STUDY GUIDE
AND
SELECTED SOLUTIONS MANUAL

Karen Timberlake

Mark Quirie

Algonquin College

General, Organic, and Biological Chemistry

STRUCTURES OF LIFE

Sixth Edition

330 Hudson Street, NY NY 10013

Courseware Portfolio Manager: Scott Dustan
Director, Courseware Portfolio Management: Jeanne Zalesky
Content Producer: Melanie Field
Managing Producer: Kristen Flathman
Courseware Analyst: Coleen Morrison
Courseware Director, Content Development: Jennifer Hart
Courseware Editorial Assistant: Fran Falk and Leslie Lee
Full-Service Vendor: SPi Global
Manufacturing Buyer: Stacey Weinberger
Cover Image Credit: Zoonar GmbH/Alamy Stock Photo

1   18

ISBN 10:   0-134-81473-8
ISBN 13: 978-0-134-81473-5

www.pearsonhighered.com

Contents

Preface

This *Study Guide and Selected Solutions Manual* is intended to accompany *General, Organic, and Biological Chemistry: Structures of Life*, sixth edition. Our purpose of this Study Guide is to provide you with additional learning resources to increase your understanding of the key concepts. Each Section in this Study Guide is correlated with a chapter Section in the text. For every Section in each chapter, there are Learning Goals, Sample Problems with Solutions, and Learning Exercises with Answers that focus on problem solving, which in turn promote an understanding of the chemical principles. A Checklist of Learning Goals and a Practice Test with Answers provide a review of the chapter content. In the Selected Solutions, we have included the answers and worked-out solutions to all the odd-numbered problems in the text.

We hope that this Study Guide and Selected Solutions Manual will help in the learning of chemistry. If you wish to make comments or corrections, or ask questions, you can send us an email message at *khemist@aol.com*

Karen Timberlake

Mark Quirie

To the Student

One must learn by doing the thing;
though you think you know it, you
have no certainty until you try.
—*Sophocles*

Here you are in a chemistry class with your textbook in front of you. Perhaps you have already been assigned some reading or some problems to do in the book. Looking through the chapter, you may see words, terms, and pictures that are new to you. This may very well be your first experience with a science class like chemistry. At this point you may have some questions about what you can do to learn chemistry. This *Study Guide and Selected Solutions Manual* is written with those considerations in mind.

Learning chemistry is similar to learning something new such as tennis or skiing or driving. If I asked you how you learn to play tennis or ski or drive a car, you would probably tell me that you would need to practice often. It is the same with learning chemistry; understanding the chemical ideas and successfully solving the problems depends on the time and effort you invest in it. If you practice every day, you will find that learning chemistry is an exciting experience and a way to understand the current issues of the environment, health, and medicine.

Manage Your Study Time

I often recommend a study system to students in which you read one Section of the text and immediately try the practice problems that are at the end of that Section. In this way, you concentrate on a small amount of information and use what you learned to solve the problems. This helps you to organize and review the information without being overwhelmed by the entire chapter. It is important to understand each Section, because they build like steps. Information presented in each chapter proceeds from the basic to the more complex. Perhaps you can only study three or four Sections of the chapter. As long as you also practice doing some problems at the same time, the information will stay with you.

Some Strategies for Learning Chemistry

Learning chemistry requires us to place new information in our long-term memory, which allows us to remember those ideas for an exam, a process called retrieval. We can develop new study habits that help us successfully recall new information by connecting it with our prior knowledge. This can be accomplished by doing a lot of practice testing. We can check how much we have learned by going back a few days later and retesting. Another useful learning strategy is to study different ideas at the same time, which allows us to connect those ideas and to differentiate between them. Although these study habits may take more time and seem more difficult, they help us find the gaps in our knowledge and connect new information with what we already know.

Form a Study Group

I highly recommend that you form a study group in the first week of your chemistry class. Working with your peers will help you use the language of chemistry. Scheduling a time to meet each week will help you study and prepare to discuss problems. You will be able to teach some things to other students in the group, and sometimes they will help you understand a topic that puzzles you. You won't always understand a concept right away. Your group will help you see your way through it. Most of all, a study group creates a strong support system when students help each other successfully complete the class.

Go to Office Hours

Try to go to your tutor's and/or professor's office hours. Your professor wants you to understand and enjoy learning this material. Often a tutor is assigned to a class or there are tutors available at your college. Don't be intimidated. Going to see a tutor or your professor is one of the best ways to clarify what you need to learn in chemistry.

Using This Study Guide and Selected Solutions Manual

Now you are ready to sit down and study chemistry. Let's go over some methods that can help you learn chemistry. This *Study Guide* is written specifically to help you understand and practice the chemical concepts that are presented in your class and in your text. The following features are part of this *Study Guide and Selected Solutions Manual*:

1. Learning Goals

The Learning Goals give you an overview of what each Section in the chapter is about and what you can expect to accomplish when you complete your study and learning of that Section.

2. Chapter Sections

Each chapter Section begins with a list of the important ideas to guide you through each of the learning activities. When you are ready to begin your study, read the matching Section in the text-book and review the *Sample Problems* in the text. Included are the *Key Math Skills* and *Core Chemistry Skills*.

3. Learning Exercises

The Learning Exercises give you an opportunity to practice problem solving related to the chemical principles in the chapter. Each set of Learning Exercises reviews one chemical principle. The answers are found immediately following each Learning Exercise. Check your answers right away. If they don't match the answers in the Study Guide, review that Section of the text again. It is important to make corrections before you go on. Chemistry involves a layering of skills such that each one must be understood before the next one can be learned.

4. Key Terms

Key Terms appear throughout each chapter in the Study Guide. As you match each Key Terms with its description, you will have an overview of the topics you have studied in that chapter. Because many of the Key Terms may be new to you, this is an opportunity to review their meaning.

5. Checklist

Use the Checklist to check your understanding of the Learning Goals. This Checklist gives you an overview of the major topics in each Section. If something does not sound familiar, go back and review it. One aspect of being a strong problem-solver is the ability to evaluate your knowledge and understanding as you study.

6. Practice Test

A Practice Test is found at the end of each chapter. When you feel you have learned the material in a chapter, you can check your understanding by taking the Practice Test. The Practice Test questions are keyed to the chapter Section, which allows you to identify the Section you may need to review. Answers are found at the end of the Practice Test. If the results of the Practice Test indicate that you know the material, you are ready to proceed to the next chapter.

7. Selected Answers and Solutions

The Selected Answers and Solutions to the odd-numbered problems for each chapter in the text follow each Practice Test for that chapter. For certain chapters, Selected Answers to the Combining Ideas for groups of chapters are included.

Resources

General, Organic, and Biological Chemistry: Structures of Life, sixth edition, provides an integrated teaching and learning package of support material for both students and professors.

Name of Supplement	Available in Print	Available Online	Instructor or Student Supplement	Description
Study Guide and Selected Solutions Manual (9780134814735)	✓		Supplement for Students	The *Study Guide and Selected Solutions Manual*, by Karen Timberlake and Mark Quirie, promotes learning through a variety of exercises with answers as well as practice tests that are connected directly to the learning goals of the textbook. Complete solutions to odd-numbered Practice Problems from the textbook are included.
Mastering™ Chemistry (www.masteringchemistry.com) (9780134787312)		✓	Supplement for Students and Instructors	Mastering™ Chemistry from Pearson is the leading online homework, tutorial, and assessment system, designed to improve results by engaging students with powerful content. Instructors ensure students arrive ready to learn by assigning educationally effective content and encourage critical thinking and retention with in-class resources such as Learning Catalytics™. Students can further master concepts through homework assignments that provide hints and answer specific feedback. The Mastering™ gradebook records scores for all automatically-graded assignments in one place, while diagnostic tools give instructors access to rich data to assess student understanding and misconceptions. http://www.masteringchemistry.com.
Mastering™ Chemistry with Pearson eText (9780134813011)		✓	Supplement for Students	The sixth edition of *General, Organic, and Biological Chemistry: Structures of Life* features a Pearson eText enhanced with media within Mastering™ Chemistry. In conjunction with Mastering™ assessment capabilities, new Interactive Videos will improve student engagement and knowledge retention. Each chapter contains a balance of interactive animations, videos, sample calculations, and self-assessments/quizzes embedded directly in the eText. Additionally, the Pearson eText offers students the power to create notes, highlight text in different colors, create bookmarks, zoom, and view single or multiple pages.
Laboratory Manual by Karen Timberlake (9780321811851)	✓		Supplement for Students	This best-selling lab manual coordinates 35 experiments with the topics in *General, Organic, and Biological Chemistry: Structures of Life*, sixth edition, uses laboratory investigations to explore chemical concepts, develop skills of manipulating equipment, reporting data, solving problems, making calculations, and drawing conclusions.
Instructor's Solutions Manual–Download Only (9780134814773)		✓	Supplement for Instructors	Prepared by Mark Quirie, the Instructor's Solutions Manual highlights chapter topics, and includes answers and solutions for all Practice Problems in the textbook.
Instructor Resource Materials–Download Only (9780134814780)		✓	Supplement for Instructors	Includes all the art, photos, and tables from the textbook in JPEG format for use in classroom projection or when creating study materials and tests. In addition, the instructors can access modifiable PowerPoint™ lecture outlines. Also available are downloadable files of the Instructor's Solutions Manual. Visit the Pearson Education catalog page for Timberlake's *General, Organic, Biological Chemistry: Structures of Life*, sixth edition, at www.pearsonhighered.com to download available instructor supplements.
TestGen Test Bank–Download Only (9780134814766)		✓	Supplement for Instructors	Prepared by William Timberlake, this resource includes more than 1600 questions in multiple-choice, matching, true/false, and short-answer format.
Online Instructor Manual for Laboratory Manual (9780321812858)		✓	Supplement for Instructors	This manual contains answers to report sheet pages for the *Laboratory Manual* and a list of the materials needed for each experiment with amounts given for 20 students working in pairs, available for download at www.pearsonhighered.com.

Chemistry in Our Lives

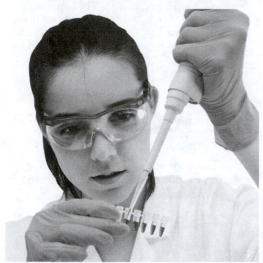

After a female victim, Gloria, is found dead in her home, samples of her blood and stomach contents are sent to Sarah, who is a forensic scientist. After using a variety of chemical tests, Sarah concludes that Gloria was poisoned when she ingested ethylene glycol. Because the initial symptoms of ethylene glycol poisoning are similar to those of alcohol intoxication, the victim was unaware of the poisoning. When ethylene glycol is oxidized, the products can cause kidney failure and may be toxic to the body. Two weeks later, police arrest Gloria's husband after finding a bottle of antifreeze containing ethylene glycol in the laundry room of the couple's home. How does this result show the use of the scientific method?

Credit: nyaivanova/Shutterstock

LOOKING AHEAD

1.1 Chemistry and Chemicals
1.2 Scientific Method: Thinking Like a Scientist
1.3 Studying and Learning Chemistry
1.4 Key Math Skills for Chemistry
1.5 Writing Numbers in Scientific Notation

The Health icon indicates a question that is related to health and medicine.

1.1   Chemistry and Chemicals

Learning Goal: Define the term chemistry, and identify chemicals.

- Chemistry is the study of the composition, structure, properties, and reactions of matter.
- A chemical is any substance that always has the same composition and properties wherever it is found.

♦ **Learning Exercise 1.1**

Is each of the following a chemical?

a. _____ ascorbic acid (vitamin C)

b. _____ time for a radioisotope to lose half its activity

c. _____ sodium fluoride in toothpaste

d. _____ amoxicillin, an antibiotic

e. _____ distance of 2.5 mi walked on a treadmill during exercise

f. _____ carbon nanotubes used to transport medication to cancer cells

Answers   **a.** yes   **b.** no   **c.** yes   **d.** yes   **e.** no   **f.** yes

1.2 Scientific Method: Thinking Like a Scientist

Learning Goal: Describe the activities that are part of the scientific method.

- The scientific method is a process of explaining natural phenomena beginning with making observations, forming a hypothesis, and performing experiments.
- When the results of the experiments are analyzed, a conclusion is made as to whether the hypothesis is *true* or *false*.
- After repeated successful experiments, a hypothesis may become a scientific theory.

Scientific Method

Observations → Law

Hypothesis — The hypothesis is modified if the results of the experiments do not support it.

Experiments

Conclusion/ Theory

The scientific method develops a conclusion or theory using observations, hypotheses, and experiments.

♦ **Learning Exercise 1.2A**

Identify each of the following as an observation, a hypothesis, an experiment, or a conclusion:

a. _____ Sunlight is necessary for the growth of plants.

b. _____ Plants in the shade are shorter than plants in the sun.

c. _____ Plant leaves are covered with aluminum foil and their growth is measured.

d. _____ Fertilizer is added to plants.

e. _____ Brown spots appear on plant leaves after exposure to ozone.

Answers   **a.** hypothesis   **b.** observation   **c.** experiment
              d. experiment   **e.** observation

♦ **Learning Exercise 1.2B**

Identify each of the following as an observation, a hypothesis, an experiment, or a conclusion:

a. _____ One half-hour after eating a wheat roll, Cynthia experiences stomach cramps.

b. _____ The next morning, Cynthia eats a piece of toast and one half-hour later has an upset stomach.

c. _____ Cynthia thinks she may be gluten intolerant.

d. _____ Cynthia tries a gluten-free roll and does not have any stomach cramps.

e. _____ Because Cynthia does not have an upset stomach after eating a gluten-free roll, she believes that she is gluten intolerant.

Answers   **a.** observation   **b.** observation   **c.** hypothesis
              d. experiment   **e.** conclusion

1.3 Studying and Learning Chemistry

Learning Goal: Identify strategies that are effective for learning. Develop a study plan for learning chemistry.

- Strategies for learning chemistry utilize features in the text that help develop a successful approach to learning chemistry.
- Components of the text that promote learning include: *Looking Ahead, Learning Goals, Review, Key Math Skills, Core Chemistry Skills, Try It First, Engage, Test, Analyze the Problem* with *Connect, Sample Problems* with *Solutions, Study Checks* with *Answers, Chemistry Link to Health, Chemistry Link to the Environment, Practice Problems, Clinical Applications, Interactive Videos, Clinical Updates, Concept Maps, Chapter Reviews, Key Terms, Review of Key Math Skills, Review of Core Chemistry Skills, Understanding the Concepts, Additional Problems, Challenge Problems, Answers to Practice Problems, Combining Ideas,* and *Glossary/Index.*

♦ **Learning Exercise 1.3**

Should each of the following activities be included in a successful study plan for learning chemistry?

a. _____ attending class occasionally

b. _____ working problems with classmates

c. _____ attending review sessions

d. _____ planning a regular study time

e. _____ not doing the assigned problems

f. _____ visiting the instructor during office hours

g. _____ attempting to work a Sample Problem before checking the Solution

h. _____ asking yourself questions as you read

i. _____ testing yourself often by working Practice Problems

Studying in a group can be beneficial to learning.

Credit: Chris Schmidt/Schmidt/E+/Getty Images

Answers   **a.** no   **b.** yes   **c.** yes   **d.** yes   **e.** no
f. yes   **g.** yes   **h.** yes   **i.** yes

1.4 Key Math Skills for Chemistry

Learning Goal: Review math concepts used in chemistry: place values, positive and negative numbers, percentages, solving equations, and interpreting graphs.

- Basic math skills are important in learning chemistry.
- In a number, we identify the place value of each digit.
- A positive number is greater than zero and has a positive sign ($+$); a negative number is less than zero and has a negative sign ($-$).
- A percentage is calculated as the parts divided by the whole, then multiplied by 100%.
- An equation is solved by rearranging it to place the unknown value on one side.
- A graph represents the relationship between two variables, which are plotted on perpendicular axes.

Study Note

On your calculator, there are four keys that are used for basic mathematical operations. The change sign $\boxed{+/-}$ key is used to change the sign of a number.

To practice these basic calculations on the calculator, work through the problem going from left to right doing the operations in the order they occur. If your calculator has a change sign $\boxed{+/-}$ key, a negative number is entered by pressing the number and then pressing the change sign $\boxed{+/-}$ key. At the end, press the equals $\boxed{=}$ key or ANS or ENTER.

Addition and Subtraction

Example 1: $15 - 8 + 2 =$

Solution: $15 \boxed{-} 8 \boxed{+} 2 \boxed{=} 9$

Example 2: $4 + (-10) - 5 =$

Solution: $4 \boxed{+} 10 \boxed{+/-} \boxed{-} 5 \boxed{=} -11$

Multiplication and Division

Example 3: $2 \times (-3) =$

Solution: $2 \boxed{\times} 3 \boxed{+/-} \boxed{=} -6$

Example 4: $\dfrac{8 \times 3}{4} =$

Solution: $8 \boxed{\times} 3 \boxed{\div} 4 \boxed{=} 6$

♦ **Learning Exercise 1.4A**

KEY MATH SKILL
Identifying Place Values

a. Identify the place value for each of the digits in the number 825.10.

Answer

Digit	Place Value
8	
2	
5	
1	
0	

Digit	Place Value
8	hundreds
2	tens
5	ones
1	tenths
0	hundredths

b. A tablet contains 0.325 g of aspirin. Identify the place value for each of the digits in the number 0.325.

Answer

Digit	Place Value
3	
2	
5	

Digit	Place Value
3	tenths
2	hundredths
5	thousandths

♦ **Learning Exercise 1.4B**

KEY MATH SKILL
Using Positive and Negative Numbers in Calculations

Evaluate each of the following:

a. $\dfrac{-14 + 22}{-4}$

b. $\dfrac{-3 + (-15)}{3}$

c. $\dfrac{-2 \times 10}{-5}$

Answers   a. -2 b. -6 c. 4

♦ **Learning Exercise 1.4C**

KEY MATH SKILL

Calculating Percentages

a. There are 20 patients on the 3rd floor of a hospital. If 4 patients were discharged today, what percentage of patients were discharged?

b. In a clinical trial with 120 participants, 36 responded to the new drug. What percentage of the participants responded?

c. Of the 12 pairs of socks in a drawer, 3 pairs are white. What percentage of the socks are white?

Answers   **a.** 20%   **b.** 30%   **c.** 25%

Study Note

1. Solve an equation for a particular variable by performing the same mathematical operation on both sides of the equation.
2. If you eliminate a symbol or number by subtracting, you need to subtract that same symbol or number on the opposite side.
3. If you eliminate a symbol or number by adding, you need to add that same symbol or number on the opposite side.
4. If you cancel a symbol or number by dividing, you need to divide both sides by that same symbol or number.
5. If you cancel a symbol or number by multiplying, you need to multiply both sides by that same symbol or number.

♦ **Learning Exercise 1.4D**

KEY MATH SKILL

Solving Equations

Solve each of the following equations for y:

  a. $2y + 14 = -2$ **b.** $3y - 8 = 22$ **c.** $-3y + 12 = -6$

Answers   **a.** $y = -8$ **b.** $y = 10$ **c.** $y = 6$

◆ **Learning Exercise 1.4E**

The dilution equation shows the relationship between the concentration (C) and volume (V) of a solution.

$$C_1 V_1 = C_2 V_2$$

Solve the dilution equation for each of the following:

a. C_1 **b.** V_2 **c.** V_1

Answers   **a.** $C_1 = \dfrac{C_2 V_2}{V_1}$ **b.** $V_2 = \dfrac{C_1 V_1}{C_2}$ **c.** $V_1 = \dfrac{C_2 V_2}{C_1}$

◆ **Learning Exercise 1.4F**

KEY MATH SKILL

Interpreting Graphs

Use the following graph to answer questions **a** and **b**:

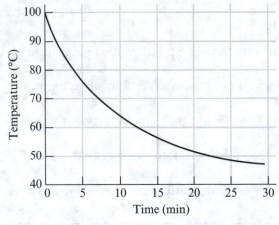

Temperature versus Time for Cooling of Coffee

a. How many minutes elapse when coffee cools from 100 °C to 65 °C?

b. What is the temperature after the coffee has cooled for 15 min?

Answers

a.

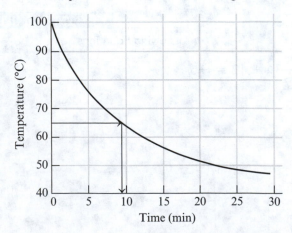

Temperature versus Time for Cooling of Coffee

Draw a horizontal line from 65 °C on the temperature axis to intersect the curved line. From the intersect, draw a vertical line to the time axis and read the time where it crosses the *x* axis, which is estimated to be 9 min.

b.

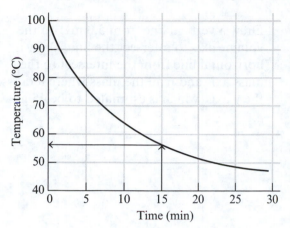

Temperature versus Time for Cooling of Coffee

Draw a vertical line from 15 min on the time axis to intersect the curved line. Draw a horizontal line from the intersect to the temperature axis and read the temperature where it crosses the *y* axis, which is estimated to be 57 °C.

♦ **Learning Exercise 1.4G**

Use the following graph to answer questions **a** and **b**:

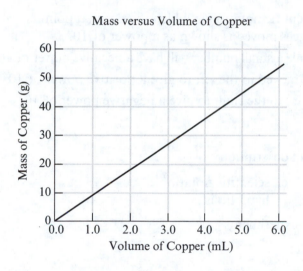

Mass versus Volume of Copper

a. What is the volume, in milliliters, of 35 g of copper?

b. What is the mass, in grams, of 5 mL of copper?

Answers

a.

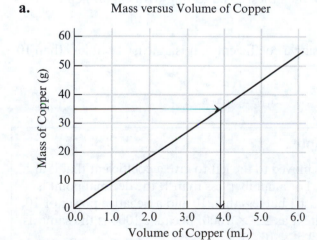

Mass versus Volume of Copper

Draw a horizontal line from 35 g on the mass axis to intersect the line. From the intersect, draw a vertical line to the volume axis and read the volume where it crosses the *x* axis, which is estimated to be 3.9 mL.

b.

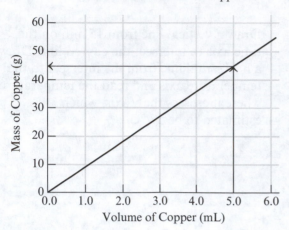

Mass versus Volume of Copper

Draw a vertical line from 5.0 mL on the volume axis to intersect the line. Draw a horizontal line from the intersect to the mass axis and read the mass where it crosses the *y* axis, which is estimated to be 45 g.

1.5 Writing Numbers in Scientific Notation

Learning Goal: Write a number in scientific notation.

- Large and small numbers can be written using scientific notation in which the decimal point is moved to give a coefficient and the number of spaces moved is shown as a power of 10.
- A large number will have a positive power of 10, while a small number will have a negative power of 10.
- For numbers greater than 10, the decimal point is moved to the left to give a positive power of 10.
- For numbers less than 1, the decimal point is moved to the right to give a negative power of 10.

Key Terms for Sections 1.1 to 1.5

Match each of the following key terms with the correct description:

- **a.** scientific method
- **b.** experiment
- **c.** scientific notation
- **d.** conclusion/theory
- **e.** chemistry
- **f.** hypothesis

1. _____ an explanation of nature validated by many experiments

2. _____ the study of substances and how they interact

3. _____ a possible explanation of a natural phenomenon

4. _____ the process of making observations, writing a hypothesis, and testing with experiments

5. _____ a procedure used to test a hypothesis

6. _____ a form of writing large and small numbers using a coefficient that is at least 1 but less than 10, followed by a power of 10

Answers   **1.** d   **2.** e   **3.** f   **4.** a   **5.** b   **6.** c

---

Study Note

1. For a number greater than 10, the decimal point is moved to the left to give a coefficient that is at least 1 but less than 10 and a positive power of 10. For a number less than 1, the decimal point is moved to the right to give a coefficient that is at least 1 but less than 10 and a negative power of 10.
2. To express 2500 in scientific notation, we can write it as 2.5×1000. Because 1000 is the same as 10^3, we can express 2500 in scientific notation as 2.5×10^3.
3. To express 0.082 in scientific notation, we move the decimal point two places to the right and write 8.2×0.01. Because 0.01 is the same as 10^{-2}, we can express 0.082 in scientific notation as 8.2×10^{-2}.

Writing a Number in Scientific Notation	
STEP 1	Move the decimal point to obtain a coefficient that is at least 1 but less than 10.
STEP 2	Express the number of places moved as a power of 10.
STEP 3	Write the product of the coefficient multiplied by the power of 10.

♦ **Learning Exercise 1.5A**

Write each of the following numbers in scientific notation:

a. 24 100 _____

b. 825 _____

c. 230 000 _____

d. 53 000 000 _____

e. 0.002 _____

f. 0.000 001 5 _____

g. 0.08 _____

h. 0.000 24 _____

Answers   **a.** 2.41×10^4 **b.** 8.25×10^2 **c.** 2.3×10^5 **d.** 5.3×10^7
   e. 2×10^{-3} **f.** 1.5×10^{-6} **g.** 8×10^{-2} **h.** 2.4×10^{-4}

♦ **Learning Exercise 1.5B**

Which number in each of the following pairs is larger?

a. 2500 or 2.5×10^2 _____

b. 0.04 or 4×10^{-3} _____

c. 65 000 or 6.5×10^5 _____

d. 0.000 35 or 3.5×10^{-3} _____

e. 300 000 or 3×10^6 _____

f. 0.002 or 2×10^{-4} _____

Answers   **a.** 2500   **b.** 0.04   **c.** 6.5×10^5
   d. 3.5×10^{-3} **e.** 3×10^6 **f.** 0.002

Checklist for Chapter 1

You are ready to take the Practice Test for Chapter 1. Be sure you have accomplished the following learning goals for this chapter. If not, review the Section listed at the end of the goal. Then apply your new skills and understanding to the Practice Test.

After studying Chapter 1, I can successfully:

_____ Identify a substance as a chemical. (1.1)

_____ Describe the steps of the scientific method. (1.2)

_____ Identify strategies that are effective for learning. (1.3)

_____ Develop a study plan for successfully learning chemistry. (1.3)

_____ Use math skills needed for chemistry: place values, positive and negative numbers, percentages, solving equations, and interpreting graphs. (1.4)

_____ Write numbers in scientific notation using a coefficient and a power of 10. (1.5)

Practice Test for Chapter 1

The chapter Sections to review are shown in parentheses at the end of each question.

1. Which of the following would be described as a chemical? (1.1)
   A. sleeping
   B. salt
   C. singing
   D. listening to a concert
   E. energy

2. Which of the following is not a chemical? (1.1)
   A. aspirin
   B. sugar
   C. feeling cold
   D. glucose
   E. vanilla

For questions 3 through 7, identify each statement as an observation (O), *a hypothesis* (H), *an experiment* (E), *or a conclusion* (C): (1.2)

3. _____ More sugar dissolves in 50 mL of hot water than in 50 mL of cold water.

4. _____ Samples containing 20 g of sugar each are placed separately in a glass of cold water and a glass of hot water.

5. _____ Sugar consists of white crystals.

6. _____ Water flows downhill due to gravity.

7. _____ Drinking 10 glasses of water a day will help me lose weight.

For questions 8 through 12, answer yes or no. (1.3)

To learn chemistry, I will:

8. _____ work Practice Problems as I read each Section of a chapter

9. _____ attend some classes but not all

10. _____ try to work each Sample Problem before I look at the Solution

11. _____ have a regular study time

12. _____ wait until the night before the exam to start studying

13. Evaluate $\dfrac{22 + 23}{-4 + 1}$ = ? (1.4)
    A. 5
    B. −5
    C. −7
    D. −15
    E. 15

14. In a big cats exhibit at the zoo, there are 8 lions, 6 tigers, and 2 leopards. What percentage of the animals in the exhibit are lions? (1.4)
    A. 50%
    B. 30%
    C. 20%
    D. 10%
    E. 5%

15. The contents found in a container at a victim's home were identified as 65.0 g of ethylene glycol in 182 g of solution. What is the percentage of ethylene glycol in the solution? (1.4)
    A. 15.3%
    B. 26.3%
    C. 35.7%
    D. 48.0%
    E. 65.0%

16. For breakfast, a person eats 55 g of eggs, 65 g of banana, and 85 g of milk. What percentage of the breakfast was milk? (1.4)
    A. 85%
    B. 65%
    C. 41%
    D. 32%
    E. 27%

17. Solve for the value of *a* in the equation $5a + 10 = 30$. (1.4)
    A. 2
    B. 4
    C. 8
    D. 20
    E. 40

18. Use the following graph to determine the solubility of carbon dioxide in water, in g CO_2/100 g water, at 40 °C: (1.4)

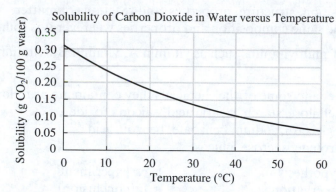

Solubility of Carbon Dioxide in Water versus Temperature

A. 0.10   B. 0.15   C. 0.20   D. 0.25   E. 0.30

19. The number 42 000 written in scientific notation is (1.5)
    A. 42   B. 42×10^3 C. 4.2×10^3 D. 4.2×10^{-3} E. 4.2×10^4

20. The number 0.005 written in scientific notation is (1.5)
    A. 5   B. 5×10^{-3} C. 5×10^{-2} D. 0.5×10^{-4} E. 5×10^3

Answers to the Practice Test

1. B   2. C   3. O   4. E   5. O
6. C   7. H   8. yes   9. no   10. yes
11. yes   12. no   13. D   14. A   15. C
16. C   17. B   18. A   19. E   20. B

Selected Answers and Solutions to Text Problems

1.1   **a.** Chemistry is the study of the composition, structure, properties, and reactions of matter.
    b. A chemical is a substance that has the same composition and properties wherever it is found.

1.3   Many chemicals are listed on a bottle of multivitamins such as vitamin A, vitamin B_3, vitamin B_{12}, vitamin C, and folic acid.

1.5   Typical items found in a medicine cabinet and some of the chemicals they contain are as follows:
    Antacid tablets: calcium carbonate, cellulose, starch, stearic acid, silicon dioxide
    Mouthwash: water, alcohol, thymol, glycerol, sodium benzoate, benzoic acid
    Cough suppressant: menthol, beta-carotene, sucrose, glucose

1.7   **a.** observation   **b.** hypothesis   **c.** experiment
    d. observation   **e.** observation   **f.** conclusion

1.9   **a.** observation   **b.** hypothesis   **c.** experiment   **d.** experiment

1.11   There are several things you can do that will help you successfully learn chemistry, including attending class regularly, forming a study group, reading the assigned pages in the textbook before class, answering the Engage questions as you read new material in the textbook, trying to solve the Sample Problems first before reading the provided Solutions, working the Study Checks and Practice Problems and checking the Answers, self-testing during and after reading each Section, retesting on new information a few days later, going to the instructor's office hours, and keeping a problem notebook.

1.13   Ways you can enhance your learning of chemistry include:
    a. forming a study group.
    c. asking yourself questions while reading the text.
    e. answering the Engage questions.

1.15   **a.** The bolded 8 is in the thousandths place.
    b. The bolded 6 is in the ones place.
    c. The bolded 6 is in the hundreds place.

1.17   **a.** $15 - (-8) = 15 + 8 = 23$
    b. $-8 + (-22) = -30$
    c. $4 \times (-2) + 6 = -8 + 6 = -2$

1.19   **a.** The graph shows the relationship between the temperature of a cup of tea and time.
    b. The vertical axis measures temperature, in °C.
    c. The values on the vertical axis range from 20 °C to 80 °C.
    d. As time increases, the temperature decreases.

1.21   **a.**
$$4a + 4 = 40$$
$$4a + \cancel{4} - \cancel{4} = 40 - 4$$
$$4a = 36$$
$$\frac{\cancel{4}a}{\cancel{4}} = \frac{36}{4}$$
$$a = 9$$

    b.
$$\frac{a}{6} = 7$$
$$\cancel{6}\left(\frac{a}{\cancel{6}}\right) = 6(7)$$
$$a = 42$$

1.23 **a.** $\dfrac{21 \text{ flu shots}}{25 \text{ patients}} \times 100\% = 84\%$ received flu shots

b. total grams of alloy = 56 g silver + 22 g copper = 78 g of alloy

$\dfrac{56 \text{ g silver}}{78 \text{ g alloy}} \times 100\% = 72\%$ silver

c. total number of coins = 11 nickels + 5 quarters + 7 dimes = 23 coins

$\dfrac{7 \text{ dimes}}{23 \text{ coins}} \times 100\% = 30\%$ dimes

1.25 **a.** The graph shows the relationship between body temperature and time since death.
b. The vertical axis measures temperature, in °C.
c. The values on the vertical axis range from 20 °C to 40 °C.
d. As time increases, the temperature decreases.

1.27 **a.** Move the decimal point four places to the left to give 5.5×10^4.
b. Move the decimal point two places to the left to give 4.8×10^2.
c. Move the decimal point six places to the right to give 5×10^{-6}.
d. Move the decimal point four places to the right to give 1.4×10^{-4}.
e. Move the decimal point three places to the right to give 7.2×10^{-3}.
f. Move the decimal point five places to the left to give 6.7×10^5.

1.29 **a.** 7.2×10^3, which is also 7200, is larger than 8.2×10^2 or 820.
b. 3.2×10^{-2}, which is also 0.032, is larger than 4.5×10^{-4} or 0.000 45.
c. 1×10^4, which is also 10 000, is larger than 1×10^{-4} or 0.0001.
d. 6.8×10^{-2}, which is also 0.068, is larger than 0.000 52.

1.31 **a.** hypothesis   **b.** conclusion   **c.** experiment   **d.** observation

1.33 $\dfrac{120 \text{ g ethylene glycol}}{450 \text{ g liquid}} \times 100\% = 27\%$ ethylene glycol

1.35 No. All of these ingredients are chemicals.

1.37 Yes. Sherlock's investigation includes making observations (gathering data), formulating a hypothesis, testing the hypothesis, and modifying it until one of the hypotheses is validated.

1.39 **a.** When two negative numbers are added, the answer has a negative sign.
b. When a positive and negative number are multiplied, the answer has a negative sign.

1.41 **a.** Describing the appearance of a patient is an observation.
b. Formulating a reason for the extinction of dinosaurs is a hypothesis.
c. Measuring a patient's blood pressure is an observation.

1.43 If experimental results do not support your hypothesis, you should:
b. modify your hypothesis.
c. do more experiments.

1.45 A successful study plan would include:
b. working the Sample Problems as you go through a chapter.
c. self-testing.

1.47 **a.** $4 \times (-8) = -32$
b. $-12 - 48 = -12 + (-48) = -60$
c. $\dfrac{-168}{-4} = 42$

1.49 total number of gumdrops = 16 orange + 8 yellow + 16 black = 40 gumdrops

a. $\dfrac{8 \text{ yellow gumdrops}}{40 \text{ total gumdrops}} \times 100\% = 20\%$ yellow gumdrops

b. $\dfrac{16 \text{ black gumdrops}}{40 \text{ total gumdrops}} \times 100\% = 40\%$ black gumdrops

1.51 **a.** Move the decimal point five places to the left to give 1.2×10^5.
b. Move the decimal point seven places to the right to give 3.4×10^{-7}.
c. Move the decimal point two places to the right to give 6.6×10^{-2}.
d. Move the decimal point three places to the left to give 2.7×10^3.

1.53 **a.** observation   **b.** hypothesis   **c.** conclusion

1.55 **a.** Self-testing allows you to check on what you understand.
b. Forming a study group can motivate you to study, fill in gaps, and correct misunderstandings by teaching and learning together.
c. Reading the assignment before class prepares you to learn new material.

1.57 **a.** observation   **b.** hypothesis   **c.** experiment   **d.** conclusion

1.59 **a.**
$$2x + 5 = 41$$
$$2x + 5 - 5 = 41 - 5$$
$$2x = 36$$
$$\frac{2x}{2} = \frac{36}{2}$$
$$x = 18$$

b.
$$\frac{5x}{3} = 40$$
$$3\left(\frac{5x}{3}\right) = 3(40)$$
$$5x = 120$$
$$\frac{5x}{5} = \frac{120}{5}$$
$$x = 24$$

1.61 **a.** The graph shows the relationship between the solubility of carbon dioxide in water and temperature.
b. The vertical axis measures the solubility of carbon dioxide in water (g CO_2/100 g water).
c. The values on the vertical axis range from 0 to 0.35 g CO_2/100 g water.
d. As temperature increases, the solubility of carbon dioxide in water decreases.

2
Chemistry and Measurements

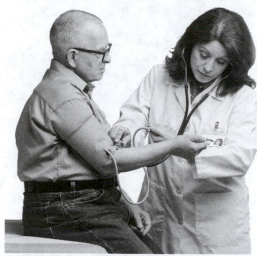

Greg is taking Inderal to treat his high blood pressure. Every two months, he has his blood pressure checked by a registered nurse. During his last visit, his blood pressure was lower, 130/85. Thus, the dosage of Inderal was reduced to 45 mg to be taken twice daily. If one tablet contains 15 mg of Inderal, how many tablets does Greg need to take in one day?

Credit: AVAVA/Shutterstock

LOOKING AHEAD

2.1 Units of Measurement	**2.4** Prefixes and Equalities	**2.6** Problem Solving Using Unit Conversion
2.2 Measured Numbers and Significant Figures	**2.5** Writing Conversion Factors	**2.7** Density
2.3 Significant Figures in Calculations		

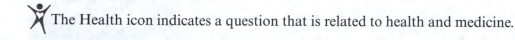 The Health icon indicates a question that is related to health and medicine.

2.1   Units of Measurement

Learning Goal: Write the names and abbreviations for the metric and SI units used in measurements of volume, length, mass, temperature, and time.

*   In science, physical quantities are described in units of the metric or International System of Units (SI).

*   Some important units of measurement are liter (L) for volume, meter (m) for length, gram (g) and kilogram (kg) for mass, degree Celsius (°C) and kelvin (K) for temperature, and second (s) for time.

Key Terms for Section 2.1

Match each of the following key terms with the correct description:

a. mass   **b.** volume
c. gram   **d.** second
e. meter   **f.** liter
g. International System
     of Units

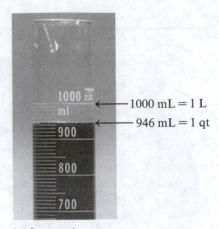

1. _____ the metric unit for volume

2. _____ the metric unit used in measurements of mass

3. _____ the amount of space occupied by a substance

4. _____ the metric and SI unit for time

5. _____ a measure of the quantity of material in an object

6. _____ the official system of measurement used by scientists and used in most countries of the world

7. _____ the metric and SI unit for length

1000 mL = 1 L
946 mL = 1 qt

In the metric system, volume is based on the liter.

Credit: Pearson Education

Answers   **1.** f   **2.** c   **3.** b   **4.** d   **5.** a   **6.** g   **7.** e

♦ **Learning Exercise 2.1**

Match the measurement with the quantity in each of the following:

a. length   **b.** volume   **c.** mass   **d.** temperature   **e.** time

1. _____ 45 g   2. _____ 8.2 m   3. _____ 215 °C   4. _____ 45 L

5. _____ 825 K   6. _____ 8.8 cm   7. _____ 140 kg   8. _____ 50 s

Answers   **1.** c   **2.** a   **3.** d   **4.** b
             5. d   **6.** a   **7.** c   **8.** e

On an electronic balance, the digital readout gives the mass of a nickel, which is 5.01 g.

Credit: Richard Megna/Fundamental Photographs, NYC

2.2   Measured Numbers and Significant Figures

Learning Goal: Identify a number as measured or exact; determine the number of significant figures in a measured number.

REVIEW
Writing Numbers in Scientific Notation (1.5)

- A measured number is obtained when you use a measuring device.

- An exact number is obtained by counting items or from a definition within the same measuring system.

- There is uncertainty in every measured number but not in exact numbers.

- Significant figures in a measured number are all the digits, including the estimated digit.

- A number is a significant figure if it is not a zero.

- A zero is significant when it occurs between nonzero digits, at the end of a decimal number, or in the coefficient of a number written in scientific notation.

- A zero is not significant if it is at the beginning of a decimal number or used as a placeholder in a large number without a decimal point.

♦ **Learning Exercise 2.2A**

Identify the number(s) in each of the following statements as measured or exact and give a reason for your answer:

a. _____ There are 7 days in 1 week.

b. _____ A pulmonary treatment lasts for 25 min.

c. _____ There are 1000 g in 1 kg.

d. _____ The chef used 12 kg of potatoes.

e. _____ There are 4 books on the shelf.

f. _____ The height of a patient is 1.8 m.

Answers   a. exact (definition)   b. measured (use a watch)
c. exact (metric definition)   d. measured (use a scale)
e. exact (counted)   f. measured (use a metric ruler)

Study Note

Significant figures (abbreviated SFs) are all the numbers reported in a measurement, including the estimated digit. Zeros are significant unless they are placeholders appearing at the beginning of a decimal number or in a large number without a decimal point.

     4.255 g (four SFs)   0.0040 m (two SFs)   46 500 L (three SFs)   150. mL (three SFs)

♦ **Learning Exercise 2.2B**

State the number of significant figures in each of the following measured numbers:

a. 35.24 g   ____   b. 8.0×10^{-5} m ____

c. 55 000 m   ____   d. 600. mL   ____

e. 5.025 L   ____   f. 0.006 kg   ____

g. 2.680×10^{5} mm ____   h. 25.0 °C   ____

CORE CHEMISTRY SKILL

Counting Significant Figures

Answers   a. 4 SFs   b. 2 SFs   c. 2 SFs   d. 3 SFs
e. 4 SFs   f. 1 SF   g. 4 SFs   h. 3 SFs

2.3 Significant Figures in Calculations

Learning Goal: Give the correct number of significant figures for a calculated answer.

REVIEW

Identifying Place Values (1.4)
Using Positive and Negative Numbers in Calculations (1.4)

• In multiplication and division, the final answer is written so that it has the same number of significant figures (SFs) as the measurement with the fewest significant figures.

- In addition and subtraction, the final answer is written so that it has the same number of decimal places as the measurement with the fewest decimal places.
- To evaluate a calculator answer, count the significant figures in the measurements and round off the calculator answer properly.
- Answers for calculations rarely use all the numbers that appear in the calculator display. Exact numbers are not included in the determination of the number of significant figures.

Study Note

1. To round off a number when the first digit to be dropped is *4 or less*, keep the proper number of digits and drop all the digits that follow.

    Round off 42.8254 to three SFs → 42.8 (drop 254)

2. To round off a number when the first digit to be dropped is *5 or greater*, keep the proper number of digits and increase the last retained digit by 1.

    Round off 8.4882 to two SFs → 8.5 (drop 882; increase the last retained digit by 1)

3. When rounding off large numbers without decimal points, maintain the value of the answer by adding nonsignificant zeros as placeholders.

    Round off 356 835 to three SFs → 357 000 (drop 835; increase the last retained digit by 1;

    add placeholder zeros)

◆ **Learning Exercise 2.3A**

KEY MATH SKILL

Rounding Off

Round off each of the following to two significant figures:

a. 88.75 m _____

b. 0.002 923 g _____

c. 50.525 s _____

d. 1.672 L _____

e. 0.001 055 8 kg _____

f. 82 080 mL _____

Answers   a. 89 m   b. 0.0029 g   c. 51 s   d. 1.7 L   e. 0.0011 kg   f. 82 000 mL

Study Note

1. An answer obtained from multiplying and dividing has the same number of significant figures as the measurement with the fewest significant figures.

    1.5   ×   32.546 = 48.819 → 49   *Answer rounded off to two SFs*

    Two SFs   Five SFs

2. An answer obtained from adding or subtracting has the same number of decimal places as the measurement with the fewest decimal places.

    82.223   +   4.1 = 86.323 → 86.3   *Answer rounded off to the tenths place*

    Thousandths place   Tenths place

♦ Learning Exercise 2.3B

CORE CHEMISTRY SKILL
Using Significant Figures in Calculations

Perform the following calculations with measured numbers. Write each answer with the correct number of significant figures or decimal places.

a. $1.3 \times 71.5 =$ **b.** $\frac{8.00}{4.00} =$ **c.** $\frac{(0.082)(25.4)}{(0.116)(3.4)} =$

d. $\frac{3.05 \times 1.86}{118.5} =$ **e.** $\frac{376}{0.0073} =$ **f.** $38.520 - 11.4 =$

g. $4.2 + 8.15 =$ **h.** $102.56 + 8.325 - 0.8825 =$

Answers **a.** 93 **b.** 2.00 **c.** 5.3 **d.** 0.0479
e. 52 000 (5.2×10^4) **f.** 27.1 **g.** 12.4 **h.** 110.00

2.4 Prefixes and Equalities

Learning Goal: Use the numerical values of prefixes to write a metric equality.

- In the metric system, larger and smaller units use prefixes to change the size of the unit by factors of 10. For example, a prefix such as *centi* or *milli* preceding the unit *meter* gives a smaller length than a meter. A prefix such as *kilo* added to *gram* gives a unit that measures a mass that is 1000 times *greater* than a gram.
- An equality contains two units that measure the *same quantity* such as length, volume, mass, or time.
- Some common metric equalities are: 1 m = 100 cm; 1 L = 1000 mL; 1 kg = 1000 g.

Metric and SI Prefixes

Prefix	Symbol	Numerical Value	Scientific Notation	Equality
Prefixes That Increase the Size of the Unit				
peta	P	1 000 000 000 000 000	10^{15}	$1\ Pg = 1 \times 10^{15}\ g$ $1\ g = 1 \times 10^{-15}\ Pg$
tera	T	1 000 000 000 000	10^{12}	$1\ Ts = 1 \times 10^{12}\ s$ $1\ s = 1 \times 10^{-12}\ Ts$
giga	G	1 000 000 000	10^{9}	$1\ Gm = 1 \times 10^{9}\ m$ $1\ m = 1 \times 10^{-9}\ Gm$
mega	M	1 000 000	10^{6}	$1\ Mg = 1 \times 10^{6}\ g$ $1\ g = 1 \times 10^{-6}\ Mg$
kilo	k	1 000	10^{3}	$1\ km = 1 \times 10^{3}\ m$ $1\ m = 1 \times 10^{-3}\ km$

Metric and SI Prefixes (continued)

Prefix	Symbol	Numerical Value	Scientific Notation	Equality
Prefixes That Decrease the Size of the Unit				
deci	d	0.1	10^{-1}	$1 \text{ dL} = 1 \times 10^{-1} \text{ L}$ $1 \text{ L} = 10 \text{ dL}$
centi	c	0.01	10^{-2}	$1 \text{ cm} = 1 \times 10^{-2} \text{ m}$ $1 \text{ m} = 100 \text{ cm}$
milli	m	0.001	10^{-3}	$1 \text{ ms} = 1 \times 10^{-3} \text{ s}$ $1 \text{ s} = 1 \times 10^{3} \text{ ms}$
micro	μ*	0.000 001	10^{-6}	$1 \text{ } \mu\text{g} = 1 \times 10^{-6} \text{ g}$ $1 \text{ g} = 1 \times 10^{6} \text{ } \mu\text{g}$
nano	n	0.000 000 001	10^{-9}	$1 \text{ nm} = 1 \times 10^{-9} \text{ m}$ $1 \text{ m} = 1 \times 10^{9} \text{ nm}$
pico	p	0.000 000 000 001	10^{-12}	$1 \text{ ps} = 1 \times 10^{-12} \text{ s}$ $1 \text{ s} = 1 \times 10^{12} \text{ ps}$
femto	f	0.000 000 000 000 001	10^{-15}	$1 \text{ fs} = 1 \times 10^{-15} \text{ s}$ $1 \text{ s} = 1 \times 10^{15} \text{ fs}$

*In medicine, the abbreviation *mc* for the prefix *micro* is used because the symbol μ may be misread, which could result in a medication error. Thus, 1 μg would be written as 1 mcg.

♦ **Learning Exercise 2.4A**

Match the items in column A with those from column B.

Using Prefixes

	A		B
1.	_____ megameter	**a.**	nanometer
2.	_____ 0.1 m	**b.**	decimeter
3.	_____ millimeter	**c.**	10^{-6} m
4.	_____ centimeter	**d.**	0.01 m
5.	_____ 10^{-9} m	**e.**	1000 m
6.	_____ micrometer	**f.**	10^{-3} m
7.	_____ kilometer	**g.**	10^{6} m

Some Normal Laboratory Test Values

Substance in Blood	Normal Range
Albumin	3.5–5.4 g/dL
Ammonia	20–70 mcg/dL
Calcium	8.5–10.5 mg/dL
Cholesterol	105–250 mg/dL
Iron (male)	80–160 mcg/dL
Protein (total)	6.0–8.5 g/dL

Laboratory results for bloodwork are often reported in mass per deciliter (dL).

Answers   **1.** g   **2.** b   **3.** f   **4.** d
            5. a   **6.** c   **7.** e

♦ **Learning Exercise 2.4B**

Complete each of the following metric relationships:

a. 1 L = _____ mL   **b.** 1 L = _____ dL

c. 1 mm = _____ cm   **d.** 1 s = _____ ms

e. 1 fg = _____ g   **f.** 1 mm = _____ m

g. 1 g = _____ mcg   **h.** 1 dL = _____ L

During a physical examination, the metric length of an infant is measured.

Credit: Eric Schrader/ Pearson Education

Answers   **a.** 1000   **b.** 10   **c.** 0.1   **d.** 1000
            e. 1×10^{-15} **f.** 0.001   **g.** 1 000 000 (10^{6})   **h.** 0.1

♦ **Learning Exercise 2.4C**

For each of the following pairs, which is the larger unit?

a. mL or mcL _____   **b.** g or kg   _____

c. nm or m   _____   **d.** centigram or decigram _____

e. kilosecond or second _____   **f.** pm or Pm   _____

Answers   **a.** mL   **b.** kg   **c.** m   **d.** decigram   **e.** kilosecond   **f.** Pm

2.5   Writing Conversion Factors

REVIEW

Calculating Percentages (1.4)

Learning Goal: Write a conversion factor for two units that describe the same quantity.

- A conversion factor represents an equality expressed in the form of a fraction.
- Two conversion factors can be written for any relationship between equal quantities.

  For the metric–U.S. equality 2.54 cm = 1 in., the corresponding conversion factors are:

  $$\frac{2.54 \text{ cm}}{1 \text{ in.}} \quad \text{and} \quad \frac{1 \text{ in.}}{2.54 \text{ cm}}$$

- A percentage (%) is written as a conversion factor by expressing matching units in the relationship as the parts of a specific substance in 100 parts of the whole.
- An equality may be stated that applies only to a specific problem.

Some Common Equalities

Quantity	Metric (SI)	U.S.	Metric–U.S.
Length	1 km = 1000 m 1 m = 1000 mm 1 cm = 10 mm	1 ft = 12 in. 1 yd = 3 ft 1 mi = 5280 ft	2.54 cm = 1 in. (exact) 1 m = 39.4 in. 1 km = 0.621 mi
Volume	1 L = 1000 mL 1 dL = 100 mL 1 mL = 1 cm³ 1 mL = 1 cc*	1 qt = 4 cups 1 qt = 2 pt 1 gal = 4 qt	946 mL = 1 qt 1 L = 1.06 qt 473 mL = 1 pt 5 mL = 1 t (tsp)* 15 mL = 1 T (tbsp)*
Mass	1 kg = 1000 g 1 g = 1000 mg 1 mg = 1000 mcg*	1 lb = 16 oz	1 kg = 2.20 lb 454 g = 1 lb
Time	1 h = 60 min 1 min = 60 s	1 h = 60 min 1 min = 60 s	

*Used in medicine.

Study Note

Metric conversion factors are obtained from metric prefixes. For example, the metric equality 1 m = 100 cm is written as two conversion factors:

$$\frac{100 \text{ cm}}{1 \text{ m}} \quad \text{and} \quad \frac{1 \text{ m}}{100 \text{ cm}}$$

♦ **Learning Exercise 2.5A**

Write the equality and two conversion factors for each of the following pairs of units:

a. millimeters and meters

b. kilograms and grams

c. kilograms and pounds

d. seconds and minutes

e. kilometers and meters

f. milliliters and quarts

g. deciliters and liters

h. square centimeters and square meters

Answers

a. $1 \text{ m} = 1000 \text{ mm}$

$$\frac{1000 \text{ mm}}{1 \text{ m}} \quad \text{and} \quad \frac{1 \text{ m}}{1000 \text{ mm}}$$

b. $1 \text{ kg} = 1000 \text{ g}$

$$\frac{1000 \text{ g}}{1 \text{ kg}} \quad \text{and} \quad \frac{1 \text{ kg}}{1000 \text{ g}}$$

c. $1 \text{ kg} = 2.20 \text{ lb}$

$$\frac{2.20 \text{ lb}}{1 \text{ kg}} \quad \text{and} \quad \frac{1 \text{ kg}}{2.20 \text{ lb}}$$

d. $1 \text{ min} = 60 \text{ s}$

$$\frac{60 \text{ s}}{1 \text{ min}} \quad \text{and} \quad \frac{1 \text{ min}}{60 \text{ s}}$$

e. $1 \text{ km} = 1000 \text{ m}$

$$\frac{1000 \text{ m}}{1 \text{ km}} \quad \text{and} \quad \frac{1 \text{ km}}{1000 \text{ m}}$$

f. $1 \text{ qt} = 946 \text{ mL}$

$$\frac{946 \text{ mL}}{1 \text{ qt}} \quad \text{and} \quad \frac{1 \text{ qt}}{946 \text{ mL}}$$

g. $1 \text{ L} = 10 \text{ dL}$

$$\frac{10 \text{ dL}}{1 \text{ L}} \quad \text{and} \quad \frac{1 \text{ L}}{10 \text{ dL}}$$

h. $(1 \text{ m})^2 = (100 \text{ cm})^2$

$$\frac{(100 \text{ cm})^2}{(1 \text{ m})^2} \quad \text{and} \quad \frac{(1 \text{ m})^2}{(100 \text{ cm})^2}$$

✗ Study Note

1. Sometimes, a statement within a problem gives an equality that is true only for that problem. Then conversion factors can be written that are true only for that problem. For example, a problem states that there are 50 mg of vitamin B in a tablet. The equality and its conversion factors are

$$1 \text{ tablet} = 50 \text{ mg of vitamin B}$$

$$\frac{50 \text{ mg vitamin B}}{1 \text{ tablet}} \quad \text{and} \quad \frac{1 \text{ tablet}}{50 \text{ mg vitamin B}}$$

Significant Figures or Exact
The 50 mg is measured: It has one significant figure. The 1 tablet is exact.

2. If a problem gives a percentage (%), it can be stated as parts per 100 parts. For example, a candy bar contains 45% by mass chocolate. This percentage (%) equality and its conversion factors are written using the same unit.

$$45 \text{ g of chocolate} = 100 \text{ g of candy bar}$$

$$\frac{45 \text{ g chocolate}}{100 \text{ g candy bar}} \quad \text{and} \quad \frac{100 \text{ g candy bar}}{45 \text{ g chocolate}}$$

Significant Figures or Exact
The 45 g is measured: It has two significant figures. The 100 g is exact.

♦ **Learning Exercise 2.5B**

Write the equality and two conversion factors for each of the following statements:

a. A cheese contains 55% fat by mass.

b. The Daily Value (DV) for zinc is 15 mg.

c. A 125-g steak contains 45 g of protein.

d. One iron tablet contains 50 mg of iron.

e. A car travels 14 miles on 1 gal of gasoline.

f. The safe drinking water standard for arsenic is 10 ppb.

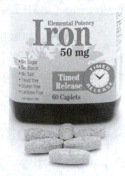

Answers

a.   55 g of fat = 100 g of cheese

$$\frac{55 \text{ g fat}}{100 \text{ g cheese}} \quad \text{and} \quad \frac{100 \text{ g cheese}}{55 \text{ g fat}}$$

b. 15 mg of zinc = 1 day

$$\frac{15 \text{ mg zinc}}{1 \text{ day}} \quad \text{and} \quad \frac{1 \text{ day}}{15 \text{ mg zinc}}$$

c. 125 g of steak = 45 g of protein

$$\frac{45 \text{ g protein}}{125 \text{ g steak}} \quad \text{and} \quad \frac{125 \text{ g steak}}{45 \text{ g protein}}$$

d.   1 tablet = 50 mg of iron

$$\frac{50 \text{ mg iron}}{1 \text{ tablet}} \quad \text{and} \quad \frac{1 \text{ tablet}}{50 \text{ mg iron}}$$

e.   1 gal = 14 mi

$$\frac{14 \text{ mi}}{1 \text{ gal}} \quad \text{and} \quad \frac{1 \text{ gal}}{14 \text{ mi}}$$

f.   1 kg of water = 10 mcg of arsenic

$$\frac{10 \text{ mcg arsenic}}{1 \text{ kg}} \quad \text{and} \quad \frac{1 \text{ kg}}{10 \text{ mcg arsenic}}$$

Each tablet contains 50 mg of iron, which is given for iron supplementation.
Credit: Editorial Image, LLC/ Alamy

♦ **Learning Exercise 2.5C**

Write the equality and two conversion factors, and identify the numbers as exact or give the number of significant figures for each of the following:

a. One tablet of calcium contains 315 mg of calcium.

b. The drug cured 37% of the patients.

c. One tablespoon of cough syrup contains 15 mL of cough syrup.

Answers

a.   1 tablet = 315 mg of calcium

$$\frac{315 \text{ mg calcium}}{1 \text{ tablet}} \quad \text{and} \quad \frac{1 \text{ tablet}}{315 \text{ mg calcium}}$$

The 315 mg is measured: It has three SFs. The 1 tablet is exact.

b. 37 cured = 100 patients

$$\frac{37 \text{ cured}}{100 \text{ patients}} \quad \text{and} \quad \frac{100 \text{ patients}}{37 \text{ cured}}$$

The 37 patients cured and the 100 patients are both exact.

c. 1 tablespoon of syrup = 15 mL of syrup

$$\frac{15 \text{ mL syrup}}{1 \text{ tablespoon syrup}} \quad \text{and} \quad \frac{1 \text{ tablespoon syrup}}{15 \text{ mL syrup}}$$

The 15 mL is measured: It has two SFs. The 1 tablespoon is exact.

2.6 Problem Solving Using Unit Conversion

Learning Goal: Use conversion factors to change from one unit to another.

- Conversion factors can be used to change a quantity expressed in one unit to a quantity expressed in another unit.
- In the process of problem solving, a given unit is multiplied by one or more conversion factors until the needed unit is obtained.

SAMPLE PROBLEM   Using Conversion Factors

A patient receives 2850 mL of saline solution. How many liters of the solution did the patient receive?

Solution:

STEP 1 State the given and needed quantities.

Analyze the Problem	Given	Need	Connect
	2850 mL	liters	metric conversion factor

STEP 2 Write a plan to convert the given unit to the needed unit.

$$\text{milliliters} \xrightarrow{\substack{\text{Metric} \\ \text{factor}}} \text{liters}$$

STEP 3 State the equalities and conversion factors.

$$1 \text{ L} = 1000 \text{ mL}$$

$$\frac{1000 \text{ mL}}{1 \text{ L}} \quad \text{and} \quad \frac{1 \text{ L}}{1000 \text{ mL}}$$

STEP 4 Set up the problem to cancel units and calculate the answer.

$$2850 \text{ mL} \times \frac{\overset{\text{Exact}}{1 \text{ L}}}{1000 \text{ mL}} = 2.85 \text{ L}$$

Three SFs   Exact   Three SFs

♦ Learning Exercise 2.6A

Use metric conversion factors to solve each of the following problems. For **e** to **h**, complete boxes for Given, Need, and Connect, and solve the problem.

CORE CHEMISTRY SKILL

Using Conversion Factors

a. 189 mL = _____ L   **b.** 2.7 cm = _____ mm

c. 0.274 m = _____ cm   **d.** 0.076 kg = _____ g

e. How many meters tall is a person whose height is 163 cm?

	Given	Need	Connect
Analyze the Problem			

f. There is 0.646 L of water in a teapot. How many milliliters is that?

	Given	Need	Connect
Analyze the Problem			

g. If a ring contains 13 500 mg of gold, how many grams of gold are in the ring?

	Given	Need	Connect
Analyze the Problem			

h. You walked a distance of 1.5 km on the treadmill at the gym. How many meters did you walk?

	Given	Need	Connect
Analyze the Problem			

Answers   **a.** 0.189 L   **b.** 27 mm   **c.** 27.4 cm   **d.** 76 g

e.

	Given	Need	Connect	Answer
Analyze the Problem	163 cm	meters	metric factor (m/cm)	1.63 m

f.

	Given	Need	Connect	Answer
Analyze the Problem	0.646 L	milliliters	metric factor (mL/L)	646 mL

g.

	Given	Need	Connect	Answer
Analyze the Problem	13 500 mg	grams	metric factor (g/mg)	13.5 g

h.

	Given	Need	Connect	Answer
Analyze the Problem	1.5 km	meters	metric factor (m/km)	1500 m

♦ **Learning Exercise 2.6B**

Use metric–U.S. conversion factors to solve each of the following problems:

a. 880 g = _____ lb

b. 4.6 qt = _____ mL

c. 1.50 ft = _____ cm

d. 8.10 in. = _____ mm

Answers   **a.** 1.9 lb   **b.** 4400 mL
            c. 45.7 cm   **d.** 206 mm

♦ **Learning Exercise 2.6C**

Use conversion factors to solve each of the following problems.
For **c** to **f**, complete boxes for Given, Need, and Connect, and solve the
problem.

a. 120 mm = _____ in.

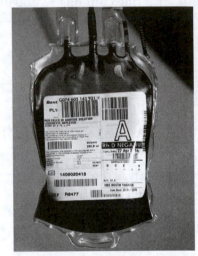

1 pt of blood contains 473 mL.

Credit: Syner-Comm/Alamy

b. 245 lb = _____ kg

c. In a triple-bypass surgery, a patient requires 3.00 pt of whole blood. How many milliliters
of blood were given?

	Given	Need	Connect
Analyze the Problem			

d. Your friend has a height of 6 ft 3 in. What is your friend's height in meters?

	Given	Need	Connect
Analyze the Problem			

e. A doctor orders 0.450 g of a sulfa drug. On hand are 150-mg tablets. How many tablets of the sulfa drug are needed?

Analyze the Problem	Given	Need	Connect

f. A mouthwash contains 27% alcohol by volume. How many milliliters of alcohol are in a 1.18-L bottle of mouthwash?

Analyze the Problem	Given	Need	Connect

Answers   **a.** 4.7 in.   **b.** 111 kg

c.

Analyze the Problem	Given	Need	Connect	Answer
	3.00 pt	milliliters	factor (mL/pt)	1420 mL

d.

Analyze the Problem	Given	Need	Connect	Answer
	6 ft 3 in.	meters	factors (in./ft, m/in.)	1.9 m

e.

Analyze the Problem	Given	Need	Connect	Answer
	0.450 g, 150-mg tablet	tablets	factors (mg/g, tablet/mg)	3 tablets

f.

Analyze the Problem	Given	Need	Connect	Answer
	1.18 L, 27% (v/v)	milliliters of alcohol	factors (mL/L, % alcohol)	320 mL

2.7   Density

Learning Goal: Calculate the density of a substance; use the density to calculate the mass or volume of a substance.

- The density of a substance is the ratio of its mass to its volume, usually in units of g/cm^3 or g/mL.

$$\text{Density} = \frac{\text{mass of substance}}{\text{volume of substance}}$$

- The volume of 1 mL is equal to the volume of $1\ cm^3$.

- Specific gravity is a relationship between the density of a substance and the density of water.

Key Terms for Sections 2.2 to 2.7

Match each of the following key terms with the correct description:

a. equality	**b.** conversion factor	**c.** measured number	**d.** exact number
e. density	**f.** prefix	**g.** significant figures	**h.** specific gravity

1. _____ a number obtained by counting

2. _____ a part of the name of a metric unit that precedes the base unit and specifies the size of the measurement

3. _____ a ratio in which the numerator and denominator are quantities from an equality

4. _____ the relationship between two units that measure the same quantity

5. _____ the digits recorded in a measurement

6. _____ the number obtained when a quantity is determined by using a measuring device

7. _____ the relationship of the mass of an object to its volume expressed as grams per cubic centimeter (g/cm^3) or grams per milliliter (g/mL)

8. _____ the relationship between the density of a substance and the density of water

Answers   **1.** d   **2.** f   **3.** b   **4.** a   **5.** g   **6.** c   **7.** e   **8.** h

♦ **Learning Exercise 2.7A**

a. Calculate the density, in grams per milliliter, of glycerol if a 200.0-mL sample has a mass of 252 g.

b. A person with diabetes may produce 5 to 12 L of urine per day. Calculate the density, in grams per milliliter, of a 100.0-mL urine sample that has a mass of 100.2 g.

c. A solid has a mass of 5.5 oz. When placed in a graduated cylinder with a water level of 55.2 mL, the object causes the water level to rise to 73.8 mL. What is the density of the object, in grams per milliliter?

Answers   **a.** 1.26 g/mL   **b.** 1.002 g/mL   **c.** 8.4 g/mL

SAMPLE PROBLEM   Using Density

Density can be used as a factor to convert between the mass (g) and volume (mL) of a substance. The density of silver is 10.5 g/mL. What is the mass, in grams, of 6.0 mL of silver?

Solution:

STEP 1 State the given and needed quantities.

Analyze the Problem	Given	Need	Connect
	6.0 mL of silver, density 10.5 g/mL	grams of silver	density conversion factor

STEP 2 Write a plan to calculate the needed quantity.

$$\text{milliliters} \xrightarrow[\text{factor}]{\text{Density}} \text{grams}$$

STEP 3 Write the equalities and their conversion factors including density.

1 mL of silver = 10.5 g of silver

$$\frac{10.5 \text{ g}}{1 \text{ mL}} \quad \text{and} \quad \frac{1 \text{ mL}}{10.5 \text{ g}}$$

STEP 4 Set up the problem to calculate the needed quantity.

$$6.0 \text{ mL silver} \times \underbrace{\frac{10.5 \text{ g silver}}{1 \text{ mL silver}}}_{\textit{Density factor}} = 63 \text{ g of silver}$$

♦ **Learning Exercise 2.7B**

Use density or specific gravity as a conversion factor to solve each of the following:

a. A sugar solution has a density of 1.20 g/mL. What is the mass, in grams, of 0.225 L of the solution?

> **CORE CHEMISTRY SKILL**
>
> Using Density as a Conversion Factor

b. A piece of pure gold weighs 0.26 lb. If gold has a density of 19.3 g/mL, what is the volume, in milliliters, of the piece of gold?

c. A urine sample has a specific gravity of 1.015 and a volume of 42.50 mL. What is the mass, in grams, of the solution?

d. A 600.-g sample of an intravenous glucose solution has a density of 1.28 g/mL. What is the volume, in liters, of the sample?

A dipstick is used to measure the specific gravity of a urine sample.
Credit: JPC-PROD/Fotolia

Answers   **a.** 270. g   **b.** 6.1 mL   **c.** 43.14 g   **d.** 0.469 L

Checklist for Chapter 2

You are ready to take the Practice Test for Chapter 2. Be sure you have accomplished the following learning goals for this chapter. If not, review the Section listed at the end of the goal. Then apply your new skills and understanding to the Practice Test.

After studying Chapter 2, I can successfully:

_____ Write the names and abbreviations for the metric (SI) units of measurement. (2.1)

_____ Identify a number as a measured number or an exact number. (2.2)

_____ Determine the number of significant figures in measured numbers. (2.2)

_____ Report a calculated answer with the correct number of significant figures. (2.3)

_____ Write an equality from the relationship between two equal quantities. (2.4)

_____ Write two conversion factors for an equality. (2.5)

_____ Use conversion factors to change from one unit to another unit. (2.6)

_____ Calculate the density or specific gravity of a substance or use density or specific gravity to calculate the mass or volume. (2.7)

Practice Test for Chapter 2

The chapter Sections to review are shown in parentheses at the end of each question.

1. Which of the following is a metric measurement of volume? (2.1)
   A. kilogram   **B.** kilowatt   **C.** kiloliter   **D.** kilometer   **E.** kiloquart

2. Which of the following is a metric measurement of length? (2.1)
   A. centiliter   **B.** centigram   **C.** centiyard   **D.** centifoot   **E.** centimeter

3. The exact number in the following is (2.2)
   A. 4.5 L   **B.** 4 syringes   **C.** 6 qt   **D.** 15 g   **E.** 2.0 m

4. The measured number in the following is (2.2)
   A. 1 book   **B.** 2 cars   **C.** 4 flowers   **D.** 5 rings   **E.** 4 g

5. The number of significant figures in 105.4 m is (2.2)
   A. 1   **B.** 2   **C.** 3   **D.** 4   **E.** 5

6. The number of significant figures in 0.000 82 g is (2.2)
   A. 1   **B.** 2   **C.** 3   **D.** 4   **E.** 5

7. The calculator answer 5.7805 rounded to two significant figures is (2.2)
   A. 5   **B.** 5.7   **C.** 5.8   **D.** 5.78   **E.** 6.0

8. The calculator answer 3486.512 rounded to three significant figures is (2.2)
   A. 4000   **B.** 3500   **C.** 349   **D.** 3487   **E.** 3490

9. The answer for $16.0 \div 8.0$ with the correct number of significant figures is (2.3)
   A. 2   **B.** 2.0   **C.** 2.00   **D.** 0.2   **E.** 5.0

10. The answer for $58.5 + 9.158$ with the correct number of decimal places is (2.3)
    A. 67   **B.** 67.6   **C.** 67.7   **D.** 67.66   **E.** 67.658

11. The answer for $\dfrac{2.5 \times 3.12}{4.6}$ with the correct number of significant figures is (2.3)

    A. 0.54   **B.** 7.8   **C.** 0.85   **D.** 1.7   **E.** 1.69

12. Which of these prefixes has the largest value? (2.4)

    A. centi   **B.** deci   **C.** milli   **D.** kilo   **E.** micro

13. What is the decimal equivalent of the prefix *centi*? (2.4)

    A. 0.001   **B.** 0.01   **C.** 0.1   **D.** 10   **E.** 100

14. Which of the following is the smallest unit of mass measurement? (2.4)

    A. gram   **B.** milligram   **C.** kilogram   **D.** decigram   **E.** centigram

15. Which of the following is the largest unit of volume measurement? (2.4)

    A. mL   **B.** dL   **C.** cm^3 **D.** L   **E.** kL

16. Which of the following is a conversion factor? (2.5)

    A. 12 in.   **B.** 3 ft = 1 yd   **C.** 20 m   **D.** $\dfrac{1000 \text{ g}}{1 \text{ kg}}$ **E.** 2 cm^3

17. Which is the correct conversion factor for milliliters and liters? (2.5)

    A. $\dfrac{1000 \text{ mL}}{1 \text{ L}}$ **B.** $\dfrac{100 \text{ mL}}{1 \text{ L}}$ **C.** $\dfrac{10 \text{ mL}}{1 \text{ L}}$ **D.** $\dfrac{0.01 \text{ mL}}{1 \text{ L}}$ **E.** $\dfrac{0.001 \text{ mL}}{1 \text{ L}}$

18. Which is the correct conversion factor for millimeters and centimeters? (2.5)

    A. $\dfrac{1 \text{ mm}}{1 \text{ cm}}$ **B.** $\dfrac{10 \text{ mm}}{1 \text{ cm}}$ **C.** $\dfrac{100 \text{ cm}}{1 \text{ mm}}$ **D.** $\dfrac{100 \text{ mm}}{1 \text{ cm}}$ **E.** $\dfrac{10 \text{ cm}}{1 \text{ mm}}$

19. A doctor orders 20 mg of prednisone. If a tablet contains 5 mg of prednisone, how many tablets should be given? (2.6)

    A. 1   **B.** 2   **C.** 3   **D.** 4   **E.** 5

20. 294 mm is equal to (2.6)

    A. 2940 m   **B.** 29.4 m   **C.** 2.94 m   **D.** 0.294 m   **E.** 0.0294 m

21. A doctor orders 200 mg of penicillin to be injected. If it is available as 1000 mg of penicillin per 5 mL, how many milliliters should be given? (2.6)

    A. 1 mL   **B.** 2 mL   **C.** 3 mL   **D.** 4 mL   **E.** 5 mL

22. What is the volume, in liters, of 800. mL of blood plasma? (2.6)

    A. 0.8 L   **B.** 0.80 L   **C.** 0.800 L   **D.** 8.0 L   **E.** 80.0 L

23. What is the mass, in kilograms, of a 22-lb toddler? (2.6)

    A. 10. kg   **B.** 48 kg   **C.** 10 000 kg   **D.** 0.048 kg   **E.** 22 000 kg

24. The number of milliliters in 2 dL of an antiseptic liquid is (2.6)

    A. 20 mL   **B.** 200 mL   **C.** 2000 mL   **D.** 20 000 mL   **E.** 500 000 mL

25. What is the height, in meters, of a person who is 5 ft 4 in. tall? (2.6)

    A. 64 m   **B.** 25 m   **C.** 14 m   **D.** 1.6 m   **E.** 1.3 m

26. How many ounces are in 1500 grams of carrots? (2.6)

    A. 94 oz   **B.** 53 oz   **C.** 24 000 oz   **D.** 33 oz   **E.** 3.3 oz

27. How many quarts of orange juice are in 255 mL of juice? (2.6)

    A. 0.255 qt   **B.** 270 qt   **C.** 236 qt   **D.** 0.270 qt   **E.** 3.71 qt

28. Your doctor places you on a 2200-kcal diet, with 19% of the kilocalories from fat. How many kilocalories are you allowed from fat? (2.6)

    A. 19 kcal   **B.** 2200 kcal   **C.** 1900 kcal   **D.** 420 kcal   **E.** 4200 kcal

29. 1.5 ft is the same length, in centimeters, as (2.6)

    A. 46 cm   **B.** 7.1 cm   **C.** 18 cm   **D.** 3.8 cm   **E.** 0.59 cm

30. How many milliliters of a salt solution with a density of 1.8 g/mL are needed to provide 450 g of salt solution? (2.7)

    A. 250 mL   **B.** 25 mL   **C.** 810 mL   **D.** 450 mL   **E.** 45 mL

31. Three liquids have densities of 1.15 g/mL, 0.79 g/mL, and 0.95 g/mL. When the liquids, which do not mix, are poured into a graduated cylinder, the liquid at the top has a density of (2.7)

    A. 1.15 g/mL   **B.** 1.00 g/mL   **C.** 0.95 g/mL   **D.** 0.79 g/mL   **E.** 0.16 g/mL

32. A sample of oil has a mass of 65 g and a volume of 80.0 mL. What is the specific gravity of the oil? (2.7)

    A. 1.5   **B.** 1.4   **C.** 1.2   **D.** 0.90   **E.** 0.81

33. What is the mass, in grams, of a 15.0-mL sample of an antihistamine solution with a density of 1.04 g/mL? (2.7)

    A. 104 g   **B.** 10.4 g   **C.** 15.6 g   **D.** 1.56 g   **E.** 14.4 g

34. Ethanol has a density of 0.785 g/mL. What is the mass, in grams, of 0.250 L of ethanol? (2.7)

    A. 196 g   **B.** 158 g   **C.** 318 g   **D.** 0.253 g   **E.** 0.160 g

35. A patient has a blood volume of 6.8 qt. If the density of blood is 1.06 g/mL, what is the mass, in grams, of the patient's blood? (2.7)

    A. 1060 g   **B.** 6400 g   **C.** 6800 g   **D.** 1500 g   **E.** 990 g

Answers to the Practice Test

1. C	**2.** E	**3.** B	**4.** E	**5.** D
6. B	**7.** C	**8.** E	**9.** B	**10.** C
11. D	**12.** D	**13.** B	**14.** B	**15.** E
16. D	**17.** A	**18.** B	**19.** D	**20.** D
21. A	**22.** C	**23.** A	**24.** B	**25.** D
26. B	**27.** D	**28.** D	**29.** A	**30.** A
31. D	**32.** E	**33.** C	**34.** A	**35.** C

Selected Answers and Solutions to Text Problems

2.1   **a.** The abbreviation for the unit gram is g.
     b. The abbreviation for the unit degree Celsius is °C.
     c. The abbreviation for the unit liter is L.
     d. The abbreviation for the unit pound is lb.
     e. The abbreviation for the unit second is s.

2.3   **a.** A liter is a unit of volume.
     b. A centimeter is a unit of length.
     c. A kilometer is a unit of length.
     d. A second is a unit of time.

2.5   **a.** The unit is a meter, which is a unit of length.
     b. The unit is a gram, which is a unit of mass.
     c. The unit is a milliliter, which is a unit of volume.
     d. The unit is a second, which is a unit of time.
     e. The unit is a degree Celsius, which is a unit of temperature.

2.7   **a.** The unit is a second, which is a unit of time.
     b. The unit is a kilogram, which is a unit of mass.
     c. The unit is a gram, which is a unit of mass.
     d. The unit is a degree Celsius, which is a unit of temperature.

2.9   **a.** All five numbers are significant figures (5 SFs).
     b. Only the two nonzero numbers are significant (2 SFs); the preceding zeros are placeholders.
     c. Only the two nonzero numbers are significant (2 SFs); the zeros that follow are placeholders.
     d. All three numbers in the coefficient of a number written in scientific notation are significant (3 SFs).
     e. All four numbers to the right of the decimal point, including the last zero in the decimal number, are significant (4 SFs).
     f. All three numbers, including the zeros at the end of a decimal number, are significant (3 SFs).

2.11   Both measurements in part **b** have three significant figures, and both measurements in part **c** have two significant figures.

2.13   **a.** Zeros at the beginning of a decimal number are not significant.
     b. Zeros between nonzero digits are significant.
     c. Zeros at the end of a decimal number are significant.
     d. Zeros in the coefficient of a number written in scientific notation are significant.
     e. Zeros used as placeholders in a large number without a decimal point are not significant.

2.15   **a.** 5000 L is the same as 5×1000 L, which is written in scientific notation as 5.0×10^3 L with two significant figures.
     b. 30 000 g is the same as $3 \times 10\,000$ g, which is written in scientific notation as 3.0×10^4 g with two significant figures.
     c. 100 000 m is the same as $1 \times 100\,000$ m, which is written in scientific notation as 1.0×10^5 m with two significant figures.
     d. 0.000 25 cm is the same as $2.5 \times \dfrac{1}{10\,000}$ cm, which is written in scientific notation as 2.5×10^{-4} cm.

2.17   Measured numbers are obtained using some type of measuring device. Exact numbers are numbers obtained by counting items or using a definition that compares two units in the same measuring system.
     a. The value 67.5 kg is a measured number; measurement of mass requires a measuring device.
     b. The value 2 tablets is obtained by counting, making it an exact number.

    c. The values in the metric definition 1 L = 1000 mL are exact numbers.

    d. The value 1720 km is a measured number; measurement of distance requires a measuring device.

2.19 Measured numbers are obtained using some type of measuring device. Exact numbers are numbers obtained by counting items or using a definition that compares two units in the same measuring system.

    a. 6 oz of hamburger meat is a measured number (3 hamburgers is a counted/exact number).

    b. Neither are measured numbers (both 1 table and 4 chairs are counted/exact numbers).

    c. Both 0.75 lb of grapes and 350 g of butter are measured numbers.

    d. Neither are measured numbers (the values in a definition are exact numbers).

2.21   **a.** 1.607 kg is a measured number and has 4 SFs.

    b. 130 mcg is a measured number and has 2 SFs.

    c. 4.02×10^6 red blood cells is a measured number and has 3 SFs.

    d. 23 babies is a counted/exact number.

2.23   **a.** 1.85 kg; the last digit is dropped since it is 4 or less.

    b. 88.2 L; since the fourth digit is 4 or less, the last three digits are dropped.

    c. 0.004 74 cm; since the fourth significant digit (the first digit to be dropped) is 5 or greater, the last retained digit is increased by 1 when the last four digits are dropped.

    d. 8810 m; since the fourth significant digit (the first digit to be dropped) is 5 or greater, the last retained digit is increased by 1 when the last digit is dropped (a nonsignificant zero is added at the end as a placeholder).

    e. 1.83×10^5 s; since the fourth digit is 4 or less, the last digit is dropped. The $\times 10^5$ is retained so that the magnitude of the answer is not changed.

2.25   **a.** To round off 56.855 m to three significant figures, drop the final digits 55 and increase the last retained digit by 1 to give 56.9 m.

    b. To round off 0.002 282 g to three significant figures, drop the final digit 2 to give 0.002 28 g.

    c. To round off 11 527 s to three significant figures, drop the final digits 27 and add two zeros as placeholders to give 11 500 s (1.15×10^4 s).

    d. To express 8.1 L to three significant figures, add a significant zero to give 8.10 L.

2.27   **a.** $45.7 \times 0.034 = 1.6$ Two significant figures are allowed since 0.034 has 2 SFs.

    b. $0.002\ 78 \times 5 = 0.01$ One significant figure is allowed since 5 has 1 SF.

    c. $\dfrac{34.56}{1.25} = 27.6$ Three significant figures are allowed since 1.25 has 3 SFs.

    d. $\dfrac{(0.2465)(25)}{1.78} = 3.5$ Two significant figures are allowed since 25 has 2 SFs.

    e. $(2.8 \times 10^4)(5.05 \times 10^{-6}) = 0.14$ or 1.4×10^{-1} Two significant figures are allowed since 2.8×10^4 has 2 SFs.

    f. $\dfrac{(3.45 \times 10^{-2})(1.8 \times 10^5)}{(8 \times 10^3)} = 0.8$ or 8×10^{-1} One significant figure is allowed since 8×10^3 has 1 SF.

2.29   **a.** $45.48 + 8.057 = 53.54$ Two decimal places are allowed since 45.48 has two decimal places.

    b. $23.45 + 104.1 + 0.025 = 127.6$ One decimal place is allowed since 104.1 has one decimal place.

    c. $145.675 - 24.2 = 121.5$ One decimal place is allowed since 24.2 has one decimal place.

    d. $1.08 - 0.585 = 0.50$ Two decimal places are allowed since 1.08 has two decimal places.

    e. $2300 + 196.11 = 2500$ Answer is rounded off to the hundreds place since 2300 is expressed to the hundreds place.

    f. $145.111 - 22.9 + 34.49 = 156.7$ One decimal place is allowed since 22.9 has one decimal place.

2.31   **a.** mg   **b.** dL   **c.** km   **d.** pg

2.33   **a.** centiliter   **b.** kilogram   **c.** millisecond   **d.** petameter

2.35   **a.** 0.01   **b.** 1 000 000 000 000 (or 1×10^{12})
        c. 0.001 (or 1×10^{-3})   **d.** 0.1

2.37   **a.** decigram   **b.** microgram   **c.** kilogram   **d.** centigram

2.39   **a.** 1 m = 100 cm   **b.** $1 \text{ m} = 1 \times 10^{9} \text{ nm}$
        c. 1 mm = 0.001 m   **d.** 1 L = 1000 mL

2.41   **a.** kilogram, since 10^{3} g is greater than 10^{-3} g
        b. milliliter, since 10^{-3} L is greater than 10^{-6} L
        c. km, since 10^{3} m is greater than 10^{0} m
        d. kL, since 10^{3} L is greater than 10^{-1} L
        e. nanometer, since 10^{-9} m is greater than 10^{-12} m

2.43   A conversion factor can be inverted to give a second conversion factor: $\dfrac{1 \text{ m}}{100 \text{ cm}}$ and $\dfrac{100 \text{ cm}}{1 \text{ m}}$

2.45   **a.** 1 m = 100 cm; $\dfrac{100 \text{ cm}}{1 \text{ m}}$ and $\dfrac{1 \text{ m}}{100 \text{ cm}}$

        b. $1 \text{ g} = 1 \times 10^{9} \text{ ng}$; $\dfrac{1 \times 10^{9} \text{ ng}}{1 \text{ g}}$ and $\dfrac{1 \text{ g}}{1 \times 10^{9} \text{ ng}}$

        c. 1 kL = 1000 L; $\dfrac{1000 \text{ L}}{1 \text{ kL}}$ and $\dfrac{1 \text{ kL}}{1000 \text{ L}}$

        d. 1 s = 1000 ms; $\dfrac{1000 \text{ ms}}{1 \text{ s}}$ and $\dfrac{1 \text{ s}}{1000 \text{ ms}}$

        e. 1 dm = 100 mm; $\dfrac{100 \text{ mm}}{1 \text{ dm}}$ and $\dfrac{1 \text{ dm}}{100 \text{ mm}}$

2.47   **a.** 1 yd = 3 ft; $\dfrac{3 \text{ ft}}{1 \text{ yd}}$ and $\dfrac{1 \text{ yd}}{3 \text{ ft}}$; the 1 yd and 3 ft are both exact (U.S. definition).

        b. 1 kg = 2.20 lb; $\dfrac{2.20 \text{ lb}}{1 \text{ kg}}$ and $\dfrac{1 \text{ kg}}{2.20 \text{ lb}}$; the 2.20 lb is measured: it has 3 SFs; the 1 kg is exact.

        c. 1 gal of gasoline = 27 mi; $\dfrac{27 \text{ mi}}{1 \text{ gal gasoline}}$ and $\dfrac{1 \text{ gal gasoline}}{27 \text{ mi}}$; the 27 mi is measured: it has 2 SFs; the 1 gal is exact.

        d. 100 g of sterling = 93 g of silver; $\dfrac{93 \text{ g silver}}{100 \text{ g sterling}}$ and $\dfrac{100 \text{ g sterling}}{93 \text{ g silver}}$; the 93 g is measured: it has 2 SFs; the 100 g is exact.

        e. 1 min = 60 sec; $\dfrac{60 \text{ sec}}{1 \text{ min}}$ and $\dfrac{1 \text{ min}}{60 \text{ sec}}$; the 1 min and 60 sec are both exact.

2.49   **a.** 1 s = 3.5 m; $\dfrac{3.5 \text{ m}}{1 \text{ s}}$ and $\dfrac{1 \text{ s}}{3.5 \text{ m}}$; the 3.5 m is measured: it has 2 SFs; the 1 s is exact.

        b. 1 day = 3.5 g of potassium; $\dfrac{3.5 \text{ g potassium}}{1 \text{ day}}$ and $\dfrac{1 \text{ day}}{3.5 \text{ g potassium}}$; the 3.5 g is measured: it has 2 SFs; the 1 day is exact.

        c. 1 L of gasoline = 26.0 km; $\dfrac{26.0 \text{ km}}{1 \text{ L gasoline}}$ and $\dfrac{1 \text{ L gasoline}}{26.0 \text{ km}}$; the 26.0 km is measured: it has 3 SFs; the 1 L is exact.

d. 1 kg of plums = 29 mcg of pesticide; $\dfrac{29 \text{ mcg pesticide}}{1 \text{ kg plums}}$ and $\dfrac{1 \text{ kg plums}}{29 \text{ mcg pesticide}}$; the 29 mcg

is measured: it has 2 SFs; the 1 kg is exact.

e. 100 g of crust = 28.2 g of silicon; $\dfrac{28.2 \text{ g silicon}}{100 \text{ g crust}}$ and $\dfrac{100 \text{ g crust}}{28.2 \text{ g silicon}}$; the 28.2 g is measured:

it has 3 SFs; the 100 g is exact.

2.51 a. 1 tablet = 630 mg of calcium; $\dfrac{630 \text{ mg calcium}}{1 \text{ tablet}}$ and $\dfrac{1 \text{ tablet}}{630 \text{ mg calcium}}$; the 630 mg is

measured: it has 2 SFs; the 1 tablet is exact.

b. 1 day = 60 mg of vitamin C; $\dfrac{60 \text{ mg vitamin C}}{1 \text{ day}}$ and $\dfrac{1 \text{ day}}{60 \text{ mg vitamin C}}$; the 60 mg is

measured: it has 1 SF; the 1 day is exact.

c. 1 tablet = 50 mg of atenolol; $\dfrac{50 \text{ mg atenolol}}{1 \text{ tablet}}$ and $\dfrac{1 \text{ tablet}}{50 \text{ mg atenolol}}$; the 50 mg is

measured: it has 1 SF; the 1 tablet is exact.

d. 1 tablet = 81 mg of aspirin; $\dfrac{81 \text{ mg aspirin}}{1 \text{ tablet}}$ and $\dfrac{1 \text{ tablet}}{81 \text{ mg aspirin}}$; the 81 mg is measured:

it has 2 SFs; the 1 tablet is exact.

2.53 a. 5 mL of syrup = 10 mg of Atarax; $\dfrac{10 \text{ mg Atarax}}{5 \text{ mL syrup}}$ and $\dfrac{5 \text{ mL syrup}}{10 \text{ mg Atarax}}$

b. 1 tablet = 0.25 g of Lanoxin; $\dfrac{0.25 \text{ g Lanoxin}}{1 \text{ tablet}}$ and $\dfrac{1 \text{ tablet}}{0.25 \text{ g Lanoxin}}$

c. 1 tablet = 300 mg of Motrin; $\dfrac{300 \text{ mg Motrin}}{1 \text{ tablet}}$ and $\dfrac{1 \text{ tablet}}{300 \text{ mg Motrin}}$

2.55 a. Given 44.2 mL   **Need** liters

     Plan mL → L $\dfrac{1 \text{ L}}{1000 \text{ mL}}$

     Set-up 44.2 mL $\times \dfrac{1 \text{ L}}{1000 \text{ mL}} = 0.0442$ L (3 SFs)

b. Given 8.65 m   **Need** nanometers

     Plan m → nm $\dfrac{1 \times 10^9 \text{ nm}}{1 \text{ m}}$

     Set-up 8.65 m $\times \dfrac{1 \times 10^9 \text{ nm}}{1 \text{ m}} = 8.65 \times 10^9$ nm (3 SFs)

c. Given 5.2×10^8 g   **Need** megagrams

     Plan g → Mg $\dfrac{1 \text{ Mg}}{1 \times 10^6 \text{ g}}$

     Set-up 5.2×10^8 g $\times \dfrac{1 \text{ Mg}}{1 \times 10^6 \text{ g}} = 5.2 \times 10^2$ Mg (2 SFs)

d. Given 0.72 ks   **Need** milliseconds

     Plan ks → s → ms $\dfrac{1000 \text{ s}}{1 \text{ ks}}$ $\dfrac{1000 \text{ ms}}{1 \text{ s}}$

     Set-up 0.72 ks $\times \dfrac{1000 \text{ s}}{1 \text{ ks}} \times \dfrac{1000 \text{ ms}}{1 \text{ s}} = 7.2 \times 10^5$ ms (2 SFs)

2.57 a. Given 3.428 lb   **Need** kilograms

Plan lb → kg $\dfrac{1 \text{ kg}}{2.20 \text{ lb}}$

Set-up 3.428 lb × $\dfrac{1 \text{ kg}}{2.20 \text{ lb}}$ = 1.56 kg (3 SFs)

b. Given 1.6 m   **Need** inches

Plan m → in. $\dfrac{39.4 \text{ in.}}{1 \text{ m}}$

Set-up 1.6 m × $\dfrac{39.4 \text{ in.}}{1 \text{ m}}$ = 63 in. (2 SFs)

c. Given 4.2 L   **Need** quarts

Plan L → qt $\dfrac{1.06 \text{ qt}}{1 \text{ L}}$

Set-up 4.2 L × $\dfrac{1.06 \text{ qt}}{1 \text{ L}}$ = 4.5 qt (2 SFs)

d. Given 0.672 ft   **Need** millimeters

Plan ft → in. → cm → mm $\dfrac{12 \text{ in.}}{1 \text{ ft}}$ $\dfrac{2.54 \text{ cm}}{1 \text{ in.}}$ $\dfrac{10 \text{ mm}}{1 \text{ cm}}$

Set-up 0.672 ft × $\dfrac{12 \text{ in.}}{1 \text{ ft}}$ × $\dfrac{2.54 \text{ cm}}{1 \text{ in.}}$ × $\dfrac{10 \text{ mm}}{1 \text{ cm}}$ = 205 mm (3 SFs)

2.59 a. Given 175 cm   **Need** meters

Plan cm → m $\dfrac{1 \text{ m}}{100 \text{ cm}}$

Set-up 175 cm × $\dfrac{1 \text{ m}}{100 \text{ cm}}$ = 1.75 m (3 SFs)

b. Given 5000 mL   **Need** liters

Plan mL → L $\dfrac{1 \text{ L}}{1000 \text{ mL}}$

Set-up 5000 mL × $\dfrac{1 \text{ L}}{1000 \text{ mL}}$ = 5 L (1 SF)

c. Given 0.0055 kg   **Need** grams

Plan kg → g $\dfrac{1000 \text{ g}}{1 \text{ kg}}$

Set-up 0.0055 kg × $\dfrac{1000 \text{ g}}{1 \text{ kg}}$ = 5.5 g (2 SFs)

d. Given 3500 cm³   **Need** liters

Plan cm³ → mL → L $\dfrac{1 \text{ mL}}{1 \text{ cm}^3}$ $\dfrac{1 \text{ L}}{1000 \text{ mL}}$

Set-up 3500 cm³ × $\dfrac{1 \text{ mL}}{1 \text{ cm}^3}$ × $\dfrac{1 \text{ L}}{1000 \text{ mL}}$ = 3.5 L (2 SFs)

2.61 a. Given 0.500 qt   **Need** milliliters

Plan qt → mL $\dfrac{946 \text{ mL}}{1 \text{ qt}}$

Set-up 0.500 qt × $\dfrac{946 \text{ mL}}{1 \text{ qt}}$ = 473 mL (3 SFs)

b. Given   175 lb   **Need**   kilograms

     Plan   lb → kg $\dfrac{1 \text{ kg}}{2.20 \text{ lb}}$

     Set-up $175 \text{ lb} \times \dfrac{1 \text{ kg}}{2.20 \text{ lb}} = 79.5 \text{ kg (3 SFs)}$

c. Given   74 kg body mass, 15% body fat   **Need**   pounds of body fat

     Plan   kg of body mass → kg of body fat → lb of body fat

         (percent equality: 100 kg of body mass = 15 g of body fat)

$$\frac{15 \text{ kg body fat}}{100 \text{ kg body mass}} \qquad \frac{2.20 \text{ lb body fat}}{1 \text{ kg body fat}}$$

     Set-up $74 \text{ kg body mass} \times \dfrac{15 \text{ kg body fat}}{100 \text{ kg body mass}} \times \dfrac{2.20 \text{ lb body fat}}{1 \text{ kg body fat}}$

         $= 24 \text{ lb of body fat (2 SFs)}$

d. Given   10.0 oz of fertilizer, 15% nitrogen   **Need**   grams of nitrogen

     Plan   oz of fertilizer → lb of fertilizer → g of fertilizer → g of nitrogen

         (percent equality: 100 g of fertilizer = 15 g of nitrogen)

$$\frac{1 \text{ lb}}{16 \text{ oz}} \quad \frac{454 \text{ g}}{1 \text{ lb}} \quad \frac{15 \text{ g nitrogen}}{100 \text{ g fertilizer}}$$

     Set-up $10.0 \text{ oz fertilizer} \times \dfrac{1 \text{ lb fertilizer}}{16 \text{ oz fertilizer}} \times \dfrac{454 \text{ g fertilizer}}{1 \text{ lb fertilizer}} \times \dfrac{15 \text{ g nitrogen}}{100 \text{ g fertilizer}}$

         $= 43 \text{ g of nitrogen (2 SFs)}$

2.63   **a. Given**   250 L of water   **Need**   gallons of water

     Plan   L → qt → gal $\dfrac{1.06 \text{ qt}}{1 \text{ L}} \quad \dfrac{1 \text{ gal}}{4 \text{ qt}}$

     Set-up $250 \text{ L} \times \dfrac{1.06 \text{ qt}}{1 \text{ L}} \times \dfrac{1 \text{ gal}}{4 \text{ qt}} = 66 \text{ gal (2 SFs)}$

b. Given   0.024 g of sulfa drug, 8-mg tablets   **Need**   number of tablets

     Plan   g of sulfa drug → mg of sulfa drug → number of tablets

$$\frac{1000 \text{ mg}}{1 \text{ g}} \qquad \frac{1 \text{ tablet}}{8 \text{ mg sulfa drug}}$$

     Set-up $0.024 \text{ g sulfa drug} \times \dfrac{1000 \text{ mg}}{1 \text{ g}} \times \dfrac{1 \text{ tablet}}{8 \text{ mg sulfa drug}} = 3 \text{ tablets (1 SF)}$

c. Given   34-lb child, 115 mg of ampicillin/kg of body mass   **Need**   milligrams of ampicillin

     Plan   lb of body mass → kg of body mass → mg of ampicillin

$$\frac{1 \text{ kg}}{2.20 \text{ lb}} \qquad \frac{115 \text{ mg ampicillin}}{1 \text{ kg body mass}}$$

     Set-up $34 \text{ lb body mass} \times \dfrac{1 \text{ kg body mass}}{2.20 \text{ lb body mass}} \times \dfrac{115 \text{ mg ampicillin}}{1 \text{ kg body mass}}$

         $= 1800 \text{ mg of ampicillin (2 SFs)}$

d. Given   4.0 oz of ointment   **Need**   grams of ointment

     Plan   oz → lb → g $\dfrac{1 \text{ lb}}{16 \text{ oz}} \quad \dfrac{454 \text{ g}}{1 \text{ lb}}$

     Set-up $4.0 \text{ oz} \times \dfrac{1 \text{ lb}}{16 \text{ oz}} \times \dfrac{454 \text{ g}}{1 \text{ lb}} = 110 \text{ g of ointment (2 SFs)}$

2.65 **a.** **Given** 500. mL of IV saline solution, 80. mL/h   **Need** infusion time in hours

     Plan mL of IV saline solution → hours $\dfrac{1\text{ h}}{80.\text{ mL saline solution}}$

     Set-up 500. ~~mL saline solution~~ $\times \dfrac{1\text{ h}}{80.\text{ ~~mL saline solution~~}} = 6.3$ h (2 SFs)

   b. **Given** 72.6-lb child, 1.5 mg of Medrol/kg of body mass, 20. mg of Medrol/mL of solution   **Need** milliliters of Medrol solution

     Plan lb of body mass → kg of body mass → mg of Medrol → mL of solution

$$\frac{1\text{ kg}}{2.20\text{ lb}} \qquad \frac{1.5\text{ mg Medrol}}{1\text{ kg body mass}} \qquad \frac{1\text{ mL solution}}{20.\text{ mg Medrol}}$$

     Set-up 72.6 ~~lb body mass~~ $\times \dfrac{1\text{ ~~kg body mass~~}}{2.20\text{ ~~lb body mass~~}} \times \dfrac{1.5\text{ ~~mg Medrol~~}}{1\text{ ~~kg body mass~~}} \times \dfrac{1\text{ mL solution}}{20.\text{ ~~mg Medrol~~}}$

     $= 2.5$ mL of Medrol solution (2 SFs)

2.67 Density is the mass of a substance divided by its volume. Density $= \dfrac{\text{mass (grams)}}{\text{volume (mL)}}$

The densities of solids and liquids are usually stated in g/mL or g/cm^3, so in some problems the units will need to be converted.

   a. Density $= \dfrac{\text{mass (grams)}}{\text{volume (mL)}} = \dfrac{24.0\text{ g}}{20.0\text{ mL}} = 1.20$ g/mL (3 SFs)

   b. **Given** 0.250 lb of butter, 130.3 mL   **Need** density (g/mL)

     Plan lb → g, then calculate density $\dfrac{454\text{ g}}{1\text{ lb}}$

     Set-up 0.250 ~~lb~~ $\times \dfrac{454\text{ g}}{1\text{ ~~lb~~}} = 113.5$ g (3 SFs allowed)

     $\therefore$ Density $= \dfrac{\text{mass}}{\text{volume}} = \dfrac{113.5\text{ g}}{130.3\text{ mL}} = 0.871$ g/mL (3 SFs)

   c. **Given** 12.00 mL initial volume, 13.45 mL final volume, 4.50 g   **Need** density (g/mL)

     Plan calculate volume by difference, then calculate density

     Set-up volume of gem: 13.45 mL total $-$ 12.00 mL water $= 1.45$ mL

     $\therefore$ Density $= \dfrac{\text{mass}}{\text{volume}} = \dfrac{4.50\text{ g}}{1.45\text{ mL}} = 3.10$ g/mL (3 SFs)

   d. Density $= \dfrac{\text{mass}}{\text{volume}} = \dfrac{3.85\text{ g}}{3.00\text{ mL}} = 1.28$ g/mL (3 SFs)

2.69 **a.** **Given** 514.1 g, 114 cm^3 **Need** density (g/mL)

     Plan convert volume cm^3 → mL, then calculate density

     Set-up 114 ~~cm^3~~ $\times \dfrac{1\text{ mL}}{1\text{ ~~cm^3~~}} = 114$ mL

     $\therefore$ Density $= \dfrac{\text{mass}}{\text{volume}} = \dfrac{514.1\text{ g}}{114\text{ mL}} = 4.51$ g/mL (3 SFs)

   b. **Given** 0.100 pt, 115.25 g initial, 182.48 g final   **Need** density (g/mL)

     Plan pt → mL, then calculate mass by difference, then calculate density

$$\frac{473\text{ mL}}{1\text{ pt}}$$

     Set-up 0.100 ~~pt~~ $\times \dfrac{473\text{ mL}}{1\text{ ~~pt~~}} = 47.3$ mL (3 SFs)

     mass of syrup $= 182.48$ g $- 115.25$ g $= 67.23$ g

     $\therefore$ Density $= \dfrac{\text{mass}}{\text{volume}} = \dfrac{67.23\text{ g}}{47.3\text{ mL}} = 1.42$ g/mL (3 SFs)

c. Given 8.51 kg, 3.15 L   **Need** density (g/mL)

Plan kg → g and L → mL, then calculate density $\dfrac{1000 \text{ g}}{1 \text{ kg}}$ $\dfrac{1000 \text{ mL}}{1 \text{ L}}$

Set-up $8.51 \text{ kg} \times \dfrac{1000 \text{ g}}{1 \text{ kg}} = 8510 \text{ g (3 SFs) and}$

$3.15 \text{ L} \times \dfrac{1000 \text{ mL}}{1 \text{ L}} = 3150 \text{ mL (3 SFs)}$

$\therefore \text{ Density} = \dfrac{\text{mass}}{\text{volume}} = \dfrac{8510 \text{ g}}{3150 \text{ mL}} = 2.70 \text{ g/mL (3 SFs)}$

2.71 In these problems, the density is used as a conversion factor.

a. Given 1.50 kg of ethanol   **Need** liters of ethanol

Plan kg → g → mL → L $\dfrac{1000 \text{ g}}{1 \text{ kg}}$ $\dfrac{1 \text{ mL}}{0.79 \text{ g}}$ $\dfrac{1 \text{ L}}{1000 \text{ mL}}$

Set-up $1.50 \text{ kg alcohol} \times \dfrac{1000 \text{ g}}{1 \text{ kg alcohol}} \times \dfrac{1 \text{ mL}}{0.79 \text{ g}} \times \dfrac{1 \text{ L}}{1000 \text{ mL}}$

$= 1.9 \text{ L of ethanol (2 SFs)}$

b. Given 6.5 mL of mercury   **Need** grams of mercury

Plan mL → g $\dfrac{13.6 \text{ g}}{1 \text{ mL}}$

Set-up $6.5 \text{ mL} \times \dfrac{13.6 \text{ g}}{1 \text{ mL}} = 88 \text{ g of mercury (2 SFs)}$

c. Given 225 cm^3 of silver   **Need** ounces of silver

Plan cm^3 → mL → g → lb → oz $\dfrac{1 \text{ mL}}{1 \text{ cm}^3}$ $\dfrac{10.5 \text{ g}}{1 \text{ cm}^3}$ $\dfrac{1 \text{ lb}}{454 \text{ g}}$ $\dfrac{16 \text{ oz}}{1 \text{ lb}}$

Set-up $225 \text{ cm}^3 \times \dfrac{1 \text{ mL}}{1 \text{ cm}^3} \times \dfrac{10.5 \text{ g}}{1 \text{ mL}} \times \dfrac{1 \text{ lb}}{454 \text{ g}} \times \dfrac{16 \text{ oz}}{1 \text{ lb}} = 83.3 \text{ oz of silver (3 SFs)}$

2.73 a. Given 74.1 cm^3 of copper   **Need** grams of copper

Plan cm^3 → mL → g $\dfrac{1 \text{ mL}}{1 \text{ cm}^3}$ $\dfrac{8.92 \text{ g}}{1 \text{ mL}}$

Set-up $74.1 \text{ cm}^3 \times \dfrac{1 \text{ mL}}{1 \text{ cm}^3} \times \dfrac{8.92 \text{ g}}{1 \text{ mL}} = 661 \text{ g of copper (3 SFs)}$

b. Given 12.0 gal of gasoline   **Need** kilograms of gasoline

Plan gal → qt → mL → g → kg $\dfrac{4 \text{ qt}}{1 \text{ gal}}$ $\dfrac{946 \text{ mL}}{1 \text{ qt}}$ $\dfrac{0.74 \text{ g}}{1 \text{ mL}}$ $\dfrac{1 \text{ kg}}{1000 \text{ g}}$

Set-up $12.0 \text{ gal} \times \dfrac{4 \text{ qt}}{1 \text{ gal}} \times \dfrac{946 \text{ mL}}{1 \text{ qt}} \times \dfrac{0.74 \text{ g}}{1 \text{ mL}} \times \dfrac{1 \text{ kg}}{1000 \text{ g}} = 34 \text{ kg of gasoline (2 SFs)}$

c. Given 27 g of ice   **Need** cubic centimeters of ice

Plan g → mL → cm^3 $\dfrac{1 \text{ mL}}{0.92 \text{ g}}$ $\dfrac{1 \text{ cm}^3}{1 \text{ mL}}$

Set-up $27 \text{ g} \times \dfrac{1 \text{ mL}}{0.92 \text{ g}} \times \dfrac{1 \text{ cm}^3}{1 \text{ mL}} = 29 \text{ cm}^3 \text{ (2 SFs)}$

2.75 Because the density of aluminum is 2.70 g/cm^3, silver is 10.5 g/cm^3, and lead is 11.3 g/cm^3, we can identify the unknown metal by calculating its density as follows:

$\text{Density} = \dfrac{\text{mass of metal}}{\text{volume of metal}} = \dfrac{217 \text{ g}}{19.2 \text{ cm}^3} = 11.3 \text{ g/cm}^3 \text{ (3 SFs)}$

$\therefore$ the metal is lead.

2.77 **a.** Specific gravity $= \dfrac{\text{density of substance}}{\text{density of water}} = \dfrac{1.030 \text{ g/mL}}{1.00 \text{ g/mL}} = 1.03$ (3 SFs)

b. Density $= \dfrac{\text{mass of glucose solution}}{\text{volume of glucose solution}} = \dfrac{20.6 \text{ g}}{20.0 \text{ mL}} = 1.03$ g/mL (3 SFs)

c. Specific gravity $= \dfrac{\text{density of substance}}{\text{density of water}}$

$\therefore$ Density of substance = specific gravity $\times$ density of water
= 0.92×1.00 g/mL = 0.92 g/mL
Mass of substance = volume of substance $\times$ density of substance
= $750 \text{ mL} \times 0.92 \text{ g/mL} = 690$ g (2 SFs)

d. Specific gravity $= \dfrac{\text{density of substance}}{\text{density of water}}$

$\therefore$ Density of substance = specific gravity $\times$ density of water
= 0.850×1.00 g/mL = 0.850 g/mL

$\therefore$ Volume of solution $= \dfrac{\text{mass of solution}}{\text{density of solution}} = \dfrac{325 \text{ g}}{0.850 \text{ g/mL}} = 382$ mL (3 SFs)

2.79 **a.** 1 dL of blood = 42 mcg of iron; $\dfrac{42 \text{ mcg iron}}{1 \text{ dL blood}}$ and $\dfrac{1 \text{ dL blood}}{42 \text{ mcg iron}}$

b. **Given** 8.0-mL blood sample, 42 mcg of iron/dL   **Need** micrograms of iron

Plan mL of blood $\rightarrow$ dL of blood $\rightarrow$ mcg of iron $\dfrac{1 \text{ dL blood}}{100 \text{ mL blood}}$ $\dfrac{42 \text{ mcg iron}}{1 \text{ dL blood}}$

Set-up $8.0 \text{ mL blood} \times \dfrac{1 \text{ dL blood}}{100 \text{ mL blood}} \times \dfrac{42 \text{ mcg iron}}{1 \text{ dL blood}} = 3.4$ mcg of iron (2 SFs)

2.81 Both measurements in part **c** have two significant figures, and both measurements in part **d** have four significant figures.

2.83 **a.** The number of legs is a counted number; it is exact.
b. The height is measured with a ruler or tape measure; it is a measured number.
c. The number of chairs is a counted number; it is exact.
d. The area is calculated from the length and width, which are measured with a ruler or tape measure; it is a measured number.

2.85 61.5 °C

2.87 **a.** length $= 38.4 \text{ in.} \times \dfrac{2.54 \text{ cm}}{1 \text{ in.}} = 97.5$ cm (3 SFs)

b. length $= 24.2 \text{ in.} \times \dfrac{2.54 \text{ cm}}{1 \text{ in.}} = 61.5$ cm (3 SFs)

c. There are three significant figures in the length measurement.
d. Area = length $\times$ width = 97.5 cm $\times$ 61.5 cm = 6.00×10^3 cm^2 (3 SFs)

2.89 **a.** Diagram 3; a cube that has a greater density than the water will sink to the bottom.
b. Diagram 4; a cube with a density of 0.80 g/mL will be about four-fifths submerged in the water.
c. Diagram 1; a cube with a density that is one-half the density of water will be one-half submerged in the water.
d. Diagram 2; a cube with the same density as water will float just at the surface of the water.

2.91 Since all three solids have a mass of 10.0 g, the one with the smallest volume must have the highest density; the one with the largest volume will have the lowest density.

A would be gold; it has the highest density (19.3 g/mL) and the smallest volume.
B would be silver; its density is intermediate (10.5 g/mL) and the volume is intermediate.
C would be aluminum; it has the lowest density (2.70 g/mL) and the largest volume.

2.93 The green cube has the same volume as the gray cube. However, the green cube has a larger mass on the scale, which means that its mass/volume ratio is larger. Thus, the density of the green cube is higher than the density of the gray cube.

2.95 **a.** To round off 0.000 012 58 L to three significant figures, drop the final digit 8 and increase the last retained digit by 1 to give 0.000 012 6 L or 1.26×10^{-5} L.

   b. To round off 3.528×10^2 kg to three significant figures, drop the final digit 8 and increase the last retained digit by 1 to give 353 kg (3.53×10^2 kg).

   c. To express 125 111 m to three significant figures, drop the final digits 111 and add three zeros as placeholders to give 125 000 m (or 1.25×10^5 m).

   d. To express 34.9673 s to three significant figures, drop the final digits 673 and increase the last retained digit by 1 to give 35.0 s.

2.97 **a.** The total mass is the sum of the individual components of the dessert.
   137.25 g + 84 g + 43.7 g = 265 g. No places to the right of the decimal point are allowed since the mass of the fudge sauce (84 g) has no digits to the right of the decimal point.

   b. **Given** grams of dessert from part **a**   **Need** pounds of dessert

   Plan g → lb $\dfrac{1 \text{ lb}}{454 \text{ g}}$

   Set-up 265 g dessert (total) $\times \dfrac{1 \text{ lb}}{454 \text{ g}}$ = 0.584 lb of dessert (3 SFs)

2.99 **Given** 1.95 euros/kg of grapes, \$1.14/euro   **Need** cost in dollars per pound

   Plan euros/kg → euros/lb → \$/lb $\dfrac{1.95 \text{ euros}}{1 \text{ kg grapes}}$ $\dfrac{1 \text{ kg}}{2.20 \text{ lb}}$ $\dfrac{\$1.14}{1 \text{ euro}}$

   Set-up $\dfrac{1.95 \text{ euros}}{1 \text{ kg grapes}} \times \dfrac{1 \text{ kg}}{2.20 \text{ lb}} \times \dfrac{\$1.14}{1 \text{ euro}}$ = \$1.01/lb of grapes (3 SFs)

2.101 **Given** 4.0 lb of onions **Need** number of onions

   Plan lb → g → number of onions $\dfrac{454 \text{ g}}{1 \text{ lb}}$ $\dfrac{1 \text{ onion}}{115 \text{ g}}$

   Set-up 4.0 lb onions $\times \dfrac{454 \text{ g}}{1 \text{ lb}} \times \dfrac{1 \text{ onion}}{115 \text{ g}}$ = 16 onions (2 SFs)

2.103 **Given** 7500 ft   **Need** minutes

   Plan ft → in. → cm → m → min $\dfrac{12 \text{ in.}}{1 \text{ ft}}$ $\dfrac{2.54 \text{ cm}}{1 \text{ in.}}$ $\dfrac{1 \text{ m}}{100 \text{ cm}}$ $\dfrac{1 \text{ min}}{55.0 \text{ m}}$

   Set-up 7500 ft $\times \dfrac{12 \text{ in.}}{1 \text{ ft}} \times \dfrac{2.54 \text{ cm}}{1 \text{ in.}} \times \dfrac{1 \text{ m}}{100 \text{ cm}} \times \dfrac{1 \text{ min}}{55.0 \text{ m}}$ = 42 min (2 SFs)

2.105 **Given** 215 mL initial, 285 mL final volume, density of lead 11.3 g/mL   **Need** grams of lead

   Plan calculate the volume by difference and mL → g $\dfrac{11.3 \text{ g}}{1 \text{ mL}}$

   Set-up The difference between the initial volume of the water and its volume with the lead object will give us the volume of the lead object: 285 mL total − 215 mL water = 70. mL of lead, then 70. mL lead $\times \dfrac{11.3 \text{ g lead}}{1 \text{ mL lead}}$ = 790 g of lead (2 SFs)

2.107 **Given** 1.2 kg of gasoline   **Need** milliliters of gasoline

   Plan kg → g → mL $\dfrac{1000 \text{ g}}{1 \text{ mL}}$ $\dfrac{1 \text{ mL}}{0.74 \text{ g}}$

   Set-up 1.2 kg $\times \dfrac{1000 \text{ g}}{1 \text{ kg}} \times \dfrac{1 \text{ mL}}{0.74 \text{ g}}$ = 1600 mL (1.6×10^3 mL) of gasoline (2 SFs)

2.109 a. Given 8.0 oz **Need** number of crackers

> **Plan** oz → number of crackers $\dfrac{6 \text{ crackers}}{0.50 \text{ oz}}$

> **Set-up** $8.0 \text{ oz} \times \dfrac{6 \text{ crackers}}{0.50 \text{ oz}} = 96$ crackers (2 SFs)

b. Given 10 crackers, 4 g of fat/serving **Need** ounces of fat

> **Plan** number of crackers → servings → g of fat → lb of fat → oz of fat
>
> $\dfrac{1 \text{ serving}}{6 \text{ crackers}}$ $\dfrac{4 \text{ g fat}}{1 \text{ serving}}$ $\dfrac{1 \text{ lb}}{454 \text{ g}}$ $\dfrac{16 \text{ oz}}{1 \text{ lb}}$

> **Set-up** $10 \text{ crackers} \times \dfrac{1 \text{ serving}}{6 \text{ crackers}} \times \dfrac{4 \text{ g fat}}{1 \text{ serving}} \times \dfrac{1 \text{ lb}}{454 \text{ g}} \times \dfrac{16 \text{ oz}}{1 \text{ lb}} = 0.2$ oz of fat (1 SF)

c. Given 2.4 g of sodium, 140 mg of sodium/serving **Need** number of servings

> **Plan** mg of sodium → servings $\dfrac{100 \text{ mg sodium}}{1 \text{ g sodium}}$ $\dfrac{1 \text{ serving}}{140 \text{ mg sodium}}$

> **Set-up** $2.4 \text{ g sodium} \times \dfrac{1000 \text{ mg sodium}}{1 \text{ g sodium}} \times \dfrac{1 \text{ serving}}{140 \text{ mg sodium}} = 17$ servings (2 SFs)

2.111 Given 10 days, 4 tablets/day, 250-mg tablets **Need** ounces of amoxicillin

> **Plan** days → tablets → mg of amoxicillin → g → lb → oz of amoxicillin
>
> $\dfrac{4 \text{ tablets}}{1 \text{ day}}$ $\dfrac{250 \text{ mg amoxicillin}}{1 \text{ tablet}}$ $\dfrac{1 \text{ g}}{1000 \text{ mg}}$ $\dfrac{1 \text{ lb}}{454 \text{ g}}$ $\dfrac{16 \text{ oz}}{1 \text{ lb}}$

> **Set-up** $10 \text{ days} \times \dfrac{4 \text{ tablets}}{1 \text{ day}} \times \dfrac{250 \text{ mg amoxicillin}}{1 \text{ tablet}} \times \dfrac{1 \text{ g}}{1000 \text{ mg}} \times \dfrac{1 \text{ lb}}{454 \text{ g}} \times \dfrac{16 \text{ oz}}{1 \text{ lb}}$
> $= 0.35$ oz of amoxicillin (2 SFs)

2.113 Given 5.0 mL of elixir, 30. mg of phenobarbital/7.5 mL
Need milligrams of phenobarbital

> **Plan** mL of elixir → mg of phenobarbital $\dfrac{30. \text{ mg phenobarbital}}{7.5 \text{ mL elixir}}$

> **Set-up** $5.0 \text{ mL elixir} \times \dfrac{30. \text{ mg phenobarbital}}{7.5 \text{ mL elixir}} = 20.$ mg of phenobarbital (2 SFs)

2.115 Because the balance can measure mass to 0.001 g, the mass should be reported to 0.001 g. You should record the mass of the object as 31.075 g.

2.117 Given 3.0-h trip **Need** gallons of gasoline

> **Plan** h → mi → km → L → qt → gal $\dfrac{55 \text{ mi}}{1 \text{ h}}$ $\dfrac{1 \text{ km}}{0.621 \text{ mi}}$ $\dfrac{1 \text{ L}}{11 \text{ km}}$ $\dfrac{1.06 \text{ qt}}{1 \text{ L}}$ $\dfrac{1 \text{ gal}}{4 \text{ qt}}$

> **Set-up** $3.0 \text{ h} \times \dfrac{55 \text{ mi}}{1 \text{ h}} \times \dfrac{1 \text{ km}}{0.621 \text{ mi}} \times \dfrac{1 \text{ L}}{11 \text{ km}} \times \dfrac{1.06 \text{ qt}}{1 \text{ L}} \times \dfrac{1 \text{ gal}}{4 \text{ qt}}$
> $= 6.4$ gal of gasoline (2 SFs)

2.119 Given 1.50 L of gasoline **Need** milliliters of olive oil

> **Plan** L of gasoline → mL of gasoline → g of gasoline → g of olive oil → mL of olive oil
> (equality from question: 1 g of olive oil = 1 g of gasoline)
>
> $\dfrac{1000 \text{ mL gasoline}}{1 \text{ L gasoline}}$ $\dfrac{0.74 \text{ g gasoline}}{1 \text{ mL gasoline}}$ $\dfrac{1 \text{ mL olive oil}}{0.92 \text{ g olive oil}}$

> **Set-up** $1.50 \text{ L gasoline} \times \dfrac{1000 \text{ mL}}{1 \text{ L}} \times \dfrac{0.74 \text{ g gasoline}}{1 \text{ mL gasoline}} = 1110$ g of gasoline

> $1110 \text{ g olive oil} \times \dfrac{1 \text{ mL olive oil}}{0.92 \text{ g olive oil}} = 1200$ mL $(1.2 \times 10^3$ mL) of olive oil (2 SFs)

2.121 a. Given   65 kg of body mass, 3.0% fat   **Need**   pounds of fat

     Plan   kg of body mass → kg of fat → lb of fat

           (percent equality: 100 kg of body mass = 3.0 kg of fat)

$$\frac{3.0 \text{ kg fat}}{100 \text{ kg body mass}} \qquad \frac{2.20 \text{ lb fat}}{1 \text{ kg fat}}$$

     Set-up $65 \text{ kg body mass} \times \dfrac{3.0 \text{ kg fat}}{100 \text{ kg body mass}} \times \dfrac{2.20 \text{ lb fat}}{1 \text{ kg fat}} = 4.3 \text{ lb of fat (2 SFs)}$

  b. Given   3.0 L of fat   **Need**   pounds of fat

     Plan   L → mL → g → lb $\dfrac{1000 \text{ mL}}{1 \text{ L}} \quad \dfrac{0.909 \text{ g}}{1 \text{ mL}} \quad \dfrac{1 \text{ lb}}{454 \text{ g}}$

     Set-up $3.0 \text{ L} \times \dfrac{1000 \text{ mL}}{1 \text{ L}} \times \dfrac{0.909 \text{ g}}{1 \text{ mL}} \times \dfrac{1 \text{ lb}}{454 \text{ g}} = 6.0 \text{ lb of fat (2 SFs)}$

3

Matter and Energy

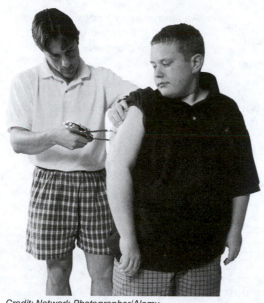

Since Charles had a visit with Daniel, a dietitian, Charles and his mother have been preparing healthier foods with smaller serving sizes. Charles has lost 6 lb. Yesterday he had a blood glucose test that showed his blood glucose level was normal. One of the meals that Charles likes consists of rice, chicken, and carrots. If Charles consumes 31 g of carbohydrate, 4 g of fat, and 24 g of protein, what is the energy content of the meal, in kilojoules and kilocalories?

Credit: Network Photographer/Alamy

LOOKING AHEAD

 The Health icon indicates a question that is related to health and medicine.

3.1 Classification of Matter

Learning Goal: Classify examples of matter as pure substances or mixtures.

- Matter is anything that has mass and occupies space.
- A pure substance, whether it is an element or compound, has a definite composition.
- Elements are the simplest type of matter.
- A compound consists of atoms of two or more elements always chemically combined in the same proportion.
- Mixtures contain two or more substances that are physically, not chemically, combined.
- Mixtures are classified as homogeneous or heterogeneous.

Key Terms for Section 3.1

Match each of the following key terms with the correct description:

a. element **b.** mixture **c.** pure substance
d. matter **e.** compound

An aluminum can consists of many
atoms of the element aluminum.
Credit: Norman Chan/Fotolia

1. _____ anything that has mass and occupies space

2. _____ an element or a compound that has a definite composition

3. _____ the physical combination of two or more substances that
does not change the identities of the substances

4. _____ a pure substance consisting of two or more elements with a definite composition that can be
broken down into simpler substances only by chemical methods

5. _____ a pure substance containing only one type of matter that cannot be broken down by chemical
methods

Answers   **1.** d   **2.** c   **3.** b   **4.** e   **5.** a

♦ **Learning Exercise 3.1A**

Identify each of the following as an element or a compound:

a. _____ iron   **b.** _____ carbon dioxide

c. _____ potassium iodide   **d.** _____ gold

e. _____ aluminum   **f.** _____ table salt (sodium chloride)

Answers   **a.** element   **b.** compound   **c.** compound
                d. element   **e.** element   **f.** compound

♦ **Learning Exercise 3.1B**

Identify each of the following as a pure substance or a mixture:

a. _____ bananas and milk **b.** _____ sulfur

c. _____ silver   **d.** _____ a bag of raisins
                                                  and nuts

e. _____ pure water   **f.** _____ sand and water

Water, H_2O, consists of two
atoms of H for one atom of O.
Credit: Pearson Education, Inc.

Answers   **a.** mixture   **b.** pure substance **c.** pure substance
                d. mixture   **e.** pure substance **f.** mixture

♦ **Learning Exercise 3.1C**

Identify each of the following mixtures as homogeneous or heterogeneous:

a. _____ chocolate milk **b.** _____ sand and water

c. _____ orange soda   **d.** _____ a bag of raisins and nuts

e. _____ air   **f.** _____ vinegar

Answers   **a.** homogeneous   **b.** heterogeneous   **c.** homogeneous
                d. heterogeneous   **e.** homogeneous   **f.** homogeneous

3.2   States and Properties of Matter

Learning Goal: Identify the states and the physical and chemical properties of matter.

- The states of matter are solid, liquid, and gas.
- Physical properties are those characteristics of a substance that can be observed or measured without affecting the identity of the substance.
- A substance undergoes a physical change when its shape, size, or state changes, but its composition does not change.
- Chemical properties are those characteristics of a substance that describe the ability of a substance to change into a new substance.
- A substance undergoes a chemical change when the original substance is converted into one or more new substances, which have different physical and chemical properties.

♦ Learning Exercise 3.2A

State whether each of the following statements describes a gas, a liquid, or a solid:

a. _____ There are no attractions among the particles.

b. _____ The particles are held close together in a definite pattern.

c. _____ This substance has a definite volume but no definite shape.

d. _____ The particles are moving extremely fast.

e. _____ This substance has no definite shape and no definite volume.

f. _____ The particles in this substance are vibrating slowly in fixed positions.

g. _____ This substance has a definite volume and a definite shape.

Answers **a.** gas   **b.** solid   **c.** liquid   **d.** gas
          e. gas   **f.** solid   **g.** solid

♦ Learning Exercise 3.2B

Classify each of the following as a physical or chemical property:

a. _____ Silver is shiny.   b. _____ Water is a liquid at 25 °C.

c. _____ Wood burns.   d. _____ Mercury is a very dense liquid.

e. _____ Helium is not reactive.   f. _____ Ice cubes float in water.

Answers **a.** physical   **b.** physical   **c.** chemical
          d. physical   **e.** chemical   **f.** physical

♦ Learning Exercise 3.2C

CORE CHEMISTRY SKILL
Identifying Physical and Chemical Changes

Classify each of the following changes as physical or chemical:

a. _____ Sodium melts at 98 °C.   b. _____ Iron forms rust in air and water.

c. _____ Water condenses on a cold window. d. _____ Fireworks explode when ignited.

e. _____ Gasoline burns in a car engine.   f. _____ Paper is cut to make confetti.

Answers **a.** physical   **b.** chemical   **c.** physical
          d. chemical   **e.** chemical   **f.** physical

3.3 Temperature

REVIEW

Using Positive and Negative
  Numbers in Calculations (1.4)
Solving Equations (1.4)
Counting Significant Figures (2.2)

Learning Goal: Given a temperature, calculate the corresponding temperature on another scale.

- Temperature is measured in degrees Celsius (°C) or kelvins (K). In the United States, the Fahrenheit scale (°F) is still in use.
- The equation $T_F = 1.8(T_C) + 32$ is used to convert a Celsius temperature to a Fahrenheit temperature.
- When rearranged for T_C, this equation is used to convert a Fahrenheit temperature to a Celsius temperature.

$$T_C = \frac{T_F - 32}{1.8}$$

- The temperature on the Celsius scale is related to the temperature on the Kelvin scale: $T_K = T_C + 273$.

Calculating Temperature	
STEP 1	State the given and needed quantities.
STEP 2	Write a temperature equation.
STEP 3	Substitute in the known values and calculate the new temperature.

♦ **Learning Exercise 3.3**

Calculate the temperature in each of the following problems:

CORE CHEMISTRY SKILL

Converting between Temperature Scales

a. To prepare yogurt, milk is warmed to 185 °F. What Celsius temperature is needed to prepare the yogurt?

b. A heat-sensitive vaccine is stored at −12 °C. What is that temperature on a Fahrenheit thermometer?

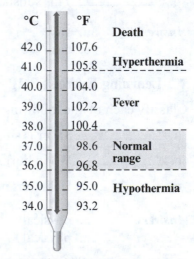

°C	°F	
		Death
42.0	107.6	
41.0	105.8	Hyperthermia
40.0	104.0	
39.0	102.2	Fever
38.0	100.4	
37.0	98.6	Normal range
36.0	96.8	
35.0	95.0	Hypothermia
34.0	93.2	

c. A patient has a temperature of 39.5 °C. What is that temperature on the Fahrenheit scale?

d. A patient in the emergency room with hyperthermia has a temperature of 105 °F. What is the temperature on the Celsius scale?

e. The temperature in an autoclave used to sterilize surgical equipment is 253 °F. What is that temperature on the Celsius scale?

Very high temperatures (hyperthermia) can lead to convulsions and brain damage.
Credit: Digital Vision/Alamy

f. Liquid nitrogen at −196 °C is used to freeze and remove precancerous skin growths. What temperature will this be on the Kelvin scale?

Answers   **a.** 85.0 °C   **b.** 10 °F   **c.** 103.1 °F   **d.** 41 °C   **e.** 123 °C   **f.** 77 K

3.4   Energy

Learning Goal: Identify energy as potential or kinetic; convert between units of energy.

- Energy is the ability to do work.
- Potential energy is stored energy, which is determined by position or composition; kinetic energy is the energy of motion.
- The SI unit of energy is the joule (J), the metric unit is the calorie (cal).

$$1 \text{ cal} = 4.184 \text{ J}$$

$$\frac{4.184 \text{ J}}{1 \text{ cal}} \quad \text{and} \quad \frac{1 \text{ cal}}{4.184 \text{ J}}$$

REVIEW

Rounding Off (2.3)
Using Significant Figures in Calculations (2.3)
Writing Conversion Factors from Equalities (2.5)
Using Conversion Factors (2.6)

Key Terms for Sections 3.2 to 3.4

Match each of the following key terms with the correct description:

    a. potential energy   **b.** joule   **c.** kinetic energy
    d. calorie   **e.** energy   **f.** physical change

1. _____ the SI unit of energy

2. _____ the energy of motion

3. _____ the amount of heat energy that raises the temperature of exactly 1 g of water exactly 1 °C

4. _____ a change in physical properties of a substance with no change in its identity

5. _____ the ability to do work

6. _____ a type of energy that is stored for future use

Answers   **1.** b   **2.** c   **3.** d   **4.** f   **5.** e   **6.** a

♦ Learning Exercise 3.4A

Indicate whether each of the following statements describes potential or kinetic energy:

a. _____ a potted plant sitting on a ledge **b.** _____ water flowing down a stream

c. _____ logs sitting in a fireplace **d.** _____ a piece of candy

e. _____ an arrow shot from a bow **f.** _____ a ski jumper standing at the top of the ski jump

g. _____ a jogger running **h.** _____ a skydiver waiting to jump

i. _____ your breakfast cereal **j.** _____ a bowling ball striking the pins

Answers   **a.** potential **b.** kinetic **c.** potential **d.** potential **e.** kinetic
    f. potential **g.** kinetic **h.** potential **i.** potential **j.** kinetic

♦ Learning Exercise 3.4B

Match the words with the definitions below.

CORE CHEMISTRY SKILL

Using Energy Units

    a. calorie   **b.** kilocalorie   **c.** joule

1. _____ the SI unit of energy

2. _____ the heat needed to raise 1 g of water by 1 °C

3. _____ 1000 cal

Answers   **1.** c   **2.** a   **3.** b

♦ **Learning Exercise 3.4C**

Convert each of the following energy units:

a. 58 000 cal to kcal

b. 3450 J to cal

c. 2.8 kJ to cal

d. 15 200 cal to kJ

Answers   **a.** 58 kcal   **b.** 825 cal   **c.** 670 cal   **d.** 63.6 kJ

3.5   Energy and Nutrition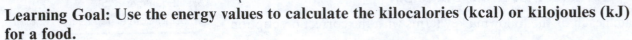

Learning Goal: Use the energy values to calculate the kilocalories (kcal) or kilojoules (kJ) for a food.

- The nutritional calorie (Cal) is the same amount of energy as 1 kcal, or 1000 calories.
- When a substance is burned in a calorimeter, the water that surrounds the reaction chamber absorbs the heat given off. The heat absorbed by the water is calculated, and the energy content for the substance (energy per gram) is determined.
- The energy values for three food types are: carbohydrate 4 kcal/g (17 kJ/g), fat 9 kcal/g (38 kJ/g), protein 4 kcal/g (17 kJ/g).
- The energy content of a food is the sum of kilocalories or kilojoules from carbohydrate, fat, and protein.

Calculating the Energy from a Food	
STEP 1	State the given and needed quantities.
STEP 2	Use the energy value for each food type to calculate the kilocalories or kilojoules, rounded off to the tens place.
STEP 3	Add the energy for each food type to give the total energy from the food.

♦ **Learning Exercise 3.5A**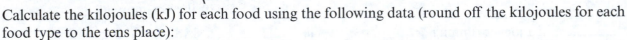

Calculate the kilojoules (kJ) for each food using the following data (round off the kilojoules for each food type to the tens place):

Food	Carbohydrate	Fat	Protein	kJ
a. green peas, cooked, 1 cup	19 g	1 g	9 g	_____
b. potato chips, 10 chips	10 g	8 g	1 g	_____
c. cream cheese, 8 oz	5 g	86 g	18 g	_____
d. lean hamburger, 3 oz	0 g	10 g	23 g	_____
e. banana, 1 medium	26 g	0 g	1 g	_____

Answers   **a.** 510 kJ   **b.** 490 kJ   **c.** 3670 kJ
          d. 770 kJ   **e.** 460 kJ

♦ **Learning Exercise 3.5B** 🏃

Use energy values to calculate each of the following:

a. Complete the following table listing ingredients for a peanut butter sandwich (round off the kilocalories for each food type to the tens place):

	Carbohydrate	Fat	Protein	kcal
Bread, 2 slices	30 g	0 g	5 g	_____
Peanut butter, 2 tbsp	10 g	13 g	8 g	_____
Jelly, 2 tsp	10 g	0 g	0 g	_____
Margarine, 1 tsp	0 g	7 g	0 g	_____

Total kcal in sandwich _____

b. How many kilocalories are in a single serving of pudding that contains 31 g of carbohydrate, 5 g of fat, and 5 g of protein (round off the kilocalories for each food type to the tens place)?

c. One bagel with light cream cheese has an energy content of 380 kcal. If there are 58 g of carbohydrate and 14 g of protein, how many grams of fat are in one bagel with light cream cheese (round off the kilocalories for each food type to the tens place)?

d. A serving of breakfast cereal provides 220 kcal. In this serving, there are 6 g of fat and 8 g of protein. How many grams of carbohydrate are in the cereal (round off the kilocalories for each food type to the tens place)?

Answers

a. bread = 140 kcal; peanut butter = 190 kcal; jelly = 40 kcal; margarine = 60 kcal
total kcal in sandwich = 430 kcal

b. carbohydrate = 120 kcal; fat = 50 kcal; protein = 20 kcal; total = 190 kcal

c. carbohydrate = 230 kcal; protein = 60 kcal
380 kcal − 230 kcal − 60 kcal = 90 kcal from fat

$$90 \text{ kcal} \times \frac{1 \text{ g fat}}{9 \text{ kcal}} = 10 \text{ g of fat}$$

d. fat = 50 kcal; protein = 30 kcal; 30 kcal + 50 kcal = 80 kcal from fat and protein
220 kcal − 80 kcal = 140 kcal from carbohydrate

$$140 \text{ kcal} \times \frac{1 \text{ g carbohydrate}}{4 \text{ kcal}} = 35 \text{ g of carbohydrate}$$

3.6 Specific Heat

Learning Goal: Use specific heat to calculate heat loss or gain.

- Specific heat is the amount of energy required to raise the temperature of 1 g of a substance by 1 °C.
- The specific heat for liquid water is 1.00 cal/g °C or 4.184 J/g °C.
- The heat lost or gained can be calculated using the mass of the substance, temperature difference, and its specific heat (SH): Heat = mass $\times \Delta T \times SH$

Calculations Using Specific Heat	
STEP 1	State the given and the needed quantities.
STEP 2	Calculate the temperature change (ΔT).
STEP 3	Write the heat equation and needed conversion factors.
STEP 4	Substitute in the given values and calculate the heat, making sure units cancel.

♦ **Learning Exercise 3.6**

The specific heat for water is 1.00 cal/g °C or 4.184 J/g °C. Calculate the kilocalories (kcal) and kilojoules (kJ) gained or released during the following:

> **CORE CHEMISTRY SKILL**
> Using the Heat Equation

a. heating 20.0 g of water from 22 °C to 77 °C

b. heating 10.0 g of water from 12.4 °C to 67.5 °C

c. cooling 0.450 kg of water from 80.0 °C to 35.0 °C

d. cooling 125 g of water from 72.0 °C to 45.0 °C

Answers **a.** 1.1 kcal, 4.6 kJ (gained) **b.** 0.551 kcal, 2.31 kJ (gained)
c. 20.3 kcal, 84.7 kJ (released) **d.** 3.38 kcal, 14.1 kJ (released)

3.7 Changes of State

> **REVIEW**
> Interpreting Graphs (1.4)

Learning Goal: Describe the changes of state between solids, liquids, and gases; calculate the energy released or absorbed.

- Melting, freezing, boiling, and condensing are typical changes of state.
- A substance melts and freezes at its melting (freezing) point. During the process of melting or freezing, the temperature remains constant.
- The heat of fusion is the energy required to change 1 g of solid to liquid at the melting point. For ice to melt at 0 °C, 80. cal/g (or 334 J/g), is required. This is also the amount of heat lost when 1 g of water freezes at 0 °C.
- A substance boils and condenses at its boiling point. During the process of boiling or condensing, the temperature remains constant.

- The heat of vaporization is the energy required to change 1 g of liquid to 1 g of gas at the boiling point. For water to boil at 100 °C, 540 cal/g (or 2260 J/g), is required to change 1 g of liquid to 1 g of gas (steam); it is also the amount of heat released when 1 g of water vapor condenses at 100 °C.

- Evaporation is a surface phenomenon, while boiling occurs throughout the liquid.

- Sublimation is the change of state from a solid directly to a gas. Deposition is the opposite process.

- A heating or cooling curve illustrates the changes in temperature and state as heat is added to or removed from a substance.

- When a substance is heated or cooled, the energy gained or released is the total of the energy involved in temperature changes as well as the energy involved in changes of state.

Key Terms for Sections 3.5 to 3.7

Match each of the following key terms with the correct description:

a. energy value   **b.** specific heat   **c.** Calorie (Cal)   **d.** sublimation   **e.** heat of vaporization

1. _____ a nutritional unit of energy equal to 1000 cal or 1 kcal

2. _____ a quantity of heat that changes the temperature of exactly 1 g of a substance by exactly 1 °C

3. _____ the number of kilocalories or kilojoules obtained per gram of carbohydrate, fat, or protein

4. _____ a substance changes directly from a solid to a gas

5. _____ the energy required to convert exactly 1 g of a liquid to vapor at its boiling point

Answers   **1.** c   **2.** b   **3.** a   **4.** d   **5.** e

♦ **Learning Exercise 3.7A**

Identify each of the following as

a. melting   **b.** freezing   **c.** sublimation   **d.** condensation

1. _____ A liquid changes to a solid.

2. _____ Dry ice in an ice cream cart changes to a gas.

3. _____ Ice forms on the surface of a lake in winter.

4. _____ Butter in a hot pan turns to liquid.

5. _____ A gas changes to a liquid.

Answers   **1.** b   **2.** c   **3.** b   **4.** a   **5.** d

Melting and freezing are reversible processes.

Credit: John A. Rizzo/Photodisc/Getty Images

Calculations Using a Heat Conversion Factor	
STEP 1	State the given and needed quantities.
STEP 2	Write a plan to convert the given quantity to the needed quantity.
STEP 3	Write the heat conversion factor and any metric factor.
STEP 4	Set up the problem and calculate the needed quantity.

♦ **Learning Exercise 3.7B**

Calculate each of the following when a substance melts or freezes:

a. How many joules of heat are needed to melt 24.0 g of ice at 0 °C?

b. How much heat, in kilojoules, is released when 325 g of water freezes at 0 °C?

c. How many kilocalories of heat are required to melt 120 g of ice in an ice bag at 0 °C?

An ice bag is used to treat a sports injury.
Credit: wsphotos/Getty Images

d. How many grams of ice would melt when 1200 cal of heat is absorbed?

Answers   **a.** 8020 J   **b.** 109 kJ   **c.** 9.6 kcal   **d.** 15 g

♦ **Learning Exercise 3.7C**

Calculate each of the following when a substance boils or condenses:

a. How many joules are needed to completely change 7.20 g of water to vapor at 100 °C?

b. How many kilocalories are released when 42 g of steam at 100 °C condenses to form liquid water at 100 °C?

c. How many grams of water can be converted to steam at 100 °C when 155 kJ of energy is absorbed?

d. How many grams of steam condense if 24 kcal is released at 100 °C?

Answers   **a.** 16 300 J   **b.** 23 kcal   **c.** 68.6 g   **d.** 44 g

♦ **Learning Exercise 3.7D**

On each heating or cooling curve, indicate the portion that corresponds to a solid, liquid, or gas and the changes of state.

1. Draw a heating curve for water that begins at −20 °C and ends at 120 °C.

2. Draw a heating curve for bromine from −25 °C to 75 °C. Bromine has a melting point of −7 °C and a boiling point of 59 °C.

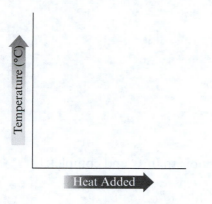

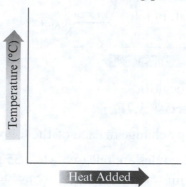

3. Draw a cooling curve for sodium from 1000 °C to 0 °C. Sodium has a freezing point of 98 °C and a boiling (condensation) point of 883 °C.

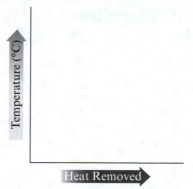

Answers

1.

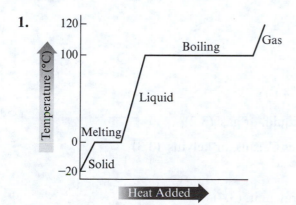

2.

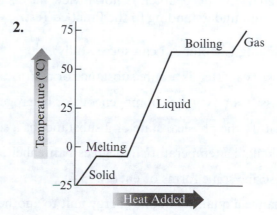

3.

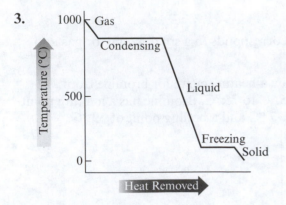

♦ **Learning Exercise 3.7E**

Calculate the energy change in each of the following:

a. How many kilojoules are released when 35 g of water at 65 °C cools to 0 °C and completely changes to solid? (*Hint*: Two steps are needed.)

b. How many kilocalories are needed to melt 15 g of ice at 0 °C, heat the water to 100 °C, and convert the water to gas at 100 °C? (*Hint*: Three steps are needed.)

Answers   **a.** 21 kJ   **b.** 11 kcal

Checklist for Chapter 3

You are ready to take the Practice Test for Chapter 3. Be sure you have accomplished the following learning goals for this chapter. If not, review the Section listed at the end of the goal. Then apply your new skills and understanding to the Practice Test.

After studying Chapter 3, I can successfully:

_____ Classify matter as a pure substance or mixture. (3.1)

_____ Classify a mixture as homogeneous or heterogeneous. (3.1)

_____ Identify the physical state of a substance as a solid, liquid, or gas. (3.2)

_____ Calculate a temperature in degrees Fahrenheit, degrees Celsius, or kelvins. (3.3)

_____ Describe some forms of energy. (3.4)

_____ Change a quantity in one energy unit to another energy unit. (3.4)

_____ Using the energy values, calculate the energy, in kilocalories (kcal) or kilojoules (kJ), for a food sample. (3.5)

_____ Given the mass of a sample, specific heat, and the temperature change, calculate the heat lost or gained. (3.6)

_____ Calculate the heat change for a change of state for a specific amount of a substance. (3.7)

_____ Calculate the total heat for a combination of change of temperature and change of state for a specific amount of a substance. (3.7)

_____ Draw heating and cooling curves using the melting and boiling points of a substance. (3.7)

Practice Test for Chapter 3

The chapter Sections to review are shown in parentheses at the end of each question.

For questions 1 through 4, classify each of the following as a pure substance (P) *or a mixture* (M): *(3.1)*

1. _____ toothpaste

2. _____ platinum

3. _____ chromium

4. _____ mouthwash

For questions 5 through 8, classify each of the following mixtures as homogeneous (Ho) *or heterogeneous* (He): *(3.1)*

5. _____ noodle soup

6. _____ salt water

7. _____ chocolate chip cookie

8. _____ mouthwash

9. Which of the following is a chemical property? (3.2)
   A. dynamite explodes
   B. a shiny metal
   C. breaking up cement
   D. a melting point of 110 °C
   E. rain on a cool day

For questions 10 through 12, answer with solid (S), *liquid* (L), *or gas* (G): *(3.2)*

10. _____ has a definite volume but takes the shape of a container

11. _____ does not have a definite shape or definite volume

12. _____ has a definite shape and a definite volume

13. Which of the following is a chemical property of silver? (3.2)
    A. density of 10.5 g/mL
    B. shiny
    C. melts at 961 °C
    D. good conductor of heat
    E. reacts to form tarnish

14. Which of the following is a physical property of silicon? (3.2)
    A. burns in chlorine
    B. has a black to gray color
    C. reacts with nitric acid
    D. used to form silicone
    E. reacts with oxygen to form sand

For questions 15 through 19, answer as physical change (P) *or chemical change* (C): *(3.2)*

15. _____ Butter melts in a hot pan.

16. _____ Iron forms rust with oxygen.

17. _____ Baking powder forms bubbles (CO_2) as a cake is baking.

18. _____ Water boils.

19. _____ Propane burns in a camp stove.

20. 105 °F = _____ °C (3.3)
    A. 73 °C
    B. 41 °C
    C. 58 °C
    D. 90 °C
    E. 189 °C

21. The melting point of gold is 1064 °C. The Fahrenheit temperature needed to melt gold would be (3.3)

    A. 129 °F   **B.** 623 °F   **C.** 1031 °F   **D.** 1913 °F   **E.** 1947 °F

22. The average daytime temperature on the planet Mercury is 683 K. What is this temperature on the Celsius scale? (3.3)

    A. 956 °C   **B.** 715 °C   **C.** 680 °C   **D.** 410 °C   **E.** 303 °C

23. Which of the following would be described as potential energy? (3.4)

    A. a car going around a racetrack   **B.** a rabbit hopping
    C. oil in an oil well   **D.** a moving merry-go-round
    E. a bouncing ball

24. Which of the following would be described as kinetic energy? (3.4)

    A. a car battery   **B.** a can of tennis balls
    C. gasoline in a car fuel tank   **D.** a box of matches
    E. a tennis ball crossing over the net

For questions 25 through 27, consider a glass of milk with an energy content of 170 kcal. In the milk, there are 12 g of carbohydrate, 9.0 g of fat, and protein: (round off the kilocalories for each food type to the tens place) (3.5)

25. The number of kilocalories provided by the carbohydrate is

    A. 4 kcal   **B.** 9 kcal   **C.** 40 kcal   **D.** 50 kcal   **E.** 80 kcal

26. The number of kilocalories provided by the fat is

    A. 9 kcal   **B.** 40 kcal   **C.** 60 kcal   **D.** 70 kcal   **E.** 80 kcal

27. Using your answers to questions 25 and 26, the number of kilocalories provided by the protein is

    A. 4 kcal   **B.** 30 kcal   **C.** 40 kcal   **D.** 90 kcal   **E.** 120 kcal

28. A patient on a diet has a lunch of 3 oz of tuna (8 g of fat, 16 g of protein), 1 cup of nonfat milk (12 g of carbohydrate, 9 g of protein), and 1 cup of raw carrots (11 g of carbohydrate, 2 g of protein). What is the total kilocalories for the lunch? (*Round off the kilocalories for each food type to the tens place.*) (3.5)

    A. 270 kcal   **B.** 200 kcal   **C.** 110 kcal   **D.** 90 kcal   **E.** 50 kcal

29. For the lunch in question 28, how many hours of swimming are needed to expend the same energy, in kilocalories, if swimming requires 2100 kJ/h? (3.5)

    A. 5.7 h   **B.** 5.0 h   **C.** 2.8 h   **D.** 1.6 h   **E.** 0.54 h

30. The number of joules needed to raise the temperature of 5.0 g of water from 25 °C to 55 °C is (3.6)

    A. 5.0 J   **B.** 36 J   **C.** 30.0 J   **D.** 335 J   **E.** 630 J

31. The number of kilojoules released when 15 g of water cools from 58 °C to 22 °C is (3.6)

    A. 0.13 kJ   **B.** 0.54 kJ   **C.** 2.3 kJ   **D.** 63 kJ   **E.** 150 kJ

For questions 32 through 36, match items A to E with one of the statements: (3.7)

    A. melting   **B.** evaporation   **C.** heat of fusion   **D.** heat of vaporization   **E.** boiling

32. _____ the energy required to convert a gram of solid to liquid

33. _____ the heat needed to convert a gram of liquid to gas

34. _____ the conversion of a liquid to gas at the surface of a liquid

35. _____ the conversion of solid to liquid

36. _____ the formation of a gas within the liquid as well as on the surface

37. The number of joules needed to convert 15.0 g of ice to liquid at 0 °C is (3.7)

    A. 73 J   **B.** 334 J   **C.** 2260 J   **D.** 4670 J   **E.** 5010 J

38. The number of calories released when 2.0 g of water at 50 °C is cooled and frozen at 0 °C is (3.7)

    A. 80. cal   **B.** 160 cal   **C.** 260 cal   **D.** 440 cal   **E.** 1100 cal

39. What is the total number of kilojoules required to convert 25 g of ice at 0 °C to gas at 100 °C? (3.7)

    A. 8.4 kJ   **B.** 19 kJ   **C.** 59 kJ   **D.** 67 kJ   **E.** 75 kJ

For questions 40 through 43, consider the heating curve for p-toluidine. Answer the following questions when heat is added to p-toluidine at −20 °C, where toluidine is below its melting point: (3.7)

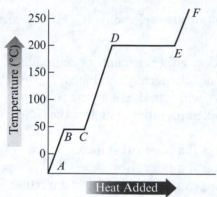

40. On the heating curve, segment **BC** indicates

    A. solid   **B.** melting   **C.** liquid   **D.** boiling   **E.** gas

41. On the heating curve, segment **CD** indicates

    A. solid   **B.** melting   **C.** liquid   **D.** boiling   **E.** gas

42. The boiling point of toluidine is

    A. 20 °C   **B.** 45 °C   **C.** 100 °C   **D.** 200 °C   **E.** 250 °C

43. On the heating curve, segment **EF** indicates

    A. solid   **B.** melting   **C.** liquid   **D.** boiling   **E.** gas

Answers to the Practice Test

  1. M   **2.** P   **3.** P   **4.** M   **5.** He
  6. Ho   **7.** He   **8.** Ho   **9.** A   **10.** L
11. G   **12.** S   **13.** E   **14.** B   **15.** P
16. C   **17.** C   **18.** P   **19.** C   **20.** B
21. E   **22.** D   **23.** C   **24.** E   **25.** D
26. E   **27.** C   **28.** A   **29.** E   **30.** E
31. C   **32.** C   **33.** D   **34.** B   **35.** A
36. E   **37.** E   **38.** C   **39.** E   **40.** B
41. C   **42.** D   **43.** E

Selected Answers and Solutions to Text Problems

3.1   *Elements* are the simplest type of pure substance, containing only one type of atom. *Compounds* contain two or more elements chemically combined in a specific proportion.
   a. A silicon chip is an element since it contains only one type of atom (Si).
   b. Hydrogen peroxide (H_2O_2) is a compound since it contains two elements (H, O) that are chemically combined.
   c. Oxygen gas (O_2) is an element since it contains only one type of atom (O).
   d. Rust (Fe_2O_3) is a compound since it contains two elements (Fe, O) that are chemically combined.
   e. Methane (CH_4) in natural gas is a compound since it contains two elements (C, H) that are chemically combined.

3.3   A *pure substance* is matter that has a fixed or definite composition: either elements or compounds. In a *mixture*, two or more substances are physically mixed but not chemically combined.
   a. Baking soda is composed of one type of matter ($NaHCO_3$), which makes it a pure substance.
   b. A blueberry muffin is composed of several substances mixed together, which makes it a mixture.
   c. Ice is composed of one type of matter (H_2O), which makes it a pure substance.
   d. Zinc is composed of one type of matter (Zn), which makes it a pure substance.
   e. Trimix is a physical mixture of oxygen, nitrogen, and helium gases, which makes it a mixture.

3.5   A *homogeneous mixture* has a uniform composition; a *heterogeneous mixture* does not have a uniform composition throughout the mixture.
   a. Vegetable soup is a heterogeneous mixture since it has chunks of vegetables.
   b. Tea is a homogeneous mixture since it has a uniform composition.
   c. Fruit salad is a heterogeneous mixture since it has chunks of fruit.
   d. Tea with ice and lemon slices is a heterogeneous mixture since it has chunks of ice and lemon.

3.7   **a.** A gas has no definite volume or shape.
   b. In a gas, the particles do not interact with each other.
   c. In a solid, the particles are held in a rigid structure.

3.9   A *physical property* is a characteristic of a substance—such as color, shape, odor, luster, size, melting point, and density—that can be observed without affecting the identity of the substance. A *chemical property* is a characteristic that indicates the ability of a substance to change into a new substance.
   a. Color and physical state are physical properties.
   b. The ability to react with oxygen is a chemical property.
   c. The temperature of a substance is a physical property.
   d. The ability to corrode iron is a chemical property.
   e. Burning butane gas in oxygen forms new substances, which makes it a chemical property.

3.11   When matter undergoes a *physical change*, its state or appearance changes, but its composition remains the same. When a *chemical change* occurs, the original substance is converted into a new substance, which has different physical and chemical properties.
   a. Water vapor condensing is a physical change since the physical form of the water changes, but the composition of the substance does not.
   b. Cesium metal reacting is a chemical change since new substances form.
   c. Gold melting is a physical change since the physical state changes, but not the composition of the substance.
   d. Cutting a puzzle is a physical change since the size and shape change, but not the composition of the substance.
   e. Grating cheese is a physical change since the size and shape change, but not the composition of the substance.

3.13 **a.** The high reactivity of fluorine is a chemical property since it allows for the formation of new substances.
b. The physical state of fluorine is a physical property.
c. The color of fluorine is a physical property.
d. The reactivity of fluorine with hydrogen is a chemical property since it allows for the formation of new substances.
e. The melting point of fluorine is a physical property.

3.15 The Fahrenheit temperature scale is still used in the United States. A normal body temperature is 98.6 °F on this scale. To convert her 99.8 °F temperature to the equivalent reading on the Celsius scale, the following calculation must be performed:

$$T_C = \frac{(99.8 - 32)}{1.8} = \frac{67.8}{1.8} = 37.7\,°C \text{ (3 SFs) (1.8 and 32 are exact numbers)}$$

Because a normal body temperature is 37.0 on the Celsius scale, her temperature of 37.7 °C would indicate a mild fever.

3.17 To convert Celsius to Fahrenheit: $T_F = 1.8(T_C) + 32$ (1.8 and 32 are exact numbers)

To convert Fahrenheit to Celsius: $T_C = \dfrac{T_F - 32}{1.8}$ (1.8 and 32 are exact numbers)

To convert Celsius to Kelvin: $T_K = T_C + 273$
To convert Kelvin to Celsius: $T_C = T_K - 273$

a. $T_F = 1.8(T_C) + 32 = 1.8(37.0) + 32 = 66.6 + 32 = 98.6\,°F$
b. $T_C = \dfrac{T_F - 32}{1.8} = \dfrac{(65.3 - 32)}{1.8} = \dfrac{33.3}{1.8} = 18.5\,°C$
c. $T_K = T_C + 273 = -27 + 273 = 246\,K$
d. $T_K = T_C + 273 = 62 + 273 = 335\,K$
e. $T_C = \dfrac{T_F - 32}{1.8} = \dfrac{(114 - 32)}{1.8} = \dfrac{82}{1.8} = 46\,°C$

3.19 **a.** $T_C = \dfrac{T_F - 32}{1.8} = \dfrac{(106 - 32)}{1.8} = \dfrac{74}{1.8} = 41\,°C$
b. $T_C = \dfrac{T_F - 32}{1.8} = \dfrac{(103 - 32)}{1.8} = \dfrac{71}{1.8} = 39\,°C$

No, there is no need to phone the doctor. The child's temperature is less than 40.0 °C.

3.21 As the roller-coaster car climbs to the top of the ramp, its kinetic energy is converted to potential energy. At the top, the car has its maximum potential energy. As the car descends, potential energy is converted into kinetic energy. At the bottom, all of its energy is kinetic.

3.23 **a.** Water at the top of the waterfall has potential energy.
b. Kicking a ball gives it kinetic energy.
c. The energy in a lump of coal is potential energy.
d. A skier at the top of a hill has potential energy.

3.25 **a.** **Given** 3500 cal   **Need** kilocalories

Plan cal → kcal $\dfrac{1\text{ kcal}}{1000\text{ cal}}$

Set-up $3500\text{ cal} \times \dfrac{1\text{ kcal}}{1000\text{ cal}} = 3.5\text{ kcal (2 SFs)}$

b. Given 415 J **Need** calories

Plan J → cal $\dfrac{1\ cal}{4.184\ J}$

Set-up $415\ \cancel{J} \times \dfrac{1\ cal}{4.184\ \cancel{J}} = 99.2\ cal\ (3\ SFs)$

c. Given 28 cal **Need** joules

Plan cal → J $\dfrac{4.184\ J}{1\ cal}$

Set-up $28\ \cancel{cal} \times \dfrac{4.184\ J}{1\ \cancel{cal}} = 120\ J\ (2\ SFs)$

d. Given 4.5 kJ **Need** calories

Plan kJ → J → cal $\dfrac{1000\ J}{1\ kJ}\quad \dfrac{1\ cal}{4.184\ J}$

Set-up $4.5\ \cancel{kJ} \times \dfrac{1000\ \cancel{J}}{1\ \cancel{kJ}} \times \dfrac{1\ cal}{4.184\ \cancel{J}} = 1100\ cal\ (2\ SFs)$

3.27 a. Given 3.0 h, 270 kJ/h **Need** joules

Plan h → kJ → J $\dfrac{1000\ J}{1\ kJ}$

Set-up $3.0\ \cancel{h} \times \dfrac{270\ \cancel{kJ}}{1.0\ \cancel{h}} \times \dfrac{1000\ J}{1\ \cancel{kJ}} = 8.1 \times 10^5\ J\ (2\ SFs)$

b. Given 3.0 h, 270 kJ/h **Need** kilocalories

Plan h → kJ → J → cal → kcal $\dfrac{1000\ J}{1\ kJ}\quad \dfrac{1\ cal}{4.184\ J}\quad \dfrac{1\ kcal}{1000\ cal}$

Set-up $3.0\ \cancel{h} \times \dfrac{270\ \cancel{kJ}}{1.0\ \cancel{h}} \times \dfrac{1000\ \cancel{J}}{1\ \cancel{kJ}} \times \dfrac{1\ \cancel{cal}}{4.184\ \cancel{J}} \times \dfrac{1\ kcal}{1000\ \cancel{cal}} = 190\ kcal\ (2\ SFs)$

3.29 a. Given 125 kJ **Need** kilocalories

Plan kJ → J → cal → kcal $\dfrac{1000\ J}{1\ kJ}\quad \dfrac{1\ cal}{4.184\ J}\quad \dfrac{1\ kcal}{1000\ cal}$

Set-up $125\ \cancel{kJ} \times \dfrac{1000\ \cancel{J}}{1\ \cancel{kJ}} \times \dfrac{1\ \cancel{cal}}{4.184\ \cancel{J}} \times \dfrac{1\ kcal}{1000\ \cancel{cal}} = 29.9\ kcal\ (3\ SFs)$

b. Given 870. kJ **Need** kilocalories

Plan kJ → J → cal → kcal $\dfrac{1000\ J}{1\ kJ}\quad \dfrac{1\ cal}{4.184\ J}\quad \dfrac{1\ kcal}{1000\ cal}$

Set-up $870.\ \cancel{kJ} \times \dfrac{1000\ \cancel{J}}{1\ \cancel{kJ}} \times \dfrac{1\ \cancel{cal}}{4.184\ \cancel{J}} \times \dfrac{1\ kcal}{1000\ \cancel{cal}} = 208\ kcal\ (3\ SFs)$

3.31 a. Given one cup of orange juice that contains 26 g of carbohydrate, 2 g of protein, and no fat
  Need total energy in kilojoules

Food Type	Mass		Energy Value		Energy (rounded off to the tens place)
Carbohydrate	26 g	×	$\dfrac{17\ kJ}{1\ g}$	=	440 kJ
Protein	2 g	×	$\dfrac{17\ kJ}{1\ g}$	=	30 kJ
			Total energy content	=	470 kJ

b. Given one apple that provides 72 kcal of energy and contains no fat or protein
   Need grams of carbohydrate

   Plan kcal → g of carbohydrate $\dfrac{1\ g\ carbohydrate}{4\ kcal}$

   Set-up $72\ kcal \times \dfrac{1\ g\ carbohydrate}{4\ kcal} = 18\ g\ of\ carbohydrate\ (2\ SFs)$

c. Given one tablespoon of vegetable oil that contains 14 g of fat, and no carbohydrate or protein
   Need total energy in kilocalories

Food Type	Mass		Energy Value		Energy (rounded off to the tens place)
Fat	14 g	×	$\dfrac{9\ kcal}{1\ g}$	=	130 kcal

d. Given one avocado that provides 410 kcal in total, and contains 13 g of carbohydrate and 5 g of protein
   Need grams of fat

Food Type	Mass		Energy Value		Energy (rounded off to the tens place)
Carbohydrate	13 g	×	$\dfrac{4\ kcal}{1\ g}$	=	50 kcal
Fat	? g	×	$\dfrac{9\ kcal}{1\ g}$	=	? kcal
Protein	5 g	×	$\dfrac{4\ kcal}{1\ g}$	=	20 kcal
			Total energy content	=	410 kcal

∴ Energy from fat = 410 kcal − (50 kcal + 20 kcal) = 340 kcal

∴ $340\ kcal \times \dfrac{1\ g\ fat}{9\ kcal} = 38\ g\ of\ fat$

3.33 Given one cup of clam chowder that contains 16 g of carbohydrate, 12 g of fat, and 9 g of protein
   Need total energy in kilocalories and kilojoules

Food Type	Mass		Energy Value		Energy (rounded off to the tens place)
Carbohydrate	16 g	×	$\dfrac{4\ kcal\ (or\ 17\ kJ)}{1\ g}$	=	60 kcal (or 270 kJ)
Fat	12 g	×	$\dfrac{9\ kcal\ (or\ 38\ kJ)}{1\ g}$	=	110 kcal (or 460 kJ)
Protein	9 g	×	$\dfrac{4\ kcal\ (or\ 17\ kJ)}{1\ g}$	=	40 kcal (or 150 kJ)
			Total energy content	=	210 kcal (or 880 kJ)

3.35 $3.2\ L\ glucose\ solution \times \dfrac{1000\ mL\ solution}{1\ L\ solution} \times \dfrac{5.0\ g\ glucose}{100.\ mL\ solution} \times \dfrac{4\ kcal}{1\ g\ glucose}$

= 640 kcal (rounded off to the tens place)

3.37 Copper, which has the lowest specific heat of the samples, would reach the highest temperature.

3.39 a. Given $SH_{water} = 1.00$ cal/g °C; $m = 8.5$ g; $\Delta T = 36\,°C - 15\,°C = 21\,°C$
   Need heat in calories
   Plan Heat $= m \times \Delta T \times SH$

   Set-up Heat $= 8.5\,\cancel{g} \times 21\,\cancel{°C} \times \dfrac{1.00\text{ cal}}{\cancel{g}\,\cancel{°C}} = 180$ cal (2 SFs)

   b. Given $SH_{water} = 4.184$ J/g °C; $m = 25$ g; $\Delta T = 86\,°C - 61\,°C = 25\,°C$
   Need heat in joules
   Plan Heat $= m \times \Delta T \times SH$

   Set-up Heat $= 25\,\cancel{g} \times 25\,\cancel{°C} \times \dfrac{4.184\text{ J}}{\cancel{g}\,\cancel{°C}} = 2600$ J (2 SFs)

   c. Given $SH_{water} = 1.00$ cal/g °C; $m = 150$ g; $\Delta T = 77\,°C - 15\,°C = 62\,°C$
   Need heat in kilocalories

   Plan Heat $= m \times \Delta T \times SH$ then cal $\rightarrow$ kcal $\dfrac{1\text{ kcal}}{1000\text{ cal}}$

   Set-up Heat $= 150\,\cancel{g} \times 62\,\cancel{°C} \times \dfrac{1.00\,\cancel{\text{cal}}}{\cancel{g}\,\cancel{°C}} \times \dfrac{1\text{ kcal}}{1000\,\cancel{\text{cal}}} = 9.3$ kcal (2 SFs)

   d. Given $SH_{copper} = 0.385$ J/g °C; $m = 175$ g; $\Delta T = 188\,°C - 28\,°C = 160.\,°C$
   Need heat in kilojoules

   Plan Heat $= m \times \Delta T \times SH$ then J $\rightarrow$ kJ $\dfrac{1\text{ kJ}}{1000\text{ J}}$

   Set-up Heat $= 175\,\cancel{g} \times 160.\,\cancel{°C} \times \dfrac{0.385\,\cancel{J}}{\cancel{g}\,\cancel{°C}} \times \dfrac{1\text{ kJ}}{1000\,\cancel{J}} = 10.8$ kJ (3 SFs)

3.41 a. Given $SH_{water} = 4.184$ J/g °C $= 1.00$ cal/g °C; $m = 25.0$ g;
   $\Delta T = 25.7\,°C - 12.5\,°C = 13.2\,°C$
   Need heat in joules and calories
   Plan Heat $= m \times \Delta T \times SH$

   Set-up Heat $= 25.0\,\cancel{g} \times 13.2\,\cancel{°C} \times \dfrac{4.184\text{ J}}{\cancel{g}\,\cancel{°C}} = 1380$ J (3 SFs)

   or Heat $= 25.0\,\cancel{g} \times 13.2\,\cancel{°C} \times \dfrac{1.00\text{ cal}}{\cancel{g}\,\cancel{°C}} = 330.$ cal (3 SFs)

   b. Given $SH_{copper} = 0.385$ J/g °C $= 0.0920$ cal/g °C; $m = 38.0$ g;
   $\Delta T = 246\,°C - 122\,°C = 124\,°C$
   Need heat in joules and calories
   Plan Heat $= m \times \Delta T \times SH$

   Set-up Heat $= 38.0\,\cancel{g} \times 124\,\cancel{°C} \times \dfrac{0.385\text{ J}}{\cancel{g}\,\cancel{°C}} = 1810$ J (3 SFs)

   or Heat $= 38.0\,\cancel{g} \times 124\,\cancel{°C} \times \dfrac{0.0920\text{ cal}}{\cancel{g}\,\cancel{°C}} = 434$ cal (3 SFs)

   c. Given $SH_{ethanol} = 2.46$ J/g °C $= 0.588$ cal/g °C; $m = 15.0$ g;
   $\Delta T = 60.5\,°C - (-42.0\,°C) = 102.5\,°C$
   Need heat in joules and calories
   Plan Heat $= m \times \Delta T \times SH$

   Set-up Heat $= 15.0\,\cancel{g} \times 102.5\,\cancel{°C} \times \dfrac{2.46\text{ J}}{\cancel{g}\,\cancel{°C}} = 3780$ J (3 SFs)

   or Heat $= 15.0\,\cancel{g} \times 102.5\,\cancel{°C} \times \dfrac{0.588\text{ cal}}{\cancel{g}\,\cancel{°C}} = 904$ cal (3 SFs)

d. Given $SH_{iron} = 0.452 \text{ J/g °C} = 0.108 \text{ cal/g °C}; m = 125 \text{ g};$
$\Delta T = 118 \text{ °C} - 55 \text{ °C} = 63 \text{ °C}$

Need heat in joules and calories

Plan Heat $= m \times \Delta T \times SH$

Set-up Heat $= 125 \text{ g} \times 63 \text{ °C} \times \dfrac{0.452 \text{ J}}{\text{g °C}} = 3600 \text{ J (2 SFs)}$

or Heat $= 125 \text{ g} \times 63 \text{ °C} \times \dfrac{0.108 \text{ cal}}{\text{g °C}} = 850 \text{ cal (2 SFs)}$

3.43 **a.** The change from solid to liquid state is melting.
b. Coffee is freeze-dried using the process of sublimation.
c. Liquid water turning to ice is freezing.
d. Ice crystals form on a package of frozen corn as a result of deposition.

3.45 **a.** $65 \text{ g ice} \times \dfrac{80. \text{ cal}}{1 \text{ g ice}} = 5200 \text{ cal (2 SFs); heat is absorbed}$

b. $17.0 \text{ g ice} \times \dfrac{334 \text{ J}}{1 \text{ g ice}} = 5680 \text{ J (3 SFs); heat is absorbed}$

c. $225 \text{ g water} \times \dfrac{80. \text{ cal}}{1 \text{ g water}} \times \dfrac{1 \text{ kcal}}{1000 \text{ cal}} = 18 \text{ kcal (2 SFs); heat is released}$

d. $50.0 \text{ g water} \times \dfrac{334 \text{ J}}{1 \text{ g water}} \times \dfrac{1 \text{ kJ}}{1000 \text{ J}} = 16.7 \text{ kJ (3 SFs); heat is released}$

3.47 **a.** Water vapor in clouds changing to rain is an example of condensation.
b. Wet clothes drying on a clothesline involves evaporation.
c. Steam forming as lava flows into the ocean involves boiling.
d. Water droplets forming on a bathroom mirror after a hot shower involves condensation.

3.49 **a.** $10.0 \text{ g water} \times \dfrac{540 \text{ cal}}{1 \text{ g water}} = 5400 \text{ cal (2 SFs); heat is absorbed}$

b. $5.00 \text{ g water} \times \dfrac{2260 \text{ J}}{1 \text{ g water}} = 11\,300 \text{ J (3 SFs); heat is absorbed}$

c. $8.0 \text{ kg steam} \times \dfrac{1000 \text{ g}}{1 \text{ kg}} \times \dfrac{540 \text{ cal}}{1 \text{ g steam}} \times \dfrac{1 \text{ kcal}}{1000 \text{ cal}} = 4300 \text{ kcal (2 SFs); heat is released}$

d. $175 \text{ g steam} \times \dfrac{2260 \text{ J}}{1 \text{ g steam}} \times \dfrac{1 \text{ kJ}}{1000 \text{ J}} = 396 \text{ kJ (3 SFs); heat is released}$

3.51

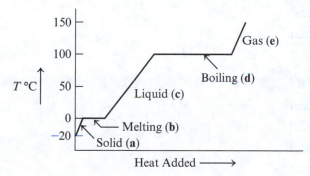

3.53 a. Two calculations are needed:

(1) ice 0 °C → water 0 °C: $50.0 \text{ g ice} \times \dfrac{334 \text{ J}}{1 \text{ g ice}} = 16\ 700 \text{ J}$

(2) water 0 °C → 65.0 °C: $\Delta T = 65.0 °C - 0 °C = 65.0 °C$

$50.0 \text{ g water} \times 65.0 °C \times \dfrac{4.184 \text{ J}}{\text{g water} °C} = 13\ 600 \text{ J}$

∴ Total heat needed $= 16\ 700 \text{ J} + 13\ 600 \text{ J} = 30\ 300 \text{ J}$ (3 SFs)

b. Two calculations are needed:

(1) steam 100 °C → water 100 °C: $15.0 \text{ g steam} \times \dfrac{540 \text{ cal}}{1 \text{ g steam}} \times \dfrac{1 \text{ kcal}}{1000 \text{ cal}} = 8.1 \text{ kcal}$

(2) water 100 °C → 0 °C: $15.0 \text{ g water} \times 100. °C \times \dfrac{1.00 \text{ cal}}{\text{g} °C} \times \dfrac{1 \text{ kcal}}{1000 \text{ cal}} = 1.50 \text{ kcal}$

∴ Total heat released $= 8.1 \text{ kcal} + 1.50 \text{ kcal} = 9.6 \text{ kcal}$ (2 SFs)

c. Three calculations are needed:

(1) ice 0 °C → water 0 °C: $24.0 \text{ g ice} \times \dfrac{334 \text{ J}}{1 \text{ g ice}} \times \dfrac{1 \text{ kJ}}{1000 \text{ J}} = 8.02 \text{ kJ}$

(2) water 0 °C → 100 °C: $24.0 \text{ g water} \times 100. °C \times \dfrac{4.184 \text{ J}}{\text{g} °C} \times \dfrac{1 \text{ kJ}}{1000 \text{ J}} = 10.0 \text{ kJ}$

(3) water 100 °C → steam 100 °C: $24.0 \text{ g water} \times \dfrac{2260 \text{ J}}{1 \text{ g water}} \times \dfrac{1 \text{ kJ}}{1000 \text{ J}} = 54.2 \text{ kJ}$

∴ Total heat needed $= 8.02 \text{ kJ} + 10.0 \text{ kJ} + 54.2 \text{ kJ} = 72.2 \text{ kJ}$ (3 SFs)

3.55 Two calculations are needed:

(1) steam 100. °C → water 100. °C: $18.0 \text{ g steam} \times \dfrac{540 \text{ cal}}{1 \text{ g steam}} \times \dfrac{1 \text{ kcal}}{1000 \text{ cal}} = 9.7 \text{ kcal}$

(2) water 100. °C → 37.0 °C: $\Delta T = 100 °C - 37.0 °C = 63 °C$

$18.0 \text{ g water} \times 63 °C \times \dfrac{1.00 \text{ cal}}{\text{g} °C} \times \dfrac{1 \text{ kcal}}{1000 \text{ cal}} = 1.1 \text{ kcal}$

∴ Total heat released $= 9.7 \text{ kcal} + 1.1 \text{ kcal} = 10.8 \text{ kcal}$ (3 SFs)

3.57 a. Using the energy totals from Table 3.8:

Breakfast	Energy
Banana, 1 medium	110 kcal
Milk, nonfat, 1 cup	90 kcal
Egg, 1 large	70 kcal
	Total = 270 kcal

Lunch	Energy
Carrots, 1 cup	50 kcal
Beef, ground, 3 oz	220 kcal
Apple, 1 medium	60 kcal
Milk, nonfat, 1 cup	90 kcal
	Total = 420 kcal

Dinner	Energy
Chicken, no skin, 6 oz	2×110 kcal
Potato, baked	100 kcal
Broccoli, 3 oz	30 kcal
Milk, nonfat, 1 cup	90 kcal
Total = 440 kcal	

b. total kilocalories for one day = 270 kcal + 420 kcal + 440 kcal = 1130 kcal

c. Since Charles will maintain his weight if he consumes 1800 kcal per day, he will lose weight on this new diet (1130 kcal/day).

d. For Charles, his energy balance is 1130 kcal/day − 1800 kcal/day = −670 kcal/day.

$$5.0 \, \cancel{\text{lb}} \times \frac{3500 \, \cancel{\text{kcal}}}{1.0 \, \cancel{\text{lb}}} \times \frac{1 \, \text{day}}{670 \, \cancel{\text{kcal}}} = 26 \, \text{days (2 SFs)}$$

3.59 a. The diagram shows two different kinds of atoms chemically combined in a definite 2:1 ratio; it represents a compound.

b. The diagram shows two different kinds of matter physically mixed, not chemically combined; it represents a mixture.

c. The diagram contains only one kind of atom; it represents an element.

3.61 A *homogeneous mixture* has a uniform composition; a *heterogeneous mixture* does not have a uniform composition throughout the mixture.

a. Lemon-flavored water is a homogeneous mixture since it has a uniform composition (as long as there are no lemon pieces).

b. Stuffed mushrooms are a heterogeneous mixture since there are mushrooms and chunks of filling.

c. Eye drops are a homogeneous mixture since they have a uniform composition.

3.63 41.5 °C

$$T_F = 1.8(T_C) + 32 = 1.8(41.5) + 32 = 74.7 + 32 = 106.7 \, °F \, (4 \, \text{SFs})$$

3.65 $T_C = \dfrac{T_F - 32}{1.8} = \dfrac{(155 - 32)}{1.8} = \dfrac{123}{1.8} = 68.3 \, °C \, (3 \, \text{SFs})$

(1.8 and 32 are exact numbers)

$$T_K = T_C + 273 = 68.3 + 273 = 341 \, K \, (3 \, \text{SFs})$$

3.67 Given 10.0 cm^3 cubes of gold (SH_{gold} = 0.129 J/g °C), and aluminum ($SH_{aluminum}$ = 0.897 J/g °C); $\Delta T = 25 \, °C - 15 \, °C = 10. \, °C$

Need energy in joules and calories

Plan cm$^3 \rightarrow$ g   Heat = $m \times \Delta T \times SH$ $\dfrac{19.3 \, \text{g gold}}{1 \, \text{cm}^3}$ $\dfrac{2.70 \, \text{g aluminum}}{1 \, \text{cm}^3}$

Set-up for gold: $10.0 \, \cancel{\text{cm}^3} \times \dfrac{19.3 \, \cancel{\text{g}}}{1 \, \cancel{\text{cm}^3}} \times 10. \, \cancel{°C} \times \dfrac{0.129 \, \text{J}}{\cancel{\text{g}} \, \cancel{°C}} = 250 \, \text{J (2 SFs)}$

$10.0 \, \cancel{\text{cm}^3} \times \dfrac{19.3 \, \cancel{\text{g}}}{1 \, \cancel{\text{cm}^3}} \times 10. \, \cancel{°C} \times \dfrac{0.129 \, \cancel{\text{J}}}{\cancel{\text{g}} \, \cancel{°C}} \times \dfrac{1 \, \text{cal}}{4.184 \, \cancel{\text{J}}} = 60. \, \text{cal (2 SFs)}$

for aluminum: $10.0 \, \cancel{\text{cm}^3} \times \dfrac{2.70 \, \cancel{\text{g}}}{1 \, \cancel{\text{cm}^3}} \times 10. \, \cancel{°C} \times \dfrac{0.897 \, \text{J}}{\cancel{\text{g}} \, \cancel{°C}} = 240 \, \text{J (2 SFs)}$

$10.0 \, \cancel{\text{cm}^3} \times \dfrac{2.70 \, \cancel{\text{g}}}{1 \, \cancel{\text{cm}^3}} \times 10. \, \cancel{°C} \times \dfrac{0.897 \, \cancel{\text{J}}}{\cancel{\text{g}} \, \cancel{°C}} \times \dfrac{1 \, \text{cal}}{4.184 \, \cancel{\text{J}}} = 58 \, \text{cal (2 SFs)}$

3.69 a. Given a meal consisting of: cheeseburger: 46 g of carbohydrate, 40. g of fat, 47 g of protein
french fries: 47 g of carbohydrate, 16 g of fat, 4 g of protein
chocolate shake: 76 g of carbohydrate, 10. g of fat, 10. g of protein

 Need energy from each food type in kilocalories
 Plan total grams of each food type, then g → kcal
 Set-up total carbohydrate = 46 g + 47 g + 76 g = 169 g
total fat = 40. g + 16 g + 10. g = 66 g
total protein = 47 g + 4 g + 10. g = 61 g

Food Type	Mass	Energy Value	Energy (rounded off to the tens place)
Carbohydrate	169 g̶ ×	$\dfrac{4\text{ kcal}}{1\text{ g̶}}$ =	680 kcal
Fat	66 g̶ ×	$\dfrac{9\text{ kcal}}{1\text{ g̶}}$ =	590 kcal
Protein	61 g̶ ×	$\dfrac{4\text{ kcal}}{1\text{ g̶}}$ =	240 kcal

 b. total energy for the meal = 680 kcal + 590 kcal + 240 kcal = 1510 kcal
 c. Given 1510 kcal from meal   **Need** hours of sleeping to "burn off"

 Plan kcal → h $\dfrac{1\text{ h sleeping}}{60\text{ kcal}}$

 Set-up 1510 k̶c̶a̶l̶ × $\dfrac{1\text{ h sleeping}}{60\text{ k̶c̶a̶l̶}}$ = 3×10^1 h of sleeping (1 SF)

 d. Given 1510 kcal from meal   **Need** hours of running to "burn off"

 Plan kcal → h $\dfrac{1\text{ h running}}{750\text{ kcal}}$

 Set-up 1510 k̶c̶a̶l̶ × $\dfrac{1\text{ h running}}{750\text{ k̶c̶a̶l̶}}$ = 2.0 h of running (2 SFs)

3.71 *Elements* are the simplest type of pure substance, containing only one type of atom.
Compounds contain two or more elements chemically combined in a definite proportion.
In a *mixture*, two or more substances are physically mixed but not chemically combined.
 a. Carbon in pencils is an element since it contains only one type of atom (C).
 b. Carbon monoxide is a compound since it contains two elements (C, O) that are chemically combined.
 c. Orange juice is composed of several substances physically mixed together (e.g., water, sugar, citric acid), which makes it a mixture.

3.73 A *homogeneous mixture* has a uniform composition; a *heterogeneous mixture* does not have a uniform composition throughout the mixture.
 a. A hot fudge sundae is a heterogeneous mixture since it has ice cream, fudge sauce, and perhaps a cherry.
 b. Herbal tea is a homogeneous mixture since it has a uniform composition.
 c. Vegetable oil is a homogeneous mixture since it has a uniform composition.

3.75 a. A vitamin tablet is a solid.   **b.** Helium in a balloon is a gas.
 c. Milk is a liquid.   **d.** Air is a mixture of gases.
 e. Charcoal is a solid.

3.77 A *physical property* is a characteristic of a substance—such as color, shape, odor, luster, size, melting point, and density—that can be observed without changing the identity of the substance. A *chemical property* is a characteristic that indicates the ability of a substance to change into a new substance.

   a. The luster of gold is a physical property.

   b. The melting point of gold is a physical property.

   c. The ability of gold to conduct electricity is a physical property.

   d. The ability of gold to form a new substance with sulfur is a chemical property.

3.79 When matter undergoes a *physical change*, its state or appearance changes, but its composition remains the same. When a *chemical change* occurs, the original substance is converted into a new substance, which has different physical and chemical properties.

   a. Plant growth produces new substances, so it is a chemical change.

   b. A change of state from solid to liquid is a physical change.

   c. Chopping wood into smaller pieces is a physical change.

   d. Burning wood, which forms new substances, results in a chemical change.

3.81 **a.** $T_C = \dfrac{T_F - 32}{1.8} = \dfrac{(134 - 32)}{1.8} = \dfrac{102}{1.8} = 56.7\,°C$

      $T_K = T_C + 273 = 56.7 + 273 = 330.\,K$

   b. $T_C = \dfrac{T_F - 32}{1.8} = \dfrac{(-69.7 - 32)}{1.8} = \dfrac{-101.7}{1.8} = -56.50\,°C$

      $T_K = T_C + 273 = -56.50 + 273 = 217\,K$

3.83 $T_C = \dfrac{T_F - 32}{1.8} = \dfrac{(-15 - 32)}{1.8} = \dfrac{-47}{1.8} = -26\,°C$

   $T_K = T_C + 273 = -26 + 273 = 247\,K$

3.85 **Given** $m = 0.50$ g of oil; 18.9 kJ produced

    Need   energy value (kcal/g)

    Plan   kJ → kcal $\dfrac{1000\ J}{1\ kJ}$ $\dfrac{1\ cal}{4.184\ J}$ $\dfrac{1\ kcal}{1000\ cal}$

          then energy value $= \dfrac{\text{heat in kcal}}{\text{mass in g}}$

    Set-up $18.9\ \cancel{kJ} \times \dfrac{1000\ \cancel{J}}{1\ \cancel{kJ}} \times \dfrac{1\ \cancel{cal}}{4.184\ \cancel{J}} \times \dfrac{1\ kcal}{1000\ \cancel{cal}} = 4.52$ kcal (3 SFs)

          ∴ energy value $= \dfrac{\text{heat produced}}{\text{mass of oil}} = \dfrac{4.52\ kcal}{0.50\ g\ oil} = 9.0$ kcal/g of vegetable oil (2 SFs)

3.87 Sand must have a lower specific heat than water since the same amount of heat causes a greater temperature change in the sand than in the water.

3.89 **a.** The melting point of chloroform is about $-60\,°C$.

   b. The boiling point of chloroform is about $60\,°C$.

   c. The diagonal line **A** represents the solid state as temperature increases. The horizontal line **B** represents the change from solid to liquid or melting of the substance. The diagonal line **C** represents the liquid state as temperature increases. The horizontal line **D** represents the change from liquid to gas or boiling of the liquid. The diagonal line **E** represents the gas state as temperature increases.

   d. At $-80\,°C$, it is solid; at $-40\,°C$, it is liquid; at $25\,°C$, it is liquid; at $80\,°C$, it is gas.

3.91

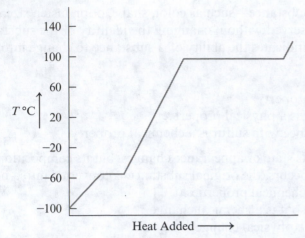

a. Dibromomethane is a solid at −75 °C.
b. At −53 °C, solid dibromomethane melts (solid → liquid).
c. Dibromomethane is a liquid at −18 °C.
d. At 110 °C, dibromomethane is a gas.
e. Both solid and liquid dibromomethane will be present at the melting temperature of −53 °C.

3.93 **Given** 1 lb of body fat; 15% (m/m) water in body fat
Need kilocalories to "burn off"
Plan Because each gram of body fat contains 15% water, a person actually loses 85 grams of fat per hundred grams of body fat. (We considered 1 lb of fat as exactly 1 lb.)

lb of body fat → g of body fat → g of fat → kcal

$$\frac{454 \text{ g body fat}}{1 \text{ lb body fat}} \qquad \frac{85 \text{ g fat}}{100 \text{ g body fat}} \qquad \frac{9 \text{ kcal}}{1 \text{ g fat}}$$

Set-up $1 \text{ lb body fat} \times \dfrac{454 \text{ g body fat}}{1 \text{ lb body fat}} \times \dfrac{85 \text{ g fat}}{100 \text{ g body fat}} \times \dfrac{9 \text{ kcal}}{1 \text{ g fat}} = 3500 \text{ kcal (2 SFs)}$

3.95 **Given** $m = 725$ g of water; $SH_{water} = 4.184$ J/g °C; $\Delta T = 65\,°C - 37\,°C = 28\,°C$
Need heat in kilojoules

Plan Heat $= m \times \Delta T \times SH$, then J → kJ $\dfrac{1 \text{ kJ}}{1000 \text{ J}}$

Set-up Heat $= 725 \text{ g} \times 28 \text{ °C} \times \dfrac{4.184 \text{ J}}{\text{g °C}} \times \dfrac{1 \text{ kJ}}{1000 \text{ J}} = 85 \text{ kJ (2 SFs)}$

∴ 85 kJ are lost from the water bottle and are transferred to the muscles.

3.97 **Given** 0.66 g of olive oil; 370 g of water; $SH_{water} = 1.00$ cal/g °C;
$\Delta T = 38.8\,°C - 22.7\,°C = 16.1\,°C$
Need energy value (in kcal/g)

Plan Heat $= m \times \Delta T \times SH$ then cal → kcal $\dfrac{1 \text{ kcal}}{1000 \text{ cal}}$

Set-up Heat $= 370 \text{ g} \times 16.1 \text{ °C} \times \dfrac{1 \text{ cal}}{\text{g °C}} \times \dfrac{1 \text{ kcal}}{1000 \text{ cal}} = 6.0 \text{ kcal}$

∴ energy value $= \dfrac{\text{heat produced}}{\text{mass of oil}} = \dfrac{6.0 \text{ kcal}}{0.66 \text{ g oil}} = 9.1 \text{ kcal/g of olive oil (2 SFs)}$

3.99 **a. Given** 2.4×10^7 J/1.0 lb of oil; $m = 150$ kg of water; $SH_{water} = 4.184$ J/g °C;
$\Delta T = 100\,°C - 22\,°C = 78\,°C$
Need kilograms of oil needed

Plan Heat $= m \times \Delta T \times SH$ then $J \rightarrow$ lb of oil $\rightarrow$ kg of oil $\dfrac{1 \text{ kg}}{2.20 \text{ lb}}$

Set-up Heat $= 150 \text{ kg water} \times \dfrac{1000 \text{ g}}{1 \text{ kg}} \times 78 \text{ °C} \times \dfrac{4.184 \text{ J}}{\text{g °C}} = 4.9 \times 10^7 \text{ J (2 SFs)}$

$\therefore m_{\text{oil}} = 4.9 \times 10^7 \text{ J} \times \dfrac{1.0 \text{ lb oil}}{2.4 \times 10^7 \text{ J}} \times \dfrac{1.0 \text{ kg oil}}{2.20 \text{ lb oil}} = 0.93 \text{ kg of oil (2 SFs)}$

b. Given $2.4 \times 10^7 \text{ J}/1.0$ lb of oil; $m = 150$ kg of water; water $100 \text{ °C} \rightarrow$ steam 100 °C
  Need kilograms of oil needed

Plan heat for water $100 \text{ °C} \rightarrow$ steam 100 °C; then $J \rightarrow$ lb of oil $\rightarrow$ kg of oil $\dfrac{1 \text{ kg}}{2.20 \text{ lb}}$

Set-up Heat $= 150 \text{ kg water} \times \dfrac{1000 \text{ g}}{1 \text{ kg}} \times \dfrac{2260 \text{ J}}{1 \text{ g water}} = 3.4 \times 10^8 \text{ J (2 SFs)}$

$\therefore m_{\text{oil}} = 3.4 \times 10^8 \text{ J} \times \dfrac{1.0 \text{ lb oil}}{2.4 \times 10^7 \text{ J}} \times \dfrac{1.0 \text{ kg oil}}{2.20 \text{ lb oil}} = 6.4 \text{ kg of oil (2 SFs)}$

3.101 Two calculations are required:

(1) ice $0.0 \text{ °C} \rightarrow$ water 0.0 °C: $275 \text{ g ice} \times \dfrac{334 \text{ J}}{1 \text{ g ice}} \times \dfrac{1 \text{ kJ}}{1000 \text{ J}} = 91.9 \text{ kJ}$

(2) water $0.0 \text{ °C} \rightarrow 24.0 \text{ °C}$: $\Delta T = 24.0 \text{ °C} - 0.0 \text{ °C} = 24.0 \text{ °C}$;

$275 \text{ g water} \times 24.0 \text{ °C} \times \dfrac{4.184 \text{ J}}{\text{g °C}} \times \dfrac{1 \text{ kJ}}{1000 \text{ J}} = 27.6 \text{ kJ}$

$\therefore$ Total heat absorbed $= 91.9 \text{ kJ} + 27.6 \text{ kJ} = 119.5 \text{ kJ (4 SFs)}$

3.103 Given

Copper	Water
$m = 70.0$ g	$m = 50.0$ g
initial temperature $= 54.0 \text{ °C}$	initial temperature $= 26.0 \text{ °C}$
final temperature $= 29.2 \text{ °C}$	final temperature $= 29.2 \text{ °C}$
$SH = ?$ J/g °C	$SH = 4.184$ J/g °C

Need specific heat (SH) of copper (J/g °C)
Plan heat lost by copper $=$ heat gained by water
  For both, heat $= m \times \Delta T \times SH$
Set-up For water:
  heat gained $= m \times \Delta T \times SH = 50.0 \text{ g} \times (29.2 \text{ °C} - 26.0 \text{ °C}) \times \dfrac{4.184 \text{ J}}{\text{g °C}} = 669 \text{ J}$

  For copper:
  $\therefore$ heat lost by copper $= 669 \text{ J}$

  $SH = \dfrac{\text{heat lost}}{m \times \Delta T} = \dfrac{669 \text{ J}}{70.0 \text{ g} \times (54.0 \text{ °C} - 29.2 \text{ °C})} = \dfrac{0.385 \text{ J}}{\text{g °C}}$ (3 SFs)

3.105 a. Given heat $= 11$ J; $m = 4.7$ g; $\Delta T = 4.5 \text{ °C}$
  Need specific heat (J/g °C)
  Plan $SH = \dfrac{\text{heat}}{m \times \Delta T}$

  Set-up $SH = \dfrac{11 \text{ J}}{4.7 \text{ g} \times 4.5 \text{ °C}} = \dfrac{0.52 \text{ J}}{\text{g °C}}$ (2 SFs)

b. By comparing this calculated value to the specific heats given in Table 3.11 we would identify the unknown metal as titanium ($SH = 0.523$ J/g °C) rather than aluminum ($SH = 0.897$ J/g °C).

Selected Answers to Combining Ideas from Chapters 1 to 3

CI.1 **a.** There are 4 significant figures in the measurement 20.17 lb.

b. $20.17 \text{ lb} \times \dfrac{1 \text{ kg}}{2.20 \text{ lb}} = 9.17 \text{ kg}$ (3 SFs)

c. $9.17 \text{ kg} \times \dfrac{1000 \text{ g}}{1 \text{ kg}} \times \dfrac{1 \text{ cm}^3}{19.3 \text{ g}} = 475 \text{ cm}^3$ (3 SFs)

d. $T_F = 1.8(T_C) + 32 = 1.8(1064) + 32 = 1947 \, °F$ (4 SFs)
$T_K = T_C + 273 = 1064 + 273 = 1337 \text{ K}$ (4 SFs)

e. $\Delta T = 1064 \, °C - 500. \, °C = 564 \, °C$

$9.17 \text{ kg} \times \dfrac{1000 \text{ g}}{1 \text{ kg}} \times \dfrac{0.129 \text{ J}}{\text{g} \, °C} \times \dfrac{1 \text{ cal}}{4.184 \text{ J}} \times 564 \, °C \times \dfrac{1 \text{ kcal}}{1000 \text{ cal}} = 159 \text{ kcal}$ (3 SFs)

f. $9.17 \text{ kg} \times \dfrac{1000 \text{ g}}{1 \text{ kg}} \times \dfrac{\$42.06}{1 \text{ g}} = \$386\,000$ (3 SFs)

CI.3 **a.** The water has its own shape in sample **B**.

b. Water sample **A** is represented by diagram **2**, which shows the particles in a random arrangement but close together.

c. Water sample **B** is represented by diagram **1**, which shows the particles fixed in a definite arrangement.

d. The state of matter indicated in diagram **1** is a <u>solid</u>; in diagram **2**, it is a <u>liquid</u>; and in diagram **3**, it is a <u>gas</u>.

e. The motion of the particles is slowest in diagram <u>**1**</u>.

f. The arrangement of particles is farthest apart in diagram <u>**3**</u>.

g. The particles fill the volume of the container in diagram <u>**3**</u>.

h. water $45 \, °C \rightarrow 0 \, °C$: $\Delta T = 45 \, °C - 0 \, °C = 45 \, °C$
Heat $= m \times \Delta T \times SH$

$= 19 \text{ g} \times 45 \, °C \times \dfrac{4.184 \text{ J}}{\text{g} \, °C} \times \dfrac{1 \text{ kJ}}{1000 \text{ J}} = 3.6 \text{ kJ}$ (2 SFs)

$\therefore 3.6 \text{ kJ}$ of heat is removed.

CI.5 **a.** $0.250 \text{ lb} \times \dfrac{454 \text{ g}}{1 \text{ lb}} \times \dfrac{1 \text{ cm}^3}{7.86 \text{ g}} = 14.4 \text{ cm}^3$ (3 SFs)

b. $30 \text{ nails} \times \dfrac{14.4 \text{ cm}^3}{75 \text{ nails}} = 5.76 \text{ cm}^3 = 5.76 \text{ mL}$ (3 SFs)

New water level $= 17.6 \text{ mL water} + 5.76 \text{ mL} = 23.4 \text{ mL}$ (3 SFs)

c. $\Delta T = 125 \, °C - 16 \, °C = 109 \, °C$
Heat $= m \times \Delta T \times SH$

$= 0.250 \text{ lb} \times \dfrac{454 \text{ g}}{1 \text{ lb}} \times 109 \, °C \times \dfrac{0.452 \text{ J}}{\text{g} \, °C} = 5590 \text{ J}$ (3 SFs)

d. iron $25 \, °C \rightarrow 1535 \, °C$: $\Delta T = 1535 \, °C - 25 \, °C = 1510. \, °C$

$1 \text{ nail} \times \dfrac{0.250 \text{ lb Fe}}{75 \text{ nails}} \times \dfrac{454 \text{ g Fe}}{1 \text{ lb Fe}} \times \dfrac{0.452 \text{ J}}{\text{g} \, °C} \times 1510. \, °C = 1030 \text{ J}$ (3 SFs)

4

Atoms and Elements

John is preparing to plant a new crop in his fields. Although the element nitrogen in the soil is important, he will also check the soil levels of the elements phosphorus and potassium, which are also important for a good yield. Tests show that the level of phosphorus in the soil is 15 mg/kg, which is low. Tests also show that the potassium level in the soil is in the optimal range from 150 to 200 mg/kg. Thus, John will apply a fertilizer that contains phosphorus but no potassium. What are the symbols of the elements potassium, phosphorus, and nitrogen found in fertilizers?

Credit: Martin Harvey/Alamy

LOOKING AHEAD

 The Health icon indicates a question that is related to health and medicine.

4.1 Elements and Symbols

Learning Goal: Given the name of an element, write its correct symbol; from the symbol, write the correct name.

- Chemical symbols are one- or two-letter abbreviations for the names of the elements; only the first letter of a chemical symbol is a capital letter.
- Element names are derived from planets, mythology, colors, minerals, and names of geographical locations and famous scientists.

♦ **Learning Exercise 4.1A** 🏃

Write the symbol for each of the following elements found in the body:

a. carbon   _C_   b. iron   _Fe_   c. sodium   _Na_

d. phosphorus _P_   e. oxygen _O_   f. nitrogen _N_

g. iodine   _I_   h. sulfur _S_   i. potassium _K_

j. cobalt _Co_   k. calcium _Ca_   l. selenium _Se_

Answers	**a.** C	**b.** Fe	**c.** Na	**d.** P	**e.** O	**f.** N
	g. I	**h.** S	**i.** K	**j.** Co	**k.** Ca	**l.** Se

♦ **Learning Exercise 4.1B**

Write the name of the element represented by each of the following symbols:

The element silver is used for dental fillings.

Credit: Pearson Science/Pearson Education

a. Mg _Magnesium_   b. Be _Beryllium_

c. H _Hydrogen_   d. Si _Silicon_

e. Ag _Silver_   f. Br _Bromine_

g. F _fluorine_   h. Zn _Zinc_

i. Cr _Chromium_   j. Al _Aluminum_

Answers	**a.** magnesium	**b.** beryllium	**c.** hydrogen	**d.** silicon	**e.** silver
	f. bromine	**g.** fluorine	**h.** zinc	**i.** chromium	**j.** aluminum

4.2   The Periodic Table

Learning Goal: Use the periodic table to identify the group and the period of an element; identify an element as a metal, a nonmetal, or a metalloid.

- The periodic table is an arrangement of the elements into vertical columns and horizontal rows.
- Each vertical column is a *group* (or *family*) of elements that have similar properties.
- A horizontal row of elements is a *period*.
- On the periodic table, the metals are located on the left of the heavy zigzag line, the nonmetals are to the right, and metalloids are found along the heavy zigzag line.
- The element hydrogen (H) located on the top left of the periodic table is a nonmetal.
- Main group, or representative, elements are found in Groups 1A (1), 2A (2), and 3A (13) to 8A (18). Transition elements are B-group elements (3 to 12).
- Of all the elements, only about 20 are essential for the well-being of the human body. Oxygen, carbon, hydrogen, and nitrogen make up 96% of human body mass.

Study Note

1. The periodic table consists of horizontal rows called *periods* and vertical columns called *groups*, which are also called *families*.
2. Group 1A (1) contains the *alkali metals*, Group 2A (2) contains the *alkaline earth metals*, Group 7A (17) contains the *halogens*, and Group 8A (18) contains the *noble gases*.

Periodic Table of Elements

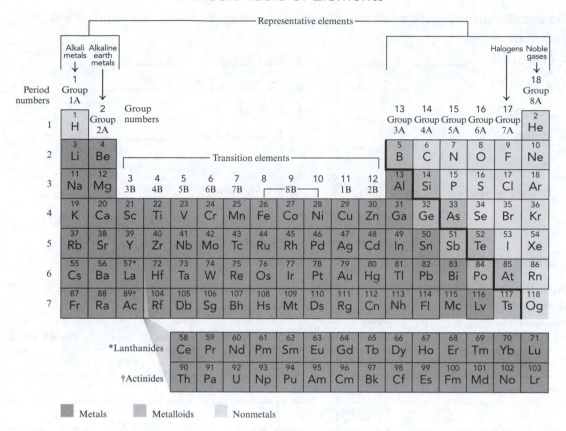

Key Terms for Sections 4.1 and 4.2

Match each of the following key terms with the correct description:

a. nonmetal   b. period   c. chemical symbol   d. metal
e. group number   f. halogen   g. periodic table   h. transition element

1. _____ a horizontal row of elements in the periodic table

2. _____ an element with little or no luster that is a poor conductor of heat and electricity

3. _____ a number that appears at the top of each vertical column (group) in the periodic table

4. _____ an element that is shiny, malleable, ductile, and a good conductor of heat and electricity

5. _____ an element in the center of the periodic table that has a group number of 3 through 12

6. _____ an arrangement of elements by increasing atomic number

7. _____ a one- or two-letter abbreviation that represents the name of an element

8. _____ an element in Group 7A (17)

Answers   **1.** b   **2.** a   **3.** e   **4.** d
           5. h   **6.** g   **7.** c   **8.** f

◆ **Learning Exercise 4.2A**

Indicate whether the following elements are part of the same group (G), the same period (P), or neither (N):

a. Li, O, and F _____   **b.** Br, Cl, and F _____

c. Al, Si, and Cl _____   **d.** Cl, Mg, and P _____

e. Mg, Ca, and Ba _____   **f.** C, S, and Br _____

g. Sb, P, and As _____   **h.** K, Ca, and Br _____

Answers   **a.** P   **b.** G   **c.** P   **d.** P
                  e. G   **f.** N   **g.** G   **h.** P

◆ **Learning Exercise 4.2B**

Complete the list of elements important in the body with the name and symbol, group number, period number, and whether it is a metal, nonmetal, or metalloid.

Name and Symbol	Group Number	Period Number	Type of Element
	2A (2)	3	
Carbon, C			
	5A (15)	2	
Sulfur, S			
	5A (15)	5	
	1A (1)	1	
		3	metalloid

Answers

Name and Symbol	Group Number	Period Number	Type of Element
Magnesium, Mg	2A (2)	3	metal
Carbon, C	4A (14)	2	nonmetal
Nitrogen, N	5A (15)	2	nonmetal
Sulfur, S	6A (16)	3	nonmetal
Antimony, Sb	5A (15)	5	metalloid
Hydrogen, H	1A (1)	1	nonmetal
Silicon, Si	4A (14)	3	metalloid

♦ **Learning Exercise 4.2C**

Identify each of the following elements as a metal (M), a nonmetal (NM), or a metalloid (ML):

a. Cl _____   **b.** P _____   **c.** Fe _____   **d.** As _____   **e.** Al _____

f. O _____   **g.** Ca _____   **h.** Zn _____   **i.** Ge _____   **j.** Ag _____

Answers   **a.** NM   **b.** NM   **c.** M   **d.** ML   **e.** M
             f. NM   **g.** M   **h.** M   **i.** ML   **j.** M

♦ **Learning Exercise 4.2D**

Match the name of each of the following chemical groups with the elements Cs, F, He, Cr, Sr, Ne, Li, Cu, and Br:

a. halogen _____

b. noble gas _____

c. alkali metal _____

d. alkaline earth metal _____

e. transition element _____

Answers   **a.** F, Br   **b.** He, Ne   **c.** Cs, Li   **d.** Sr   **e.** Cr, Cu

4.3   The Atom

Learning Goal: Describe the electrical charge of a proton, neutron, and electron and identify their location within an atom.

- An atom is the smallest particle that retains the characteristics of an element.
- Atoms are composed of three types of subatomic particles. Protons have a positive charge (+), electrons carry a negative charge (−), and neutrons are electrically neutral.
- The protons and neutrons, with masses of about 1 amu, are found in the tiny, dense nucleus.
- The electrons are located outside the nucleus and have masses much smaller than the mass of a neutron or proton.

Dalton's Atomic Theory

1. All matter is made up of tiny particles called atoms.
2. All atoms of a given element are the same and different from atoms of other elements.
3. Atoms of two or more different elements combine to form compounds. A particular compound is always made up of the same kinds of atoms and the same number of each kind of atom.
4. A chemical reaction involves the rearrangement, separation, or combination of atoms. Atoms are neither created nor destroyed during a chemical reaction.

♦ **Learning Exercise 4.3A**

True (T) or False (F): Each of the following statements is consistent with current atomic theory:

a. _____ All matter is composed of atoms.

b. _____ All atoms of an element are similar.

c. _____ Atoms can combine to form compounds.

d. _____ Most of the mass of the atom is outside of the nucleus.

Answers   **a.** T   **b.** T   **c.** T   **d.** F

Aluminum foil consists of atoms of aluminum.

Credit: Russ Lappa/Pearson Education/ Prentice Hall

♦ **Learning Exercise 4.3B**

Match each of the following terms with the correct description:

   a. proton   **b.** neutron   **c.** electron   **d.** nucleus

1. _____ is found in the nucleus of an atom

2. _____ has a 1− charge

3. _____ is found outside the nucleus

4. _____ has a mass of about 1 amu

5. _____ is the small, dense center of the atom

6. _____ is neutral

7. _____ is attracted to a proton

Answers   **1.** a and b   **2.** c   **3.** c   **4.** a and b   **5.** d   **6.** b   **7.** c

4.4   Atomic Number and Mass Number

Learning Goal: Given the atomic number and the mass number of an atom, state the number of protons, neutrons, and electrons.

REVIEW

Using Positive and Negative Numbers in Calculations (1.4)

- The atomic number, which can be found on the periodic table, is the number of protons in a given atom of an element.
- In a neutral atom, the number of protons is equal to the number of electrons.
- The mass number is the total number of protons and neutrons in an atom.

Study Note

1. The _atomic number_ is the number of protons in every atom of an element. In neutral atoms, the number of electrons equals the number of protons.
2. The _mass number_ is the total number of neutrons and protons in the nucleus of an atom.
3. The number of neutrons is _mass number − atomic number._

Example: How many protons and neutrons are in the nucleus of a strontium atom with a mass number of 89?

Solution: The atomic number of Sr is 38. Thus, an atom of Sr has 38 protons.

Mass number − atomic number = number of neutrons: 89 − 38 = 51 neutrons

---

♦ **Learning Exercise 4.4A**

CORE CHEMISTRY SKILL
Counting Protons and Neutrons

Determine the number of protons in each of the following neutral atoms:

a. an atom of carbon ____
b. an atom of the element with atomic number 15 ____
c. an atom with a mass number of 40 and atomic number 19 ____
d. an atom with 9 neutrons and a mass number of 19 ____
e. a neutral atom that has 18 electrons ____

Answers **a.** 6 **b.** 15 **c.** 19 **d.** 10 **e.** 18

♦ **Learning Exercise 4.4B**

Determine the number of neutrons in each of the following neutral atoms:

a. a mass number of 33 and atomic number 15 ____
b. a mass number of 10 and 5 protons ____
c. a selenium atom with a mass number of 80 ____
d. a mass number of 9 and atomic number 4 ____
e. a mass number of 22 and 10 protons ____
f. a titanium atom with a mass number of 46 ____

Answers **a.** 18 **b.** 5 **c.** 46 **d.** 5 **e.** 12 **f.** 24

4.5 Isotopes and Atomic Mass

REVIEW
Calculating Percentages (1.4)

Learning Goal: Determine the number of protons, neutrons, and electrons in one or more of the isotopes of an element; calculate the atomic mass of an element using the percent abundance and mass of its naturally occurring isotopes.

- *Isotopes* are atoms of the same element that have the same number of protons but different numbers of neutrons.
- The atomic mass of an element is the average mass of all the isotopes in a naturally occurring sample of that element.

Key Terms for Sections 4.3 to 4.5

Match each of the following key terms with the correct description:

a. electron **b.** atom **c.** atomic mass **d.** mass number **e.** neutron
f. isotope **g.** atomic number **h.** nucleus **i.** proton

1. ____ the total number of neutrons and protons in the nucleus of an atom
2. ____ the smallest particle of an element that retains the characteristics of the element
3. ____ a negatively charged subatomic particle having a minute mass
4. ____ a neutral subatomic particle found in the nucleus of an atom

5. _____ a number that is equal to the number of protons in an atom

6. _____ the compact, extremely dense center of an atom where the protons and neutrons are located

7. _____ a positively charged subatomic particle found in the nucleus of an atom

8. _____ the weighted average mass of all the naturally occurring isotopes of an element

9. _____ an atom that differs in mass number from another atom of the same element

Answers   **1.** d   **2.** b   **3.** a   **4.** e   **5.** g
            6. h   **7.** i   **8.** c   **9.** f

Study Note

In the atomic symbol for a particular atom, the mass number appears in the upper left corner and the atomic number in the lower left corner of the atomic symbol. $\begin{array}{l}\text{Mass number} \rightarrow \\ \text{Atomic number} \rightarrow\end{array} {}^{32}_{16}\text{S}$

This isotope is also written several other ways including sulfur-32, S-32, and ^{32}S.

♦ **Learning Exercise 4.5A**

Complete the following table for neutral atoms:

Atomic Symbol	Atomic Number	Mass Number	Number of Protons	Number of Neutrons	Number of Electrons
	15			16	
			47	60	
		55		30	
	35			45	
		35	17		
$^{120}_{50}$Sn					

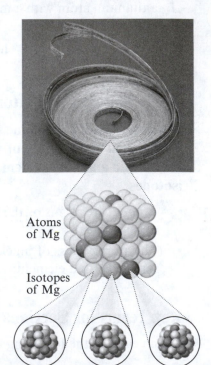

Atoms of Mg

Isotopes of Mg

$^{24}_{12}$Mg $^{25}_{12}$Mg $^{26}_{12}$Mg

Answers

Atomic Symbol	Atomic Number	Mass Number	Number of Protons	Number of Neutrons	Number of Electrons
$^{31}_{15}$P	15	31	15	16	15
$^{107}_{47}$Ag	47	107	47	60	47
$^{55}_{25}$Mn	25	55	25	30	25
$^{80}_{35}$Br	35	80	35	45	35
$^{35}_{17}$Cl	17	35	17	18	17
$^{120}_{50}$Sn	50	120	50	70	50

The nuclei of three naturally occurring magnesium isotopes have the same number of protons but different numbers of neutrons.

Credit: Pearson Science/Pearson Education

♦ **Learning Exercise 4.5B**

Identify the sets of atoms that are isotopes.

    A. $^{20}_{10}X$ **B.** $^{20}_{11}X$ **C.** $^{21}_{11}X$ **D.** $^{19}_{10}X$ **E.** $^{19}_{9}X$

Answer   Atoms A and D are isotopes (atomic number 10); atoms B and C are isotopes (atomic number 11).

♦ **Learning Exercise 4.5C**

Copper has two naturally occurring isotopes, $^{63}_{29}Cu$ and $^{65}_{29}Cu$. If that is the case, why is the atomic mass of copper listed as 63.55 on the periodic table?

Answer   Copper in nature consists of two isotopes that have different atomic masses. The atomic mass is the weighted average of the masses of the two isotopes, which takes into consideration their percent abundance in the sample. The atomic mass does not represent the mass of any individual atom.

Study Note

To calculate the atomic mass of an element, multiply the mass of each isotope by its percent abundance. Then add these values together. For example, a large sample of naturally occurring chlorine atoms consists of 75.76% of Cl-35 atoms (mass 34.97 amu) and 24.24% of Cl-37 atoms (mass 36.97 amu).

Atomic Symbol	Mass (amu)		Percent Abundance		Contribution to the Atomic Mass
Cl-35	34.97	×	$\dfrac{75.76}{100}$	=	26.49 amu
Cl-37	36.97	×	$\dfrac{24.24}{100}$	=	8.962 amu
			Atomic mass of Cl	=	35.45 amu

♦ **Learning Exercise 4.5D**

Rubidium is the element with atomic number 37.

  1. What is the symbol for rubidium? _____

  2. What is the group number for rubidium? _____

  3. What is the name of the family that includes rubidium? _____

  4. Rubidium has two naturally occurring isotopes, $^{85}_{37}Rb$ and $^{87}_{37}Rb$. How many protons, neutrons, and electrons are in each isotope?

5. In naturally occurring rubidium, 72.17% is $^{85}_{37}$Rb with a mass of 84.91 amu; 27.83% is $^{87}_{37}$Rb with a mass of 86.91 amu. Calculate the atomic mass for rubidium using the weighted average mass method.

Answers

1. Rb
2. Group 1A (1)
3. alkali metals
4. The isotope $^{85}_{37}$Rb has 37 protons, 48 neutrons, and 37 electrons. The isotope $^{87}_{37}$Rb has 37 protons, 50 neutrons, and 37 electrons.
5. 84.91 amu (72.17/100) + 86.91 amu (27.83/100) = 61.28 amu + 24.19 amu = 85.47 amu for the atomic mass of Rb.

4.6   Electron Energy Levels

Learning Goal: Describe the energy levels, sublevels, and orbitals for the electrons in an atom.

- Energy levels, which are indicated by the principal quantum number n, contain electrons of similar energies, and increase in order of energy ($n = 1 < 2 < 3 < 4 < 5 < 6 < 7$).
- Within each energy level, electrons with identical energy are grouped in sublevels s, p, d, f.
- An orbital is a region around the nucleus in which an electron with a specific energy is most likely to be found. Each orbital holds a maximum of two electrons.
- An s orbital is spherical, and a p orbital has two lobes along an axis. The d and f orbitals have more complex shapes.
- An s sublevel contains one s orbital; a p sublevel contains three p orbitals; a d sublevel contains five d orbitals; and an f sublevel contains seven f orbitals.

♦ **Learning Exercise 4.6**

State the maximum number of electrons for each of the following:

a. 3p sublevel _____   **b.** 3d sublevel _____   **c.** 2s orbital _____

d. energy level 4 _____   **e.** 1s sublevel _____   **f.** 4p orbital _____

g. 5p sublevel _____   **h.** 4f sublevel _____

Answers   **a.** 6   **b.** 10   **c.** 2   **d.** 32
              e. 2   **f.** 2   **g.** 6   **h.** 14

4.7   Electron Configurations

Learning Goal: Draw the orbital diagram, and write the electron configuration for an element.

- The orbital diagram for an element shows the orbitals as boxes that are occupied by paired and unpaired electrons.
- Within a sublevel, electrons enter orbitals in the same energy level one at a time until all the orbitals are half-filled.
- Additional electrons enter with opposite spins until the orbitals in that sublevel are filled with two electrons each.
- The electron configuration for an element such as silicon shows the number of electrons in each sublevel: $1s^2 2s^2 2p^6 3s^2 3p^2$.

- An abbreviated electron configuration for an element such as silicon places the symbol of a noble gas in brackets to represent the filled sublevels: $[Ne]3s^23p^2$.
- The periodic table consists of s, p, d, and f sublevel blocks.
- Beginning with $1s$, an electron configuration is obtained by writing the sublevel blocks in order going across each period on the periodic table until the element is reached.

Drawing Orbital Diagrams	
STEP 1	Draw boxes to represent the occupied orbitals.
STEP 2	Place a pair of electrons with opposite spins in each filled orbital.
STEP 3	Place the remaining electrons in the last occupied sublevel in separate orbitals.

♦ **Learning Exercise 4.7A**

Draw the orbital diagram for each of the following elements:

a. beryllium _____

b. carbon _____

c. sodium _____

d. nitrogen _____

e. fluorine _____

f. magnesium _____

Answers

a. $1s$ [↑↓] $2s$ [↑↓]

b. $1s$ [↑↓] $2s$ [↑↓] $2p$ [↑][↑][ ]

c. $1s$ [↑↓] $2s$ [↑↓] $2p$ [↑↓][↑↓][↑↓] $3s$ [↑]

d. $1s$ [↑↓] $2s$ [↑↓] $2p$ [↑][↑][↑]

e. $1s$ [↑↓] $2s$ [↑↓] $2p$ [↑↓][↑↓][↑]

f. $1s$ [↑↓] $2s$ [↑↓] $2p$ [↑↓][↑↓][↑↓] $3s$ [↑↓]

♦ **Learning Exercise 4.7B**

CORE CHEMISTRY SKILL

Writing Electron Configurations

Write the electron configuration ($1s^22s^22p^6$, etc.) for each of the following elements:

a. carbon _____

b. magnesium _____

c. iron _____

d. sulfur _____

e. chlorine _____

f. phosphorus _____

Answers
a. $1s^22s^22p^2$
b. $1s^22s^22p^63s^2$
c. $1s^22s^22p^63s^23p^64s^23d^6$
d. $1s^22s^22p^63s^23p^4$
e. $1s^22s^22p^63s^23p^5$
f. $1s^22s^22p^63s^23p^3$

♦ **Learning Exercise 4.7C**

Write the abbreviated electron configuration for each of the elements in Learning Exercise 4.7B:

a. carbon _____

b. magnesium _____

c. iron _____

d. sulfur _____

e. chlorine _____

f. phosphorus _____

Answers
a. $[He]2s^22p^2$
b. $[Ne]3s^2$
c. $[Ar]4s^23d^6$
d. $[Ne]3s^23p^4$
e. $[Ne]3s^23p^5$
f. $[Ne]3s^23p^3$

♦ **Learning Exercise 4.7D**

Name the element with an electron configuration that ends with each of the following notations:

a. $3p^5$ _____

b. $2s^1$ _____

c. $3d^8$ _____

d. $4p^1$ _____

e. $5p^5$ _____

f. $3p^2$ _____

g. $1s^1$ _____

h. $6s^2$ _____

Answers   **a.** chlorine   **b.** lithium   **c.** nickel   **d.** gallium
           e. iodine   **f.** silicon   **g.** hydrogen   **h.** barium

Writing Electron Configurations Using Sublevel Blocks	
STEP 1	Locate the element on the periodic table.
STEP 2	Write the filled sublevels in order, going across each period.
STEP 3	Complete the electron configuration by counting the electrons in the last occupied sublevel block.

♦ **Learning Exercise 4.7E**

CORE CHEMISTRY SKILL

Using the Periodic Table to Write
Electron Configurations

Use the sublevel blocks on the periodic table to write the electron configuration for each of the following elements:

a. S _____

b. As _____

c. Tc _____

d. Br _____

e. Sr _____

f. V _____

Answers   **a.** S $1s^2 2s^2 2p^6 3s^2 3p^4$

b. As $1s^2 2s^2 2p^6 3s^2 3p^6 4s^2 3d^{10} 4p^3$

c. Tc $1s^2 2s^2 2p^6 3s^2 3p^6 4s^2 3d^{10} 4p^6 5s^2 4d^5$

d. Br $1s^2 2s^2 2p^6 3s^2 3p^6 4s^2 3d^{10} 4p^5$

e. Sr $1s^2 2s^2 2p^6 3s^2 3p^6 4s^2 3d^{10} 4p^6 5s^2$

f. V $1s^2 2s^2 2p^6 3s^2 3p^6 4s^2 3d^3$

4.8   Trends in Periodic Properties

CORE CHEMISTRY SKILL

Identifying Trends in Periodic
Properties

Learning Goal: Use the electron configurations of elements to explain the trends in periodic properties.

- The physical and chemical properties of the elements change in a periodic manner going across each period and are repeated in each successive period.

- The group number of a representative element gives the number of valence electrons in the atoms of that group.

- A Lewis symbol is a convenient way to represent the valence electrons, which are shown as dots on the sides, top, or bottom of the symbol for an element. For beryllium, a Lewis symbol is:

$\dot{Be}\cdot$

- The atomic size of representative elements increases going down a group and decreases going across a period.
- The ionization energy of representative elements decreases going down a group and increases going across a period.
- The metallic character of representative elements increases going down a group and decreases going across a period.

Key Terms for Sections 4.6 to 4.8

Match each of the following key terms with the correct description:

a. *f* block   **b.** *s* block   **c.** sublevel   **d.** electron configuration
e. *d* block   **f.** orbital   **g.** ionization energy   **h.** valence electrons
i. group number   **j.** Lewis symbol

1. _____ the block of 14 elements in the rows at the bottom of the periodic table

2. _____ a list of the number of electrons in each sublevel within an atom, arranged by increasing energy

3. _____ the elements in groups 1A(1) and 2A(2) in which electrons fill the *s* orbitals

4. _____ contains electrons of identical energy

5. _____ the block of 10 elements from Groups 3B(3) to 2B(12)

6. _____ the representation of an atom that shows valence electrons as dots around the symbol of the element

7. _____ a measure of how easily an element loses a valence electron

8. _____ the electrons in the highest energy level of an atom

9. _____ the number at the top of each vertical column in the periodic table

10. _____ a region around the nucleus where electrons are most likely to be found

Answers   **1.** a   **2.** d   **3.** b   **4.** c   **5.** e
              6. j   **7.** g   **8.** h   **9.** i   **10.** f

♦ **Learning Exercise 4.8A**

CORE CHEMISTRY SKILL
Drawing Lewis Symbols

State the group number, the number of valence electrons, and draw the Lewis symbol for each of the following:

Element	Group Number	Valence Electrons	Lewis Symbol
Sulfur			
Silicon			
Magnesium			
Hydrogen			
Fluorine			
Aluminum			

Answers

Element	Group Number	Valence Electrons	Lewis Symbol
Sulfur	6A (16)	6 e^-	$\cdot\ddot{\underset{\cdot\cdot}{S}}\colon$
Silicon	4A (14)	4 e^-	$\cdot\underset{\cdot}{Si}\cdot$
Magnesium	2A (2)	2 e^-	$\overset{\cdot}{Mg}\cdot$
Hydrogen	1A (1)	1 e^-	H $\cdot$
Fluorine	7A (17)	7 e^-	$\cdot\ddot{\underset{\cdot\cdot}{F}}\colon$
Aluminum	3A (13)	3 e^-	$\cdot\overset{\cdot}{Al}\cdot$

♦ **Learning Exercise 4.8B**

Select the larger atom in each pair.

a. _____ Mg or Ca   b. _____ Si or Cl

c. _____ Sr or Rb   d. _____ Br or Cl

e. _____ Li or Cs   f. _____ Li or N

g. _____ N or P   h. _____ As or Ca

Answers   a. Ca   b. Si   c. Rb   d. Br
              e. Cs   f. Li   g. P   h. Ca

♦ **Learning Exercise 4.8C**

Select the element in each pair with the lower ionization energy.

a. _____ Mg or Na   b. _____ P or Cl

c. _____ K or Rb   d. _____ Br or F

e. _____ Li or O   f. _____ Sb or N

g. _____ K or Br   h. _____ S or Na

Answers   a. Na   b. P   c. Rb   d. Br
              e. Li   f. Sb   g. K   h. Na

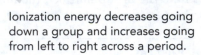

Ionization energy decreases going down a group and increases going from left to right across a period.

♦ **Learning Exercise 4.8D**

Select the element in each pair that has more metallic character.

a. _____ K or Na   b. _____ O or Se

c. _____ Ca or Br   d. _____ I or F

e. _____ Li or N   f. _____ Pb or Ba

Answers   a. K   b. Se   c. Ca
              d. I   e. Li   f. Ba

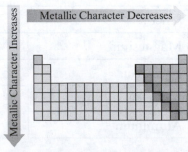

Metallic character increases going down a group and decreases going from left to right across a period.

Checklist for Chapter 4

You are ready to take the Practice Test for Chapter 4. Be sure that you have accomplished the following learning goals for this chapter. If not, review the Section listed at the end of each goal. Then apply your new skills and understanding to the Practice Test.

After studying Chapter 4, I can successfully:

_____ Write the correct symbol or name for an element. (4.1)

_____ Use the periodic table to identify the group and period of an element. (4.2)

_____ Identify an element as a metal, nonmetal, or metalloid. (4.2)

_____ State the electrical charge, mass, and location of the protons, neutrons, and electrons in an atom. (4.3)

_____ Given the atomic number and mass number of an atom, state the number of protons, neutrons, and electrons. (4.4)

_____ State the number of protons, neutrons, and electrons in an isotope of an element. (4.5)

_____ Calculate the atomic mass of an element using the abundance and mass of its naturally occurring isotopes. (4.5)

_____ Describe the energy levels, sublevels, and orbitals for the electrons in an atom. (4.6)

_____ State the maximum number of electrons in orbitals and sublevels. (4.6)

_____ Draw the orbital diagram for an element. (4.7)

_____ Write the electron configuration for an element using sublevel notation or abbreviated notation. (4.7)

_____ Predict the periodic trends in valence electrons, atomic size, ionization energy, and metallic character going down a group or across a period. (4.8)

_____ Draw the Lewis symbol for a representative element. (4.8)

Practice Test for Chapter 4

The chapter Sections to review are shown in parentheses at the end of each question.

For questions 1 through 5, write the correct symbol for each of the elements listed: (4.1)

1. zinc _____   2. barium _____   3. radon _____

4. bromine _____   5. lead _____

For questions 6 through 10, write the correct element name for each of the symbols listed: (4.1)

6. Fe _____   7. Cu _____   8. Cl _____

9. Pt _____   10. Au _____

11. The important elements in the body C, N, and O are part of a (4.2)
    A. period   B. group   C. neither

12. The important elements in the body N, P, and As are part of a (4.2)
    A. period   B. group   C. neither

13. What is the classification of an atom with 15 protons and 17 neutrons? (4.2)
    A. metal   B. nonmetal   C. halogen   D. noble gas   E. transition element

14. What is the group number of the element with atomic number 3? (4.2)
    A. 1A (1)   B. 2A (2)   C. 3B (3)   D. 3A (13)   E. 5A (15)

For questions 15 through 17, indicate whether the statement is True or False: (4.3)

15. _____ The electrons of an atom are located in the nucleus.

16. _____ Most of the mass of an atom is due to the protons and neutrons.

17. _____ All matter is composed of atoms.

For questions 18 through 21, consider an atom with 12 protons and 13 neutrons: (4.4)

18. This atom has an atomic number of
    A. 12   **B.** 13   **C.** 23   **D.** 24   **E.** 25

19. This atom has a mass number of
    A. 12   **B.** 13   **C.** 23   **D.** 24   **E.** 25

20. This is an atom of
    A. carbon   **B.** sodium   **C.** magnesium   **D.** aluminum   **E.** manganese

21. The number of electrons in this atom is
    A. 12   **B.** 13   **C.** 23   **D.** 24   **E.** 25

For questions 22 through 25, consider an atom of calcium with a mass number of 44: (4.4)

22. This atom of calcium has an atomic number of
    A. 20   **B.** 24   **C.** 40   **D.** 41   **E.** 44

23. The number of protons in this atom of calcium is
    A. 20   **B.** 24   **C.** 40   **D.** 41   **E.** 44

24. The number of neutrons in this atom of calcium is
    A. 20   **B.** 24   **C.** 40   **D.** 41   **E.** 44

25. The number of electrons in this atom of calcium is
    A. 20   **B.** 24   **C.** 40   **D.** 41   **E.** 44

26. A platinum atom, $^{195}_{78}Pt$, has (4.4)
    A. $78\ p^+, 78\ e^-, 78\ n^0$ **B.** $195\ p^+, 195\ e^-, 195\ n^0$
    C. $78\ p^+, 78\ e^-, 195\ n^0$ **D.** $78\ p^+, 78\ e^-, 117\ n^0$
    E. $78\ p^+, 117\ e^-, 117\ n^0$

For questions 27 and 28, use the following list of atoms: (4.5)

    $^{14}_{7}G$ $^{16}_{8}J$ $^{19}_{9}M$ $^{16}_{7}Q$ $^{18}_{8}R$

27. Which atom(s) is (are) isotopes of an atom with 8 protons and 9 neutrons?
    A. J   **B.** J, R   **C.** M, Q   **D.** M   **E.** Q

28. Which atom(s) are isotopes of an atom with 7 protons and 8 neutrons?
    A. G   **B.** J   **C.** G, Q   **D.** J, R   **E.** none

29. A hypothetical element has two naturally occurring isotopes. Isotope A has a mass of 52.0 amu and a percent abundance of 20.0%. Isotope B has a mass of 54.0 amu and a percent abundance of 80.0%. The atomic mass is (4.5)
    A. 52.0 amu   **B.** 52.4 amu   **C.** 53.0 amu   **D.** 53.6 amu   **E.** 54.0 amu

30. How many sublevels are in the $n = 3$ energy level? (4.6)
    A. 1   **B.** 2   **C.** 3   **D.** 4   **E.** 5

31. How many *d* orbitals are in each *d* sublevel? (4.6)
    A. 1   **B.** 2   **C.** 3   **D.** 4   **E.** 5

32. The electron configuration for oxygen is (4.7)
    A. $2s^2 2p^4$ **B.** $1s^2 2s^4 2p^4$ **C.** $1s^2 2s^6$ **D.** $1s^2 2s^2 2p^2 3s^2$ **E.** $1s^2 2s^2 2p^4$

33. The electron configuration for aluminum is (4.7)

    A. $1s^2 2s^2 2p^9$ **B.** $1s^2 2s^2 2p^6 3p^5$ **C.** $1s^2 2s^2 2p^6 3s^2 3p^1$ **D.** $1s^2 2s^2 2p^8 3p^1$ **E.** $1s^2 2s^2 2p^6 3p^3$

For questions 34 through 38, match the final notation in the electron configuration with the following elements: (4.7)

    A. As   **B.** Rb   **C.** Na   **D.** N   **E.** Xe

34. $4p^3$ _____   **35.** $5s^1$ _____   **36.** $3s^1$ _____

37. $5p^6$ _____   **38.** $2p^3$ _____

39. The number of valence electrons in gallium is (4.8)

    A. 1   **B.** 2   **C.** 3   **D.** 13   **E.** 31

40. An atom of oxygen is larger than an atom of (4.8)

    A. lithium   **B.** sulfur   **C.** fluorine   **D.** boron   **E.** argon

41. Of Li, Na, K, Rb, and Cs, the element with the most metallic character is (4.8)

    A. Li   **B.** Na   **C.** K   **D.** Rb   **E.** Cs

42. Of C, Si, Ge, Sn, and Pb, the element with the highest ionization energy is (4.8)

    A. C   **B.** Si   **C.** Ge   **D.** Sn   **E.** Pb

43. The Lewis symbol •X• would be correct for an element in (4.8)

    A. Group 1A (1)   **B.** Group 2A (2)   **C.** Group 4A (14)

    D. Group 6A (16)   **E.** Group 7A (17)

44. The Lewis symbol X• would be correct for (4.8)

    A. magnesium   **B.** chlorine   **C.** sulfur

    D. cesium   **E.** nitrogen

45. Which element has a larger atomic size, Mg or P? (4.8)

46. Which element has a larger atomic size, Ar or Xe? (4.8)

47. Which element has a higher ionization energy, N or F? (4.8)

48. Which element has a lower ionization energy, Br or F? (4.8)

49. Which element is more metallic, Li or K? (4.8)

50. Which element is more metallic, Mg or P? (4.8)

Answers to the Practice Test

1. Zn	**2.** Ba	**3.** Rn	**4.** Br	**5.** Pb
6. iron	**7.** copper	**8.** chlorine	**9.** platinum	**10.** gold
11. A	**12.** B	**13.** B	**14.** A	**15.** False
16. True	**17.** True	**18.** A	**19.** E	**20.** C
21. A	**22.** A	**23.** A	**24.** B	**25.** A
26. D	**27.** B	**28.** C	**29.** D	**30.** C
31. E	**32.** E	**33.** C	**34.** A	**35.** B
36. C	**37.** E	**38.** D	**39.** C	**40.** C
41. E	**42.** A	**43.** B	**44.** D	**45.** Mg
46. Xe	**47.** F	**48.** Br	**49.** K	**50.** Mg

Selected Answers and Solutions to Text Problems

4.1 **a.** Cu is the symbol for copper.
    b. Pt is the symbol for platinum.
    c. Ca is the symbol for calcium.
    d. Mn is the symbol for manganese.
    e. Fe is the symbol for iron.
    f. Ba is the symbol for barium.
    g. Pb is the symbol for lead.
    h. Sr is the symbol for strontium.

4.3 **a.** Carbon is the element with the symbol C.
    b. Chlorine is the element with the symbol Cl.
    c. Iodine is the element with the symbol I.
    d. Selenium is the element with the symbol Se.
    e. Nitrogen is the element with the symbol N.
    f. Sulfur is the element with the symbol S.
    g. Zinc is the element with the symbol Zn.
    h. Cobalt is the element with the symbol Co.

4.5 **a.** Sodium (Na) and chlorine (Cl) are in NaCl.
    b. Calcium (Ca), sulfur (S), and oxygen (O) are in $CaSO_4$.
    c. Carbon (C), hydrogen (H), chlorine (Cl), nitrogen (N), and oxygen (O) are in $C_{15}H_{22}ClNO_2$.
    d. Lithium (Li), carbon (C), and oxygen (O) are in Li_2CO_3.

4.7 **a.** C, N, and O are in Period 2.
    b. He is the element at the top of Group 8A (18).
    c. The alkali metals are the elements in Group 1A (1).
    d. Period 2 is the horizontal row of elements that ends with neon (Ne).

4.9 **a.** C is Group 4A (14), Period 2.
    b. He is the noble gas in Period 1.
    c. Na is the alkali metal in Period 3.
    d. Ca is Group 2A (2), Period 4.
    e. Al is Group 3A (13), Period 3.

4.11 On the periodic table, *metals* are located to the left of the heavy zigzag line, *nonmetals* are elements to the right, and *metalloids* (B, Si, Ge, As, Sb, Te, Po, At, and Ts) are located along the line.
    a. Calcium (Ca) is a metal.
    b. Sulfur (S) is a nonmetal.
    c. Metals are shiny.
    d. An element that is a gas at room temperature is a nonmetal.
    e. Group 8A (18) elements are nonmetals.
    f. Bromine (Br) is a nonmetal.
    g. Boron (B) is a metalloid.
    h. Silver (Ag) is a metal.

4.13 **a.** Ca, an alkaline earth metal, is needed for bones and teeth, muscle contraction, and nerve impulses.
    b. Fe, a transition element, is a component of the oxygen carrier hemoglobin.
    c. K, an alkali metal, is the most common ion (K^+) in cells, and is needed for muscle contraction and nerve impulses.
    d. Cl, a halogen, is the most common negative ion (Cl^-) in fluids outside cells, and is a component of stomach acid (HCl).

4.15 a. The *macrominerals*—Ca, P, K, Cl, S, Na, and Mg—are elements located in Period 3 and Period 4 of the periodic table. They are present in lower amounts than the major elements, so that smaller amounts are required in our daily diets.
  b. Sulfur is a component of proteins, vitamin B$_1$, and insulin.
  c. 86 g is a typical amount of sulfur in a 60.-kg adult.

4.17 a. The electron has the smallest mass.
  b. The proton has a 1+ charge.
  c. The electron is found outside the nucleus.
  d. The neutron is electrically neutral.

4.19 Rutherford determined that an atom contains a small, compact nucleus that is positively charged.

4.21 a. True
  b. True
  c. True
  d. False; since a neutron has no charge, it is not attracted to a proton (a proton is attracted to an electron).

4.23 In the process of brushing your hair, strands of hair become charged with like charges that repel each other.

4.25 a. The atomic number is the same as the number of protons in an atom.
  b. Both are needed since the number of neutrons is the (mass number) − (atomic number).
  c. The mass number is the number of particles (protons + neutrons) in the nucleus.
  d. The atomic number is the same as the number of electrons in a neutral atom.

4.27 The atomic number defines the element and is found above the symbol of the element in the periodic table.
  a. Lithium, Li, has an atomic number of 3.
  b. Fluorine, F, has an atomic number of 9.
  c. Calcium, Ca, has an atomic number of 20.
  d. Zinc, Zn, has an atomic number of 30.
  e. Neon, Ne, has an atomic number of 10.
  f. Silicon, Si, has an atomic number of 14.
  g. Iodine, I, has an atomic number of 53.
  h. Oxygen, O, has an atomic number of 8.

4.29 The atomic number gives the number of protons in the nucleus of an atom. Since atoms are neutral, the atomic number also gives the number of electrons in the neutral atom.
  a. There are 18 protons and 18 electrons in a neutral argon atom.
  b. There are 25 protons and 25 electrons in a neutral manganese atom.
  c. There are 53 protons and 53 electrons in a neutral iodine atom.
  d. There are 48 protons and 48 electrons in a neutral cadmium atom.

4.31 The atomic number is the same as the number of protons in an atom and the number of electrons in a neutral atom; the atomic number defines the element. The number of neutrons is the (mass number) − (atomic number).

Name of the Element	Symbol	Atomic Number	Mass Number	Number of Protons	Number of Neutrons	Number of Electrons
Zinc	**Zn**	30	**66**	30	66 − 30 = 36	30
Magnesium	Mg	**12**	12 + 12 = 24	12	**12**	12
Potassium	K	19	19 + 20 = 39	19	**20**	19
Sulfur	S	16	16 + 15 = 31	**16**	15	16
Iron	Fe	26	**56**	26	56 − 26 = 30	**26**

4.33 a. Since the atomic number of strontium is 38, every Sr atom has 38 protons. An atom of strontium (mass number 89) has 51 neutrons ($89 - 38 = 51\ n$). Neutral atoms have the same number of protons and electrons. Therefore, 38 protons, 51 neutrons, 38 electrons.

b. Since the atomic number of chromium is 24, every Cr atom has 24 protons. An atom of chromium (mass number 52) has 28 neutrons ($52 - 24 = 28\ n$). Neutral atoms have the same number of protons and electrons. Therefore, 24 protons, 28 neutrons, 24 electrons.

c. Since the atomic number of sulfur is 16, every S atom has 16 protons. An atom of sulfur (mass number 34) has 18 neutrons ($34 - 16 = 18\ n$). Neutral atoms have the same number of protons and electrons. Therefore, 16 protons, 18 neutrons, 16 electrons.

d. Since the atomic number of bromine is 35, every Br atom has 35 protons. An atom of bromine (mass number 81) has 46 neutrons ($81 - 35 = 46\ n$). Neutral atoms have the same number of protons and electrons. Therefore, 35 protons, 46 neutrons, 35 electrons.

4.35 a. Since the number of protons is 15, the atomic number is 15 and the element symbol is P. The mass number is the sum of the number of protons and the number of neutrons, $15 + 16 = 31$. The atomic symbol for this isotope is $^{31}_{15}\text{P}$.

b. Since the number of protons is 35, the atomic number is 35 and the element symbol is Br. The mass number is the sum of the number of protons and the number of neutrons, $35 + 45 = 80$. The atomic symbol for this isotope is $^{80}_{35}\text{Br}$.

c. Since the number of electrons is 50, there must be 50 protons in a neutral atom. Since the number of protons is 50, the atomic number is 50, and the element symbol is Sn. The mass number is the sum of the number of protons and the number of neutrons, $50 + 72 = 122$. The atomic symbol for this isotope is $^{122}_{50}\text{Sn}$.

d. Since the element is chlorine, the element symbol is Cl, the atomic number is 17, and the number of protons is 17. The mass number is the sum of the number of protons and the number of neutrons, $17 + 18 = 35$. The atomic symbol for this isotope is $^{35}_{17}\text{Cl}$.

e. Since the element is mercury, the element symbol is Hg, the atomic number is 80, and the number of protons is 80. The mass number is the sum of the number of protons and the number of neutrons, $80 + 122 = 202$. The atomic symbol for this isotope is $^{202}_{80}\text{Hg}$.

4.37 a. Since the element is argon, the element symbol is Ar, the atomic number is 18, and the number of protons is 18. The atomic symbols for the isotopes with mass numbers of 36, 38, and 40 are $^{36}_{18}\text{Ar}$, $^{38}_{18}\text{Ar}$, and $^{40}_{18}\text{Ar}$, respectively.

b. They all have the same atomic number (the same number of protons and electrons).

c. They have different numbers of neutrons, which gives them different mass numbers.

d. The atomic mass of argon listed on the periodic table is the weighted average mass of all the naturally occurring isotopes of argon.

e. The isotope Ar-40 ($^{40}_{18}\text{Ar}$) is the most abundant in a sample of argon because its mass is closest to the average atomic mass of argon listed on the periodic table (39.95 amu).

4.39 Since the atomic mass of copper (63.55 amu) is closer to 63 amu, there are more atoms of $^{63}_{29}\text{Cu}$ in a sample of copper.

4.41 Since the atomic mass of thallium (204.4 amu) is closest to 205 amu, the more abundant isotope of thallium is $^{205}_{81}\text{Tl}$.

4.43 $^{69}_{31}\text{Ga}$ $\quad 68.93 \text{ amu} \times \dfrac{60.11}{100} = 41.43 \text{ amu}$

$^{71}_{31}\text{Ga}$ $\quad 70.92 \text{ amu} \times \dfrac{39.89}{100} = 28.29 \text{ amu}$

$$\text{Atomic mass of Ga} = 69.72 \text{ amu (4 SFs)}$$

4.45 a. A $1s$ orbital is spherical.

b. A $2p$ orbital has two lobes.

c. A $5s$ orbital is spherical.

4.47 **a. 1** and **2**; 1*s* and 2*s* orbitals have the same shape (spherical) and can contain a maximum of two electrons.

  b. 3; 3*s* and 3*p* sublevels are in the same energy level ($n = 3$).

  c. 1 and **2**; 3*p* and 4*p* sublevels contain orbitals with the same two-lobed shape and can contain the same maximum number of electrons (six).

  d. 1, 2, and **3**; all of the 3*p* orbitals have two lobes, can contain a maximum of two electrons, and are in energy level $n = 3$.

4.49 **a.** There are five orbitals in the 3*d* sublevel.

  b. There is one sublevel in the $n = 1$ energy level.

  c. There is one orbital in the 6*s* sublevel.

  d. There are nine orbitals in the $n = 3$ energy level: one 3*s* orbital, three 3*p* orbitals, and five 3*d* orbitals.

4.51 **a.** Any orbital can hold a maximum of two electrons. Thus, a 2*p* orbital can have a maximum of two electrons.

  b. The 3*p* sublevel contains three *p* orbitals, each of which can hold a maximum of two electrons, which gives a maximum of six electrons in the 3*p* sublevel.

  c. The $n = 4$ energy level has 4 sublevels: 4*s* which can hold two electrons, 4*p* which can hold six electrons, 4*d* which can hold 10 electrons, and 4*f* which can hold 14 electrons. This gives a total of $2 + 6 + 10 + 14 = 32$ electrons.

  d. The 5*d* sublevel contains five *d* orbitals, each of which can hold a maximum of two electrons, which gives a maximum of 10 electrons in the 5*d* sublevel.

4.53 The electron configuration shows the number of electrons in each sublevel of an atom. The abbreviated electron configuration uses the symbol of the preceding noble gas to show completed sublevels.

4.55 Determine the number of electrons and then fill orbitals in the following order: $1s^2 2s^2 2p^6 3s^2 3p^6$

  a. Boron is atomic number 5 and so it has 5 electrons. Fill orbitals as

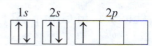

  b. Aluminum is atomic number 13 and so it has 13 electrons. Fill orbitals as

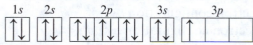

  c. Phosphorus is atomic number 15 and so it has 15 electrons. Fill orbitals as

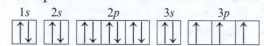

  d. Argon is atomic number 18 and so it has 18 electrons. Fill orbitals as

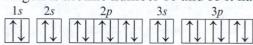

4.57 **a.** B $\quad 1s^2 2s^2 2p^1$

  b. Na $\quad 1s^2 2s^2 2p^6 3s^1$

  c. Li $\quad 1s^2 2s^1$

  d. Mg $\quad 1s^2 2s^2 2p^6 3s^2$

4.59 **a.** C $\quad [He] 2s^2 2p^2$

  b. Zr $\quad [Kr] 5s^2 4d^2$

  c. Sb $\quad [Kr] 5s^2 4d^{10} 5p^3$

  d. F $\quad [He] 2s^2 2p^5$

4.61 **a.** Li is the element with one electron in the 2*s* sublevel.
    b. S is the element with two electrons in the 3*s* and four electrons in the 3*p* sublevels.
    c. Si is the element with two electrons in the 3*s* and two electrons in the 3*p* sublevels.
    d. F is the element with two electrons in the 2*s* and five electrons in the 2*p* sublevels.

4.63 **a.** Al has three electrons in the energy level $n = 3$, $3s^2 3p^1$.
    b. C has two 2*p* electrons, $2p^2$.
    c. Ar completes the 3*p* sublevel, $3p^6$.
    d. Be completes the 2*s* sublevel, $2s^2$.

4.65 On the periodic table, the *s* sublevel block is on the left, the *p* sublevel block is on the right, and the *d* sublevel block is in the center between the *s* and *p* blocks.
    a. As $1s^2 2s^2 2p^6 3s^2 3p^6 4s^2 3d^{10} 4p^3$ $[\text{Ar}]4s^2 3d^{10} 4p^3$
    b. Fe $1s^2 2s^2 2p^6 3s^2 3p^6 4s^2 3d^6$ $[\text{Ar}]4s^2 3d^6$
    c. In $1s^2 2s^2 2p^6 3s^2 3p^6 4s^2 3d^{10} 4p^6 5s^2 4d^{10} 5p^1$ $[\text{Kr}]5s^2 4d^{10} 5p^1$
    d. Br $1s^2 2s^2 2p^6 3s^2 3p^6 4s^2 3d^{10} 4p^5$ $[\text{Ar}]4s^2 3d^{10} 4p^5$
    e. Ti $1s^2 2s^2 2p^6 3s^2 3p^6 4s^2 3d^2$ $[\text{Ar}]4s^2 3d^2$

4.67 **a.** Ga has three electrons in energy level $n = 4$; two are in the 4*s* block, and one is in the 4*p* block.
    b. N is the third element in the 2*p* block; it has three 2*p* electrons.
    c. Xe is the final element (6 electrons) in the 5*p* block.
    d. Zr is the second element in the 4*d* block; it has two 4*d* electrons.

4.69 **a.** Zn is the tenth element in the 3*d* block; it has 10 3*d* electrons.
    b. Na has one electron in the 3*s* block; the 2*p* block in Na is complete with six electrons.
    c. As is the third element in the 4*p* block; it has three 4*p* electrons.
    d. Rb is the first element in the 5*s* block; Rb has one 5*s* electron.

4.71 **a.** An element with two valence electrons is in Group 2A (2).
    b. An element with five valence electrons is in Group 5A (15).
    c. An element with two valence electrons and five electrons in the *d* block is in Group 7B (7).
    d. An element with six valence electrons is in Group 6A (16).

4.73 **a.** Alkali metals, which are in Group 1A (1), have a valence electron configuration ns^1.
    b. Elements in Group 4A (14) have a valence electron configuration $ns^2 np^2$.
    c. Elements in Group 7A (17), have a valence electron configuration $ns^2 np^5$.
    d. Elements in Group 5A (15) have a valence electron configuration $ns^2 np^3$.

4.75 **a.** Aluminum in Group 3A (13) has three valence electrons.
    b. Any element in Group 5A (15) has five valence electrons.
    c. Barium in Group 2A (2) has two valence electrons.
    d. Each halogen in Group 7A (17) has seven valence electrons.

4.77 The number of dots is equal to the number of valence electrons, as indicated by the group number.
    a. Sulfur is in Group 6A (16); $\cdot \ddot{\underset{\cdot\cdot}{\text{S}}} :$

    b. Nitrogen is in Group 5A (15); $\cdot \ddot{\text{N}} \cdot$

    c. Calcium is in Group 2A (2); $\dot{\text{Ca}} \cdot$

    d. Sodium is in Group 1A (1); Na·

    e. Gallium is in Group 3A (13); $\cdot \dot{\text{Ga}} \cdot$

4.79 The atomic size of representative elements decreases going across a period from Group 1A to 8A and increases going down a group.
    a. In Period 3, Na, which is on the left, is larger than Cl.
    b. In Group 1A (1), Rb, which is farther down the group, is larger than Na.

    c. In Period 3, Na, which is on the left, is larger than Mg.

    d. In Period 5, Rb, which is on the left, is larger than I.

4.81 a. The atomic size of representative elements decreases from Group 1A to 8A: Mg, Al, Si.

    b. The atomic size of representative elements decreases going up a group: I, Br, Cl.

    c. The atomic size of representative elements decreases from Group 1A to 8A: Sr, Sb, I.

    d. The atomic size of representative elements decreases from Group 1A to 8A: Na, Si, P.

4.83 a. In Br, the valence electrons are closer to the nucleus, so Br has a higher ionization energy than I.

    b. In Mg, the valence electrons are closer to the nucleus, so Mg has a higher ionization energy than Sr.

    c. Attraction for the valence electrons increases going from left to right across a period, giving P a higher ionization energy than Si.

    d. Attraction for the valence electrons increases going from left to right across a period, giving Xe a higher ionization energy than I.

4.85 a. Br, Cl, F; ionization energy increases going up a group.

    b. Na, Al, Cl; going across a period from left to right, ionization energy increases.

    c. Cs, K, Na; ionization energy increases going up a group.

    d. Ca, As, Br; going across a period from left to right, ionization energy increases.

4.87 Ca, Ga, Ge, Br; since metallic character decreases from left to right across a period, Ca, in Group 2A (2), will be the most metallic, followed by Ga, then Ge, and finally Br, in Group 7A (17).

4.89 Sr has a <u>lower</u> ionization energy and is <u>more</u> metallic than Sb.

4.91 Going down Group 6A (16),

    a. 1; the ionization energy <u>decreases</u>

    b. 2; the atomic size <u>increases</u>

    c. 2; the metallic character <u>increases</u>

    d. 3; the number of valence electrons <u>remains the same</u>

4.93 a. False; N has a smaller atomic size than Li.

    b. True

    c. True

    d. False; N has less metallic character than Li.

    e. True

4.95 a. Group 1A (1), alkali metals

    b. metal

    c. 19 protons

    d. Since the atomic mass of potassium is 39.10 amu, the most abundant isotope is K-39, $^{39}_{19}$K.

4.97 a. False; according to Dalton's atomic theory, all atoms of a given element are identical to one another but different from atoms of other elements.

    b. True

    c. True

    d. False; atoms are never created or destroyed during a chemical reaction.

4.99 a. The atomic mass is the weighted average of the masses of all of the naturally occurring isotopes of the element. Isotope masses are based on the masses of all subatomic particles in the atom: protons (1), neutrons (2), and electrons (3), although almost all of that mass comes from the protons (1) and neutrons (2).

    b. The number of protons (1) is the atomic number.

    c. The protons (1) are positively charged.

    d. The electrons (3) are negatively charged.

    e. The number of neutrons (2) is the (mass number) − (atomic number).

4.101 a. $^{16}_{8}X$, $^{17}_{8}X$, and $^{18}_{8}X$ all have an atomic number of 8, so all have 8 protons.

b. $^{16}_{8}X$, $^{17}_{8}X$, and $^{18}_{8}X$ all have an atomic number of 8, so all are isotopes of oxygen.

c. $^{16}_{8}X$ and $^{16}_{9}X$ have mass numbers of 16, whereas $^{18}_{8}X$ and $^{18}_{10}X$ have mass numbers of 18.

4.103

	Atomic Symbol		
	$^{70}_{32}Ge$	$^{73}_{32}Ge$	$^{76}_{32}Ge$
Atomic Number	32	32	32
Mass Number	70	73	76
Number of Protons	32	32	32
Number of Neutrons	70 − 32 = 38	73 − 32 = 41	76 − 32 = 44
Number of Electrons	32	32	32

4.105 a. The diagram shows 4 protons and 5 neutrons in the nucleus of the element. Since the number of protons is 4, the atomic number is 4, and the element symbol is Be. The mass number is the sum of the number of protons and the number of neutrons, 4 + 5 = 9. The atomic symbol is $^{9}_{4}Be$.

b. The diagram shows 5 protons and 6 neutrons in the nucleus of the element. Since the number of protons is 5, the atomic number is 5, and the element symbol is B. The mass number is the sum of the number of protons and the number of neutrons, 5 + 6 = 11. The atomic symbol is $^{11}_{5}B$.

c. The diagram shows 6 protons and 7 neutrons in the nucleus of the element. Since the number of protons is 6, the atomic number is 6, and the element symbol is C. The mass number is the sum of the number of protons and the number of neutrons, 6 + 7 = 13. The atomic symbol is $^{13}_{6}C$.

d. The diagram shows 5 protons and 5 neutrons in the nucleus of the element. Since the number of protons is 5, the atomic number is 5, and the element symbol is B. The mass number is the sum of the number of protons and the number of neutrons, 5 + 5 = 10. The atomic symbol is $^{10}_{5}B$.

e. The diagram shows 6 protons and 6 neutrons in the nucleus of the element. Since the number of protons is 6, the atomic number is 6, and the element symbol is C. The mass number is the sum of the number of protons and the number of neutrons, 6 + 6 = 12. The atomic symbol is $^{12}_{6}C$.

Representations **B** ($^{11}_{5}B$) and **D** ($^{10}_{5}B$) are isotopes of boron; **C** ($^{11}_{6}C$) and **E** ($^{12}_{6}C$) are isotopes of carbon.

4.107 a. This orbital diagram is possible. The orbitals up to $3s^2$ are filled with electrons having opposite spins. The element represented would be magnesium.

b. Not possible. The $2p$ sublevel would fill completely before the $3s$, and only 2 electrons are allowed in an orbital (not 3 as shown in the $3s$).

4.109 Atomic size increases going down a group. Li is **D** because it would be smallest. Na is **A**, K is **C**, and Rb is **B**.

4.111 a. Na has the largest atomic size.

b. Cl is a halogen.

c. Si has the electron configuration $1s^2 2s^2 2p^6 3s^2 3p^2$.

d. Ar, a noble gas, has the highest ionization energy.

e. S is in Group 6A (16).

f. Na, in Group 1A (1), has the most metallic character.

g. Mg, in Group 2A (2), has two valence electrons.

4.113 a. Bromine (Br) is in Group 7A (17), Period 4.
  b. Argon (Ar) is in Group 8A (18), Period 3.
  c. Lithium (Li) is in Group 1A (1), Period 2.
  d. Radium (Ra) is in Group 2A (2), Period 7.

4.115 On the periodic table, *metals* are located to the left of the heavy zigzag line, *nonmetals* are located to the right of the line, and *metalloids* (B, Si, Ge, As, Sb, Te, Po, At, and Ts) are located along the line.
  a. Zinc (Zn) is a metal.
  b. Cobalt (Co) is a metal.
  c. Manganese (Mn) is a metal.
  d. Iodine (I) is a nonmetal.

4.117 a. False; the proton is a positively charged particle.
  b. False; the neutron has about the same mass as a proton.
  c. True
  d. False; the nucleus is the tiny, dense central core of an atom.
  e. True

4.119 a. Since the atomic number of cadmium is 48, every Cd atom has 48 protons. An atom of cadmium (mass number 114) has 66 neutrons ($114 - 48 = 66\,n$). Neutral atoms have the same number of protons and electrons. Therefore, 48 protons, 66 neutrons, 48 electrons.
  b. Since the atomic number of technetium is 43, every Tc atom has 43 protons. An atom of technetium (mass number 98) has 55 neutrons ($98 - 43 = 55\,n$). Neutral atoms have the same number of protons and electrons. Therefore, 43 protons, 55 neutrons, 43 electrons.
  c. Since the atomic number of gold is 79, every Au atom has 79 protons. An atom of gold (mass number 199) has 120 neutrons ($199 - 79 = 120\,n$). Neutral atoms have the same number of protons and electrons. Therefore, 79 protons, 120 neutrons, 79 electrons.
  d. Since the atomic number of radon is 86, every Rn atom has 86 protons. An atom of radon (mass number 222) has 136 neutrons ($222 - 86 = 136\,n$). Neutral atoms have the same number of protons and electrons. Therefore, 86 protons, 136 neutrons, 86 electrons.
  e. Since the atomic number of xenon is 54, every Xe atom has 54 protons. An atom of xenon (mass number 136) has 82 neutrons ($136 - 54 = 82\,n$). Neutral atoms have the same number of protons and electrons. Therefore, 54 protons, 82 neutrons, 54 electrons.

4.121

Name of the Element	Atomic Symbol	Number of Protons	Number of Neutrons	Number of Electrons
Selenium	$^{80}_{34}\text{Se}$	34	$80 - 34 = 46$	34
Nickel	$^{28+34}_{28}\text{Ni}$ or $^{62}_{28}\text{Ni}$	**28**	**34**	28
Magnesium	$^{12+14}_{12}\text{Mg}$ or $^{26}_{12}\text{Mg}$	12	**14**	12
Radium	$^{228}_{88}\text{Ra}$	88	$228 - 88 = 140$	88

4.123 a. The $3p$ sublevel starts to fill after completion of the $3s$ sublevel.
  b. The $5s$ sublevel starts to fill after completion of the $4p$ sublevel.
  c. The $4p$ sublevel starts to fill after completion of the $3d$ sublevel.
  d. The $4s$ sublevel starts to fill after completion of the $3p$ sublevel.

4.125 a. Iron is the sixth element in the $3d$ block; iron has six $3d$ electrons.
  b. Barium has a completely filled $5p$ sublevel, which is six $5p$ electrons.
  c. Iodine has a completely filled $4d$ sublevel, which is 10 $4d$ electrons.
  d. Radium has a filled $7s$ sublevel, which is two $7s$ electrons.

4.127 a. Phosphorus in Group 5A (15) has an electron configuration that ends with $3s^23p^3$.
 b. Lithium is the alkali metal that is highest in Group 1A (1) and has the smallest atomic size (H in Group 1A (1) is a nonmetal).
 c. Cadmium in Period 5 has a complete $4d$ sublevel with 10 electrons.
 d. Nitrogen at the top of Group 5A (15) has the highest ionization energy in that group.
 e. Sodium, the first element in Period 3, has the largest atomic size of that period.

4.129 a. Na on the far left of the heavy zigzag line on the periodic table is a metal.
 b. P is in Group 5A (15).
 c. F at the top of Group 7A (17) and to the far right in Period 2 has the highest ionization energy.
 d. Na has the lowest ionization energy and loses an electron most easily.
 e. Cl is found in Period 3 in Group 7A (17).

4.131 a. Since the element is lead (Pb), the atomic number is 82, and the number of protons is 82. In a neutral atom, the number of electrons is equal to the number of protons, so there will be 82 electrons. The number of neutrons is the (mass number) − (atomic number) = 208 − 82 = 126 n. Therefore, 82 protons, 126 neutrons, 82 electrons.
 b. Since the element is lead (Pb), the atomic number is 82, and the number of protons is 82. The mass number is the sum of the number of protons and the number of neutrons, 82 + 132 = 214. The atomic symbol for this isotope is $^{214}_{82}\text{Pb}$.
 c. Since the mass number is 214 (as in part **b**) and the number of neutrons is 131, the number of protons is the (mass number) − (number of neutrons) = 214 − 131 = 83 p. Since there are 83 protons, the atomic number is 83, and the element symbol is Bi (bismuth). The atomic symbol for this isotope is $^{214}_{83}\text{Bi}$.

4.133 a. Atomic size decreases going up a group, which gives O the smallest atomic size in Group 6A (16).
 b. Atomic size decreases going from left to right across a period, which gives Ar the smallest atomic size in Period 3.
 c. Ionization energy increases going up a group, which gives C the highest ionization energy in Group 4A (14).
 d. Ionization energy decreases going from right to left across a period, which gives Na the lowest ionization energy in Period 3.
 e. Metallic character increases going down a group, which gives Ra the most metallic character in Group 2A (2).

4.135 $^{28}_{14}\text{Si}$ 27.977 amu $\times \dfrac{92.23}{100} = 25.80$ amu

 $^{29}_{14}\text{Si}$ 28.976 amu $\times \dfrac{4.68}{100} = 1.36$ amu

 $^{30}_{14}\text{Si}$ 29.974 amu $\times \dfrac{3.09}{100} = 0.926$ amu

 ————————————————————————————

 Atomic mass of Si = 28.09 amu (4 SFs)

4.137 a. X is a metal; Y and Z are nonmetals.
 b. X has the largest atomic size.
 c. Because Y and Z have six valence electrons (ns^2np^4) and are in Group 6A (16), they will have similar properties.
 d. Y has the highest ionization energy.
 e. Y has the smallest atomic size.

5

Nuclear Chemistry

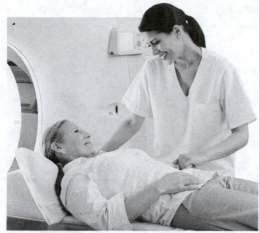

During a nuclear stress test, Simone walks on a treadmill until she reaches the maximum level. Then she is given an injection of Tl-201 and placed under a scanner that takes images of her heart muscle. The comparison of her scans at rest and under stress shows that she has normal blood flow to her heart muscle. If the half-life of Tl-201 is 3.0 days, how many days will pass for the activity to decrease to one-fourth the activity of the dose she received?

Credit: Tyler Olson/Shutterstock

LOOKING AHEAD

5.1 Natural Radioactivity

5.2 Nuclear Reactions

5.3 Radiation Measurement

5.4 Half-Life of a Radioisotope

5.5 Medical Applications Using Radioactivity

5.6 Nuclear Fission and Fusion

 The Health icon indicates a question that is related to health and medicine.

5.1 Natural Radioactivity

Learning Goal: Describe alpha, beta, positron, and gamma radiation.

REVIEW

Writing Atomic Symbols for Isotopes (4.5)

- Radioactive isotopes have unstable nuclei that break down (decay), spontaneously emitting alpha (α), beta (β), positron (β^+), or gamma (γ) radiation.

- An alpha particle is the same as a helium nucleus; it contains two protons and two neutrons.

- A beta particle is a high-energy electron, and a positron is a high-energy positive particle. A gamma ray is high-energy radiation.

- Because radiation can damage cells in the body, proper protection must be used: shielding, limiting time of exposure, and distance.

Study Note

It is important to learn the symbols for the radiation particles in order to describe the different types of radiation:

^4_2He or α	$^0_{-1}e$ or β	$^0_{+1}e$ or β^+	$^0_0\gamma$ or γ	^1_1H or p	^1_0n or n
Alpha particle	Beta particle	Positron	Gamma ray	Proton	Neutron

---

♦ **Learning Exercise 5.1A**

Match the items in column A with the descriptions in column B.

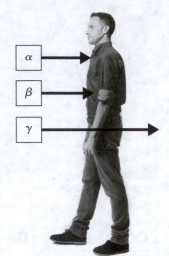

A	**B**
1. _____ $^{18}_8\text{O}$	**a.** symbol for a beta particle
2. _____ $^0_0\gamma$	**b.** symbol for an alpha particle
3. _____ $^0_{+1}e$	**c.** an atom that emits radiation
4. _____ radioisotope	**d.** symbol for a positron
5. _____ ^4_2He	**e.** symbol for an isotope of oxygen
6. _____ $^0_{-1}e$	**f.** symbol for gamma radiation

Answers   **1.** e   **2.** f   **3.** d   **4.** c   **5.** b   **6.** a

Different types of radiation penetrate the body to different depths.
Credit: Celig/Shutterstock

♦ **Learning Exercise 5.1B** 🏃

Discuss some things you can do to minimize the amount of radiation received if you work with a radioactive substance. Describe how each method helps to limit the amount of radiation you would receive.

Answer

Three ways to minimize exposure to radiation are (1) use shielding, (2) keep time short in the radioactive area, and (3) keep as much distance as possible from the radioactive materials. Shielding materials such as clothing and gloves stop alpha and beta particles from reaching your skin, whereas lead or concrete will absorb gamma rays. Limiting the time spent near radioactive samples reduces exposure time. Increasing the distance from a radioactive source reduces the intensity of radiation. Wearing a film badge will monitor the amount of radiation you receive.

♦ **Learning Exercise 5.1C**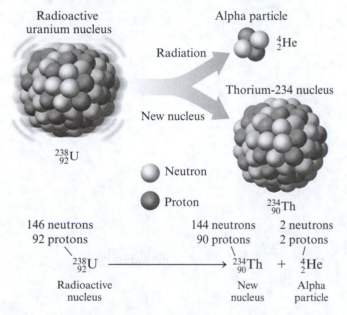

What type(s) of radiation (alpha, beta, and/or gamma) would each of the following shielding materials protect you from?

a. heavy clothing _____

b. skin _____

c. paper _____

d. concrete _____

e. lead wall _____

Answers **a.** alpha, beta **b.** alpha **c.** alpha
d. alpha, beta, gamma **e.** alpha, beta, gamma

5.2 Nuclear Reactions

Learning Goal: Write a balanced nuclear equation for radioactive decay, showing mass numbers and atomic numbers.

> **REVIEW**
> Using Positive and Negative Numbers in Calculations (1.4)
> Solving Equations (1.4)
> Counting Protons and Neutrons (4.4)

- A balanced nuclear equation represents the changes in the nuclei of radioisotopes.

- In all nuclear equations, the sum of the mass numbers and the sum of the atomic numbers on one side of the arrow must equal the sum of the mass numbers and the sum of the atomic numbers on the other side.

Radioactive uranium nucleus

Alpha particle

Radiation

$_{2}^{4}\text{He}$

Thorium-234 nucleus

New nucleus

$_{92}^{238}\text{U}$

○ Neutron

● Proton

$_{90}^{234}\text{Th}$

| 146 neutrons | 144 neutrons | 2 neutrons |
| 92 protons | 90 protons | 2 protons |

$$_{92}^{238}\text{U} \longrightarrow\ _{90}^{234}\text{Th}\ +\ _{2}^{4}\text{He}$$

Radioactive nucleus New nucleus Alpha particle

In the nuclear equation for alpha decay, the mass number of the new nucleus decreases by 4 and its atomic number decreases by 2.

- The new isotopes and the type of radiation emitted can be determined from the symbols that show the mass numbers and atomic numbers of the isotopes in the nuclear reaction.

$$\text{mass number} = \text{sum of mass numbers}$$
$$11\ =\ 7\ +\ 4$$
$$_{6}^{11}\text{C} \longrightarrow\ _{4}^{7}\text{Be}\ +\ _{2}^{4}\text{He}$$
$$6\ =\ 4\ +\ 2$$
$$\text{atomic number} = \text{sum of atomic numbers}$$

- A new radioactive isotope is produced when a nonradioactive isotope is bombarded by a small particle such as an alpha particle, a proton, a neutron, or a small nucleus.

$$_{2}^{4}\text{He} \qquad + \qquad _{5}^{10}\text{B} \qquad \longrightarrow \qquad _{7}^{13}\text{N} \qquad + \qquad _{0}^{1}n$$

Bombarding particle (α)   *Stable nucleus*   *New nucleus*   *Neutron emitted*

Writing a Nuclear Equation	
STEP 1	Write the incomplete nuclear equation.
STEP 2	Determine the missing mass number.
STEP 3	Determine the missing atomic number.
STEP 4	Determine the symbol of the new nucleus and complete the nuclear equation.

♦ **Learning Exercise 5.2A**

Complete each of the following nuclear equations:

a. $^{213}_{83}\text{Bi} \longrightarrow {}^{209}_{81}\text{Tl} + ?$ _____

b. $^{127}_{53}\text{I} \longrightarrow ? + {}^{1}_{0}n$ _____

c. $^{238}_{92}\text{U} \longrightarrow ? + {}^{4}_{2}\text{He}$ _____

d. $^{24}_{11}\text{Na} \longrightarrow ? + {}^{0}_{-1}e$ _____

e. $? \longrightarrow {}^{30}_{14}\text{Si} + {}^{0}_{-1}e$ _____

Answers   a. $^{4}_{2}\text{He}$ b. $^{126}_{53}\text{I}$ c. $^{234}_{90}\text{Th}$ d. $^{24}_{12}\text{Mg}$ e. $^{30}_{13}\text{Al}$

♦ **Learning Exercise 5.2B**

Complete each of the following bombardment reactions:

a. $? + {}^{40}_{20}\text{Ca} \longrightarrow {}^{40}_{19}\text{K} + {}^{1}_{1}\text{H}$ _____

b. $^{1}_{0}n + {}^{27}_{13}\text{Al} \longrightarrow {}^{24}_{11}\text{Na} + ?$ _____

c. $^{1}_{0}n + {}^{10}_{5}\text{B} \longrightarrow ? + {}^{4}_{2}\text{He}$ _____

d. $? + {}^{23}_{11}\text{Na} \longrightarrow {}^{23}_{12}\text{Mg} + {}^{1}_{0}n$ _____

e. $^{1}_{1}\text{H} + {}^{197}_{79}\text{Au} \longrightarrow ? + {}^{1}_{0}n$ _____

Answers   a. $^{1}_{0}n$ b. $^{4}_{2}\text{He}$ c. $^{7}_{3}\text{Li}$ d. $^{1}_{1}\text{H}$ e. $^{197}_{80}\text{Hg}$

5.3   Radiation Measurement

Learning Goal: Describe how radiation is detected and measured.

- A Geiger counter may be used to detect radiation. When radiation passes through the gas in the counter tube, atoms of gas are ionized, producing an electrical current.

- The activity of a radioactive sample measures the number of nuclear transformations per second. The curie (Ci) is equal to 3.7×10^{10} disintegrations/s. The SI unit is the becquerel (Bq), which is equal to 1 disintegration/s.

- The radiation dose absorbed by a gram of body tissue is measured in the unit rad (radiation absorbed dose) or the SI unit gray (Gy), which is equal to 100 rad.
- The biological damage to the body caused by different types of radiation is measured in the unit rem (radiation equivalent) or the SI unit sievert (Sv), which is equal to 100 rem.

♦ **Learning Exercise 5.3A**

Match each type of measurement unit with the radiation process measured.

     a. curie   **b.** becquerel   **c.** rad   **d.** gray   **e.** rem

1. _____ an activity of 1 disintegration/s

2. _____ the amount of radiation absorbed by 1 g of material

3. _____ an activity of 3.7×10^{10} disintegrations/s

4. _____ a unit measuring the biological damage caused by different kinds of radiation

5. _____ a unit that measures the absorbed dose of radiation, which is equal to 100 rad

Answers   **1.** b   **2.** c, d   **3.** a   **4.** e   **5.** d

♦ **Learning Exercise 5.3B**

A sample of Ir-192 used in brachytherapy to treat breast and prostate cancers has an activity of 35 μCi.

  a. What is its activity in becquerels?

  b. If an absorbed dose is 15 mrem, what is the absorbed dose in sieverts?

Answers   **a.** 1.3×10^{6} Bq   **b.** 1.5×10^{-4} Sv

5.4   Half-Life of a Radioisotope

REVIEW

Learning Goal: Given the half-life of a radioisotope, calculate the amount of radioisotope remaining after one or more half-lives.

Interpreting Graphs (1.4)

Using Conversion Factors (2.6)

- The half-life of a radioactive sample is the time required for one-half of the sample to decay (emit radiation).
- Most radioisotopes used in medicine, such as Tc-99m and I-131, have short half-lives. By comparison, many naturally occurring radioisotopes, such as C-14, Ra-226, and U-238, have long half-lives. For example, potassium-42 has a half-life of 12 h, whereas potassium-40 takes 1.3×10^{9} yr for one-half of the radioactive sample to decay.

SAMPLE PROBLEM   Using Half-Lives of a Radioisotope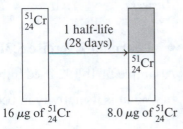

Chromium-51, which is used to measure blood volume, has a half-life of 28 days. How many micrograms of a 16-μg sample of chromium-51 will remain active after 84 days?

Solution:

STEP 1 State the given and needed quantities.

	Given	Need	Connect
Analyze the Problem	16 μg of Cr-51, 84 days elapsed, half-life = 28 days	micrograms of Cr-51 remaining	number of half-lives

STEP 2 Write a plan to calculate the unknown quantity.

84 days $\xrightarrow{\text{Half-life}}$ number of half-lives

16 micrograms of Cr-51 $\xrightarrow{\text{Number of half-lives}}$ micrograms of Cr-51 remaining

STEP 3 Write the half-life equality and conversion factors.

1 half-life of Cr-51 = 28 days

$$\frac{28 \text{ days}}{1 \text{ half-life}} \quad \text{and} \quad \frac{1 \text{ half-life}}{28 \text{ days}}$$

STEP 4 Set up the problem to calculate the needed quantity.

$$\text{Number of half-lives} = 84 \text{ days} \times \frac{1 \text{ half-life}}{28 \text{ days}} = 3.0 \text{ half-lives}$$

Now, we can calculate the quantity of Cr-51 that remains active.

16 μg $\xrightarrow{\text{1 half-life}}$ 8.0 μg $\xrightarrow{\text{2 half-lives}}$ 4.0 μg $\xrightarrow{\text{3 half-lives}}$ 2.0 μg

16 μg of $^{51}_{24}$Cr   8.0 μg of $^{51}_{24}$Cr

♦ **Learning Exercise 5.4A**

An 80.-mg sample of iodine-125 is used to treat prostate cancer. If iodine-125 has a half-life of 60. days, how many milligrams are radioactive

CORE CHEMISTRY SKILL
Using Half-Lives

a. after one half-life?

b. after two half-lives?

c. after 240 days?

Answers   **a.** 40. mg   **b.** 20. mg   **c.** 5.0 mg

♦ **Learning Exercise 5.4B**

a. $^{99m}_{43}$Tc, which is used to image the heart, brain, and lungs, has a half-life of 6.0 h. If a technician picked up a 16-mg sample at 8 A.M., how many milligrams of the radioactive sample will remain active at 8 P.M. that same day?

b. Samarium-153, which is used to treat bone cancer, has a half-life of 46 h. How many micrograms of a 240-μg sample will be radioactive after 46 h?

c. Gallium-68, which is used for the detection of pancreatic cancer, has a half-life of 68 min. How many minutes will it take for 44 mg of Ga-68 to decay to 11 mg?

d. An archaeologist finds some pieces of a wooden boat at an ancient site. When a sample of the wood is analyzed for C-14, one-eighth of the original amount of C-14 remains. If the half-life of carbon-14 is 5730 yr, how many years ago was the boat made?

Answers   **a.** 4.0 mg   **b.** 120 μg   **c.** 136 min   **d.** 17 000 yr

5.5   Medical Applications Using Radioactivity

Learning Goal: Describe the use of radioisotopes in medicine.

- Nuclear medicine uses radioactive isotopes that go to specific sites in the body.
- For diagnostic work, radioisotopes are used that emit gamma rays and produce nonradioactive products.
- By detecting the radiation emitted by medical radioisotopes, evaluations can be made about the location and extent of an injury, disease, tumor, blood flow, or level of function of a particular organ.

♦ **Learning Exercise 5.5**

Write the atomic symbol for each of the following radioactive isotopes:

a. _____ bismuth-213, used to treat leukemia

b. _____ strontium-85, used to detect bone lesions

c. _____ sodium-24, used to determine blood flow and to locate a blood clot or embolism

d. _____ nitrogen-13, used in positron emission tomography

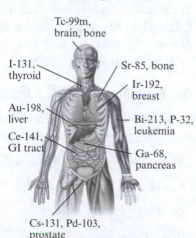

Tc-99m, brain, bone
I-131, thyroid
Sr-85, bone
Ir-192, breast
Au-198, liver
Bi-213, P-32, leukemia
Ce-141, GI tract
Ga-68, pancreas
Cs-131, Pd-103, prostate

Different radioisotopes are used to diagnose and treat a number of diseases.

Answers   **a.** $^{213}_{83}$Bi   **b.** $^{85}_{38}$Sr   **c.** $^{24}_{11}$Na   **d.** $^{13}_{7}$N

5.6   Nuclear Fission and Fusion

Learning Goal: Describe the processes of nuclear fission and fusion.

- In fission, a large nucleus breaks apart into smaller nuclei, releasing one or more types of radiation and a great amount of energy.
- A chain reaction is a fission reaction that will continue once initiated.
- In fusion, small nuclei combine to form a larger nucleus, which releases great amounts of energy.
- Nuclear fission is currently used to produce electrical power while nuclear fusion applications are still in the experimental stage.

Key Terms for Sections 5.1 to 5.6

Match each of the following key terms with the correct description:

a. radiation	**b.** half-life	**c.** curie	**d.** fission
e. alpha particle	**f.** positron	**g.** rem	

1. _____ a particle identical to a helium nucleus produced in a radioactive nucleus

2. _____ the time required for one-half of a radioactive sample to undergo radioactive decay

3. _____ a unit of radiation measurement equal to 3.7×10^{10} disintegrations per second

4. _____ a process in which large nuclei split into smaller nuclei with the release of energy

5. _____ energy or particles released by radioactive atoms

6. _____ a measure of biological damage caused by radiation

7. _____ a particle produced when a proton is transformed into a neutron

Answers   1. e   2. b   3. c   4. d   5. a   6. g   7. f

◆ Learning Exercise 5.6A

Discuss the nuclear processes of fission and fusion for the production of energy.

Answer

Nuclear fission is a splitting of the atom into smaller nuclei accompanied by the release of large amounts of energy and radiation. In the process of *nuclear fusion*, two or more nuclei combine to form a heavier nucleus and release a large amount of energy. However, fusion requires a considerable amount of energy to initiate the process. Because the extremely high temperatures are difficult to maintain, fusion technology is still in the experimental stage.

◆ Learning Exercise 5.6B

Balance each of the following nuclear equations by writing the atomic symbol for the missing particle and state if each is a fission or fusion reaction:

a. $^{1}_{0}n + ^{235}_{92}U \longrightarrow ^{143}_{54}Xe + ? + 3^{1}_{0}n$ _____

b. $^{2}_{1}H + ^{2}_{1}H \longrightarrow ? + ^{1}_{0}n$ _____

c. $^{3}_{1}H + ? \longrightarrow ^{4}_{2}He + ^{1}_{0}n$ _____

d. $^{1}_{0}n + ^{235}_{92}U \longrightarrow ^{91}_{36}Kr + ? + 3^{1}_{0}n$ _____

Answers   a. $^{90}_{38}Sr$, fission   b. $^{3}_{2}He$, fusion   c. $^{2}_{1}H$, fusion   d. $^{142}_{56}Ba$, fission

Checklist for Chapter 5

You are ready to take the Practice Test for Chapter 5. Be sure you have accomplished the following learning goals for this chapter. If not, review the Section listed at the end of the goal. Then apply your new skills and understanding to the Practice Test.

After studying Chapter 5, I can successfully:

_____ Describe alpha, beta, positron, and gamma radiation. (5.1)

_____ Write a balanced nuclear equation showing mass numbers and atomic numbers for radioactive decay. (5.2)

_____ Write a balanced nuclear equation for the formation of a radioactive isotope. (5.2)

_____ Describe the detection and measurement of radiation. (5.3)

_____ Calculate the amount of radioisotope remaining after one or more half-lives. (5.4)

_____ Calculate the amount of time for a specific amount of a radioisotope to decay given its half-life. (5.4)

_____ Describe the uses of radioisotopes in medicine. (5.5)

_____ Describe the processes of nuclear fission and fusion. (5.6)

Practice Test for Chapter 5

The chapter Sections to review are shown in parentheses at the end of each question.

1. The correctly written symbol for an isotope of sulfur is (5.1)

   A. $^{30}_{16}\text{Su}$ **B.** $^{14}_{30}\text{Si}$ **C.** $^{30}_{16}\text{S}$ **D.** $^{30}_{16}\text{Si}$ **E.** $^{16}_{30}\text{S}$

2. Alpha particles are composed of (5.1)

   A. protons   **B.** neutrons   **C.** electrons
   D. protons and electrons   **E.** protons and neutrons

3. Gamma radiation is a type of radiation that (5.1)

   A. originates in the electron shells
   B. is most dangerous
   C. is least dangerous
   D. is the heaviest
   E. travels the shortest distance

4. The charge on an alpha particle is (5.1)

   A. 1−   **B.** 1+   **C.** 2−   **D.** 2+   **E.** 4+

5. Beta particles are (5.1)

   A. protons   **B.** neutrons   **C.** electrons
   D. protons and electrons   **E.** protons and neutrons

For questions 6 through 10, select from the following: (5.1)

   A. $^{0}_{-1}\text{X}$ **B.** $^{4}_{2}\text{X}$ **C.** $^{1}_{1}\text{X}$ **D.** $^{1}_{0}\text{X}$ **E.** $^{0}_{0}\text{X}$

6. an alpha particle

7. a beta particle

8. a gamma ray

9. a proton

10. a neutron

11. Shielding from gamma rays is provided by (5.1)
    A. skin   B. paper   C. clothing   D. lead   E. air

12. The skin will provide shielding from (5.1)
    A. alpha particles   B. beta particles   C. gamma rays
    D. ultraviolet rays   E. X-rays

13. When an atom emits an alpha particle, its mass number will (5.2)
    A. increase by 1   B. increase by 2   C. increase by 4
    D. decrease by 4   E. not change

14. When a nucleus emits a positron, the atomic number of the new nucleus (5.2)
    A. increases by 1   B. increases by 2   C. decreases by 1
    D. decreases by 2   E. does not change

15. When a nucleus emits a gamma ray, the atomic number of the new nucleus (5.2)
    A. increases by 1   B. increases by 2   C. decreases by 1
    D. decreases by 2   E. does not change

For questions 16 through 19, select the particle that completes each of the following equations: (5.2)

    A. neutron   B. alpha particle   C. beta particle   D. gamma ray

16. $^{126}_{50}Sn \longrightarrow ^{126}_{51}Sb + ?$

17. $^{99m}_{43}Tc \longrightarrow ^{99}_{43}Tc + ?$

18. $^{69}_{30}Zn \longrightarrow ^{69}_{31}Ga + ?$

19. $^{149}_{62}Sm \longrightarrow ^{145}_{60}Nd + ?$

20. What symbol completes the following reaction? (5.2)
    $$^{1}_{0}n + ^{14}_{7}N \longrightarrow ? + ^{1}_{1}H$$

    A. $^{15}_{8}O$ B. $^{15}_{6}C$ C. $^{14}_{8}O$ D. $^{14}_{6}C$ E. $^{15}_{7}N$

21. To complete this nuclear equation, you need to write (5.2)
    $$? + ^{54}_{26}Fe \longrightarrow ^{57}_{28}Ni + ^{1}_{0}n$$

    A. an alpha particle   B. a beta particle   C. gamma
    D. neutron   E. proton

22. The name of the unit used to measure the number of disintegrations per second is (5.3)
    A. curie   B. rad   C. rem   D. gray   E. sievert

23. The rem and the sievert are units used to measure (5.3)
    A. activity of a radioactive sample
    B. biological damage of different types of radiation
    C. radiation absorbed
    D. background radiation
    E. half-life of a radioactive sample

24. Radiation can cause (5.3)
    A. nausea   B. a lower white cell count   C. fatigue
    D. hair loss   E. all of these

25. The time required for a radioisotope to decay is measured by its (5.4)
    A. half-life   B. protons   C. activity   D. fusion   E. radioisotope

26. Oxygen-15, used in imaging brain function and blood flow, has a half-life of 2 min. How many half-lives have occurred in the 10 min it takes to prepare the sample? (5.4)
    A. 2   B. 3   C. 4   D. 5   E. 6

27. Iodine-131, used to treat Graves' disease as well as thyroid and prostate cancers, has a half-life of 8.0 days. How many days will it take for a 80.-mg sample to decay to 10. mg? (5.4)

    A. 8.0 days   **B.** 16 days   **C.** 24 days   **D.** 32 days   **E.** 40. days

28. Phosphorus-32, used to treat leukemia and pancreatic cancer, has a half-life of 14.3 days. After 28.6 days, how many milligrams of a 120-mg sample will still be radioactive? (5.4)

    A. 240 mg   **B.** 120 mg   **C.** 60. mg   **D.** 30. mg   **E.** 15 mg

29. The radioisotope iodine-131 is used as a radioactive tracer for studying thyroid activity. The correctly written symbol for iodine-131 is (5.5)

    A. $_{53}I$ **B.** $_{131}I$ **C.** $^{131}_{53}I$ **D.** $^{53}_{131}I$ **E.** $^{78}_{53}I$

30. Radioisotopes used in medical diagnosis (5.5)

    A. have short half-lives   **B.** emit only gamma rays   **C.** locate in specific organs
    D. produce nonradioactive nuclei   **E.** all of these

31. The "splitting" of a large nucleus to form smaller particles accompanied by a release of energy is called (5.6)
    A. radioisotope   **B.** nuclear fission   **C.** nuclear fusion   **D.** bombardment   **E.** half-life

32. The process of combining small nuclei to form larger nuclei is (5.6)
    A. radioisotope   **B.** nuclear fission   **C.** nuclear fusion   **D.** bombardment   **E.** half-life

33. A fusion reaction (5.6)
    A. occurs in the Sun
    B. forms larger nuclei from smaller nuclei
    C. requires extremely high temperatures
    D. releases a large amount of energy
    E. all of these

Answers to the Practice Test

1. C	**2.** E	**3.** B	**4.** D	**5.** C
6. B	**7.** A	**8.** E	**9.** C	**10.** D
11. D	**12.** A	**13.** D	**14.** C	**15.** E
16. C	**17.** D	**18.** C	**19.** B	**20.** D
21. A	**22.** A	**23.** B	**24.** E	**25.** A
26. D	**27.** C	**28.** D	**29.** C	**30.** E
31. B	**32.** C	**33.** E		

Selected Answers and Solutions to Text Problems

5.1 **a.** ^4_2He is the symbol for an alpha particle.
  b. $^0_{+1}e$ is the symbol for a positron.
  c. $^0_0\gamma$ is the symbol for gamma radiation.

5.3 **a.** Since the element is potassium, the element symbol is K and the atomic number is 19. The potassium isotopes with mass number 39, 40, and 41 will have the atomic symbol $^{39}_{19}K$, $^{40}_{19}K$, and $^{41}_{19}K$, respectively.
  b. Each isotope has 19 protons and 19 electrons, but they differ in the number of neutrons present. Potassium-39 has $39 - 19 = 20$ neutrons, potassium-40 has $40 - 19 = 21$ neutrons, and potassium-41 has $41 - 19 = 22$ neutrons.

5.5 **a.** beta particle (β, $^0_{-1}e$)
  b. alpha particle (α, ^4_2He)
  c. neutron (n, ^1_0n)
  d. argon-38 ($^{38}_{18}Ar$)
  e. carbon-14 ($^{14}_6C$)

5.7 **a.** Since the element is copper, the element symbol is Cu and the atomic number is 29. The copper isotope with mass number 64 will have the atomic symbol $^{64}_{29}Cu$.
  b. Since the element is selenium, the element symbol is Se and the atomic number is 34. The selenium isotope with mass number 75 will have the atomic symbol $^{75}_{34}Se$.
  c. Since the element is sodium, the element symbol is Na and the atomic number is 11. The sodium isotope with mass number 24 will have the atomic symbol $^{24}_{11}Na$.
  d. Since the element is nitrogen, the element symbol is N and the atomic number is 7. The nitrogen isotope with mass number 15 will have the atomic symbol $^{15}_7N$.

5.9

Medical Use	Atomic Symbol	Mass Number	Number of Protons	Number of Neutrons
Heart imaging	$^{201}_{81}Tl$	201	81	120
Radiation therapy	$^{60}_{27}Co$	**60**	**27**	33
Abdominal scan	$^{67}_{31}Ga$	67	**31**	**36**
Hyperthyroidism	$^{131}_{53}I$	131	53	78
Leukemia treatment	$^{32}_{15}P$	**32**	15	**17**

5.11 **a.** 1. Alpha particles do not penetrate the skin.
  b. 3. Gamma radiation requires shielding protection that includes lead or thick concrete.
  c. 1. Alpha particles can be very harmful if ingested.

5.13 The mass number of the radioactive atom is reduced by four when an alpha particle (^4_2He) is emitted. The unknown product will have an atomic number that is two less than the atomic number of the radioactive atom.
  a. $^{208}_{84}Po \longrightarrow ^{204}_{82}Pb + ^4_2He$
  b. $^{232}_{90}Th \longrightarrow ^{228}_{88}Ra + ^4_2He$
  c. $^{251}_{102}No \longrightarrow ^{247}_{100}Fm + ^4_2He$
  d. $^{220}_{86}Rn \longrightarrow ^{216}_{84}Po + ^4_2He$

5.15 The mass number of the radioactive atom is not changed when a beta particle ($_{-1}^{0}e$) is emitted. The unknown product will have an atomic number that is one greater than the atomic number of the radioactive atom.

a. $_{11}^{25}\text{Na} \longrightarrow {}_{12}^{25}\text{Mg} + {}_{-1}^{0}e$

b. $_{8}^{20}\text{O} \longrightarrow {}_{9}^{20}\text{F} + {}_{-1}^{0}e$

c. $_{38}^{92}\text{Sr} \longrightarrow {}_{39}^{92}\text{Y} + {}_{-1}^{0}e$

d. $_{26}^{60}\text{Fe} \longrightarrow {}_{27}^{60}\text{Co} + {}_{-1}^{0}e$

5.17 The mass number of the radioactive atom is not changed when a positron ($_{+1}^{0}e$) is emitted. The unknown product will have an atomic number that is one less than the atomic number of the radioactive atom.

a. $_{14}^{26}\text{Si} \longrightarrow {}_{13}^{26}\text{Al} + {}_{+1}^{0}e$

b. $_{27}^{54}\text{Co} \longrightarrow {}_{26}^{54}\text{Fe} + {}_{+1}^{0}e$

c. $_{37}^{77}\text{Rb} \longrightarrow {}_{36}^{77}\text{Kr} + {}_{+1}^{0}e$

d. $_{45}^{93}\text{Rh} \longrightarrow {}_{44}^{93}\text{Ru} + {}_{+1}^{0}e$

5.19 Balance the mass numbers and the atomic numbers in each nuclear equation.

a. $_{13}^{28}\text{Al} \longrightarrow {}_{14}^{28}\text{Si} + {}_{-1}^{0}e$ $? = {}_{14}^{28}\text{Si}$ beta decay

b. $_{73}^{180m}\text{Ta} \longrightarrow {}_{73}^{180}\text{Ta} + {}_{0}^{0}\gamma$ $? = {}_{0}^{0}\gamma$ gamma emission

c. $_{29}^{66}\text{Cu} \longrightarrow {}_{30}^{66}\text{Zn} + {}_{-1}^{0}e$ $? = {}_{-1}^{0}e$ beta decay

d. $_{92}^{238}\text{U} \longrightarrow {}_{90}^{234}\text{Th} + {}_{2}^{4}\text{He}$ $? = {}_{92}^{238}\text{U}$ alpha decay

e. $_{80}^{188}\text{Hg} \longrightarrow {}_{79}^{188}\text{Au} + {}_{+1}^{0}e$ $? = {}_{79}^{188}\text{Au}$ positron emission

5.21 Balance the mass numbers and the atomic numbers in each nuclear equation.

a. $_{0}^{1}n + {}_{4}^{9}\text{Be} \longrightarrow {}_{4}^{10}\text{Be}$ $? = {}_{4}^{10}\text{Be}$

b. $_{0}^{1}n + {}_{52}^{131}\text{Te} \longrightarrow {}_{53}^{132}\text{I} + {}_{-1}^{0}e$ $? = {}_{53}^{132}\text{I}$

c. $_{0}^{1}n + {}_{13}^{27}\text{Al} \longrightarrow {}_{11}^{24}\text{Na} + {}_{2}^{4}\text{He}$ $? = {}_{13}^{27}\text{Al}$

d. $_{2}^{4}\text{He} + {}_{7}^{14}\text{N} \longrightarrow {}_{8}^{17}\text{O} + {}_{1}^{1}\text{H}$ $? = {}_{8}^{17}\text{O}$

5.23 **a.** 2. Absorbed dose can be measured in rad.

b. 3. Biological damage can be measured in mrem.

c. 1. Activity can be measured in mCi.

d. 2. Absorbed dose can be measured in Gy.

5.25 $8 \text{ mGy} \times \dfrac{1 \text{ Gy}}{1000 \text{ mGy}} \times \dfrac{100 \text{ rad}}{1 \text{ Gy}} = 0.8 \text{ rad (1 SF)}$

Thus, a technician exposed to a 5-rad dose of radiation received more radiation than one exposed to 8 mGy (0.8 rad) of radiation.

5.27 **a.** $70.0 \text{ kg body mass} \times \dfrac{4.20 \text{ } \mu\text{Ci}}{1 \text{ kg body mass}} = 294 \text{ } \mu\text{Ci (3 SFs)}$

b. $50 \text{ rad} \times \dfrac{1 \text{ Gy}}{100 \text{ rad}} = 0.5 \text{ Gy (1 SF)}$

5.29 **a.** 2. two half-lives:

$34 \text{ days} \times \dfrac{1 \text{ half-life}}{17 \text{ days}} = 2.0 \text{ half-lives}$

b. 1. one half-life:

$20 \text{ min} \times \dfrac{1 \text{ half-life}}{20 \text{ min}} = 1 \text{ half-life}$

c. 3. three half-lives:

$$21 \cancel{K} \times \frac{1 \text{ half-life}}{7 \cancel{K}} = 3 \text{ half-lives}$$

5.31 a. After one half-life, one-half of the sample would be radioactive:

80.0 mg of $^{99m}_{43}\text{Tc}$ $\xrightarrow{\text{1 half-life}}$ 40.0 mg of $^{99m}_{43}\text{Tc}$ (3 SFs)

b. After two half-lives, one-fourth of the sample would still be radioactive:

80.0 mg of $^{99m}_{43}\text{Tc}$ $\xrightarrow{\text{1 half-life}}$ 40.0 mg of $^{99m}_{43}\text{Tc}$ $\xrightarrow{\text{2 half-lives}}$ 20.0 mg of $^{99m}_{43}\text{Tc}$ (3 SFs)

c. $18 \cancel{K} \times \dfrac{1 \text{ half-life}}{6.0 \cancel{K}} = 3.0 \text{ half-lives}$

80.0 mg of $^{99m}_{43}\text{Tc}$ $\xrightarrow{\text{1 half-life}}$ 40.0 mg of $^{99m}_{43}\text{Tc}$ $\xrightarrow{\text{2 half-lives}}$ 20.0 mg of $^{99m}_{43}\text{Tc}$ $\xrightarrow{\text{3 half-lives}}$

10.0 mg of $^{99m}_{43}\text{Tc}$ (3 SFs)

d. $1.5 \cancel{\text{days}} \times \dfrac{24 \cancel{K}}{1 \cancel{\text{day}}} \times \dfrac{1 \text{ half-life}}{6.0 \cancel{K}} = 6.0 \text{ half-lives}$

80.0 mg of $^{99m}_{43}\text{Tc}$ $\xrightarrow{\text{1 half-life}}$ 40.0 mg of $^{99m}_{43}\text{Tc}$ $\xrightarrow{\text{2 half-lives}}$ 20.0 mg of $^{99m}_{43}\text{Tc}$ $\xrightarrow{\text{3 half-lives}}$

10.0 mg of $^{99m}_{43}\text{Tc}$ $\xrightarrow{\text{4 half-lives}}$ 5.00 mg of $^{99m}_{43}\text{Tc}$ $\xrightarrow{\text{5 half-lives}}$ 2.50 mg of $^{99m}_{43}\text{Tc}$ $\xrightarrow{\text{6 half-lives}}$

1.25 mg of $^{99m}_{43}\text{Tc}$ (3 SFs)

5.33 The radiation level in a radioactive sample is cut in half with each half-life; the half-life of Sr-85 is 65 days.

a. For the radiation level to drop to one-fourth of its original level, $\frac{1}{4} = \frac{1}{2} \times \frac{1}{2}$ or two half-lives

$$2 \cancel{\text{half-lives}} \times \frac{65 \text{ days}}{1 \cancel{\text{half-life}}} = 130 \text{ days (2 SFs)}$$

b. For the radiation level to drop to one-eighth of its original level,
$\frac{1}{8} = \frac{1}{2} \times \frac{1}{2} \times \frac{1}{2}$ or three half-lives

$$3 \cancel{\text{half-lives}} \times \frac{65 \text{ days}}{1 \cancel{\text{half-life}}} = 195 \text{ days (2 SFs)}$$

5.35 a. Because the elements calcium and phosphorus are part of bone, any calcium or phosphorus atom, regardless of isotope, will be carried to and become part of the bony structures of the body. Once there, the radiation emitted by any radioactive isotope can be used to diagnose or treat bone diseases.

b. Strontium (Sr) acts much like calcium (Ca) because both are Group 2A (2) elements. The body will accumulate radioactive strontium in bones in the same way that it incorporates calcium. Radioactive strontium is harmful to children because the radiation it produces causes more damage in cells that are dividing rapidly.

5.37 $4.0 \cancel{\text{mL solution}} \times \dfrac{45 \,\mu\text{Ci}}{1 \cancel{\text{mL solution}}} = 180 \,\mu\text{Ci of selenium-75 (2 SFs)}$

5.39 a. $^{68}_{31}\text{Ga} \longrightarrow {}^{68}_{30}\text{Mn} + {}^{0}_{+1}e$

b. First, calculate the number of half-lives that have elapsed:

$$136 \cancel{\text{min}} \times \frac{1 \text{ half-life}}{68 \cancel{\text{min}}} = 2.0 \text{ half-lives}$$

Now we can calculate the number of micrograms of gallium-68 that remain:

64 mcg of $^{68}_{31}\text{Ga}$ $\xrightarrow{\text{1 half-life}}$ 32 mcg of $^{68}_{31}\text{Ga}$ $\xrightarrow{\text{2 half-lives}}$ 16 mcg of $^{68}_{31}\text{Ga}$

5.41 Nuclear fission is the splitting of a large atom into smaller fragments with a simultaneous release of a large amount of energy.

5.43 $^{1}_{0}n + ^{235}_{92}U \longrightarrow ^{131}_{50}Sn + ^{103}_{42}Mo + 2^{1}_{0}n + energy$? $= ^{103}_{42}Mo$

5.45 **a.** Neutrons bombard a nucleus in the <u>fission</u> process.
b. The nuclear process that occurs in the Sun is <u>fusion</u>.
c. <u>Fission</u> is the process in which a large nucleus splits into smaller nuclei.
d. <u>Fusion</u> is the process in which small nuclei combine to form larger nuclei.

5.47 **a.** $74 \text{ MBq} \times \dfrac{1 \times 10^6 \text{ Bq}}{1 \text{ MBq}} \times \dfrac{1 \text{ Ci}}{3.7 \times 10^{10} \text{ Bq}} = 2.0 \times 10^{-3} \text{ Ci (2 SFs)}$

b. $74 \text{ MBq} \times \dfrac{1 \times 10^6 \text{ Bq}}{1 \text{ MBq}} \times \dfrac{1 \text{ Ci}}{3.7 \times 10^{10} \text{ Bq}} \times \dfrac{1 \times 10^3 \text{ mCi}}{1 \text{ Ci}} = 2.0 \text{ mCi (2 SFs)}$

5.49 For the activity to drop to one-eighth of its original level, $\frac{1}{8} = \frac{1}{2} \times \frac{1}{2} \times \frac{1}{2}$ or three half-lives

$3 \text{ half-lives} \times \dfrac{3.0 \text{ days}}{1 \text{ half-life}} = 9.0 \text{ days (2 SFs)}$

5.51

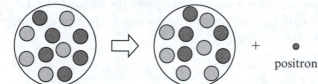

● proton
◐ neutron

+ ● positron

5.53

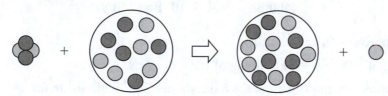

5.55 Half of a radioactive sample decays with each half-life:

$6.4 \ \mu\text{Ci of } ^{14}_{6}\text{C} \xrightarrow{\text{1 half-life}} 3.2 \ \mu\text{Ci of } ^{14}_{6}\text{C} \xrightarrow{\text{2 half-lives}} 1.6 \ \mu\text{Ci of } ^{14}_{6}\text{C} \xrightarrow{\text{3 half-lives}}$
$0.80 \ \mu\text{Ci of } ^{14}_{6}\text{C}$

∴ The activity of carbon-14 drops to 0.80 μCi in three half-lives or 3 × 5730 yr, which makes the age of the painting 17 000 yr (2 SFs).

5.57 **a.** $^{25}_{11}\text{Na}$ has 11 protons and $25 - 11 = 14$ neutrons.
b. $^{61}_{28}\text{Ni}$ has 28 protons and $61 - 28 = 33$ neutrons.
c. $^{84}_{37}\text{Rb}$ has 37 protons and $84 - 37 = 47$ neutrons.
d. $^{110}_{47}\text{Ag}$ has 47 protons and $110 - 47 = 63$ neutrons.

5.59 **a.** gamma emission
b. positron emission
c. alpha decay

5.61 **a.** $^{225}_{90}\text{Th} \longrightarrow ^{221}_{88}\text{Ra} + ^{4}_{2}\text{He}$
b. $^{210}_{83}\text{Bi} \longrightarrow ^{206}_{81}\text{Tl} + ^{4}_{2}\text{He}$
c. $^{137}_{55}\text{Cs} \longrightarrow ^{137}_{56}\text{Ba} + ^{0}_{-1}e$
d. $^{126}_{50}\text{Sn} \longrightarrow ^{126}_{51}\text{Sb} + ^{0}_{-1}e$
e. $^{18}_{9}\text{F} \longrightarrow ^{18}_{8}\text{O} + ^{0}_{+1}e$

5.63 **a.** $^4_2He + ^{14}_7N \longrightarrow ^{17}_8O + ^1_1H$ $? = ^{17}_8O$

   b. $^4_2He + ^{27}_{13}Al \longrightarrow ^{30}_{14}Si + ^1_1H$ $? = ^1_1H$

   c. $^1_0n + ^{235}_{92}U \longrightarrow ^{90}_{38}Sr + 3^1_0n + ^{143}_{54}Xe$ $? = ^{143}_{54}Xe$

   d. $^{23m}_{12}Mg \longrightarrow ^{23}_{12}Mg + ^0_0\gamma$ $? = ^{23}_{12}Mg$

5.65 **a.** $^{16}_8O + ^{16}_8O \longrightarrow ^{28}_{14}Si + ^4_2He$

   b. $^{18}_8O + ^{249}_{98}Cf \longrightarrow ^{263}_{106}Sg + 4^1_0n$

   c. $^{222}_{86}Rn \longrightarrow ^{218}_{84}Po + ^4_2He$

   d. $^{80}_{38}Sr \longrightarrow ^{80}_{37}Rb + ^0_{+1}e$

5.67 First, calculate the number of half-lives that have elapsed:

$$24 \cancel{h} \times \frac{1 \text{ half-life}}{6.0 \cancel{h}} = 4.0 \text{ half-lives}$$

Now we can calculate the number of milligrams of technicium-99m that remain:

120 mg of $^{99m}_{43}Tc \xrightarrow{\text{1 half-life}}$ 60. mg of $^{99m}_{43}Tc \xrightarrow{\text{2 half-lives}}$ 30. mg of $^{99m}_{43}Tc \xrightarrow{\text{3 half-lives}}$

15 mg of $^{99m}_{43}Tc \xrightarrow{\text{4 half-lives}}$ 7.5 mg of $^{99m}_{43}Tc$ (2 SFs)

5.69 In the fission process, an atom splits into smaller nuclei with a simultaneous release of a large amount of energy. In fusion, two (or more) small nuclei combine (fuse) to form a larger nucleus, with a simultaneous release of a large amount of energy.

5.71 Fusion occurs naturally in the Sun and other stars.

5.73 $120 \text{ } n\cancel{Ci} \times \dfrac{1 \times 10^{-9} \text{ } \cancel{Ci}}{1 \text{ } n\cancel{Ci}} \times \dfrac{3.7 \times 10^{10} \text{ Bq}}{1 \text{ } \cancel{Ci}} = 4400 \text{ Bq} = 4.4 \times 10^3 \text{ Bq (2 SFs)}$

5.75 Half of a radioactive sample decays with each half-life:

1.2 mg of $^{32}_{15}P \xrightarrow{\text{1 half-life}}$ 0.60 mg of $^{32}_{15}P \xrightarrow{\text{2 half-lives}}$ 0.30 mg of $^{32}_{15}P$

∴ Two half-lives must have elapsed during this time (28.6 days), yielding the half-life for phosphorus-32:

$$\frac{28.6 \text{ days}}{2 \text{ half-lives}} = 14.3 \text{ days/half-life (3 SFs)}$$

5.77 **a.** $^{47}_{20}Ca \longrightarrow ^{47}_{21}Sc + ^0_{-1}e$

   b. First, calculate the number of half-lives that have elapsed:

$$18 \cancel{\text{days}} \times \frac{1 \text{ half-life}}{4.5 \cancel{\text{days}}} = 4.0 \text{ half-lives}$$

   Now we can calculate the number of milligrams of calcium-47 that remain:

16 mg of $^{47}_{20}Ca \xrightarrow{\text{1 half-life}}$ 8.0 mg of $^{47}_{20}Ca \xrightarrow{\text{2 half-lives}}$ 4.0 mg of $^{47}_{20}Ca \xrightarrow{\text{3 half-lives}}$

2.0 mg of $^{47}_{20}Ca \xrightarrow{\text{4 half-lives}}$ 1.0 mg of $^{47}_{20}Ca$

   c. Half of a radioactive sample decays with each half-life:

4.8 mg of $^{47}_{20}Ca \xrightarrow{\text{1 half-life}}$ 2.4 mg of $^{47}_{20}Ca \xrightarrow{\text{2 half-lives}}$ 1.2 mg of $^{47}_{20}Ca$

   ∴ Two half-lives have elapsed.

$$2 \cancel{\text{half-lives}} \times \frac{4.5 \text{ days}}{1 \cancel{\text{half-life}}} = 9.0 \text{ days (2 SFs)}$$

5.79 **a.** $^{180}_{80}Hg \longrightarrow ^{176}_{78}Pt + ^4_2He$

   b. $^{198}_{79}Au \longrightarrow ^{198}_{80}Hg + ^0_{-1}e$

   c. $^{82}_{37}Rb \longrightarrow ^{82}_{36}Kr + ^0_{+1}e$

5.81 $^1_0n + ^{238}_{92}U \longrightarrow ^{239}_{93}Np + ^0_{-1}e$

5.83 Half of a radioactive sample decays with each half-life:

64 μCi of $^{201}_{81}$Tl $\xrightarrow{\text{1 half-life}}$ 32 μCi of $^{201}_{81}$Tl $\xrightarrow{\text{2 half-lives}}$ 16 μCi of $^{201}_{81}$Tl $\xrightarrow{\text{3 half-lives}}$

8.0 μCi of $^{201}_{81}$Tl $\xrightarrow{\text{4 half-lives}}$ 4.0 μCi of $^{201}_{81}$Tl

$\therefore$ Four half-lives must have elapsed during this time (12 days), yielding the half-life for thallium-201:

$$\frac{12 \text{ days}}{4 \text{ half-lives}} = 3.0 \text{ days/half-life (2 SFs)}$$

5.85 $^{48}_{20}$Ca + $^{244}_{94}$Pu $\longrightarrow$ $^{292}_{114}$Fl

5.87 First, calculate the number of half-lives that have elapsed:

$$27 \text{ days} \times \frac{1 \text{ half-life}}{4.5 \text{ days}} = 6.0 \text{ half-lives}$$

Because the activity of a radioactive sample is cut in half with each half-life, the activity must have been double its present value before each half-life. For 6.0 half-lives, we need to double the value six times:

1.0 μCi of $^{47}_{20}$Ca $\xleftarrow{\text{1 half-life}}$ 2.0 μCi of $^{47}_{20}$Ca $\xleftarrow{\text{2 half-lives}}$ 4.0 μCi of $^{47}_{20}$Ca $\xleftarrow{\text{3 half-lives}}$ 8.0 μCi of $^{47}_{20}$Ca

$\xleftarrow{\text{4 half-lives}}$ 16 μCi of $^{47}_{20}$Ca $\xleftarrow{\text{5 half-lives}}$ 32 μCi of $^{47}_{20}$Ca $\xleftarrow{\text{6 half-lives}}$ 64 μCi of $^{47}_{20}$Ca

$\therefore$ The initial activity of the sample was 64 μCi (2 SFs).

5.89 First, calculate the number of half-lives that have elapsed since the technician was exposed:

$$36 \text{ h} \times \frac{1 \text{ half-life}}{12 \text{ h}} = 3.0 \text{ half-lives}$$

Because the activity of a radioactive sample is cut in half with each half-life, the activity must have been double its present value before each half-life. For 3.0 half-lives, we need to double the value three times:

2.0 μCi of $^{42}_{19}$K $\xleftarrow{\text{1 half-life}}$ 4.0 μCi of $^{42}_{19}$K $\xleftarrow{\text{2 half-lives}}$ 8.0 μCi of $^{42}_{19}$K $\xleftarrow{\text{3 half-lives}}$ 16 μCi of $^{42}_{19}$K

$\therefore$ The activity of the sample when the technician was exposed was 16 μCi (2 SFs).

6

Ionic and Molecular Compounds

Sarah, a pharmacist, talked with her customer regarding questions about stomach upset from the low-dose aspirin, $C_9H_8O_4$, which he is taking to prevent a heart attack or stroke. Sarah explained that when he takes a low-dose aspirin, he needs to drink a glass of water. If his stomach is still upset, he can take the aspirin with food or milk. A few days later, Sarah talked with her customer again. He told her he followed her suggestions and that he no longer had any upset stomach from the low-dose aspirin. Is aspirin an ionic or a molecular compound?

Credit: Design Pics Inc./Alamy

LOOKING AHEAD

6.1 Ions: Transfer of Electrons
6.2 Ionic Compounds
6.3 Naming and Writing Ionic Formulas
6.4 Polyatomic Ions

6.5 Molecular Compounds: Sharing Electrons
6.6 Lewis Structures for Molecules and Polyatomic Ions

6.7 Electronegativity and Bond Polarity
6.8 Shapes and Polarity of Molecules
6.9 Intermolecular Forces in Compounds

 The Health icon indicates a question that is related to health and medicine.

6.1 Ions: Transfer of Electrons

Learning Goal: Write the symbols for the simple ions of the representative elements.

REVIEW

Using Positive and Negative Numbers in Calculations (1.4)
Writing Electron Configurations (4.7)

- The stability of the noble gases is associated with a filled valence electron energy level of eight electrons, an octet. Helium is stable with two electrons in the $n = 1$ energy level.
- Atoms of elements other than the noble gases achieve stability by losing, gaining, or sharing valence electrons with other atoms in the formation of compounds.

- Metals in Groups 1A (1), 2A (2), and 3A (13) achieve a stable electron configuration by losing their valence electron(s) to form positively charged cations with a charge of 1+, 2+, or 3+, respectively.
- Nonmetals in Groups 5A (15), 6A (16), and 7A (17) achieve a stable electron configuration by gaining valence electron(s) to form negatively charged anions with a charge of 3−, 2−, or 1−, respectively.
- *Ionic bonds* occur when the valence electrons of atoms of a metal are transferred to atoms of nonmetals. For example, sodium atoms lose electrons and chlorine atoms gain electrons to form the ionic compound NaCl.
- *Covalent bonds* form when atoms of nonmetals share valence electrons. In the molecular compounds H_2O and C_3H_8, atoms share electrons.

Study Note

When an atom loses or gains electrons, it acquires the electron configuration of the nearest noble gas. For example, lithium loses one electron, which gives the Li^+ ion the same electron configuration as helium. Oxygen gains two electrons to give an oxide ion, O^{2-}, the same electron configuration as neon.

♦ **Learning Exercise 6.1A**

The following metals, which are essential to health, lose electrons when they form ions. Indicate the group number, the number of electrons lost, and the ion (symbol and charge) for each of the following:

CORE CHEMISTRY SKILL
Writing Positive and Negative Ions

Metal	Group Number	Electrons Lost	Ion Formed
Magnesium	2A	2	Mg^{2+}
Sodium	1A	1	Na^+
Calcium	2A	2	Ca^{2+}
Potassium	1A	1	K^+

Answers

Metal	Group Number	Electrons Lost	Ion Formed
Magnesium	2A (2)	2	Mg^{2+}
Sodium	1A (1)	1	Na^+
Calcium	2A (2)	2	Ca^{2+}
Potassium	1A (1)	1	K^+

♦ **Learning Exercise 6.1B**

The following nonmetals, which are essential to health, gain electrons when they form ions. Indicate the group number, the number of electrons gained, and the ion (symbol and charge) for each of the following:

Nonmetal	Group Number	Electrons Gained	Ion Formed
Chlorine	7 A	1	Cl⁻
Nitrogen	5 A	3	N³⁻
Iodine	7A	1	I⁻
Sulfur	6 A	2	S²⁻

Answers

Nonmetal	Group Number	Electrons Gained	Ion Formed
Chlorine	7A (17)	1	Cl^-
Nitrogen	5A (15)	3	N^{3-}
Iodine	7A (17)	1	I^-
Sulfur	6A (16)	2	S^{2-}

6.2 Ionic Compounds

Learning Goal: Using charge balance, write the correct formula for an ionic compound.

- In the formulas of ionic compounds, the total positive charge is equal to the total negative charge. For example, the compound magnesium chloride, $MgCl_2$, contains Mg^{2+} and two Cl^-. The sum of the charges is zero: $1(2+) + 2(1-) = 0$.
- When two or more ions are needed for charge balance, that number is indicated by subscripts in the formula.
- A subscript of 1 is understood and not written.

For the following exercises, you may want to cut pieces of paper that represent typical positive and negative ions, as shown. To determine an ionic formula, place the smallest number of positive and negative pieces together to complete a geometric shape. Write the number of positive ions and negative ions as the subscripts for the formula.

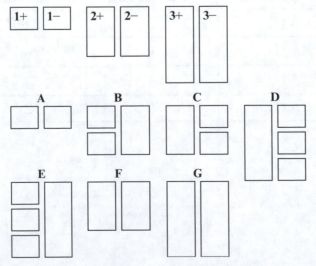

♦ Learning Exercise 6.2A

CORE CHEMISTRY SKILL
Writing Ionic Formulas

Write the letter (A to F) that matches the combination of ions in each of the following compounds:

Compound	Combination	Compound	Combination
a. $MgCl_2$	C	b. Na_2S	B
c. LiCl	A	d. CaO	F
e. K_3N	E	f. $AlBr_3$	D
g. MgS	F	h. AlN	G

Answers a. C b. B c. A d. F e. E f. D g. F h. G

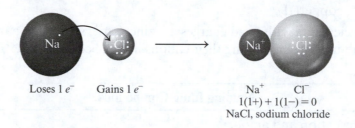

Na
:Cl:
Na+
:Cl:
Loses 1 e−
Gains 1 e−
Na+ Cl−
1(1+) + 1(1−) = 0
NaCl, sodium chloride

Study Note

You can check that the ionic formula you write is electrically neutral by multiplying each ionic charge by its subscript. When added together, their sum should equal zero. For example, the formula Na_2O gives $2(1+) + 1(2-) = (2+) + (2-) = 0$.

♦ Learning Exercise 6.2B

Write the correct formula for the ionic compound formed from each of the following pairs of ions:

a. Na^+ and Cl^-	NaCl	b. K^+ and S^{2-}	K_2S
c. Al^{3+} and O^{2-}	Al_2O_3	d. Sr^{2+} and Cl^-	$SrCl_2$
e. Ca^{2+} and S^{2-}	CaS	f. Al^{3+} and Cl^-	$AlCl_3$
g. Li^+ and N^{3-}	Li_3N	h. Ba^{2+} and P^{3-}	Ba_3P_2

Answers a. NaCl b. K_2S c. Al_2O_3 d. $SrCl_2$
e. CaS f. $AlCl_3$ g. Li_3N h. Ba_3P_2

♦ Learning Exercise 6.2C

Write the symbols for the ions and the correct formula for the ionic compound formed by each of the following:

a. lithium and sulfur b. potassium and phosphorus
c. barium and fluorine d. gallium and sulfur

Answers a. Li^+ and S^{2-}, Li_2S b. K^+ and P^{3-}, K_3P
c. Ba^{2+} and F^-, BaF_2 d. Ga^{3+} and S^{2-}, Ga_2S_3

6.3   Naming and Writing Ionic Formulas

Learning Goal: Given the formula of an ionic compound, write the correct name; given the name of an ionic compound, write the correct formula.

*   In naming ionic compounds, the positive ion is named first, followed by the name of the negative ion. The name of a representative metal ion in Group 1A (1), 2A (2), or 3A (13) is the same as its element name. The name of a nonmetal ion is obtained by using the first syllable of its element name followed by *ide*.

*   Most transition elements form cations with two or more ionic charges. Then the ionic charge must be written as a Roman numeral in parentheses after the name of the metal. For example, the cations of iron, Fe^{2+} and Fe^{3+}, are named iron(II) and iron(III). The ions of copper are Cu^+, copper(I), and Cu^{2+}, copper(II).

*   The only transition elements with fixed charges are zinc, Zn^{2+}; silver, Ag^+; and cadmium, Cd^{2+}; no Roman numerals are used when naming their cations in ionic compounds.

	Naming Ionic Compounds
STEP 1	Identify the cation and anion.
STEP 2	Name the cation by its element name.
STEP 3	Name the anion by using the first syllable of its element name followed by *ide*.
STEP 4	Write the name for the cation first and the name for the anion second.

♦   **Learning Exercise 6.3A**

Write the symbols for the ions and the correct name for each of the following ionic compounds:

Formula	Ions		Name
a. Cs_2O	Cs^+	O^{-2}	cesium oxide
b. $BaBr_2$	Ba^{2+}	Br^-	barium bromide
c. Ca_3P_2	Ca^{2+}	P^{3-}	calcium phosphide
d. Na_2S	Na^+	S^{2-}	sodium sulfide

Answers   **a.** Cs^+, O^{2-}, cesium oxide   **b.** Ba^{2+}, Br^-, barium bromide
c. Ca^{2+}, P^{3-}, calcium phosphide   **d.** Na^+, S^{2-}, sodium sulfide

	Naming Ionic Compounds with Variable Charge Metal Ions
STEP 1	Determine the charge of the cation from the anion.
STEP 2	Name the cation by its element name, and use a Roman numeral in parentheses for the charge.
STEP 3	Name the anion by using the first syllable of its element name followed by *ide*.
STEP 4	Write the name for the cation first and the name for the anion second.

♦ **Learning Exercise 6.3B**

Write the name for each of the following ions:

a. Cl^- _Chlorine_   **b.** Fe^{2+} _iron(II)_   **c.** Cu^+ _Copper (I)_

d. Ag^+ _Silver_   **e.** O^{2-} _Oxygen_   **f.** Ca^{2+} _Calcium_

g. S^{2-} _Sulphide_   **h.** Al^{3+} _Aluminium_   **i.** Fe^{3+} _iron(III)_

j. Ba^{2+} _Barium_   **k.** Cu^{2+} _Copper(II)_   **l.** N^{3-} _Nitrogen_

Answers		
a. chloride	**b.** iron(II)	**c.** copper(I)
d. silver	**e.** oxide	**f.** calcium
g. sulfide	**h.** aluminum	**i.** iron(III)
j. barium	**k.** copper(II)	**l.** nitride

♦ **Learning Exercise 6.3C**

Write the symbol for each of the following ions:

Name	**Symbol**
a. chromium(III) ion	Cr^{3+}
b. cobalt(II) ion	Co^{2+}
c. zinc ion	Zn^+
d. lead(IV) ion	Pb^{4+}
e. gold(I) ion	Au^+
f. silver ion	Ag^+
g. potassium ion	K^+
h. nickel(III) ion	Ni^{3+}

Answers		
a. Cr^{3+}	**b.** Co^{2+}	**c.** Zn^{2+}
d. Pb^{4+}	**e.** Au^+	**f.** Ag^+
g. K^+	**h.** Ni^{3+}	

Some Metals That Form More Than One Positive Ion

Element	Positive Ions	Name of Ion
Bismuth	Bi^{3+}	Bismuth(III)
	Bi^{5+}	Bismuth(V)
Chromium	Cr^{2+}	Chromium(II)
	Cr^{3+}	Chromium(III)
Cobalt	Co^{2+}	Cobalt(II)
	Co^{3+}	Cobalt(III)
Copper	Cu^+	Copper(I)
	Cu^{2+}	Copper(II)
Gold	Au^+	Gold(I)
	Au^{3+}	Gold(III)
Iron	Fe^{2+}	Iron(II)
	Fe^{3+}	Iron(III)
Lead	Pb^{2+}	Lead(II)
	Pb^{4+}	Lead(IV)
Manganese	Mn^{2+}	Manganese(II)
	Mn^{3+}	Manganese(III)
Mercury	Hg_2^{2+}	Mercury(I)*
	Hg^{2+}	Mercury(II)
Nickel	Ni^{2+}	Nickel(II)
	Ni^{3+}	Nickel(III)
Tin	Sn^{2+}	Tin(II)
	Sn^{4+}	Tin(IV)

*Mercury(I) ions form an ion pair with a 2+ charge.

♦ **Learning Exercise 6.3D**

Write the symbols for the ions and the correct name for each of the following ionic compounds:

> **CORE CHEMISTRY SKILL**
> Naming Ionic Compounds

Formula		**Ions**	**Name**
a. $CrCl_2$	Cr^{2+}	Cl^-	Chromium (II) chloride
b. $SnBr_4$	Sn^{4+}	Br^-	Tin (IV) bromide
c. Na_3P	Na^+	P^{3-}	Sodium phosphide
d. Ni_2O_3	Ni^{3+}	O^{2-}	Nickel (III) oxide
e. CuO	Cu^{2+}	O^{2-}	Copper (II) oxide
f. Mg_3N_2	Mg^{2+}	N^{3-}	Magnesium nitride

Answers   **a.** Cr^{2+}, Cl^-, chromium(II) chloride   **b.** Sn^{4+}, Br^-, tin(IV) bromide
   c. Na^+, P^{3-}, sodium phosphide   **d.** Ni^{3+}, O^{2-}, nickel(III) oxide
   e. Cu^{2+}, O^{2-}, copper(II) oxide   **f.** Mg^{2+}, N^{3-}, magnesium nitride

Writing Formulas for Ionic Compounds	
STEP 1	Identify the cation and anion.
STEP 2	Balance the charges.
STEP 3	Write the formula, cation first, using subscripts from the charge balance.

♦ **Learning Exercise 6.3E**

Write the symbols for the ions and the correct formula for each of the following ionic compounds:

Compound	Cation	Anion	Formula of Compound
Aluminum sulfide	Al^{3+}	S^{2-}	Al_2S_3
Lead(II) chloride	Pb^{2+}	Cl^-	$PbCl_2$
Barium oxide	Ba^{2+}	O^{2-}	BaO
Gold(III) bromide	Au^{3+}	Br^-	$AuBr_3$
Silver oxide	Ag^+	O^{2-}	Ag_2O

Answers

Compound	Cation	Anion	Formula of Compound
Aluminum sulfide	Al^{3+}	S^{2-}	Al_2S_3
Lead(II) chloride	Pb^{2+}	Cl^-	$PbCl_2$
Barium oxide	Ba^{2+}	O^{2-}	BaO
Gold(III) bromide	Au^{3+}	Br^-	$AuBr_3$
Silver oxide	Ag^+	O^{2-}	Ag_2O

6.4   Polyatomic Ions

Learning Goal: Write the name and formula for an ionic compound containing a polyatomic ion.

- A polyatomic ion is a group of nonmetal atoms that carries an electrical charge, usually negative, 1−, 2−, or 3−. The polyatomic ion NH_4^+, ammonium ion, has a positive charge.

- Polyatomic ions cannot exist alone but are combined with an ion of the opposite charge.

- The names of ionic compounds containing polyatomic anions often end with *ate* or *ite*.

- When there is more than one polyatomic ion in the formula for a compound, the entire polyatomic ion formula is enclosed in parentheses and the subscript written outside the parentheses.

Fertilizer
NH_4NO_3

NH_4^+ NO_3^-
Ammonium ion Nitrate ion
Many products contain poly-
atomic ions, which are groups of
atoms that have an ionic charge.
Credit: Pearson Education/Pearson Science

Study Note

By learning the most common polyatomic ions such as nitrate NO_3^-, carbonate CO_3^{2-}, sulfate SO_4^{2-}, and phosphate PO_4^{3-}, you can derive their related polyatomic ions. For example, the nitrite ion, NO_2^-, has one oxygen atom less than the nitrate ion, NO_3^-.

Names and Formulas of Some Common Polyatomic Ions

Nonmetal	Formula of Ion*	Name of Ion
Hydrogen	**OH⁻**	Hydroxide
Nitrogen	NH_4^+	Ammonium
	NO_3^-	**Nitrate**
	NO_2^-	Nitrite
Chlorine	ClO_4^-	Perchlorate
	ClO_3^-	**Chlorate**
	ClO_2^-	Chlorite
	ClO^-	Hypochlorite
Carbon	**CO_3^{2-}**	**Carbonate**
	HCO_3^-	Hydrogen carbonate (or bicarbonate)
	CN^-	Cyanide
	$C_2H_3O_2^-$	Acetate
Sulfur	**SO_4^{2-}**	**Sulfate**
	HSO_4^-	Hydrogen sulfate (or bisulfate)
	SO_3^{2-}	Sulfite
	HSO_3^-	Hydrogen sulfite (or bisulfite)
Phosphorus	**PO_4^{3-}**	**Phosphate**
	HPO_4^{2-}	Hydrogen phosphate
	$H_2PO_4^-$	Dihydrogen phosphate
	PO_3^{3-}	Phosphite

*Formulas and names in bold type indicate the most common polyatomic ion for that element.

♦ **Learning Exercise 6.4A**

Write the polyatomic ion (symbol and charge) for each of the following:

a. sulfate ion _____

b. hydroxide ion _____

c. carbonate ion _____

d. sulfite ion _____

e. ammonium ion _____

f. phosphate ion _____

g. nitrate ion _____

h. nitrite ion _____

Answers

a. SO_4^{2-}

b. OH^-

c. CO_3^{2-}

d. SO_3^{2-}

e. NH_4^+

f. PO_4^{3-}

g. NO_3^-

h. NO_2^-

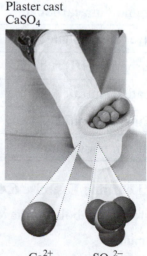

Plaster cast
$CaSO_4$

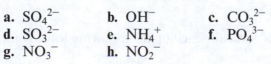

Ca^{2+} SO_4^{2-}
Sulfate ion

The sulfate ion in $CaSO_4$ is a polyatomic ion, which is a group of atoms that have an ionic charge.

Credit: Pearson Education/Pearson Science

Writing Formulas Containing Polyatomic Ions	
STEP 1	Identify the cation and polyatomic ion (anion).
STEP 2	Balance the charges.
STEP 3	Write the formula, cation first, using the subscripts from charge balance.

♦ **Learning Exercise 6.4B**

Write the formulas for the ions and the correct formula for each of the following compounds:

Compound	Cation	Anion	Formula
Sodium phosphate	Na^+	PO_4^{3-}	Na_3PO_4
Iron(II) hydroxide	Fe^{2+}	OH^-	$Fe(OH)_2$
Ammonium carbonate			
Silver bicarbonate			
Chromium(III) sulfate			
Lead(II) nitrate			
Potassium sulfite			
Barium phosphate			

Answers

Compound	Cation	Anion	Formula
Sodium phosphate	Na^+	PO_4^{3-}	Na_3PO_4
Iron(II) hydroxide	Fe^{2+}	OH^-	$Fe(OH)_2$
Ammonium carbonate	NH_4^+	CO_3^{2-}	$(NH_4)_2CO_3$
Silver bicarbonate	Ag^+	HCO_3^-	$AgHCO_3$
Chromium(III) sulfate	Cr^{3+}	SO_4^{2-}	$Cr_2(SO_4)_3$
Lead(II) nitrate	Pb^{2+}	NO_3^-	$Pb(NO_3)_2$
Potassium sulfite	K^+	SO_3^{2-}	K_2SO_3
Barium phosphate	Ba^{2+}	PO_4^{3-}	$Ba_3(PO_4)_2$

♦ **Learning Exercise 6.4C**

Write the symbols for the ions and the correct name for each of the following ionic compounds:

Formula	Ions		Name
a. $Ba(NO_3)_2$	_____	_____	_____
b. $Fe_2(SO_4)_3$	_____	_____	_____
c. Na_3PO_3	_____	_____	_____

d. $Al(ClO_3)_3$ ___ ___ _____

e. NH_4NO_2 ___ ___ _____

f. $Cr(OH)_2$ ___ ___ _____

Answers   **a.** Ba^{2+}, NO_3^-, barium nitrate   **b.** Fe^{3+}, SO_4^{2-}, iron(III) sulfate
c. Na^+, PO_3^{3-}, sodium phosphite   **d.** Al^{3+}, ClO_3^-, aluminum chlorate
e. NH_4^+, NO_2^-, ammonium nitrite   **f.** Cr^{2+}, OH^-, chromium(II) hydroxide

6.5 Molecular Compounds: Sharing Electrons

Learning Goal: Given the formula of a molecular compound, write its correct name; given the name of a molecular compound, write its formula.

- Molecular compounds are composed of nonmetals bonded together to give discrete units called *molecules*.
- The name of a molecular compound is written using the element name of the first nonmetal and the first syllable of the second nonmetal name followed by the suffix *ide*.
- Prefixes are used to indicate the number of atoms of each nonmetal in the formula: mono (1), di (2), tri (3), tetra (4), penta (5), hexa (6), hepta (7), octa (8), nona (9), deca (10).
- In the name of a molecular compound, the prefix *mono* is usually omitted, as in NO, nitrogen oxide. Traditionally, however, CO is named carbon monoxide.
- If the vowels *o* and *o* or *a* and *o* appear together, the first vowel is omitted.
- The formula of a molecular compound is written using the symbols of the nonmetals in the name followed by subscripts determined from the prefixes.

Prefixes Used in Naming Molecular Compounds

1 mono	6 hexa
2 di	7 hepta
3 tri	8 octa
4 tetra	9 nona
5 penta	10 deca

Key Terms for Sections 6.1 to 6.5

Match each of the following key terms with the correct description:

a. ion   **b.** cation   **c.** anion
d. polyatomic ion   **e.** octet   **f.** covalent bond

1. _b_ a positively charged ion

2. _c_ a negatively charged ion

3. _f_ the attraction between atoms formed by sharing electrons

4. _d_ a group of covalently bonded nonmetal atoms that has an overall charge

5. _a_ an atom or group of atoms having an electrical charge

6. _e_ a set of eight valence electrons

Answers   **1.** b   **2.** c   **3.** f   **4.** d   **5.** a   **6.** e

Naming Molecular Compounds	
STEP 1	Name the first nonmetal by its element name.
STEP 2	Name the second nonmetal by using the first syllable of its element name followed by *ide*.
STEP 3	Add prefixes to indicate the number of atoms (subscripts).

♦ **Learning Exercise 6.5A**

Name each of the following molecular compounds:

a. CS_2 _____

b. PCl_3 _____

c. CO _____

d. SO_3 _____

e. N_2O_4 _____

f. CCl_4 _____

g. P_4S_6 _____

h. IF_7 _____

i. ClO_2 _____

j. S_2O _____

Answers

a. carbon disulfide

c. carbon monoxide

e. dinitrogen tetroxide

g. tetraphosphorus hexasulfide

i. chlorine dioxide

b. phosphorus trichloride

d. sulfur trioxide

f. carbon tetrachloride

h. iodine heptafluoride

j. disulfur oxide

Writing Formulas for Molecular Compounds	
STEP 1	Write the symbols in the order of the elements in the name.
STEP 2	Write any prefixes as subscripts.

♦ **Learning Exercise 6.5B**

Write the formula for each of the following molecular compounds:

a. dinitrogen oxide _____

b. silicon tetrabromide _____

c. bromine pentafluoride _____

d. carbon dioxide _____

e. sulfur hexafluoride _____

f. oxygen difluoride _____

g. phosphorus octasulfide _____

h. selenium hexafluoride _____

i. iodine chloride _____

j. xenon trioxide _____

Answers

a. N_2O

b. $SiBr_4$

c. BrF_5

d. CO_2

e. SF_6

f. OF_2

g. PS_8

h. SeF_6

i. ICl

j. XeO_3

Summary of Writing Formulas and Names

- In ionic compounds containing two different elements, the first takes its element name. The ending of the name of the second element is replaced by *ide*. For example, $BaCl_2$ is named *barium chloride*. If the metal forms two or more positive ions, a Roman numeral is added to its name to indicate the ionic charge in the compound. For example, $FeCl_3$ is named *iron*(III) *chloride*.

- In ionic compounds with three or more elements, a group of atoms is named as a polyatomic ion. The names of negative polyatomic ions end in *ate* or *ite*, except for hydroxide and cyanide. For example, Na_2SO_4 is named *sodium sulfate*. No prefixes are used.

- When a polyatomic ion occurs two or more times in a formula, its formula is placed inside parentheses, and the number of ions is shown as a subscript after the parentheses as in $Ca(NO_3)_2$. If the subscript is 1, it is not written.

- In naming molecular compounds, a prefix before the name of one or both nonmetal names indicates the numerical value of a subscript. For example, N_2O_3 is named *dinitrogen trioxide*.

♦ **Learning Exercise 6.5C**

Indicate the type of compound (ionic or molecular). Then give the formula or name for each.

Formula	Ionic or Molecular?	Name of Compound
$MgCl_2$	Ionic	Magnisium chloride
NCl_3	molecular	nitrogen trichloride
K_2SO_4	Ionic	potassium sulfate
Li_2O	Ionic	Lithium oxide
CBr_4	Molecular	Carbon tetrabromide
Na_3PO_4	Ionic	Sodium Phosphate
H_2S	Molecular	dihydrogen sulfide
$Ca(HCO_3)_2$	Ionic	calcium hydrogen carbonate (bicarbonate)

Answers

Formula	Ionic or Molecular?	Name of Compound
$MgCl_2$	ionic	magnesium chloride
NCl_3	molecular	nitrogen trichloride
K_2SO_4	ionic	potassium sulfate
Li_2O	ionic	lithium oxide
CBr_4	molecular	carbon tetrabromide
Na_3PO_4	ionic	sodium phosphate
H_2S	molecular	dihydrogen sulfide
$Ca(HCO_3)_2$	ionic	calcium hydrogen carbonate (bicarbonate)

6.6 Lewis Structures for Molecules and Polyatomic Ions

Learning Goal: Draw the Lewis structures for molecular compounds and polyatomic ions.

REVIEW

Drawing Lewis Symbols (4.8)

- In a covalent bond, atoms of nonmetals share electrons to achieve an octet (two for H).
- In a double bond, two pairs of electrons are shared between the same two atoms to complete octets.
- In a triple bond, three pairs of electrons are shared to complete octets.
- Bonding pairs of electrons are shared between atoms.
- Nonbonding electrons (lone pairs) are not shared between atoms.

♦ **Learning Exercise 6.6A**

List the number of bonds typically formed by the following atoms in molecular compounds:

a. N __3__   **b.** S __2__   **c.** P __3__   **d.** C __4__

e. Cl __1__   **f.** O __2__   **g.** H __1__   **h.** F __1__

Answers   **a.** 3   **b.** 2   **c.** 3   **d.** 4   **e.** 1   **f.** 2   **g.** 1   **h.** 1

SAMPLE PROBLEM   Drawing Lewis Structures

Draw the Lewis structures for SCl_2, one with electron dots and the other with lines for bonding pairs.

Solution:

Analyze the Problem	Given	Need	Connect
	SCl_2	Lewis structure	total valence electrons

STEP 1 Determine the arrangement of atoms.

  Cl S Cl

STEP 2 Determine the total number of valence electrons.

$$1 \text{ S atom} \times 6\,e^- = 6\,e^-$$
$$2 \text{ Cl atoms} \times 7\,e^- = \underline{14\,e^-}$$
Total valence electrons for SCl_2 = $20\,e^-$

STEP 3 Attach each bonded atom to the central atom with a pair of electrons. Four electrons are used to form two single bonds to connect the S atom and each Cl atom. Each bonding pair can be represented by a pair of electrons or a bond line.

  Cl:S:Cl   or   Cl—S—Cl $20\,e^- - 4\,e^- = 16$ remaining valence electrons (8 pairs)

STEP 4 Use the remaining electrons to complete octets. The remaining 16 electrons are placed as lone pairs around the S and Cl atoms to complete octets for all the atoms.

  :C̈l:S̈:C̈l: or   :C̈l—S̈—C̈l:

♦ **Learning Exercise 6.6B**

Draw the Lewis structures for each of the following, one with electron dots and the other with lines for bonding pairs:

a. H_2

  H:H

b. NCl_3 $5 + 7 \times 3 = 26$

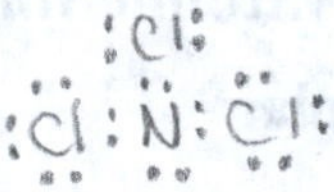

  :C̈l:
  :C̈l:N̈:C̈l:

c. HCl $1 + 7 = 8$

  H:C̈l:

d. Cl_2

  :C̈l:C̈l:

e. H_2S $2 + 6 = 8$

  H:S̈:H

f. CBr_4 $4 + 7 \times 4 = 32$

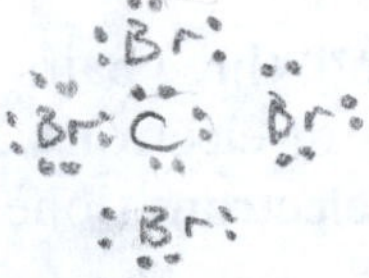

  :B̈r:
  :B̈r:C:B̈r:
  :B̈r:

Answers

a. H:H or H—H

b. :C̈l:N:C̈l: or :C̈l—N—C̈l:
         :C̈l: :C̈l:

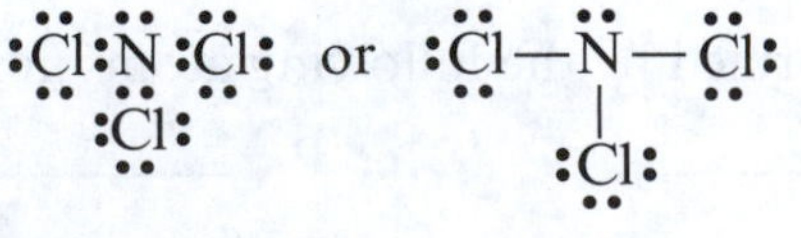

c. H:C̈l: or H—C̈l:

d. :C̈l:C̈l: or :C̈l—C̈l:

e. H:S̈: or H—S̈:
         H   H

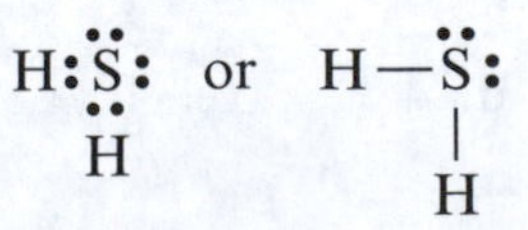

f. :B̈r:C:B̈r: or :B̈r—C—B̈r:
     :B̈r: :B̈r:
   with :B̈r: above and :B̈r: below

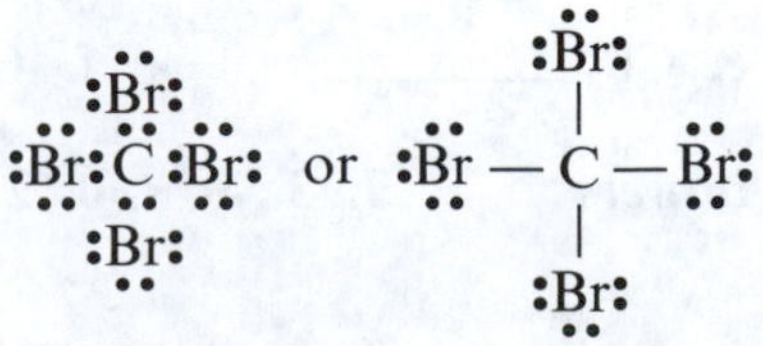

SAMPLE PROBLEM   Drawing Lewis Structures with Multiple Bonds

Draw the Lewis structures for sulfur dioxide, SO_2, in which the central atom is S.

Solution:

Analyze the Problem	Given	Need	Connect
	SO_2	Lewis structure	total valence electrons

STEP 1 Determine the arrangement of atoms.

O   S   O

STEP 2 Determine the total number of valence electrons.

$$1\ S\ atom\ \times 6\ e^- = 6\ e^-$$
$$2\ O\ atoms \times 6\ e^- = \underline{12\ e^-}$$
Total valence electrons for $SO_2 = 18\ e^-$

STEP 3 Attach each bonded atom to the central atom with a pair of electrons. Each bonding pair can be represented by a pair of electrons or a bond line.

O : S : O or O—S—O $18\ e^- - 4\ e^- = 14$ remaining valence electrons (7 pairs)

STEP 4 Use the remaining electrons to complete octets, using multiple bonds if needed.

:Ö:S:Ö: or :Ö—S—Ö:

Because the octet for S is not complete, a lone pair from one O atom is shared with the central S atom to form a double bond.

:Ö:S::Ö: or :Ö—S=Ö:

♦ **Learning Exercise 6.6C**

Draw the Lewis structures for each of the following molecules, one with electron dots and the other with lines for bonding pairs:

a. CS_2 $4 + 6(2) = 16$

:S̈: C :S̈:   :S̈=C=S̈:

b. HCN

c. H_2CCH_2

d. HONO

Answers   **a.** :S̈::C::S̈: or :S̈=C=S̈:   **b.** H:C⦂⦂N: or H—C≡N:

c. H:C̈::C̈:H or

$$\begin{array}{cc} H & H \\ | & | \\ H—C{=}C—H \end{array}$$

d. H:Ö:N::Ö: or H—Ö—N̈=Ö:

SAMPLE PROBLEM   Drawing Lewis Structures for Polyatomic ions

Draw the Lewis structures for the polyatomic ion bromite, BrO_2^-.

Solution:

Analyze the Problem	Given	Need	Connect
	BrO_2^-	Lewis structure	total valence electrons

STEP 1   Determine the arrangement of atoms.

[O   Br   O]⁻

STEP 2   Determine the total number of valence electrons.

1 Br and 2 O = $1(7\ e^-) + 2(6\ e^-) + 1\ e^-$ (negative charge) = 20 valence electrons

STEP 3   Attach each bonded atom to the central atom with a pair of electrons. Each bonding pair can be represented by a pair of electrons or a bond line.

[O:Br:O]⁻   or   [O—Br—O]⁻

STEP 4   Place the remaining electrons using single or multiple bonds to complete octets.

20 valence e^- − 4 bonding e^- = 16 remaining valence electrons (8 pairs)
Thus, 8 lone pairs of electrons are placed on the Br and O atoms.

$$\left[:\ddot{\underset{..}{O}}:\ddot{\underset{..}{Br}}:\ddot{\underset{..}{O}}:\right]^-\quad \text{or}\quad \left[:\ddot{\underset{..}{O}}-\ddot{\underset{..}{Br}}-\ddot{\underset{..}{O}}:\right]^-$$

♦   **Learning Exercise 6.6D**

Draw a Lewis structures for each of the following polyatomic ions, one with electron dots and the other with lines for bonding pairs:

   a. CO_3^{2-} **b.** NO_3^-

Answers

a. $\left[:\ddot{\underset{..}{O}}-C=\ddot{\underset{..}{O}}:\;\;{\overset{\displaystyle :\ddot{O}:}{|}}\right]^{2-}$ or $\left[:\ddot{\underset{..}{O}}:\overset{\displaystyle :\ddot{O}:}{C}::\ddot{\underset{..}{O}}:\right]^{2-}$

b. $\left[:\ddot{\underset{..}{O}}=N-\ddot{\underset{..}{O}}:\;\;{\overset{\displaystyle :\ddot{O}:}{|}}\right]^{-}$ or $\left[:\ddot{\underset{..}{O}}:\overset{\displaystyle :\ddot{O}:}{N}:\ddot{\underset{..}{O}}:\right]^{-}$

6.7   Electronegativity and Bond Polarity

Learning Goal: Use electronegativity to determine the polarity of a bond.

- Electronegativity values indicate the ability of an atom to attract electrons in a chemical bond.
- Electronegativity values are low for metals and high for nonmetals.

Electronegativity Increases →

							18 8A
H 2.1							

1 1A	2 2A		13 3A	14 4A	15 5A	16 6A	17 7A
Li 1.0	Be 1.5		B 2.0	C 2.5	N 3.0	O 3.5	F 4.0
Na 0.9	Mg 1.2		Al 1.5	Si 1.8	P 2.1	S 2.5	Cl 3.0
K 0.8	Ca 1.0		Ga 1.6	Ge 1.8	As 2.0	Se 2.4	Br 2.8
Rb 0.8	Sr 1.0		In 1.7	Sn 1.8	Sb 1.9	Te 2.1	I 2.5
Cs 0.7	Ba 0.9		Tl 1.8	Pb 1.9	Bi 1.9	Po 2.0	At 2.1

Electronegativity Decreases ↓

- When atoms sharing electrons have the same or similar electronegativity values, electrons are shared equally and the bond is *nonpolar covalent*.
- Electrons are shared unequally in *polar covalent* bonds because they are attracted to the more electronegative atom.
- An electronegativity difference of 0.0 to 0.4 indicates a nonpolar covalent bond, whereas a difference of 0.5 to 1.8 indicates a polar covalent bond.
- An electronegativity difference of 1.9 to 3.3 indicates a bond that is ionic.
- A polar bond has a charge separation, which is called a *dipole*; the positive end of the dipole is marked as δ^+ and the negative end as δ^-.

$$\overset{\delta^+}{H} \!-\! \overset{\delta^-}{Cl}$$

Electrons are shared unequally in the polar covalent bond of HCl.

Credit: Kameel4u/Shutterstock

◆ **Learning Exercise 6.7A**

Using the electronegativity values, determine the following:

1. the electronegativity difference for each pair

2. the type of bonding as ionic (I), polar covalent (PC), or nonpolar covalent (NC)

CORE CHEMISTRY SKILL
Using Electronegativity

Elements	Electronegativity Difference	Type of Bonding	Elements	Electronegativity Difference	Type of Bonding
a. H and O	1.4	Polar	b. N and S	0.5	P
c. Al and O	2	I	d. Cl and Cl	0.0	N
e. H and Cl	0.9	P	f. Li and F	3.0	I
g. S and F	1.5	P	h. H and C	0.4	N

Answers a. 1.4, PC   b. 0.5, PC   c. 2.0, I   d. 0.0, NC
e. 0.9, PC   f. 3.0, I   g. 1.5, PC   h. 0.4, NC

♦ **Learning Exercise 6.7B**

For each of the following, write the symbols δ^+ and δ^- over the appropriate atoms.

a. H—O

b. N—H

c. C—Cl

d. O—F

e. N—F

f. P—Cl

Answers

a. $H^{\delta+}—O^{\delta-}$

b. $N^{\delta-}—H^{\delta+}$

c. $C^{\delta+}—Cl^{\delta-}$

d. $O^{\delta+}—F^{\delta-}$

e. $N^{\delta+}—F^{\delta-}$

f. $P^{\delta+}—Cl^{\delta-}$

6.8   Shapes and Polarity of Molecules

Learning Goal: Predict the three-dimensional structure of a molecule, and classify it as polar or nonpolar.

- The VSEPR theory predicts the geometry of a molecule by placing the electron groups around a central atom as far apart as possible.
- One electron group may consist of the following: a lone pair, a single bond, a double bond, or a triple bond.
- The atoms bonded to the central atom determine the three-dimensional geometry of a molecule.
- A central atom with two electron groups has a linear geometry (180°); three electron groups give a trigonal planar geometry (120°); and four electron pairs give a tetrahedral geometry (109°).
- A linear molecule has a central atom bonded to two atoms and no lone pairs.

Linear shape

- A trigonal planar molecule has a central atom bonded to three atoms and no lone pairs. A bent molecule with a bond angle of 120° has a central atom bonded to two atoms and one lone pair.

Trigonal planar shape   Bent shape (120°)

- A tetrahedral molecule has a central atom bonded to four atoms and no lone pairs. In a trigonal pyramidal molecule, a central atom is bonded to three atoms and one lone pair. In a bent molecule with a bond angle of 109°, a central atom is bonded to two atoms and two lone pairs.

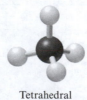

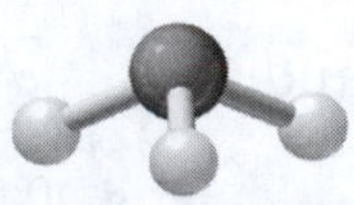

Tetrahedral   Trigonal pyramidal   Bent shape (109°)
shape   shape

- Nonpolar molecules can have polar bonds if the dipoles are in a symmetric arrangement and the bond dipoles cancel each other.
- In polar molecules, the bond dipoles do not cancel each other.

♦ **Learning Exercise 6.8A**

Match the shape of a molecule with the following descriptions:

    a. linear   **b.** trigonal planar   **c.** tetrahedral
    d. trigonal pyramidal   **e.** bent (120°)   **f.** bent (109°)

1. has a central atom with three electron groups and three bonded atoms   _____

2. has a central atom with two electron groups and two bonded atoms   _____

3. has a central atom with four electron groups with three bonded atoms   _____

4. has a central atom with three electron groups and two bonded atoms   _____

5. has a central atom with four electron groups and four bonded atoms   _____

6. has a central atom with four electron groups and two bonded atoms   _____

Answers   **1.** b   **2.** a   **3.** d   **4.** e   **5.** c   **6.** f

Predicting Shape of Molecules	
STEP 1	Draw the Lewis structure.
STEP 2	Arrange the electron groups around the central atom to minimize repulsion.
STEP 3	Use the atoms bonded to the central atom to determine the shape.

♦ **Learning Exercise 6.8B**

For each molecule or ion, state the total number of valence electrons, draw the Lewis structure, state the number of electron groups and bonded atoms, and predict the shape.

CORE CHEMISTRY SKILL
Predicting Shape

Molecule or Ion	Valence Electrons	Lewis Structure	Electron Groups	Bonded Atoms	Shape
CH_4					
H_2CO (C is the central atom)					
SO_3^{2-}					
H_2S					
SeO_2					

Answers

Molecule or Ion	Valence Electrons	Lewis Structure	Electron Groups	Bonded Atoms	Shape
CH₄	8	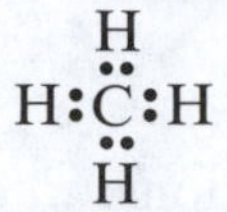 H:C:H (with H above and below)	4	4	Tetrahedral
H₂CO	12	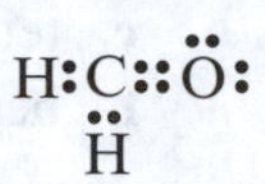 H:C::Ö: (with H below)	3	3	Trigonal planar
SO₃²⁻	26	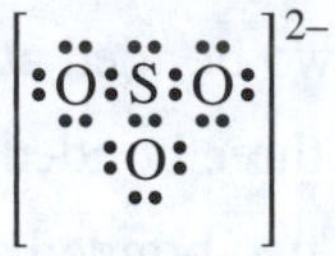 [:Ö:S:Ö: with :Ö: below]²⁻	4	3	Trigonal pyramidal
H₂S	8	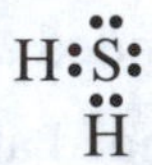 H:S: (with H below)	4	2	Bent, 109°
SeO₂	18	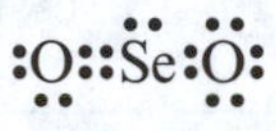 :Ö::Se:Ö:	3	2	Bent, 120°

SAMPLE PROBLEM   Polarity of Molecules

Determine if NH_3 is polar or nonpolar.

Solution:

Analyze the Problem	Given	Need	Connect
	NH₃	polarity	Lewis structure, bond polarity

STEP 1   Determine if the bonds are polar covalent or nonpolar covalent. N and H have an electronegativity difference of 0.9 (3.0 − 2.1 = 0.9), which makes the N—H bonds polar covalent.

STEP 2   If the bonds are polar covalent, draw the Lewis structure and determine if the dipoles cancel. The Lewis structure for NH_3 has one N atom with four electron groups and three bonded H atoms. The molecule has a trigonal pyramidal shape in which the dipoles of the N—H bonds do not cancel. The NH_3 molecule is polar.

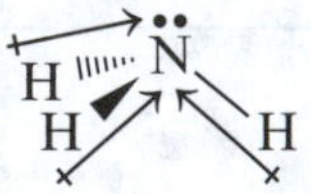

♦ **Learning Exercise 6.8C**

Determine the shape, bond dipoles, and whether each of the following is polar or nonpolar:

a. CF₄   b. HBr   c. SF₂

> CORE CHEMISTRY SKILL
>
> Identifying Polarity of Molecules

Answers

a. Tetrahedral, bond dipoles cancel, nonpolar

b. 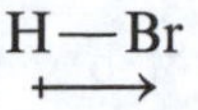 H—Br
Linear, bond dipole does not cancel, polar

c.  Bent, 109°, bond dipoles do not cancel, polar

6.9   Intermolecular Forces in Compounds

Learning Goal: Describe the intermolecular forces between ions, polar covalent molecules, and nonpolar covalent molecules.

- The intermolecular forces between particles in solids and liquids determine their melting points.
- Ionic solids have high melting points due to strong ionic interactions between positive and negative ions.
- In polar substances, dipole–dipole attractions occur between the positive end of one molecule and the negative end of another.

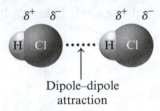

Dipole–dipole attraction

- Hydrogen bonding, a particularly strong type of dipole–dipole attraction, occurs between partially positive hydrogen atoms and the strongly electronegative atoms N, O, or F.

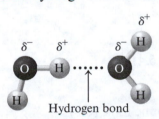

Hydrogen bond

- Dispersion forces occur when temporary dipoles form within nonpolar molecules, causing weak attractions to other nonpolar molecules. Dispersion forces are the only intermolecular forces that occur between nonpolar molecules.

Key Terms for Sections 6.6 to 6.9

Match each of the following key terms with the correct description:

    a. electronegativity   **b.** double bond   **c.** VSEPR theory
    d. nonpolar covalent bond   **e.** dipole–dipole attraction   **f.** dispersion forces

1. __d__ the equal sharing of valence electrons by two atoms

2. __a__ the ability of an atom to attract the shared electrons in a chemical bond

3. __c__ the atoms in a molecule are arranged to minimize repulsion between electrons

4. __b__ the sharing of two pairs of electrons by two atoms

5. _____ interaction between the positive end of one polar molecule and the negative end of another

6. _____ results from momentary polarization of nonpolar molecules

Answers   **1.** d   **2.** a   **3.** c   **4.** b   **5.** e   **6.** f

♦ **Learning Exercise 6.9**

Identify the major type of intermolecular forces between the molecules of each of the following compounds:

    a. dipole–dipole attractions   **b.** hydrogen bonding   **c.** dispersion forces

1. _____ NF_3 **2.** _____ PF_3 **3.** _____ Cl_2

4. _____ HF   **5.** _____ H_2O **6.** _____ C_4H_{10}

Answers   **1.** a   **2.** a   **3.** c
           4. b   **5.** b   **6.** c

Checklist for Chapter 6

You are ready to take the Practice Test for Chapter 6. Be sure you have accomplished the following learning goals for this chapter. If not, review the Section listed at the end of the goal. Then apply your new skills and understanding to the Practice Test.

After studying Chapter 6, I can successfully:

_____ Illustrate the octet rule for the formation of ions. (6.1)

_____ Write the formulas for ionic compounds containing the ions of metals and nonmetals of representative elements. (6.2)

_____ Use charge balance to write an ionic formula. (6.2)

_____ Write the name for an ionic compound. (6.3)

_____ Write the formula for a compound containing a polyatomic ion. (6.4)

_____ Write the names and formulas for molecular compounds. (6.5)

_____ Draw the Lewis structures for molecular compounds and polyatomic ions. (6.6)

_____ Classify a bond as nonpolar covalent, polar covalent, or ionic. (6.7)

_____ Use the VSEPR theory to determine the shape of a molecule. (6.8)

_____ Classify a molecule as polar or nonpolar. (6.8)

_____ Identify the intermolecular forces between particles. (6.9)

Practice Test for Chapter 6

The chapter Sections to review are shown in parentheses at the end of each question.

For questions 1 through 4, consider an atom of phosphorus: (6.1)

1. Phosphorus is in Group
    A. 2A (2)   **B.** 3A (13)   **C.** 5A (15)   **D.** 7A (17)   **E.** 8A (18)

2. How many valence electrons does an atom of phosphorus have?
    A. 2   **B.** 3   **C.** 5   **D.** 8   **E.** 15

3. To achieve an octet, a phosphorus atom in an ionic compound will
   A. lose 1 electron   B. lose 2 electrons   C. lose 5 electrons
   D. gain 2 electrons   E. gain 3 electrons

4. A phosphide ion has a charge of
   A. 1+   B. 2+   C. 5+   D. 2−   E. 3−

5. To achieve an octet, a calcium atom (6.1)
   A. loses 1 electron   B. loses 2 electrons   C. loses 3 electrons
   D. gains 1 electron   E. gains 2 electrons

6. The correct ionic charge for a calcium ion is (6.1)
   A. 1+   B. 2+   C. 1−   D. 2−   E. 3−

7. Another name for a positive ion is (6.1)
   A. anion   B. cation   C. proton   D. positron   E. sodium

8. To achieve an octet, a chlorine atom (6.1)
   A. loses 1 electron   B. loses 2 electrons   C. loses 3 electrons
   D. gains 1 electron   E. gains 2 electrons

9. The correct ionic charge for a cesium ion is (6.1)
   A. 1+   B. 2+   C. 1−   D. 2−   E. 3−

10. The correct ionic charge for a fluoride ion is (6.1)
    A. 1+   B. 2+   C. 1−   D. 2−   E. 3−

11. When the elements magnesium and sulfur react, (6.2)
    A. an ionic compound forms
    B. a molecular compound forms
    C. no reaction occurs
    D. the two repel each other and will not combine
    E. none of the above

12. An ionic bond typically occurs between (6.2)
    A. two different nonmetals   B. two of the same type of nonmetals
    C. two noble gases   D. two different metals
    E. a metal and a nonmetal

13. The correct formula for a compound between sodium and sulfur is (6.2)
    A. SoS   B. NaS   C. Na_2S D. NaS_2 E. Na_2SO_4

14. The correct formula for a compound between aluminum and oxygen is (6.2)
    A. AlO   B. Al_2O C. AlO_3 D. Al_2O_3 E. Al_3O_2

15. The correct formula for a compound between barium and sulfur is (6.2)
    A. BaS   B. Ba_2S C. BaS_2 D. Ba_2S_2 E. $BaSO_4$

16. The correct formula for iron(III) fluoride is (6.2, 6.3)
    A. FeF   B. FeF_2 C. Fe_2F D. Fe_3F E. FeF_3

17. The correct formula for copper(II) chloride is (6.2, 6.3)
    A. CoCl   B. CuCl   C. $CoCl_2$ D. $CuCl_2$ E. Cu_2Cl

18. The correct formula for silver oxide is (6.2, 6.3)
    A. AgO   B. Ag_2O C. AgO_2 D. Ag_3O_2 E. Ag_3O

19. The correct ionic charge for a phosphate ion is (6.4)
    A. 1+   B. 2+   C. 1−   D. 2−   E. 3−

20. The correct ionic charge for a sulfate ion is (6.4)

    A. 1+   **B.** 2+   **C.** 1−   **D.** 2−   **E.** 3−

21. The correct formula for ammonium sulfate is (6.2, 6.4)

    A. AmS   **B.** $AmSO_4$ **C.** $(NH_4)_2S$ **D.** NH_4SO_4 **E.** $(NH_4)_2SO_4$

22. The correct formula for lithium phosphate is (6.2, 6.4)

    A. $LiPO_4$ **B.** Li_2PO_4 **C.** Li_3PO_4 **D.** $Li_2(PO_4)_3$ **E.** $Li_3(PO_4)_2$

23. The correct formula for magnesium carbonate, used as a dietary magnesium supplement, is (6.2, 6.4)

    A. $MgCO_3$ **B.** Mg_2CO_3 **C.** $Mg(CO_3)_2$ **D.** MgCO   **E.** $Mg_2(CO_3)_3$

24. The correct formula for copper(I) sulfate, used as a fungicide, is (6.3, 6.4)

    A. $CuSO_3$ **B.** $CuSO_4$ **C.** Cu_2SO_3 **D.** $Cu(SO_4)_2$ **E.** Cu_2SO_4

25. The correct name of $Al_2(HPO_4)_3$ is (6.3, 6.4)

    A. aluminum hydrogen phosphite
    B. aluminum hydrogen phosphate
    C. aluminum hydrogen phosphorus
    D. aluminum hydrogen phosphorus oxide
    E. trialuminum dihydrogen phosphate

26. The correct name of CoS is (6.3, 6.4)

    A. copper sulfide   **B.** cobalt(II) sulfate   **C.** cobalt(I) sulfide
    D. cobalt sulfide   **E.** cobalt(II) sulfide

27. The correct name of $MnCl_2$, used as a supplement in intravenous solutions, is (6.3, 6.4)

    A. magnesium chloride   **B.** manganese(II) chlorine   **C.** manganese(II) chloride
    D. manganese chlorine   **E.** manganese(III) chloride

28. The correct name of $ZnCO_3$ is (6.3, 6.4)

    A. zinc(III) carbonate   **B.** zinc(II) carbonate   **C.** zinc bicarbonate
    D. zinc carbon trioxide   **E.** zinc carbonate

29. The correct name of Al_2O_3 is (6.3)

    A. aluminum oxide   **B.** aluminum(II) oxide   **C.** aluminum trioxide
    D. dialuminum trioxide   **E.** aluminum oxygenate

30. The correct name of $Cr_2(SO_3)_3$ is (6.3, 6.4)

    A. chromium sulfite   **B.** dichromium trisulfite   **C.** chromium(III) sulfite
    D. chromium(III) sulfate   **E.** chromium sulfate

31. The correct name of PF_5 is (6.5)

    A. potassium pentafluoride   **B.** phosphorus pentafluoride   **C.** phosphorus pentafluorine
    D. potassium pentafluorine   **E.** phosphorus heptafluoride

32. The correct name of NCl_3 is (6.5)

    A. nitrogen chloride   **B.** nitrogen trichloride   **C.** trinitrogen chloride
    D. nitrogen chlorine three   **E.** nitrogen trichlorine

33. The correct name of CO is (6.5)

    A. carbon monoxide   **B.** carbonic oxide   **C.** carbon oxide
    D. carbonious oxide   **E.** carboxide

34. The correct Lewis structure for CS_2 is: (6.6)

    A. S̈=C̈=S̈   **B.** S̈=C=S̈   **C.** :S̈—C̈—S̈:

    D. :S—C—S: **E.** :S̈—C≡S̈:

35. A polar covalent bond occurs between (6.7)
 A. two different nonmetals
 B. two of the same type of nonmetals
 C. two noble gases
 D. two different metals
 E. a metal and a nonmetal

For questions 36 through 41, indicate the type of bonding expected to occur between the following elements: (6.7)

  ionic (I)   polar covalent (PC)   nonpolar covalent (NC)

36. ____ silicon and oxygen   **37.** ____ barium and chlorine

38. ____ aluminum and fluorine   **39.** ____ bromine and bromine

40. ____ sulfur and oxygen   **41.** ____ nitrogen and oxygen

For questions 42 through 47, match each of the following shapes with one of the descriptions: (6.8)

 A. linear   **B.** trigonal planar   **C.** tetrahedral
 D. trigonal pyramidal   **E.** bent (120°)   **F.** bent (109°)

42. ____ has a central atom with three electron groups and three bonded atoms

43. ____ has a central atom with two electron groups and two bonded atoms

44. ____ has a central atom with four electron groups and three bonded atoms

45. ____ has a central atom with three electron groups and two bonded atoms

46. ____ has a central atom with four electron groups and four bonded atoms

47. ____ has a central atom with four electron groups and two bonded atoms

For questions 48 through 51, determine the shape for each of the following molecules: (6.8)

 A. linear   **B.** trigonal planar   **C.** bent, 120°
 D. tetrahedral   **E.** trigonal pyramidal   **F.** bent, 109°

48. PBr_3 **49.** CBr_4 **50.** H_2S **51.** CO

For questions 52 through 55, determine if each of the following molecules is polar (P) *or nonpolar* (N): (6.8)

52. PBr_3 **53.** CBr_4 **54.** H_2S **55.** CO

For questions 56 through 61, indicate the major type of intermolecular forces (A to D) *between the molecules of each of the following:* (6.9)

 A. dipole–dipole attractions   **B.** hydrogen bonds   **C.** dispersion forces

56. ____ NCl_3 **57.** ____ SF_2

58. ____ Br_2 **59.** ____ H_2O_2

60. ____ NH_3 **61.** ____ C_3H_8

For questions 62 through 66, draw the Lewis structure for CO₂: (6.6, 6.7, 6.8)

62. The number of valence electrons needed in the Lewis structure of CO_2 is
    A. 10   **B.** 12   **C.** 14   **D.** 16   **E.** 18

63. The number of single bonds is
    A. 0   **B.** 1   **C.** 2   **D.** 3   **E.** 4

64. The number of double bonds is
    A. 0   **B.** 1   **C.** 2   **D.** 3   **E.** 4

65. The molecule is
    A. nonpolar   **B.** polar   **C.** ionic

66. The shape of the molecule is
    A. linear   **B.** bent (120°)   **C.** bent (109°)
    D. trigonal planar   **E.** tetrahedral

Answers to the Practice Test

1. C	**2.** C	**3.** E	**4.** E	**5.** B
6. B	**7.** B	**8.** D	**9.** A	**10.** C
11. A	**12.** E	**13.** C	**14.** D	**15.** A
16. E	**17.** D	**18.** B	**19.** E	**20.** D
21. E	**22.** C	**23.** A	**24.** E	**25.** B
26. E	**27.** C	**28.** E	**29.** A	**30.** C
31. B	**32.** B	**33.** A	**34.** B	**35.** A
36. PC	**37.** I	**38.** I	**39.** NC	**40.** PC
41. PC	**42.** B	**43.** A	**44.** D	**45.** E
46. C	**47.** F	**48.** E	**49.** D	**50.** F
51. A	**52.** P	**53.** N	**54.** N	**55.** P
56. C	**57.** A	**58.** C	**59.** B	**60.** B
61. C	**62.** D	**63.** A	**64.** C	**65.** A
66. A				

Selected Answers and Solutions to Text Problems

6.1 Atoms with one, two, or three valence electrons will lose those electrons to acquire a noble gas electron configuration.
   a. Li loses 1 e^-.   **b.** Ca loses 2 e^-.
   c. Ga loses 3 e^-.   **d.** Cs loses 1 e^-.
   e. Ba loses 2 e^-.

6.3 Atoms form ions by losing or gaining valence electrons to achieve the same electron configuration as the nearest noble gas. Elements in Groups 1A (1), 2A (2), and 3A (13) lose valence electrons, whereas elements in Groups 5A (15), 6A (16), and 7A (17) gain valence electrons to complete octets.
   a. Sr loses 2 e^-.   **b.** P gains 3 e^-.
   c. Elements in Group 7A (17) gain 1 e^-.   **d.** Na loses 1 e^-.
   e. Br gains 1 e^-.

6.5 **a.** The element with 3 protons is lithium. In a lithium ion with 2 electrons, the ionic charge would be 1+, $(3+) + (2-) = 1+$. The lithium ion is written as Li^+.
   b. The element with 9 protons is fluorine. In a fluorine ion with 10 electrons, the ionic charge would be 1−, $(9+) + (10-) = 1-$. The fluoride ion is written as F^-.
   c. The element with 12 protons is magnesium. In a magnesium ion with 10 electrons, the ionic charge would be 2+, $(12+) + (10-) = 2+$. The magnesium ion is written as Mg^{2+}.
   d. The element with 27 protons is cobalt. In a cobalt ion with 24 electrons, the ionic charge would be 3+, $(27+) + (24-) = 3+$. This cobalt ion is written as Co^{3+}.

6.7 **a.** Cu has an atomic number of 29, which means it has 29 protons. In a neutral atom, the number of electrons equals the number of protons. A copper ion with a charge of 2+ has lost $2\ e^-$ to have $29 - 2 = 27$ electrons. ∴ in a Cu^{2+} ion, there are 29 protons and 27 electrons.
   b. Se has an atomic number of 34, which means it has 34 protons. In a neutral atom, the number of electrons equals the number of protons. A selenium ion with a charge of 2− has gained $2\ e^-$ to have $34 + 2 = 36$ electrons. ∴ in a Se^{2-} ion, there are 34 protons and 36 electrons.
   c. Br has an atomic number of 35, which means it has 35 protons. In a neutral atom, the number of electrons equals the number of protons. A bromine ion with a charge of 1− has gained $1\ e^-$ to have $35 + 1 = 36$ electrons. ∴ in a Br^- ion, there are 35 protons and 36 electrons.
   d. Fe has an atomic number of 26, which means it has 26 protons. In a neutral atom, the number of electrons equals the number of protons. An iron ion with a charge of 2+ has lost $2\ e^-$ to have $26 - 2 = 24$ electrons. ∴ in a Fe^{2+} ion, there are 26 protons and 24 electrons.

6.9 **a.** Chlorine in Group 7A (17) gains one electron to form chloride ion, Cl^-.
   b. Cesium in Group 1A (1) loses one electron to form cesium ion, Cs^+.
   c. Nitrogen in Group 5A (15) gains three electrons to form nitride ion, N^{3-}.
   d. Radium in Group 2A (2) loses two electrons to form radium ion, Ra^{2+}.

6.11 **a.** Li^+ is called the lithium ion.   **b.** Ca^{2+} is called the calcium ion.
   c. Ga^{3+} is called the gallium ion.   **d.** P^{3-} is called the phosphide ion.

6.13 **a.** There are 8 protons and 10 electrons in the O^{2-} ion; $(8+) + (10-) = 2-$.
   b. There are 19 protons and 18 electrons in the K^+ ion; $(19+) + (18-) = 1+$.
   c. There are 53 protons and 54 electrons in the I^- ion; $(53+) + (54-) = 1-$.
   d. There are 11 protons and 10 electrons in the Na^+ ion; $(11+) + (10-) = 1+$.

6.15 **a** (Li and Cl) and **c** (K and O) will form ionic compounds.

6.17 **a.** Na^+ and $O^{2-} \rightarrow Na_2O$ **b.** Al^{3+} and $Br^- \rightarrow AlBr_3$
       c. Ba^{2+} and $N^{3-} \rightarrow Ba_3N_2$ **d.** Mg^{2+} and $F^- \rightarrow MgF_2$
       e. Al^{3+} and $S^{2-} \rightarrow Al_2S_3$

6.19 **a.** Ions: K^+ and $S^{2-} \rightarrow K_2S$ Check: $2(1+) + 1(2-) = 0$
       b. Ions: Na^+ and $N^{3-} \rightarrow Na_3N$ Check: $3(1+) + 1(3-) = 0$
       c. Ions: Al^{3+} and $I^- \rightarrow AlI_3$ Check: $1(3+) + 3(1-) = 0$
       d. Ions: Ga^{3+} and $O^{2-} \rightarrow Ga_2O_3$ Check: $2(3+) + 3(2-) = 0$

6.21 **a.** aluminum fluoride   **b.** calcium chloride
       c. sodium oxide   **d.** magnesium phosphide
       e. potassium iodide   **f.** barium fluoride

6.23 **a.** iron(II)   **b.** copper(II)
       c. zinc   **d.** lead(IV)
       e. chromium(III)   **f.** manganese(II)

6.25 **a.** Ions: Sn^{2+} and $Cl^- \rightarrow$ tin(II) chloride
       b. Ions: Fe^{2+} and $O^{2-} \rightarrow$ iron(II) oxide
       c. Ions: Cu^+ and $S^{2-} \rightarrow$ copper(I) sulfide
       d. Ions: Cu^{2+} and $S^{2-} \rightarrow$ copper(II) sulfide
       e. Ions: Cd^{2+} and $Br^- \rightarrow$ cadmium bromide
       f. Ions: Hg^{2+} and $Cl^- \rightarrow$ mercury(II) chloride

6.27 **a.** Au^{3+} **b.** Fe^{3+}
       c. Pb^{4+} **d.** Al^{3+}

6.29 **a.** Ions: Mg^{2+} and $Cl^- \rightarrow MgCl_2$ **b.** Ions: Na^+ and $S^{2-} \rightarrow Na_2S$
       c. Ions: Cu^+ and $O^{2-} \rightarrow Cu_2O$ **d.** Ions: Zn^{2+} and $P^{3-} \rightarrow Zn_3P_2$
       e. Ions: Au^{3+} and $N^{3-} \rightarrow AuN$ **f.** Ions: Co^{3+} and $F^- \rightarrow CoF_3$

6.31 **a.** Ions: Co^{3+} and $Cl^- \rightarrow CoCl_3$ **b.** Ions: Pb^{4+} and $O^{2-} \rightarrow PbO_2$
       c. Ions: Ag^+ and $I^- \rightarrow AgI$ **d.** Ions: Ca^{2+} and $N^{3-} \rightarrow Ca_3N_2$
       e. Ions: Cu^+ and $P^{3-} \rightarrow Cu_3P$ **f.** Ions: Cr^{2+} and $Cl^- \rightarrow CrCl_2$

6.33 **a.** Ions: K^+ and $P^{3-} \rightarrow K_3P$ **b.** Ions: Cu^{2+} and $Cl^- \rightarrow CuCl_2$
       c. Ions: Fe^{3+} and $Br^- \rightarrow FeBr_3$ **d.** Ions: Mg^{2+} and $O^{2-} \rightarrow MgO$

6.35 **a.** HCO_3^- **b.** NH_4^+ **c.** PO_3^{3-} **d.** ClO_3^-

6.37 **a.** sulfate   **b.** carbonate
       c. hydrogen sulfite (or bisulfite)   **d.** nitrate

6.39

	NO_2^-	CO_3^{2-}	HSO_4^-	PO_4^{3-}
Li^+	$LiNO_2$ Lithium nitrite	Li_2CO_3 Lithium carbonate	$LiHSO_4$ Lithium hydrogen sulfate	Li_3PO_4 Lithium phosphate
Cu^{2+}	$Cu(NO_2)_2$ Copper(II) nitrite	$CuCO_3$ Copper(II) carbonate	$Cu(HSO_4)_2$ Copper(II) hydrogen sulfate	$Cu_3(PO_4)_2$ Copper(II) phosphate
Ba^{2+}	$Ba(NO_2)_2$ Barium nitrite	$BaCO_3$ Barium carbonate	$Ba(HSO_4)_2$ Barium hydrogen sulfate	$Ba_3(PO_4)_2$ Barium phosphate

6.41 **a.** Ions: Ba^{2+} and $OH^- \rightarrow Ba(OH)_2$ **b.** Ions: Na^+ and $HSO_4^- \rightarrow NaHSO_4$
       c. Ions: Fe^{2+} and $NO_2^- \rightarrow Fe(NO_2)_2$ **d.** Ions: Zn^{2+} and $PO_4^{3-} \rightarrow Zn_3(PO_4)_2$
       e. Ions: Fe^{3+} and $CO_3^{2-} \rightarrow Fe_2(CO_3)_3$

6.43 **a.** CO_3^{2-}, sodium carbonate
b. NH_4^+, ammonium sulfide
c. OH^-, calcium hydroxide
d. NO_2^-, tin(II) nitrite

6.45 **a.** Ions: $Zn^{2+} + C_2H_3O_2^- \rightarrow$ zinc acetate
b. Ions: Mg^{2+} and $PO_4^{3-} \rightarrow$ magnesium phosphate
c. Ions: $NH_4^+ + Cl^- \rightarrow$ ammonium chloride
d. Ions: Na^+ and $HCO_3^- \rightarrow$ sodium bicarbonate or sodium hydrogen carbonate
e. Ions: Na^+ and $NO_2^- \rightarrow$ sodium nitrite

6.47 When naming molecular compounds, prefixes are used to indicate the number of each atom as shown in the subscripts of the formula. The first nonmetal is named by its elemental name; the second nonmetal is named by using its elemental name with the ending changed to *ide*.
a. 1 P and 3 Br $\rightarrow$ phosphorus tribromide
b. 2 Cl and 1 O $\rightarrow$ dichlorine oxide
c. 1 C and 4 Br $\rightarrow$ carbon tetrabromide
d. 1 H and 1 F $\rightarrow$ hydrogen fluoride
e. 1 N and 3 F $\rightarrow$ nitrogen trifluoride

6.49 When naming molecular compounds, prefixes are used to indicate the number of each atom as shown in the subscripts of the formula. The first nonmetal is named by its elemental name; the second nonmetal is named by using its elemental name with the ending changed to *ide*.
a. 2 N and 3 O $\rightarrow$ dinitrogen trioxide
b. 2 Si and 6 Br $\rightarrow$ disilicon hexabromide
c. 4 P and 3 S $\rightarrow$ tetraphosphorus trisulfide
d. 1 P and 5 Cl $\rightarrow$ phosphorus pentachloride
e. 1 Se and 6 F $\rightarrow$ selenium hexafluoride

6.51 **a.** 1 C and 4 Cl $\rightarrow CCl_4$
b. 1 C and 1 O $\rightarrow$ CO
c. 1 P and 3 F $\rightarrow PF_3$
d. 2 N and 4 O $\rightarrow N_2O_4$

6.53 **a.** 1 O and 2 F $\rightarrow OF_2$
b. 1 B and 3 Cl $\rightarrow BCl_3$
c. 2 N and 3 O $\rightarrow N_2O_3$
d. 1 S and 6 F $\rightarrow SF_6$

6.55 **a.** Ions: Al^{3+} and $SO_4^{2-} \rightarrow$ aluminum sulfate
b. Ions: Ca^{2+} and $CO_3^{2-} \rightarrow$ calcium carbonate
c. 2 N and 1 O $\rightarrow$ dinitrogen oxide
d. Ions: $Mg^{2+} + OH^- \rightarrow$ magnesium hydroxide

6.57 **a.** $2\,H(1\,e^-) + 1\,S(6\,e^-) = 2 + 6 = 8$ valence electrons
b. $2\,I(7\,e^-) = 14$ valence electrons
c. $1\,C(4\,e^-) + 4\,Cl(7\,e^-) = 4 + 28 = 32$ valence electrons
d. $1\,O(6\,e^-) + 1\,H(1\,e^-) + 1\,e^-$ (negative charge) $= 6 + 1 + 1 = 8$ valence electrons

6.59 **a.** $1\,H(1\,e^-) + 1\,F(7\,e^-) = 1 + 7 = 8$ valence electrons

$$H\!:\!\ddot{\underset{\ddot{}}{F}}\!: \quad \text{or} \quad H\!-\!\ddot{\underset{\ddot{}}{F}}\!:$$

b. $1\,S(6\,e^-) + 2\,F(7\,e^-) = 6 + 14 = 20$ valence electrons

$$:\!\ddot{F}\!:\!\ddot{S}\!:\!\ddot{F}\!: \quad \text{or} \quad :\!\ddot{F}\!-\!\ddot{S}\!-\!\ddot{F}\!:$$

c. $1\,N(5\,e^-) + 3\,Br(7\,e^-) = 5 + 21 = 26$ valence electrons

$$
\begin{array}{c}
:\!\ddot{Br}\!: \\
:\!\ddot{Br}\!:\!N\!:\!\ddot{Br}\!:
\end{array}
\quad \text{or} \quad
\begin{array}{c}
:\!\ddot{Br}\!: \\
| \\
:\!\ddot{Br}\!-\!N\!-\!\ddot{Br}\!:
\end{array}
$$

d. $1\,B(3\,e^-) + 4\,H(1\,e^-) + 1\,e^-$ (negative charge) $= 3 + 4 + 1 = 8$ valence electrons

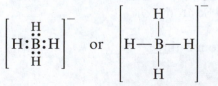

6.61 If complete octets cannot be formed by using all the valence electrons, it is necessary to draw multiple bonds.

6.63 **a.** $1\,C(4\,e^-) + 1\,O(6\,e^-) = 4 + 6 = 10$ valence electrons

:C::O: or :C≡O:

b. $1\,N(5\,e^-) + 3\,O(6\,e^-) + 1\,e^-$ (negative charge) $= 5 + 18 + 1 = 24$ valence electrons

$$\left[\begin{array}{c} \ddot{\text{:O:}} \\ \ddot{\text{:O:N:O:}} \end{array}\right]^- \quad \text{or} \quad \left[\begin{array}{c} \text{:O:} \\ \| \\ \ddot{\text{:O}}—\text{N}—\ddot{\text{O:}} \end{array}\right]^-$$

c. $2\,H(1\,e^-) + 1\,C(4\,e^-) + 1\,O(6\,e^-) = 2 + 4 + 6 = 12$ valence electrons

$$\begin{array}{cc} \ddot{\text{:O:}} & \text{:O:} \\ \| \\ \text{H:C:H} & \text{or} \quad \text{H}—\text{C}—\text{H} \end{array}$$

6.65 **a.** The electronegativity values increase going from left to right across Period 2 from B to F.
b. The electronegativity values decrease going down Group 2A (2) from Mg to Ba.
c. The electronegativity values decrease going down Group 7A (17) from F to I.

6.67 **a.** Electronegativity increases going up a group: K, Na, Li.
b. Electronegativity increases going left to right across a period: Na, P, Cl.
c. Electronegativity increases going across a period and at the top of a group: Ca, Se, O.

6.69 a; A nonpolar covalent bond would result from an electronegativity difference from 0.0 to 0.4.

6.71 **a.** electronegativity difference: $2.8 - 1.8 = 1.0$, polar covalent
b. electronegativity difference: $4.0 - 1.0 = 3.0$, ionic
c. electronegativity difference: $4.0 - 2.8 = 1.2$, polar covalent
d. electronegativity difference: $2.5 - 2.5 = 0.0$, nonpolar covalent
e. electronegativity difference: $3.0 - 2.1 = 0.9$, polar covalent
f. electronegativity difference: $2.5 - 2.1 = 0.4$, nonpolar covalent

6.73 A dipole arrow points from the atom with the lower electronegativity value (more positive) to the atom in the bond that has the higher electronegativity value (more negative).

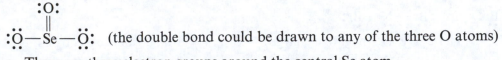

a. $\overset{\delta^+\quad\delta^-}{\text{N—F}}$ **b.** $\overset{\delta^+\quad\delta^-}{\text{Si—Br}}$ **c.** $\overset{\delta^+\quad\delta^-}{\text{C—O}}$

d. $\overset{\delta^+\quad\delta^-}{\text{P—Br}}$ **e.** $\overset{\delta^-\quad\delta^+}{\text{N—P}}$

6.75 **a.** 6, tetrahedral; Four electron groups around a central atom have a tetrahedral electron-group geometry. With four bonded atoms, the shape of the molecule is tetrahedral.
b. 5, trigonal pyramidal; Four electron groups around a central atom have a tetrahedral electron-group geometry. With three bonded atoms (and one lone pair of electrons), the shape of the molecule is trigonal pyramidal.
c. 3, trigonal planar; Three electron groups around a central atom have a trigonal planar electron-group geometry. With three bonded atoms, the shape of the molecule is trigonal planar.

6.77 SeO_3 $1\,Se(6\,e^-) + 3\,O(6\,e^-) = 6 + 18 = 24$ valence electrons

$$\begin{array}{c} \text{:O:} \\ \| \\ \ddot{\text{:O}}—\text{Se}—\ddot{\text{O:}} \end{array}$$ (the double bond could be drawn to any of the three O atoms)

a. There are <u>three</u> electron groups around the central Se atom.
b. The electron-group geometry is <u>trigonal planar</u>.
c. The number of atoms attached to the central Se atom is <u>three</u>.
d. The shape of the molecule is <u>trigonal planar</u>.

6.79 In PH_3 and NH_3, the four electron groups on the central P and N atoms have a tetrahedral geometry, with 109° bond angles. The shapes of the molecules are determined by the number of bonded atoms: since both PH_3 and NH_3 have three bonded atoms and one lone pair of electrons, the shapes of both are trigonal pyramidal.

$$H-\overset{\displaystyle \cdot\cdot}{\underset{\displaystyle |}{P}}-H \qquad H-\overset{\displaystyle \cdot\cdot}{\underset{\displaystyle |}{N}}-H$$
$$\qquad H \qquad\qquad\quad H$$

6.81 **a.** The central Se atom has four electron groups with two bonded atoms and two lone pairs of electrons, which gives $SeBr_2$ a bent shape (109°).

$$:\overset{\displaystyle \cdot\cdot}{\underset{\displaystyle |}{Br}}:$$
$$:\overset{\cdot\cdot}{\underset{\cdot\cdot}{Br}}-\underset{\cdot\cdot}{Se}:$$

b. The central C atom has four electron groups bonded to four chlorine atoms; CCl_4 has a tetrahedral shape.

$$:\overset{\displaystyle \cdot\cdot}{\underset{\displaystyle |}{Cl}}:$$
$$:\overset{\cdot\cdot}{\underset{\cdot\cdot}{Cl}}-C-\overset{\cdot\cdot}{\underset{\cdot\cdot}{Cl}}:$$
$$:\overset{\displaystyle \cdot\cdot}{\underset{\displaystyle |}{Cl}}:$$

c. The central O atom has four electron groups, but only two are bonded to bromine atoms; the shape of OBr_2 is bent (109°).

$$:\overset{\displaystyle \cdot\cdot}{\underset{\displaystyle |}{Br}}:$$
$$:\overset{\cdot\cdot}{\underset{\cdot\cdot}{Br}}-\underset{\cdot\cdot}{O}:$$

6.83 To find the total valence electrons for an ion, add the total valence electrons for each atom and add the number of electrons indicated by a negative charge, or subtract for a positive charge.

a. $1\ Al(3\ e^-) + 4\ H(1\ e^-) + 1\ e^-$(negative charge) $= 3 + 4 + 1 = 8$ valence electrons

$$\left[H-\overset{\displaystyle H}{\underset{\displaystyle H}{\underset{\displaystyle |}{\overset{\displaystyle |}{Al}}}}-H \right]^-$$ four electron groups around Al bonded to four atoms; tetrahedral shape

b. $1\ S(6\ e^-) + 4\ O(6\ e^-) + 2\ e^-$(negative charge) $= 6 + 24 + 2 = 32$ valence electrons

$$\left[:\overset{\cdot\cdot}{\underset{\cdot\cdot}{O}}-\overset{\displaystyle :\overset{\cdot\cdot}{O}:}{\underset{\displaystyle :\overset{\cdot\cdot}{O}:}{\underset{\displaystyle |}{\overset{\displaystyle |}{S}}}}-\overset{\cdot\cdot}{\underset{\cdot\cdot}{O}}: \right]^{2-}$$ four electron groups around S bonded to four atoms; tetrahedral shape

c. $1\ N(5\ e^-) + 4\ H(1\ e^-) - 1\ e^-$(positive charge) $= 5 + 4 - 1 = 8$ valence electrons

$$\left[H-\overset{\displaystyle H}{\underset{\displaystyle H}{\underset{\displaystyle |}{\overset{\displaystyle |}{N}}}}-H \right]^+$$ four electron groups around N bonded to four atoms; tetrahedral shape

d. $1\ N(5\ e^-) + 2\ O(6\ e^-) - 1\ e^-$(positive charge) $= 5 + 12 - 1 = 16$ valence electrons

$$\left[:\overset{\cdot\cdot}{O}=N=\overset{\cdot\cdot}{O}: \right]^+$$ two electron groups around N bonded to two atoms; linear shape

6.85 Since molecular shape is determined by the electron groups around the <u>central</u> atom, the structures below have been simplified by not showing the lone pairs on the atoms bonded to the central atom.

   a. The molecule HBr contains the polar covalent H—Br bond; this single dipole makes HBr a polar molecule.

   H—Br
   ├────→
    polar

   b. The molecule NF_3 contains three polar covalent N—F bonds and a lone pair of electrons on the central N atom. This asymmetric trigonal pyramidal shape makes NF_3 a polar molecule.

   polar

   c. In the molecule CHF_3, there are three polar covalent C—F bonds and one nonpolar covalent C—H bond, which makes CHF_3 a polar molecule.

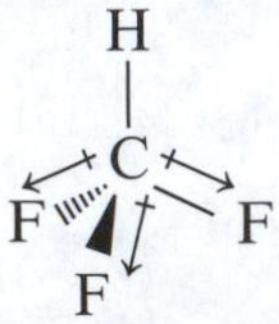

   polar

6.87 **a.** BrF is a polar molecule. An attraction between the positive end of one polar molecule and the negative end of another polar molecule is called a dipole–dipole attraction.

   b. An ionic bond is an attraction between a positive and negative ion, as in KCl.

   c. NF_3 is a polar molecule. An attraction between the positive end of one polar molecule and the negative end of another polar molecule is called a dipole–dipole attraction.

   d. Cl_2 is a nonpolar molecule. The weak attractions that occur between temporary dipoles in nonpolar molecules are called dispersion forces.

6.89 **a.** CH_3OH is a polar molecule. Hydrogen bonds are strong dipole–dipole attractions that occur between a partially positive hydrogen atom of one molecule and one of the strongly electronegative atoms N, O, or F in another, as is seen with CH_3OH molecules.

   b. CO is a polar molecule. Dipole–dipole attractions occur between dipoles in polar molecules.

   c. CF_4 is a nonpolar molecule. The weak attractions that occur between temporary dipoles in nonpolar molecules are called dispersion forces.

   d. CH_3CH_3 is a nonpolar molecule. The weak attractions that occur between temporary dipoles in nonpolar molecules are called dispersion forces.

6.91 **a.** HF would have a higher melting point than HBr; the hydrogen bonds in HF are stronger than the dipole–dipole attractions in HBr.

   b. NaF would have a higher melting point than HF; the ionic bonds in NaF are stronger than the hydrogen bonds in HF.

   c. $MgBr_2$ would have a higher melting point than PBr_3; the ionic bonds in $MgBr_2$ are stronger than the dipole–dipole attractions in PBr_3.

   d. CH_3OH would have a higher melting point than CH_4; the hydrogen bonds in CH_3OH are stronger than the dispersion forces in CH_4.

6.93 **a.** C—C electronegativity difference $2.5 - 2.5 = 0.0$
   b. C—H electronegativity difference $2.5 - 2.1 = 0.4$
   c. C—O electronegativity difference $3.5 - 2.5 = 1.0$

6.95 **a.** nonpolar covalent   **b.** nonpolar covalent   **c.** polar covalent

6.97 a. By losing two valence electrons from the third energy level, magnesium achieves an octet in the second energy level.
b. The magnesium ion Mg^{2+} has the same electron configuration as the noble gas Ne $1s^2 2s^2 2p^6$.
c. Group 1A (1) and 2A (2) elements achieve octets by losing electrons when they form compounds. Group 8A (18) elements already have a stable octet of valence electrons (or two electrons for helium), so they are not normally found in compounds.

6.99 a. The element with 15 protons is phosphorus. In an ion of phosphorus with 18 electrons, the ionic charge would be 3−, (15+) + (18−) = 3−. The phosphide ion is written P^{3-}.
b. The element with 8 protons is oxygen. Since there are also 8 electrons, this is an oxygen (O) atom.
c. The element with 30 protons is zinc. In an ion of zinc with 28 electrons, the ionic charge would be 2+, (30+) + (28−) = 2+. The zinc ion is written Zn^{2+}.
d. The element with 26 protons is iron. In an ion of iron with 23 electrons, the ionic charge would be 3+, (26+) + (23−) = 3+. This iron ion is written Fe^{3+}.

6.101 a. X is in Group 1A (1); Y is in Group 6A (16). **b.** ionic
c. X^+, Y^{2-} **d.** X_2Y
e. X^+ and $S^{2-} \rightarrow X_2S$ **f.** 1 Y and 2 Cl $\rightarrow YCl_2$
g. molecular

6.103

Electron Configurations		Cation	Anion	Formula of Compound	Name of Compound
$1s^2 2s^2 2p^6 3s^2$	$1s^2 2s^2 2p^3$	Mg^{2+}	N^{3-}	Mg_3N_2	Magnesium nitride
$1s^2 2s^2 2p^6 3s^2 3p^6 4s^1$	$1s^2 2s^2 2p^4$	K^+	O^{2-}	K_2O	Potassium oxide
$1s^2 2s^2 2p^6 3s^2 3p^1$	$1s^2 2s^2 2p^6 3s^2 3p^5$	Al^{3+}	Cl^-	$AlCl_3$	Aluminum chloride

6.105 a. 2 valence electrons, 1 bonding pair, no lone pairs
b. 8 valence electrons, 1 bonding pair, 3 lone pairs
c. 14 valence electrons, 1 bonding pair, 6 lone pairs

6.107 a. 2; trigonal pyramidal shape, polar molecule
b. 1; bent shape (109°), polar molecule
c. 3; tetrahedral shape, nonpolar molecule

6.109 a. C—O (EN 3.5 − 2.5 = 1.0) and N—O (EN 3.5 − 3.0 = 0.5) are polar covalent bonds.
b. O—O (EN 3.5 − 3.5 = 0.0) is a nonpolar covalent bond.
c. Ca—O (EN 3.5 − 1.0 = 2.5) and K—O (EN 3.5 − 0.8 = 2.7) are ionic bonds.
d. C—O, N—O, O—O

6.111 a. PH_3 is a nonpolar molecule. Dispersion forces occur between temporary dipoles in nonpolar molecules.
b. NO_2 is a polar molecule. Dipole–dipole attractions occur between dipoles in polar molecules.
c. Hydrogen bonds are strong dipole–dipole attractions that occur between a partially positive hydrogen atom of one molecule and one of the strongly electronegative atoms N, O, or F in another.
d. Dispersion forces occur between temporary dipoles in Ar atoms.

6.113 a. N^{3-} is called the nitride ion. **b.** Mg^{2+} is called the magnesium ion.
c. O^{2-} is called the oxide ion. **d.** Al^{3+} is called the aluminum ion.

6.115 a. An element that forms an ion with a 2+ charge would be in Group 2A (2).
b. The Lewis symbol for an element in Group 2A (2) is $\dot{X}\cdot$
c. Mg is the Group 2A (2) element in Period 3.
d. Ions: X^{2+} and $N^{3-} \rightarrow X_3N_2$

6.117 a. Tin(IV) is Sn^{4+}.
    b. The Sn^{4+} ion has 50 protons and $50 - 4 = 46$ electrons.
    c. Ions: Sn^{4+} and $O^{2-} \rightarrow SnO_2$
    d. Ions: Sn^{4+} and $PO_4^{3-} \rightarrow Sn_3(PO_4)_4$

6.119 a. Ions: Sn^{2+} and $S^{2-} \rightarrow SnS$ **b.** Ions: Pb^{4+} and $O^{2-} \rightarrow PbO_2$
    c. Ions: Ag^+ and $Cl^- \rightarrow AgCl$ **d.** Ions: Ca^{2+} and $N^{3-} \rightarrow Ca_3N_2$
    e. Ions: Cu^+ and $P^{3-} \rightarrow Cu_3P$ **f.** Ions: Cr^{2+} and $Br^- \rightarrow CrBr_2$

6.121 a. 1 N and 3 Cl $\rightarrow$ nitrogen trichloride
    b. 2 N and 3 S $\rightarrow$ dinitrogen trisulfide
    c. 2 N and 1 O $\rightarrow$ dinitrogen oxide
    d. 1 I and 1 F $\rightarrow$ iodine fluoride
    e. 1 B and 3 F $\rightarrow$ boron trifluoride
    f. 2 P and 5 O $\rightarrow$ diphosphorus pentoxide

6.123 a. 1 C and 1 S $\rightarrow$ CS   **b.** 2 P and 5 O $\rightarrow P_2O_5$
    c. 2 H and 1 S $\rightarrow H_2S$ **d.** 1 S and 2 Cl $\rightarrow SCl_2$

6.125 a. Ionic, ions are Fe^{3+} and $Cl^- \rightarrow$ iron(III) chloride
    b. Ionic, ions are Na^+ and $SO_4^{2-} \rightarrow$ sodium sulfate
    c. Molecular, 1 N and 2 O $\rightarrow$ nitrogen dioxide
    d. Ionic, ions are Rb^+ and $S^{2-} \rightarrow$ rubidium sulfide
    e. Molecular, 1 P and 5 F $\rightarrow$ phosphorus pentafluoride
    f. Molecular, 1 C and 4 F $\rightarrow$ carbon tetrafluoride

6.127 a. Ions: Sn^{2+} and $CO_3^{2-} \rightarrow SnCO_3$ **b.** Ions: Li^+ and $P^{3-} \rightarrow Li_3P$
    c. Molecular, 1 Si and 4 Cl $\rightarrow SiCl_4$ **d.** Ions: Mn^{3+} and $O^{2-} \rightarrow Mn_2O_3$
    e. Molecular, 4 P and 3 Se $\rightarrow P_4Se_3$ **f.** Ions: Ca^{2+} and $Br^- \rightarrow CaBr_2$

6.129 a. $1\,H(1\,e^-) + 1\,N(5\,e^-) + 2\,O(6\,e^-) = 1 + 5 + 12 = 18$ valence electrons
    b. $2\,C(4\,e^-) + 4\,H(1\,e^-) + 1\,O(6\,e^-) = 8 + 4 + 6 = 18$ valence electrons
    c. $1\,C(4\,e^-) + 5\,H(1\,e^-) + 1\,N(5\,e^-) = 4 + 5 + 5 = 14$ valence electrons

6.131 a. $2\,Cl(7\,e^-) + 1\,O(6\,e^-) = 14 + 6 = 20$ valence electrons

    $:\!\ddot{C}l\!:\!\ddot{O}\!:\!\ddot{C}l\!:$ or $:\!\ddot{C}l\!-\!\ddot{O}\!-\!\ddot{C}l\!:$

    b. $1\,N(5\,e^-) + 3\,H(1\,e^-) + 1\,O(6\,e^-) = 5 + 3 + 6 = 14$ valence electrons

$$\begin{array}{ccc} & & H \\ & & | \\ H & & \\ H\!:\!\ddot{N}\!:\!\ddot{O}\!:\!H & \text{or} & H\!-\!N\!-\!\ddot{O}\!-\!H \end{array}$$

    c. $2\,C(4\,e^-) + 2\,H(1\,e^-) + 2\,Cl(7\,e^-) = 8 + 2 + 14 = 24$ valence electrons

$$\begin{array}{ccc} H \quad :\!\ddot{C}l\!: & & H \quad :\!\ddot{C}l\!: \\ & & |\quad\quad| \\ H\!:\!C\!:\!:\!C\!:\!\ddot{C}l\!: & \text{or} & H\!-\!C\!=\!C\!-\!\ddot{C}l\!: \end{array}$$

    d. $1\,B(3\,e^-) + 4\,F(7\,e^-) + 1\,e^-$ (negative charge) $= 3 + 28 + 1 = 32$ valence electrons

$$\left[\begin{array}{c} :\!\ddot{F}\!: \\ :\!\ddot{F}\!:\!B\!:\!\ddot{F}\!: \\ :\!\ddot{F}\!: \end{array}\right]^- \quad \text{or} \quad \left[\begin{array}{c} :\!\ddot{F}\!: \\ | \\ :\!\ddot{F}\!-\!B\!-\!\ddot{F}\!: \\ | \\ :\!\ddot{F}\!: \end{array}\right]^-$$

6.133 a. Electronegativity increases going up a group: I, Cl, F
    b. Electronegativity increases going left to right across a period and at the top of a group: K, Li, S, Cl
    c. Electronegativity increases going up a group: Ba, Sr, Mg, Be

6.135 a. C—O $(3.5 - 2.5 = 1.0)$ is more polar than C—N $(3.5 - 3.0 = 0.5)$.
  b. N—F $(4.0 - 3.0 = 1.0)$ is more polar than N—Br $(3.0 - 2.8 = 0.2)$.
  c. S—Cl $(3.0 - 2.5 = 0.5)$ is more polar than Br—Cl $(3.0 - 2.8 = 0.2)$.
  d. Br—I $(2.8 - 2.5 = 0.3)$ is more polar than Br—Cl $(3.0 - 2.8 = 0.2)$.
  e. N—F $(4.0 - 3.0 = 1.0)$ is more polar than N—O $(3.5 - 3.0 = 0.5)$.

6.137 A dipole arrow points from the atom with the lower electronegativity value (more positive) to the atom in the bond that has the higher electronegativity value (more negative).

  a. $\overset{\delta^+}{\text{Si}}-\overset{\delta^-}{\text{Cl}}$ $\longrightarrow$

  b. $\overset{\delta^+}{\text{C}}-\overset{\delta^-}{\text{N}}$ $\longrightarrow$

  c. $\overset{\delta^-}{\text{F}}-\overset{\delta^+}{\text{Cl}}$ $\longleftarrow$

  d. $\overset{\delta^+}{\text{C}}-\overset{\delta^-}{\text{F}}$ $\longrightarrow$

  e. $\overset{\delta^+}{\text{N}}-\overset{\delta^-}{\text{O}}$ $\longrightarrow$

6.139 a. electronegativity difference: $3.0 - 1.8 = 1.2$, polar covalent
  b. electronegativity difference: $2.5 - 2.5 = 0.0$, nonpolar covalent
  c. electronegativity difference: $3.0 - 0.9 = 2.1$, ionic
  d. electronegativity difference: $2.5 - 2.1 = 0.4$, nonpolar covalent
  e. electronegativity difference: $4.0 - 4.0 = 0.0$, nonpolar covalent

6.141 a. $1 \text{ N}(5 \ e^-) + 3 \text{ F}(7 \ e^-) = 5 + 21 = 26$ valence electrons

$$:\!\overset{..}{\underset{..}{\text{F}}}\!-\!\overset{..}{\underset{..}{\text{N}}}\!-\!\overset{..}{\underset{..}{\text{F}}}\!:$$
$$|$$
$$:\!\overset{..}{\underset{..}{\text{F}}}\!:$$

The central atom N has four electron groups with three bonded atoms and one lone pair of electrons, which gives NF_3 a trigonal pyramidal shape.

  b. $1 \text{ Si}(4 \ e^-) + 4 \text{ Br}(7 \ e^-) = 4 + 28 = 32$ valence electrons

$$:\!\overset{..}{\underset{..}{\text{Br}}}\!:$$
$$|$$
$$:\!\overset{..}{\underset{..}{\text{Br}}}\!-\!\text{Si}\!-\!\overset{..}{\underset{..}{\text{Br}}}\!:$$
$$|$$
$$:\!\overset{..}{\underset{..}{\text{Br}}}\!:$$

The central atom Si has four electron groups bonded to four bromine atoms; $SiBr_4$ has a tetrahedral shape.

  c. $1 \text{ C}(4 \ e^-) + 2 \text{ Se}(6 \ e^-) = 4 + 12 = 16$ valence electrons

$$:\!\overset{..}{\underset{..}{\text{Se}}}\!=\!\text{C}\!=\!\overset{..}{\underset{..}{\text{Se}}}\!:$$

The central atom C has two electron groups bonded to two selenium atoms; CSe_2 has a linear shape.

6.143 a. $1 \text{ Br}(7 \ e^-) + 2 \text{ O}(6 \ e^-) + 1 \ e^- \text{(negative charge)} = 7 + 12 + 1 = 20$ valence electrons

$$\left[:\!\overset{..}{\underset{..}{\text{O}}}\!-\!\overset{..}{\underset{..}{\text{Br}}}\!-\!\overset{..}{\underset{..}{\text{O}}}\!:\right]^-$$

The central atom Br has four electron groups with two bonded atoms and two lone pairs of electrons, which gives BrO_2^- a bent shape $(109°)$.

  b. $1 \text{ O}(6 \ e^-) + 2 \text{ H}(1 \ e^-) = 6 + 2 = 8$ valence electrons

$$\text{H}$$
$$|$$
$$:\!\overset{}{\underset{..}{\text{O}}}\!-\!\text{H}$$

The central atom O has four electron groups with two bonded atoms and two lone pairs of electrons, which gives H_2O a bent shape $(109°)$.

c. 1 C($4\,e^-$) + 4 Br($7\,e^-$) = 4 + 28 = 32 valence electrons

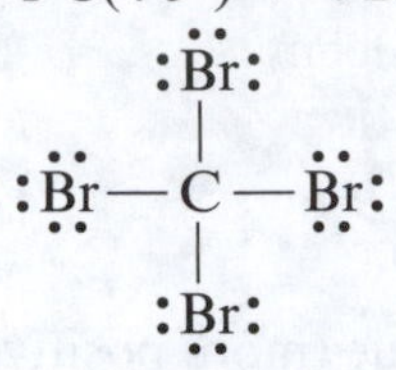

$$
\begin{array}{c}
\ddot{\text{Br}}\!: \\
| \\
:\!\ddot{\text{Br}}\!-\!\text{C}\!-\!\ddot{\text{Br}}\!: \\
| \\
:\!\ddot{\text{Br}}\!:
\end{array}
$$

The central atom C has four electron groups bonded to four bromine atoms, which gives CBr_4 a tetrahedral shape.

d. 1 P($5\,e^-$) + 3 O($6\,e^-$) + 3 e^- (negative charge) = 5 + 18 + 3 = 26 valence electrons

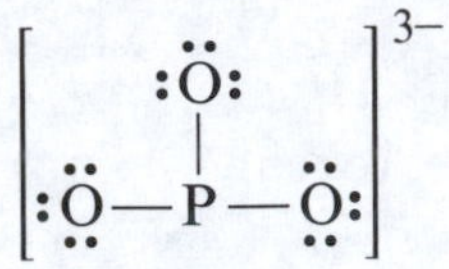

$$
\left[
\begin{array}{c}
:\!\ddot{\text{O}}\!: \\
| \\
:\!\ddot{\text{O}}\!-\!\text{P}\!-\!\ddot{\text{O}}\!: \\
\end{array}
\right]^{3-}
$$

The central atom P has four electron groups with three bonded atoms and one lone pair of electrons, which gives PO_3^{3-} a trigonal pyramidal shape.

6.145 a. A molecule that has a central atom with three bonded atoms and one lone pair will have a trigonal pyramidal shape. This asymmetric shape means the dipoles do not cancel, and the molecule will be polar.

b. A molecule that has a central atom with two bonded atoms and two lone pairs will have a bent shape (109°). This asymmetric shape means the dipoles do not cancel, and the molecule will be polar.

6.147 Since molecular shape is determined by the electron groups around the <u>central</u> atom, the structures below have been simplified by not showing the lone pairs on the atoms bonded to the central atom.

a. The molecule HBr contains only a single polar covalent H—Br bond (EN 2.8 − 2.1 = 0.7); it is a polar molecule.

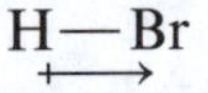

H—Br

b. The molecule SiO_2 contains two polar covalent Si—O bonds (EN 3.5 − 1.8 = 1.7) and has a linear shape. The two equal dipoles directed away from each other at 180° will cancel, resulting in a nonpolar molecule.

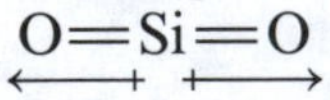

O=Si=O

c. The molecule NCl_3 contains three nonpolar covalent N—Cl bonds (EN 3.0 − 3.0 = 0.0) and a lone pair of electrons on the central N atom, which gives the molecule a trigonal pyramidal shape. Since the molecule contains only nonpolar bonds, NCl_3 is a nonpolar molecule.

d. The molecule CH_3Cl consists of a central atom, C, with three nonpolar covalent C—H bonds (EN 2.5 − 2.1 = 0.4) and one polar covalent C—Cl bond (EN 3.0 − 2.5 = 0.5). The molecule has a tetrahedral shape, but the single dipole makes CH_3Cl a polar molecule.

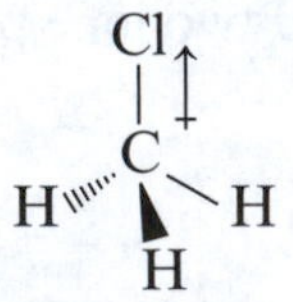

e. The molecule NI_3 contains three polar covalent N—I bonds (EN 3.0 − 2.5 = 0.5) and a lone pair of electrons on the central N atom, which gives the molecule a trigonal pyramidal shape. This asymmetric shape does not allow the dipoles to cancel, which makes NI_3 a polar molecule.

f. The molecule H_2O contains two polar covalent O—H bonds (EN 3.5 − 2.1 = 1.4) and two lone pairs of electrons on the central O atom. The molecule has a bent shape (109°) and the asymmetric geometry of the O—H dipoles makes H_2O a polar molecule.

$$\overset{\displaystyle \ddot{\text{O}}{\cdot}}{H \nearrow \,\, \nwarrow H}$$

6.149 a. NF_3 is a polar molecule. Dipole–dipole attractions (2) occur between dipoles in polar molecules.
b. ClF is a polar molecule (EN 4.0 − 3.0 = 1.0). Dipole–dipole attractions (2) occur between dipoles in polar molecules.
c. Dispersion forces (4) occur between temporary dipoles in nonpolar Br_2 molecules.
d. Ionic bonds (1) are strong attractions between positive and negative ions, as in Cs_2O.
e. Dispersion forces (4) occur between temporary dipoles in nonpolar C_4H_{10} molecules.
f. CH_3OH is a polar molecule and contains the polar O—H bond. Hydrogen bonding (3) involves strong dipole–dipole attractions that occur between a partially positive hydrogen atom of one polar molecule and one of the strongly electronegative atoms F, O, or N in another, as is seen with CH_3OH molecules.

6.151

Atom or Ion	Number of Protons	Number of Electrons	Electrons Lost/Gained
K^+	$19\ p^+$	$18\ e^-$	$1\ e^-$ lost
Mg^{2+}	$12\ p^+$	$10\ e^-$	$2\ e^-$ lost
O^{2-}	$8\ p^+$	$10\ e^-$	$2\ e^-$ gained
Al^{3+}	$13\ p^+$	$10\ e^-$	$3\ e^-$ lost

6.153 a. X as a X^{3+} ion would be in Group 3A (13).
b. X as a X^{2-} ion would be in Group 6A (16).
c. X as a X^{2+} ion would be in Group 2A (2).

6.155 Compounds with a metal and nonmetal are classified as ionic; compounds with two nonmetals are molecular.
a. Ionic, ions are Li^+ and $HPO_4^{2-} \rightarrow$ lithium hydrogen phosphate
b. Molecular, 1 Cl and 3 F $\rightarrow$ chlorine trifluoride
c. Ionic, ions are Mg^{2+} and $ClO_2^- \rightarrow$ magnesium chlorite
d. Molecular, 1 N and 3 F $\rightarrow$ nitrogen trifluoride
e. Ionic, ions are Ca^{2+} and $HSO_4^- \rightarrow$ calcium bisulfate or calcium hydrogen sulfate
f. Ionic, ions are K^+ and $ClO_4^- \rightarrow$ potassium perchlorate
g. Ionic, ions are Au^{3+} and $SO_3^{2-} \rightarrow$ gold(III) sulfite

6.157 a. $3\ H(1\ e^-) + 1\ N(5\ e^-) + 1\ C(4\ e^-) + 1\ O(6\ e^-) = 3 + 5 + 4 + 6 = 18$ valence electrons

$$\begin{array}{ccc} \text{H} & & {:}\ddot{\text{O}}{:} \\ | & & \| \\ \text{H}\!-\!\underset{\displaystyle ..}{\text{N}}\!-\!\text{C}\!-\!\text{H} \end{array}$$

b. $1\ Cl(7\ e^-) + 2\ H(1\ e^-) + 2\ C(4\ e^-) + 1\ N(5\ e^-) = 7 + 2 + 8 + 5 = 22$ valence electrons

$$\begin{array}{c} \text{H} \\ | \\ {:}\ddot{\ddot{\text{Cl}}}\!-\!\text{C}\!-\!\text{C}\!\equiv\!\text{N}{:} \\ | \\ \text{H} \end{array}$$

c. $2\ H(1\ e^-) + 2\ N(5\ e^-) = 2 + 10 = 12$ valence electrons

$$\text{H}\!-\!\ddot{\text{N}}\!=\!\ddot{\text{N}}\!-\!\text{H}$$

   d. 1 Cl(7 e^-) + 2 C(4 e^-) + 2 O(6 e^-) + 3 H(1 e^-) = 7 + 8 + 12 + 3 = 30 valence electrons

6.159 a. The central atom N has four electron groups with three bonded atoms and one lone pair of electrons, which gives NH_2Cl a trigonal pyramidal shape.

   b. The central Te atom has three electron groups with two bonded atoms and one lone pair of electrons, which gives TeO_2 a bent shape (120°).

(the double bond could be drawn to either of the O atoms)

Selected Answers to Combining Ideas from Chapters 4 to 6

CI.7 **a.** Y has the higher electronegativity, since it is a nonmetal in Group 7A (17).

b. X^{2+}, Y^-

c. X has an electron configuration of $1s^2 2s^2 2p^6 3s^2$.
Y has an electron configuration of $1s^2 2s^2 2p^6 3s^2 3p^5$.

d. X^{2+} has an electron configuration of $1s^2 2s^2 2p^6$.
Y^- has an electron configuration of $1s^2 2s^2 2p^6 3s^2 3p^6$.

e. X^{2+} has the same electron configuration as the noble gas neon (Ne).
Y^- has the same electron configuration as the noble gas argon (Ar).

f. $MgCl_2$, magnesium chloride

CI.9 **a.**

Isotope	Number of Protons	Number of Neutrons	Number of Electrons
$^{28}_{14}Si$	14	14	14
$^{29}_{14}Si$	14	15	14
$^{30}_{14}Si$	14	16	14

b. Electron configuration of $1s^2 2s^2 2p^6 3s^2 3p^2$

c. Since the atomic mass for Si is 28.02, the most abundant isotope of silicon is Si-28.

d. $^{31}_{14}Si \longrightarrow {}^{31}_{15}P + {}^{0}_{-1}e$

e. Half of a radioactive sample decays with each half-life:

16 μCi of $^{31}_{14}Si$ $\xrightarrow{\text{1 half-life}}$ 8.0 μCi of $^{31}_{14}Si$ $\xrightarrow{\text{2 half-lives}}$ 4.0 μCi of $^{31}_{14}Si$ $\xrightarrow{\text{3 half-lives}}$ 2.0 μCi of $^{31}_{14}Si$

Therefore, 3 half-lives have passed.

$$3 \text{ half-lives} \times \frac{2.6 \text{ h}}{1 \text{ half-life}} = 7.8 \text{ h (2 SFs)}$$

f.

$$:\!\ddot{C}l\!:$$
$$|$$
$$:\!\ddot{C}l\!-\!Si\!-\!\ddot{C}l\!:$$
$$|$$
$$:\!\ddot{C}l\!:$$

The central atom Si has four electron groups bonded to four chlorine atoms; $SiCl_4$ has a tetrahedral shape.

CI.11 **a.** $^{40}_{19}K \longrightarrow {}^{40}_{20}Ca + {}^{0}_{-1}e$, beta particle   and $^{40}_{19}K \longrightarrow {}^{40}_{18}Ar + {}^{0}_{+1}e$, positron

b. $1.6 \text{ oz K} \times \dfrac{1 \text{ lb K}}{16 \text{ oz K}} \times \dfrac{454 \text{ g K}}{1 \text{ lb K}} \times \dfrac{0.012 \text{ g } ^{40}_{19}K}{100 \text{ g K}} \times \dfrac{7.0 \text{ } \mu Ci}{1 \text{ g } ^{40}_{19}K} \times \dfrac{1 \text{ mCi}}{1000 \text{ } \mu Ci}$

$= 3.8 \times 10^{-5} \text{ mCi (2 SFs)}$

$1.6 \text{ oz K} \times \dfrac{1 \text{ lb K}}{16 \text{ oz K}} \times \dfrac{454 \text{ g K}}{1 \text{ lb K}} \times \dfrac{0.012 \text{ g } ^{40}_{19}K}{100 \text{ g K}} \times \dfrac{7.0 \text{ } \mu Ci}{1 \text{ g } ^{40}_{19}K} \times \dfrac{1 \text{ Ci}}{10^6 \text{ } \mu Ci} \times \dfrac{3.7 \times 10^{10} \text{ Bq}}{1 \text{ Ci}}$

$= 1.4 \times 10^3 \text{ Bq (2 SFs)}$

7
Chemical Reactions and Quantities

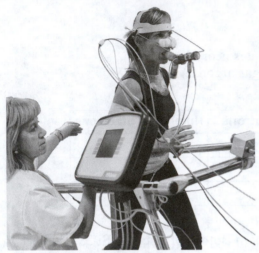

Credit: Javier Larrea/age fotostock

Since Natalie's diagnosis of mild emphysema caused by secondhand cigarette smoke, she is working with Angela, an exercise physiologist. Natalie's workout starts with low-intensity exercises that use smaller muscles. Now that Natalie has increased the oxygen concentration in her blood, Angela adds some exercises that use larger muscles, which need more O_2. Today Natalie had a stress test and an ECG to assess and evaluate her progress. Her results show that her heart is functioning more efficiently and that she needs less oxygen. Her doctor tells her that she has begun to reverse the progression of emphysema. How many moles of O_2 are in 48.0 g of O_2?

LOOKING AHEAD

7.1 Equations for Chemical Reactions

7.2 Types of Chemical Reactions

7.3 Oxidation–Reduction Reactions

7.4 The Mole

7.5 Molar Mass

7.6 Calculations Using Molar Mass

7.7 Mole Relationships in Chemical Equations

7.8 Mass Calculations for Chemical Reactions

7.9 Limiting Reactants and Percent Yield

7.10 Energy in Chemical Reactions

 The Health icon indicates a question that is related to health and medicine.

7.1 Equations for Chemical Reactions

Learning Goal: Write a balanced chemical equation from the formulas of the reactants and products for a reaction; determine the number of atoms in the reactants and products.

REVIEW

Writing Ionic Formulas (6.2)

Naming Ionic Compounds (6.3)

Writing the Names and Formulas for Molecular Compounds (6.5)

- When new substances form, a chemical reaction has taken place.

- Chemical change may be indicated by a formation of a gas (bubbles), change in color, formation of a solid, and/or heat produced or heat absorbed.

- A chemical equation shows the formulas of the reactants on the left side of the arrow and the formulas of the products on the right side, with + to separate two or more formulas.

- In a balanced equation, numbers called *coefficients*, which appear in front of the symbols or formulas, provide the same number of atoms for each kind of element on the reactant and product sides.

Bubbles (gas) form when $CaCO_3$ reacts with acid.

Credit: Sciencephotos/Alamy

Copyright © 2019 Pearson Education, Inc.

- Each formula in an equation is followed by an abbreviation, in parentheses, that gives the physical state of the substance: solid (*s*), liquid (*l*), or gas (*g*), and, if dissolved in water, an aqueous solution (*aq*).
- The Greek letter delta (Δ) over the arrow in an equation represents the application of heat to the reaction.

♦ **Learning Exercise 7.1A**

Indicate the number of atoms of each element on the reactant side and on the product side for each of the following balanced equations:

a. $CaCO_3(s) \longrightarrow CaO(s) + CO_2(g)$

Element	Atoms on Reactant Side	Atoms on Product Side
Ca	1	1
C	1	1
O	3	3

$2H_2(g) + O_2(g) \longrightarrow 2H_2O(g)$

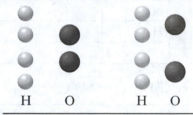

H   O   H   O

Reactant atoms = Product atoms

When there are the same numbers of each type of atom in the reactants as in the products, the equation is balanced.

b. $2Na(s) + H_2O(l) \longrightarrow Na_2O(s) + H_2(g)$

Element	Atoms on Reactant Side	Atoms on Product Side
Na	2	2
H	2	2
O	1	1

c. $C_5H_{12}(g) + 8O_2(g) \xrightarrow{\Delta} 5CO_2(g) + 6H_2O(g)$

Element	Atoms on Reactant Side	Atoms on Product Side
C	5	5
H	12	12
O	16	16

d. $2AgNO_3(aq) + K_2S(aq) \longrightarrow Ag_2S(s) + 2KNO_3(aq)$

Element	Atoms on Reactant Side	Atoms on Product Side
Ag		
N		
O		
K		
S		

e. $2Al(OH)_3(s) + 3H_2SO_4(aq) \longrightarrow 6H_2O(l) + Al_2(SO_4)_3(aq)$

Element	Atoms on Reactant Side	Atoms on Product Side
Al		
O		
H		
S		

Answers

a. $CaCO_3(s) \longrightarrow CaO(s) + CO_2(g)$

Element	Atoms on Reactant Side	Atoms on Product Side
Ca	1	1
C	1	1
O	3	3

b. $2Na(s) + H_2O(l) \longrightarrow Na_2O(s) + H_2(g)$

Element	Atoms on Reactant Side	Atoms on Product Side
Na	2	2
H	2	2
O	1	1

c. $C_5H_{12}(g) + 8O_2(g) \xrightarrow{\Delta} 5CO_2(g) + 6H_2O(g)$

Element	Atoms on Reactant Side	Atoms on Product Side
C	5	5
H	12	12
O	16	16

d. $2AgNO_3(aq) + K_2S(aq) \longrightarrow Ag_2S(s) + 2KNO_3(aq)$

Element	Atoms on Reactant Side	Atoms on Product Side
Ag	2	2
N	2	2
O	6	6
K	2	2
S	1	1

e. $2Al(OH)_3(s) + 3H_2SO_4(aq) \longrightarrow 6H_2O(l) + Al_2(SO_4)_3(aq)$

Element	Atoms on Reactant Side	Atoms on Product Side
Al	2	2
O	18	18
H	12	12
S	3	3

SAMPLE PROBLEM   Balancing a Chemical Equation

CORE CHEMISTRY SKILL
Balancing a Chemical Equation

Balance the following equation:

$$N_2(g) + 3H_2(g) \longrightarrow 2NH_3(g)$$

Solution:

Analyze the Problem	Given	Need	Connect
	reactants, products	balanced equation	equal numbers of atoms in reactants and products

STEP 1 **Write an equation using the correct formulas for the reactants and products.**

$$N_2(g) + H_2(g) \longrightarrow NH_3(g)$$

STEP 2 **Count the atoms of each element in the reactants and products.**

$$N_2(g) + H_2(g) \longrightarrow NH_3(g)$$

2N 2H   1N, 3H

STEP 3 **Use coefficients to balance each element.** Balance the N atoms by placing a coefficient of 2 in front of NH_3. (This increases the number of H atoms, too.) Recheck the number of N atoms and the number of H atoms.

$$N_2(g) + H_2(g) \longrightarrow 2NH_3(g)$$

2N 2H   2N, 6H

Balance the H atoms by placing a coefficient of 3 in front of H_2.

$$N_2(g) + 3H_2(g) \longrightarrow 2NH_3(g)$$

STEP 4 **Check the final equation to confirm it is balanced.**

$$N_2(g) + 3H_2(g) \longrightarrow 2NH_3(g)$$

2N 6H   2N, 6H Balanced

♦ **Learning Exercise 7.1B**

Balance each of the following equations by placing coefficients in front of the formulas as needed:

a. __2__ $MgO(s) \longrightarrow$ __2__ $Mg(s) +$ ____ $O_2(g)$

b. ____ $Zn(s) +$ __2__ $HCl(aq) \longrightarrow$ ____ $H_2(g) +$ ____ $ZnCl_2(aq)$

c. __2__ $Al(s) +$ __3__ $CuSO_4(aq) \longrightarrow$ __3__ $Cu(s) +$ ____ $Al_2(SO_4)_3(aq)$

d. ____ $Al_2S_3(s) +$ __6__ $H_2O(l) \longrightarrow$ __2__ $Al(OH)_3(s) +$ __3__ $H_2S(g)$

e. ____ $BaCl_2(aq) +$ ____ $Na_2SO_4(aq) \longrightarrow$ ____ $BaSO_4(s) +$ ____ $NaCl(aq)$

f. ____ $CO(g) +$ ____ $Fe_2O_3(s) \longrightarrow$ ____ $Fe(s) +$ ____ $CO_2(g)$

g. ____ $K(s) +$ ____ $H_2O(l) \longrightarrow$ ____ $H_2(g) +$ ____ $K_2O(aq)$

h. ____ $Fe(OH)_3(s) \longrightarrow$ ____ $H_2O(l) +$ ____ $Fe_2O_3(s)$

i. Galactose, $C_6H_{12}O_6(aq)$, is found in milk. In the body, galactose reacts with O_2 gas to form CO_2 gas and liquid water. Write a balanced chemical equation for the reaction.

$$C_6H_{12}O_6(aq) + 6O_2(g) \longrightarrow 6CO_2(g) + 6H_2O(l)$$

j. Maltose, $C_{12}H_{22}O_{11}(aq)$, is found in corn syrup. In the body, maltose reacts with O_2 gas to form CO_2 gas and liquid water. Write a balanced chemical equation for the reaction.

$$C_{12}H_{22}O_{11}(aq) + 12O_2(g) \longrightarrow 12CO_2(g) + 11H_2O(l)$$

Answers

a. $2MgO(s) \longrightarrow 2Mg(s) + O_2(g)$

b. $Zn(s) + 2HCl(aq) \longrightarrow H_2(g) + ZnCl_2(aq)$

c. $2Al(s) + 3CuSO_4(aq) \longrightarrow 3Cu(s) + Al_2(SO_4)_3(aq)$

d. $Al_2S_3(s) + 6H_2O(l) \longrightarrow 2Al(OH)_3(s) + 3H_2S(g)$

e. $BaCl_2(aq) + Na_2SO_4(aq) \longrightarrow BaSO_4(s) + 2NaCl(aq)$

f. $3CO(g) + Fe_2O_3(s) \longrightarrow 2Fe(s) + 3CO_2(g)$

g. $2K(s) + H_2O(l) \longrightarrow H_2(g) + K_2O(aq)$

h. $2Fe(OH)_3(s) \longrightarrow 3H_2O(l) + Fe_2O_3(s)$

i. $C_6H_{12}O_6(aq) + 6O_2(g) \longrightarrow 6CO_2(g) + 6H_2O(l)$

j. $C_{12}H_{22}O_{11}(aq) + 12O_2(g) \longrightarrow 12CO_2(g) + 11H_2O(l)$

7.2 Types of Chemical Reactions

Learning Goal: Identify a chemical reaction as a combination, decomposition, single replacement, double replacement, or combustion.

- Reactions are classified as combination, decomposition, single replacement, double replacement, or combustion.

- In a *combination* reaction, two or more reactants form one product:

    $A + B \longrightarrow AB$

- In a *decomposition* reaction, a reactant splits into two or more simpler products:

    $AB \longrightarrow A + B$

- In a *single replacement* reaction, an uncombined element takes the place of an element in a compound:

    $A + BC \longrightarrow AC + B$

- In a *double replacement* reaction, the positive ions in the reactants switch places:

    $AB + CD \longrightarrow AD + CB$

- In a *combustion* reaction, a compound of carbon and hydrogen reacts with oxygen to form CO_2, H_2O, and energy.

    $$C_XH_Y(g \text{ or } l) + ZO_2(g) \xrightarrow{\Delta} XCO_2(g) + \frac{Y}{2}H_2O(g) + energy$$

♦ **Learning Exercise 7.2A**

CORE CHEMISTRY SKILL
Classifying Types of Chemical Reactions

Match each of the following reactions with the type of reaction:

    a. combination   **b.** decomposition

    c. single replacement   **d.** double replacement   **e.** combustion

1. _____ $N_2(g) + 3H_2(g) \longrightarrow 2NH_3(g)$

2. _____ $CaCl_2(aq) + K_2CO_3(aq) \longrightarrow CaCO_3(s) + 2KCl(aq)$

3. _____ $2K_2O_2(aq) \longrightarrow 2K_2O(aq) + O_2(g)$

4. _____ $CuO(s) + H_2(g) \longrightarrow Cu(s) + H_2O(l)$

5. _____ $N_2(g) + O_2(g) \longrightarrow 2NO(g)$

6. ____ $C_2H_4(g) + 3O_2(g) \xrightarrow{\Delta} 2CO_2(g) + 2H_2O(g) +$ energy

7. ____ $PbCO_3(s) \longrightarrow PbO(s) + CO_2(g)$

8. ____ $2Al(s) + Fe_2O_3(s) \longrightarrow 2Fe(s) + Al_2O_3(s)$

Answers	**1.** a	**2.** d	**3.** b	**4.** c
	5. a	**6.** e	**7.** b	**8.** c

In a combustion reaction, a candle burns using the oxygen in the air.

Credit: Sergiy Zavgorodny/ Shutterstock

♦ **Learning Exercise 7.2B**

One way to remove tarnish from silver is to place the silver object on a piece of aluminum foil and add boiling water and some baking soda. The unbalanced chemical equation is the following:

$$Al(s) + Ag_2S(s) \longrightarrow Ag(s) + Al_2S_3(s)$$

a. Write the balanced chemical equation for the reaction.

b. What type of reaction takes place?

Answers   **a.** $2Al(s) + 3Ag_2S(s) \longrightarrow 6Ag(s) + Al_2S_3(s)$
b. single replacement

♦ **Learning Exercise 7.2C**

Octane, C_8H_{18}, a liquid compound in gasoline, burns in oxygen gas to produce the gases carbon dioxide and water, and energy.

a. Write the balanced chemical equation for the reaction.

b. What type of reaction takes place?

Answers   **a.** $2C_8H_{18}(l) + 25O_2(g) \xrightarrow{\Delta} 16CO_2(g) + 18H_2O(g) +$ energy
b. combustion

♦ **Learning Exercise 7.2D**

Predict the products of each of the following reactions and balance:

a. decomposition: $Al_2O_3(s) \longrightarrow$

b. single replacement: $Mg(s) + H_3PO_4(aq) \longrightarrow$

c. combustion: $C_6H_{12}O_2(l) + O_2(g) \xrightarrow{\Delta}$

d. combination: $Ca(s) + N_2(g) \longrightarrow$

Answers　**a.** decomposition: $2Al_2O_3(s) \longrightarrow 4Al(s) + 3O_2(g)$

b. single replacement: $6Mg(s) + 2H_3PO_4(aq) \longrightarrow 3H_2(g) + 2Mg_3(PO_4)_2(aq)$

c. combustion: $C_6H_{12}O_2(l) + 8O_2(g) \xrightarrow{\Delta} 6CO_2(g) + 6H_2O(g) + energy$

d. combination: $3Ca(s) + N_2(g) \longrightarrow Ca_3N_2(s)$

7.3　Oxidation–Reduction Reactions

Learning Goal: Define the terms oxidation and reduction; identify the reactants oxidized and reduced.

- In an oxidation–reduction reaction, there is a loss and gain of electrons. In an oxidation, electrons are lost. In a reduction, electrons are gained.
- An oxidation must always be accompanied by a reduction. The number of electrons lost in the oxidation reaction is equal to the number of electrons gained in the reduction reaction.
- In biological systems, the term *oxidation* describes the gain of oxygen or the loss of hydrogen. The term *reduction* is used to describe a loss of oxygen or a gain of hydrogen.

Key Terms for Sections 7.1 to 7.3

Match each of the following key terms with the correct description:

　a. combustion reaction　　**b.** chemical equation　　**c.** combination reaction

　d. decomposition reaction　　**e.** single replacement reaction　　**f.** oxidation–reduction reaction

1. _d_ a reaction in which a single reactant splits into two or more simpler substances

2. _e_ a reaction in which one element replaces a different element in a compound

3. _a_ a reaction in which a carbon-containing compound burns in oxygen from the air to produce carbon dioxide, water, and energy

4. _c_ a reaction in which reactants combine to form a single product

5. _b_ a shorthand method of writing a chemical reaction with the formulas of the reactants written on the left side of an arrow and the formulas of the products written on the right side

6. _f_ a reaction in which electrons are lost from one reactant and gained by another reactant

Answers　　**1.** d　　**2.** e　　**3.** a　　**4.** c　　**5.** b　　**6.** f

♦ **Learning Exercise 7.3**

For each of the following reactions, indicate whether the underlined element is *oxidized* or *reduced*:

> **CORE CHEMISTRY SKILL**
> Identifying Oxidized and Reduced Substances

a. $\underline{Fe}^{3+}(aq) + e^- \longrightarrow Fe^{2+}(aq)$　　　　Fe^{3+} is ___Reduced___

b. $2\underline{Cl}^-(aq) \longrightarrow Cl_2(g) + 2 e^-$　　　　Cl^- is ___Oxidized___

c. $\underline{Ca}O(s) + H_2(g) \longrightarrow Ca(s) + H_2O(l)$　　　Ca^{2+} is ___Reduced___

d. $4\underline{Al}(s) + 3O_2(g) \longrightarrow 2Al_2O_3(s)$　　　　Al is ___Oxidized___

e. $\underline{Cu}Cl_2(aq) + Zn(s) \longrightarrow ZnCl_2(aq) + Cu(s)$　　Cu^{2+} is ___Reduced___

f. $2H\underline{Br}(aq) + Cl_2(g) \longrightarrow 2HCl(aq) + Br_2(l)$ Br^- is   <u>Oxidized</u>

g. $2\underline{Na}(s) + Cl_2(g) \longrightarrow 2NaCl(s)$ Na is   <u>Oxidized</u>

Answers   **a.** reduced   **b.** oxidized   **c.** reduced   **d.** oxidized
            e. reduced   **f.** oxidized   **g.** oxidized

7.4 The Mole

REVIEW

Writing Numbers in Scientific Notation (1.5)
Counting Significant Figures (2.2)
Writing Conversion Factors from Equalities (2.5)
Using Conversion Factors (2.6)

Learning Goal: Use Avogadro's number to determine the number of particles in a given number of moles.

- One mole of any compound contains Avogadro's number, 6.02×10^{23}, of particles (atoms, molecules, ions, formula units).
- Avogadro's number provides conversion factors between moles and the number of particles:

$$1 \text{ mole} = 6.02 \times 10^{23} \text{ particles}$$

$$\frac{6.02 \times 10^{23} \text{ particles}}{1 \text{ mole}} \quad \text{and} \quad \frac{1 \text{ mole}}{6.02 \times 10^{23} \text{ particles}}$$

Calculating the Number of Atoms or Molecules	
STEP 1	State the given and needed quantities.
STEP 2	Write a plan to convert moles to atoms or molecules.
STEP 3	Use Avogadro's number to write conversion factors.
STEP 4	Set up the problem to calculate the number of particles.

♦ **Learning Exercise 7.4A**

CORE CHEMISTRY SKILL
Converting Particles to Moles

Use Avogadro's number to calculate each of the following:

a. number of Ca atoms in 3.00 moles of Ca

$$3 \text{ mol} \times \frac{6.02 \times 10^{23}}{1 \text{ mol}} = 1.8 \times 10^{24}$$

b. number of Zn atoms in 0.250 mole of Zn

$$0.250 \times \frac{6.02 \times 10^{23}}{1 \text{ mol}} = 1.51 \times 10^{23}$$

c. number of SO_2 molecules in 0.118 mole of SO_2

$$7.10 \times 10^{22}$$

d. number of moles of Ag in 4.88×10^{23} atoms of Ag

$$4.88 \times 10^{23} \times \frac{1 \text{ mol}}{6.02 \times 10^{23}} = 0.811$$

e. number of moles of NH_3 (ammonia) in 7.52×10^{23} molecules of NH_3

$$1.25$$

Answers

a. 1.81×10^{24} atoms of Ca
b. 1.51×10^{23} atoms of Zn
c. 7.10×10^{22} molecules of SO_2
d. 0.811 mole of Ag
e. 1.25 moles of NH_3

Study Note

The subscripts in the formula of a compound indicate the number of moles of each element in 1 mole of that compound. For example, we can write the equalities and conversion factors for the moles of elements in the formula Mg_3N_2 as

1 mole of Mg_3N_2 = 3 moles of Mg atoms

$\dfrac{3 \text{ moles Mg atoms}}{1 \text{ mole } Mg_3N_2}$ and $\dfrac{1 \text{ mole } Mg_3N_2}{3 \text{ moles Mg atoms}}$

1 mole of Mg_3N_2 = 2 moles of N atoms

$\dfrac{2 \text{ moles N atoms}}{1 \text{ mole } Mg_3N_2}$ and $\dfrac{1 \text{ mole } Mg_3N_2}{2 \text{ moles N atoms}}$

Calculating the Moles of an Element in a Compound	
STEP 1	State the given and needed quantities.
STEP 2	Write a plan to convert moles of compound to moles of an element.
STEP 3	Write the equalities and conversion factors using subscripts.
STEP 4	Set up the problem to calculate the moles of an element.

♦ **Learning Exercise 7.4B**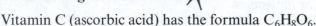

Vitamin C (ascorbic acid) has the formula $C_6H_8O_6$.

a. How many moles of C are in 2.0 moles of vitamin C?

2.0 mol $C_6H_8O_6 \times \dfrac{6 \text{ mol } C}{1 \text{ mol } C_6H_8O_6} = 12$ moles

b. How many moles of H are in 5.0 moles of vitamin C?

5.0 mol $C_6H_8O_6 \times \dfrac{8 \text{ mol } H}{1 \text{ mol } C_6H_8O_6} = 40$ moles

c. How many moles of O are in 1.5 moles of vitamin C?

1.5 mol $C_6H_8O_6 \times \dfrac{6 \text{ mol } O}{1 \text{ mol } C_6H_8O_6}$

Answers   a. 12 moles of C   b. 40. moles of H   c. 9.0 moles of O

♦ **Learning Exercise 7.4C**

For the compound ibuprofen ($C_{13}H_{18}O_2$) used in Advil and Motrin, determine each of the following:

a. moles of C in 2.20 moles of ibuprofen

2.20 mol $C_{13}H_{18}O_2 \times \dfrac{13 \text{ mol } C}{1 \text{ mol } C_{13}H_{18}O_2} = 28.6$ moles of C

b. moles of O in 0.750 mole of ibuprofen

1.50

c. moles of ibuprofen that contain 15 moles of H

$$15 \text{ mol } H \times \frac{1 \text{ mol } C_{13}H_{18}O_2}{18 \text{ mol } H} = 0.83$$

Answers   **a.** 28.6 moles of C   **b.** 1.50 moles of O   **c.** 0.83 mole of ibuprofen

7.5   Molar Mass

Learning Goal: Given the chemical formula of a substance, calculate its molar mass.

- The molar mass (g/mole) of an element is numerically equal to its atomic mass in grams.
- The molar mass (g/mole) of a compound is the sum of the molar mass (grams) for each element multiplied by the subscript in the formula.
- Round the molar mass to the hundredths place (0.01) or use at least four significant figures for calculations.

SAMPLE PROBLEM   Calculating Molar Mass

What is the molar mass of silver nitrate, $AgNO_3$?

Solution:

Analyze the Problem	Given	Need	Connect
	formula $AgNO_3$	molar mass of $AgNO_3$	periodic table

STEP 1 Obtain the molar mass of each element.

$$\frac{107.9 \text{ g Ag}}{1 \text{ mole Ag}} \qquad \frac{14.01 \text{ g N}}{1 \text{ mole N}} \qquad \frac{16.00 \text{ g O}}{1 \text{ mole O}}$$

STEP 2 Multiply each molar mass by the number of moles (subscript) in the formula.

$$1 \text{ mole Ag} \times \frac{107.9 \text{ g Ag}}{1 \text{ mole Ag}} = 107.9 \text{ g of Ag}$$

$$1 \text{ mole N} \times \frac{14.01 \text{ g N}}{1 \text{ mole N}} = 14.01 \text{ g of N}$$

$$3 \text{ moles O} \times \frac{16.00 \text{ g O}}{1 \text{ mole O}} = 48.00 \text{ g of O}$$

STEP 3 Calculate the molar mass by adding the masses of the elements.

Molar mass of $AgNO_3$ = 107.9 g + 14.01 g + 48.00 g = 169.9 g

♦ **Learning Exercise 7.5**

Calculate the molar mass for each of the following:

a. NaF, prevents dental caries

$$22.990 + 18.998 = 41.988 g$$

b. $FeC_4H_2O_4$, iron supplement

$$55.845 + 2 \times 1.008$$
$$4 \times 12.011 + 4 \times 15.999 = 169.901 g$$

c. $C_{13}H_{18}O_2$, ibuprofen, anti-inflammatory

d. $KMnO_4$, used to treat fungus

$C_{12}H_{22}O_{11}$

1 mole of sucrose contains 342.3 g of sucrose (table sugar).

Credit: Pearson Education/ Pearson Science

e. CaF_2, source of fluoride in drinking water   f. $C_6H_8O_6$, vitamin C

Answers   **a.** 41.99 g   **b.** 169.91 g   **c.** 206.27 g
                  d. 158.04 g   **e.** 78.08 g   **f.** 176.12 g

7.6   Calculations Using Molar Mass

Learning Goal: Use molar mass to convert between grams and moles.

- The molar mass is used as a conversion factor to change a given quantity in moles to grams or grams to moles.
- The two conversion factors for the molar mass of NaOH (40.00 g/mole) are:

$$\frac{40.00 \text{ g NaOH}}{1 \text{ mole NaOH}} \quad \text{and} \quad \frac{1 \text{ mole NaOH}}{40.00 \text{ g NaOH}}$$

SAMPLE PROBLEM   Converting Between Moles and Grams

What is the mass, in grams, of 0.254 mole of Na_2CO_3?

Solution:

STEP 1 State the given and needed quantities.

Analyze the Problem	Given	Need	Connect
	0.254 mole of Na_2CO_3	grams of Na_2CO_3	molar mass

STEP 2 Write a plan to convert moles to grams.

moles of Na_2CO_3 $\xrightarrow{\text{Molar mass}}$ grams of Na_2CO_3

STEP 3 Determine the molar mass and write conversion factors.

1 mole of Na_2CO_3 = 105.99 g of Na_2CO_3

$$\frac{105.99 \text{ g } Na_2CO_3}{1 \text{ mole } Na_2CO_3} \quad \text{and} \quad \frac{1 \text{ mole } Na_2CO_3}{105.99 \text{ g } Na_2CO_3}$$

STEP 4 Set up the problem to convert moles to grams.

$$0.254 \text{ mole } Na_2CO_3 \times \frac{105.99 \text{ g } Na_2CO_3}{1 \text{ mole } Na_2CO_3} = 26.9 \text{ g of } Na_2CO_3$$

♦ **Learning Exercise 7.6A**

Calculate the mass, in grams, for each of the following:

CORE CHEMISTRY SKILL
Using Molar Mass as a Conversion Factor

a. 0.75 mole of S   **b.** 3.18 moles of K_2SO_4

$$0.75 \text{ mol S} \times \frac{32.06 \text{ g S}}{1 \text{ mol S}} = 24 \text{ g}$$

c. 2.50 moles of NH_4Cl **d.** 4.08 moles of $FeCl_3$

e. 2.28 moles of PCl_3 **f.** 0.815 mole of $Mg(NO_3)_2$

Answers   **a.** 24 g   **b.** 554 g   **c.** 134 g   **d.** 662 g   **e.** 313 g   **f.** 121 g

◆ **Learning Exercise 7.6B**

Calculate the number of moles in each of the following quantities:

a. 108 g of C_5H_{12}

$$108g \times \frac{1 \, mol}{72,151} = 1.50$$

b. 6.12 g of CO_2

$$6.12g \times \frac{1 \, mol}{44.009} = 0.139$$

c. 38.7 g of $CaBr_2$ **d.** 236 g of Cl_2

e. 128 g of $Mg(OH)_2$ **f.** 172 g of Al_2O_3

g. The methane used to heat a home has a formula of CH_4. If 725 g of methane is used in one day, 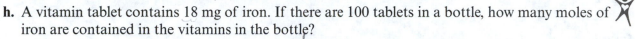 how many moles of methane are used during this time?

$$725 \times \frac{1 \, mol}{16.043} = 45.2$$

h. A vitamin tablet contains 18 mg of iron. If there are 100 tablets in a bottle, how many moles of iron are contained in the vitamins in the bottle?

$$0.018g \times \frac{1 \, mol}{55.845g} \times 100 = 0.032$$

Answers   **a.** 1.50 moles   **b.** 0.139 mole   **c.** 0.194 mole   **d.** 3.33 moles
                e. 2.19 moles   **f.** 1.69 moles   **g.** 45.2 moles   **h.** 0.032 mole

◆ **Learning Exercise 7.6C**

Octane, C_8H_{18}, is a component of gasoline.

a. How many moles of octane are in 34.7 g of octane?

$$34.7g \times \frac{1 \, mol}{114.232} = 0.304$$

b. How many grams of octane are in 0.737 mole of octane?

$$0.737 \, mol \, C_8H_{18} \times \frac{114.232g}{1 \, mol} = 84.2g$$

c. How many grams of C are in 135 g of octane?

d. How many grams of H are in 0.654 g of octane?

Answers   **a.** 0.304 mole   **b.** 84.2 g   **c.** 114 g   **d.** 0.104 g

7.7 Mole Relationships in Chemical Equations

Learning Goal: Use a mole–mole factor from a balanced chemical equation to calculate the number of moles of another substance in the reaction.

- The Law of Conservation of Mass states that there is no change in the total mass of the substances reacting in a chemical reaction.
- The coefficients in a balanced chemical equation indicate the moles of reactants and products in a reaction.
- Using the coefficients, mole–mole conversion factors can be written to relate any two substances in an equation.
- For the reaction of oxygen forming ozone, $3O_2(g) \longrightarrow 2O_3(g)$, the mole–mole factors are the following:

$$\frac{3 \text{ moles } O_2}{2 \text{ moles } O_3} \quad \text{and} \quad \frac{2 \text{ moles } O_3}{3 \text{ moles } O_2}$$

♦ **Learning Exercise 7.7A**

Write all of the possible mole–mole factors for the following equation:

$$N_2(g) + O_2(g) \longrightarrow 2NO(g)$$

Answers

For N_2 and O_2:

$$\frac{1 \text{ mole } N_2}{1 \text{ mole } O_2} \quad \text{and} \quad \frac{1 \text{ mole } O_2}{1 \text{ mole } N_2}$$

For N_2 and NO:

$$\frac{2 \text{ moles NO}}{1 \text{ mole } N_2} \quad \text{and} \quad \frac{1 \text{ mole } N_2}{2 \text{ moles NO}}$$

For O_2 and NO:

$$\frac{2 \text{ moles NO}}{1 \text{ mole } O_2} \quad \text{and} \quad \frac{1 \text{ mole } O_2}{2 \text{ moles NO}}$$

SAMPLE PROBLEM Calculating Moles of a Reactant or Product

For the reaction $N_2(g) + O_2(g) \longrightarrow 2NO(g)$, calculate the number of moles of NO produced from 3.0 moles of N_2.

Solution:

STEP 1 State the given and needed quantities (moles).

	Given	Need	Connect
Analyze the Problem	3.0 moles of N_2	moles of NO	mole–mole factor
	Equation		
	$N_2(g) + O_2(g) \longrightarrow 2NO(g)$		

STEP 2 Write a plan to convert the given to the needed quantity (moles).

$$\text{moles of } N_2 \xrightarrow{\text{Mole–mole factor}} \text{moles of NO}$$

STEP 3 Use coefficients to write mole–mole factors.

$$1 \text{ mole of } N_2 = 2 \text{ moles of NO}$$

$$\frac{2 \text{ moles NO}}{1 \text{ mole } N_2} \quad \text{and} \quad \frac{1 \text{ mole } N_2}{2 \text{ moles NO}}$$

STEP 4 Set up the problem to give the needed quantity (moles).

$$3.0 \text{ \sout{moles N}}_2 \times \frac{2 \text{ moles NO}}{1 \text{ \sout{mole N}}_2} = 6.0 \text{ moles of NO}$$

♦ Learning Exercise 7.7B

CORE CHEMISTRY SKILL
Using Mole–Mole Factors

Use the balanced chemical equation to answer the following questions:

$$C_3H_8(g) + 5O_2(g) \xrightarrow{\Delta} 3CO_2(g) + 4H_2O(g)$$

a. How many moles of O_2 are needed to react with 2.00 moles of C_3H_8?

$2.00 \text{ mol } C_3H_8 \times \frac{5 \text{ mol } O_2}{1 \text{ mol } C_3H_8} = 10.00 \text{ moles of } O_2$

b. How many moles of CO_2 are produced when 4.00 moles of O_2 reacts?

$4.00 \text{ mol } O_2 \times \frac{3 \text{ mol } CO_2}{5 \text{ mol } O_2} = 2.40 \text{ moles } CO_2$

c. How many moles of C_3H_8 react with 3.00 moles of O_2?

d. How many moles of H_2O are produced from 0.500 mole of C_3H_8?

Answers **a.** 10.0 moles of O_2 **b.** 2.40 moles of CO_2 **c.** 0.600 mole of C_3H_8
d. 2.00 moles of H_2O

7.8 Mass Calculations for Chemical Reactions

REVIEW
Using Significant Figures in Calculations (2.3)

Learning Goal: Given the mass in grams of a substance in a reaction, calculate the mass in grams of another substance in the reaction.

- The grams of a substance in an equation are converted to grams of another substance using their molar masses and mole–mole factors.

Substance A **Substance B**

$$\text{grams of A} \xrightarrow{\text{Molar mass A}} \text{moles of A} \xrightarrow{\text{Mole–mole factor B/A}} \text{moles of B} \xrightarrow{\text{Molar mass B}} \text{grams of B}$$

SAMPLE PROBLEM Calculating Mass of a Product

How many grams of H_2O are produced when 45.0 g of O_2 reacts with NH_3?

$$4NH_3(g) + 3O_2(g) \longrightarrow 2N_2(g) + 6H_2O(g)$$

Solution:

STEP 1 State the given and needed quantities (grams).

Analyze the Problem	**Given**	**Need**	**Connect**
	45.0 g of O_2	grams of H_2O	molar masses, mole–mole factor
	Equation		
	$4NH_3(g) + 3O_2(g) \longrightarrow 2N_2(g) + 6H_2O(g)$		

STEP 2 Write a plan to convert the given to the needed quantity (grams).

grams of O_2 $\xrightarrow{\text{Molar mass } O_2}$ moles of O_2 $\xrightarrow{\text{Mole–mole factor}}$ moles of H_2O $\xrightarrow{\text{Molar mass } H_2O}$ grams of H_2O

STEP 3 Use coefficients to write mole–mole factors; write molar masses.

1 mole of O_2 = 32.00 g of O_2 1 mole of H_2O = 18.02 g of H_2O

$$\frac{32.00 \text{ g } O_2}{1 \text{ mole } O_2} \text{ and } \frac{1 \text{ mole } O_2}{32.00 \text{ g } O_2} \qquad \frac{18.02 \text{ g } H_2O}{1 \text{ mole } H_2O} \text{ and } \frac{1 \text{ mole } H_2O}{18.02 \text{ g } H_2O}$$

3 moles of O_2 = 6 moles of H_2O

$$\frac{3 \text{ moles } O_2}{6 \text{ moles } H_2O} \text{ and } \frac{6 \text{ moles } H_2O}{3 \text{ moles } O_2}$$

STEP 4 Set up the problem to give the needed quantity (grams).

$$45.0 \text{ g } O_2 \times \underset{\text{4 SFs}}{\frac{1 \text{ mole } O_2}{32.00 \text{ g } O_2}} \times \frac{6 \text{ moles } H_2O}{3 \text{ moles } O_2} \times \underset{\text{Exact}}{\frac{18.02 \text{ g } H_2O}{1 \text{ mole } H_2O}} = 50.7 \text{ g of } H_2O$$

3 SFs   4 SFs   Exact   Exact   3 SFs

♦ **Learning Exercise 7.8**

CORE CHEMISTRY SKILL
Converting Grams to Grams

Use the balanced chemical equation to answer the questions below:

$$2C_2H_6(g) + 7O_2(g) \xrightarrow{\Delta} 4CO_2(g) + 6H_2O(g)$$

a. How many grams of O_2 are needed to react with 120. g of C_2H_6?

$$120 \text{g } C_2H_6 \times \frac{1 \text{ mol } C_2H_6}{30.07 \text{g } C_2H_6} \times \frac{7 \text{ mol } O_2}{2 \text{ mol } C_2H_6} \times \frac{31.998 \text{g } O_2}{1 \text{ mol } O_2} = 447 \text{g } O_2$$

b. How many grams of C_2H_6 are needed to react with 115 g of O_2?

$$115 \text{g } O_2 \times \frac{1 \text{ mol } O_2}{31.998 \text{g } O_2} \times \frac{2 \text{ mol } C_2H_6}{7 \text{ mol } O_2} \times \frac{30.07 \text{g } C_2H_6}{1 \text{ mol } C_2H_6} = 30.9 \text{ g } C_2H_6$$

c. How many grams of C_2H_6 react if 2.00 g of CO_2 is produced?

$$2.00 \text{g } CO_2 \times \frac{1 \text{ mol } CO_2}{44.009 \text{g } CO_2} \times \frac{2 \text{ mol } C_2H_6}{4 \text{ mol } CO_2} \times \frac{30.07 \text{g } C_2H_6}{1 \text{ mol } C_2H_6} = 0.683 \text{g } C_2H_6$$

d. How many grams of CO_2 are produced when 60.0 g of C_2H_6 reacts with sufficient oxygen?

$$60.0 \text{g } C_2H_6 \times \frac{1 \text{ mol } C_2H_6}{30.07 \text{g } C_2H_6} \times \frac{4 \text{ mol } CO_2}{2 \text{ mol } C_2H_6} \times \frac{44.009 \text{g } CO_2}{1 \text{ mol } CO_2} = 176 \text{g } CO_2$$

e. How many grams of H_2O are produced when 82.5 g of O_2 reacts with sufficient C_2H_6?

Answers   **a.** 447 g of O_2 **b.** 30.9 g of C_2H_6 **c.** 0.683 g of C_2H_6
              d. 176 g of CO_2 **e.** 39.8 g of H_2O

7.9 Limiting Reactants and Percent Yield

Learning Goal: Identify a limiting reactant when given the quantities of two reactants; calculate the amount of product formed from the limiting reactant. Given the actual quantity of product, calculate the percent yield for a reaction.

- In a limiting reactant problem, the availability of one of the reactants limits the amount of product.
- The reactant that is used up is the limiting reactant; the reactant that remains is the excess reactant.
- The limiting reactant produces the smaller number of moles of product.
- Theoretical yield is the maximum amount of product calculated for a given amount of a reactant.
- Percent yield is the ratio of the actual amount (yield) of product obtained to the theoretical yield.

SAMPLE PROBLEM   Calculating the Grams of Product from a Limiting Reactant

In the reaction $S(l) + 3F_2(g) \longrightarrow SF_6(g)$, how many grams of SF_6 can be produced when 16.1 g of S is reacted with 45.6 g of F_2?

Solution:

REVIEW
Calculating Percentages (1.4)

STEP 1   State the given and needed quantities.

	Given	Need	Connect
Analyze the Problem	16.1 g of S, 45.6 g of F_2	grams of SF_6	molar masses, mole–mole factors
	Equation		
	$S(l) + 3F_2(g) \longrightarrow SF_6(g)$		

STEP 2   Write a plan to convert the quantity (moles) of each reactant to quantity (moles) of product.

grams of S $\longrightarrow$ moles of S $\longrightarrow$ moles of SF_6 $\longrightarrow$ grams of SF_6

grams of F_2 $\longrightarrow$ moles of F_2 $\longrightarrow$ moles of SF_6 $\longrightarrow$ grams of SF_6

STEP 3   Use coefficients to write mole–mole factors; write molar mass factors.

1 mole of S = 1 mole of SF_6 3 moles of F_2 = 1 mole of SF_6

$$\frac{1 \text{ mole S}}{1 \text{ mole SF}_6} \text{ and } \frac{1 \text{ mole SF}_6}{1 \text{ mole S}} \qquad \frac{3 \text{ moles F}_2}{1 \text{ mole SF}_6} \text{ and } \frac{1 \text{ mole SF}_6}{3 \text{ moles F}_2}$$

1 mole of S = 32.07 g of S   1 mole of F_2 = 38.00 g of F_2 1 mole of SF_6 = 146.07 g of SF_6

$$\frac{32.07 \text{ g S}}{1 \text{ mole S}} \text{ and } \frac{1 \text{ mole S}}{32.07 \text{ g S}} \qquad \frac{38.00 \text{ g F}_2}{1 \text{ mole F}_2} \text{ and } \frac{1 \text{ mole F}_2}{38.00 \text{ g F}_2} \qquad \frac{146.07 \text{ g SF}_6}{1 \text{ mole SF}_6} \text{ and } \frac{1 \text{ mole SF}_6}{146.07 \text{ g SF}_6}$$

STEP 4   Calculate the quantity (grams) of product from each reactant, and select the smaller quantity (grams) as the limiting reactant.

$$16.1 \text{ g S} \times \frac{1 \text{ mole S}}{32.07 \text{ g S}} \times \frac{1 \text{ mole SF}_6}{1 \text{ mole S}} \times \frac{146.07 \text{ g SF}_6}{1 \text{ mole SF}_6} = 73.3 \text{ g of SF}_6$$

$$45.6 \text{ g F}_2 \times \frac{1 \text{ mole F}_2}{38.00 \text{ g F}_2} \times \frac{1 \text{ mole SF}_6}{3 \text{ moles F}_2} \times \frac{146.07 \text{ g SF}_6}{1 \text{ mole SF}_6} = 58.4 \text{ g of SF}_6 \text{ from the limiting reactant } F_2 \text{ (smaller amount of product)}$$

♦ Learning Exercise 7.9A

CORE CHEMISTRY SKILL
Calculating Quantity of Product from a Limiting Reactant

a. How many grams of Co_2S_3 can be produced from the reaction of 250. g of Co and 250. g of S?

$$2Co(s) + 3S(s) \longrightarrow Co_2S_3(s)$$

$$250\,g\ Co \times \frac{1\,mol\ Co}{58.933\,g\ Co} \times \frac{1\,mol\ Co_2S_3}{2\,mol\ Co} \times \frac{214.046\,g\ Co_2S_3}{1\,mol\ Co_2S_3} = 454\,g$$

$$250\,g\ S \times \frac{1\,mol\ S}{32.06\,g\ S} \times \frac{1\,mol\ Co_2S_3}{3\,mol\ S} \times \frac{214.046\,g\ Co_2S_3}{1\,mol\ Co_2S_3} = 556\,g$$

b. How many grams of NO_2 can be produced from the reaction of 32.0 g of NO and 24.0 g of O_2?

$$2NO(g) + O_2(g) \longrightarrow 2NO_2(g)$$

$$32.0\,g\ NO \times \frac{1\,mol\ NO}{30.006\,g\ NO} \times \frac{2\,mol\ NO_2}{2\,mol\ NO} \times \frac{46.007\,g\ NO_2}{1\,mol\ NO_2} = 49.1\,g$$

$$24.0\ O_2 \times \frac{1\,mol\ O_2}{31.998\,g\ O_2} \times \frac{2\,mol\ NO_2}{1\,mol\ O_2} \times \frac{46.007\,g\ NO_2}{1\,mol\ NO_2} = 69.0\,g$$

Answers **a.** 454 g **b.** 49.1 g

Study Note

The percent yield is the ratio of the actual yield obtained, which is given, to the theoretical yield, which is calculated for a given amount of starting reactant. If we calculate that a reaction can theoretically produce 43.1 g of NH_3, but the actual yield is 26.0 g of NH_3, the percent yield is:

$$\frac{26.0\text{ g } NH_3 \text{ (actual mass produced)}}{43.1\text{ g } NH_3 \text{ (theoretical mass)}} \times 100\% = 60.3\%$$

Calculating Percent Yield	
STEP 1	State the given and needed quantities.
STEP 2	Write a plan to calculate the theoretical yield and the percent yield.
STEP 3	Use coefficients to write mole–mole factors; write molar mass factors.
STEP 4	Calculate the percent yield by dividing the actual yield (given) by the theoretical yield and multiplying the result by 100%.

♦ Learning Exercise 7.9B

CORE CHEMISTRY SKILL
Calculating Percent Yield

Consider the following reaction:

$$2H_2S(g) + 3O_2(g) \longrightarrow 2SO_2(g) + 2H_2O(g)$$

a. If 60.0 g of H_2S reacts with sufficient oxygen and produces 45.5 g of SO_2, what is the percent yield of SO_2?

$$60.0\ H_2S \times \frac{1\ \text{mol}\ H_2S}{34.076\text{g}\ H_2S} \times \frac{2\ \text{mol}\ SO_2}{2\ \text{mol}\ H_2S} \times \frac{64.058\text{g}\ SO_2}{1\ \text{mol}\ SO_2} = 112\text{g} \qquad \frac{45.5}{112} \times 100 = 40.6\%$$

b. If 25.0 g of O_2 reacting with H_2S produces 18.6 g of SO_2, what is the percent yield of SO_2?

$$25.0\text{g}\ O_2 \times \frac{1\ \text{mol}\ O_2}{31.998\text{g}\ O_2} \times \frac{2\ \text{mol}\ SO_2}{3\ \text{mol}\ O_2} \times \frac{64.058\text{g}\ SO_4}{1\ \text{mol}\ SO_4} = 33.4\text{g} \qquad \frac{18.6}{33.4} \times 100 = 54.8\%$$

Consider the following reaction:

$$\underset{\text{Ethane}}{2C_2H_6(g)} + 7O_2(g) \xrightarrow{\Delta} 4CO_2(g) + 6H_2O(g)$$

c. If 125 g of C_2H_6 reacting with sufficient oxygen produces 175 g of CO_2, what is the percent yield of CO_2?

$$125\text{g}\ C_2H_6 \times \frac{1\ \text{mol}\ C_2H_6}{30.07\text{g}\ C_2H_6} \times \frac{4\ \text{mol}\ CO_2}{2\ \text{mol}\ C_2H_6} \times \frac{44.007\text{g}\ CO_2}{1\ \text{mol}\ CO_2} = 365 \qquad \frac{175}{365} \times 100 = 47.9\%$$

d. When 35.0 g of O_2 reacts with sufficient ethane to produce 12.5 g of H_2O, what is the percent yield of H_2O?

$$35.0\text{g}\ O_2 \times \frac{1\ \text{mol}\ O_2}{31.998\text{g}\ O_2} \times \frac{2\ \text{mol}\ C_2H_6}{7\ \text{mol}\ O_2} \times \frac{30.07\text{g}\ C_2H_6}{1\ \text{mol}\ C_2H_6} = 9.40$$

Answers **a.** 40.3% **b.** 55.7% **c.** 47.8% **d.** 74.0%

♦ Learning Exercise 7.9C

Valine, $C_5H_{11}NO_2$, an amino acid, reacts in the body according to the following reaction:

$$\underset{\text{Valine}}{2C_5H_{11}NO_2(aq)} + 12O_2(g) \longrightarrow 9CO_2(g) + 9H_2O(l) + \underset{\text{Urea}}{CH_4N_2O(aq)}$$

a. How many grams of O_2 will react with 15.0 g of valine?

$$15.0\text{g}\ C_5H_{11}NO_2 \times \frac{1\ \text{mol}\ C_5H_{11}NO_2}{117.146\ C_5H_{11}NO_2} \times \frac{12\ \text{mol}}{2\ \text{mol}} \times \frac{31.998\text{g}\ O_2}{1\ \text{mol}\ O_2} = 98.3\text{g}$$

b. How many grams of CO_2 are formed when 100. g of valine and 200. g of O_2 react?

100g $C_5H_{11}NO_2$ ✗

c. If 45.0 g of valine reacts to give 6.75 g of urea, what is the percent yield of urea?

Galactose, $C_6H_{12}O_6(aq)$, found in milk products, reacts in the body by the following reaction:

$$C_6H_{12}O_6(aq) + 6O_2(g) \longrightarrow 6CO_2(g) + 6H_2O(l)$$

d. How many grams of O_2 will react with 46.4 g of galactose?

e. How many grams of CO_2 are formed when 77.0 g of galactose and 100. g of O_2 react?

f. If 23.3 g of galactose reacts to give 26.3 g of CO_2, what is the percent yield of CO_2?

Answers	**a.** 24.6 g	**b.** 169 g	**c.** 58.7%
	d. 49.4 g	**e.** 113 g	**f.** 77.1%

7.10 Energy in Chemical Reactions

Learning Goal: Given the heat of reaction, calculate the loss or gain of heat for an exothermic or endothermic reaction.

- The heat of reaction is the energy released or absorbed when the reaction occurs.
- The heat of reaction is the energy difference between the reactants and the products.

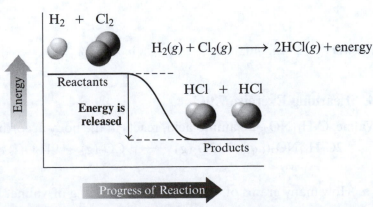

$$H_2(g) + Cl_2(g) \longrightarrow 2HCl(g) + energy$$

In an exothermic reaction, the energy of the products is lower than the energy of the reactants.

- In exothermic reactions, the energy of the products is lower than that of the reactants, which means that energy is released. The heat of reaction for an exothermic reaction is written as a $-\Delta H$ value with a negative sign ($-$), or by writing the heat of reaction as a product.
- In endothermic reactions, the energy of the products is higher than that of the reactants, which means that energy is absorbed. The heat of reaction for an endothermic reaction is written as a $+\Delta H$ value with a positive sign (+), or by writing the heat of reaction as a reactant.

Key Terms for Sections 7.4 to 7.10

Match each of the following key terms with the correct description:

    a. exothermic reaction   **b.** endothermic reaction   **c.** molar mass
    d. mole–mole factor   **e.** Avogadro's number   **f.** conservation of mass

1. _____ a reaction in which the energy of the reactants is less than that of the products

2. _____ the sum of the molar mass (grams) for each element in a compound multiplied by the subscript in the formula

3. _____ a reaction in which the energy of the reactants is greater than that of the products

4. _____ the law that states that there is no change in the mass of substances reacting

5. _____ the number of items in a mole, equal to 6.02×10^{23}

6. _____ a factor that relates the number of moles of two compounds derived from the coefficients in a balanced equation

Answers   **1.** b   **2.** c   **3.** a   **4.** f   **5.** e   **6.** d

♦ Learning Exercise 7.10A

Indicate whether each of the following is an endothermic or exothermic reaction:

1. $2H_2(g) + O_2(g) \longrightarrow 2H_2O(g) + 582 \text{ kJ}$ _____

2. $C_2H_4(g) + 176 \text{ kJ} \longrightarrow H_2(g) + C_2H_2(g)$ _____

3. $2C(s) + O_2(g) \longrightarrow 2CO(g) + 220 \text{ kJ}$ _____

4. $C_6H_{12}O_6(s) + 6O_2(g) \longrightarrow 6CO_2(g) + 6H_2O(l) + 670. \text{ kcal}$ _____

5. $C_2H_6O(l) + 21 \text{ kcal} \longrightarrow C_2H_4(g) + H_2O(g)$ _____

Answers   **1.** exothermic   **2.** endothermic   **3.** exothermic   **4.** exothermic   **5.** endothermic

Calculating Heat in a Reaction	
STEP 1	State the given and needed quantities.
STEP 2	Write a plan using the heat of reaction and any molar mass needed.
STEP 3	Write the conversion factors including heat of reaction.
STEP 4	Set up the problem to calculate the heat.

♦ **Learning Exercise 7.10B**

Consider the following reaction:

$$C_6H_{12}O_6(s) + 6O_2(g) \longrightarrow 6CO_2(g) + 6H_2O(l) + 670. \text{ kcal}$$

How much heat, in kilocalories and kilojoules, is produced when 50.0 g of CO_2 is formed?

Answers   127 kcal; 531 kJ

♦ **Learning Exercise 7.10C**

CORE CHEMISTRY SKILL
Using the Heat of Reaction

Consider each of the following reactions, and calculate the number of kilojoules absorbed or released for each:

1. $C_2H_4(g) + 176 \text{ kJ} \longrightarrow H_2(g) + C_2H_2(g)$

    a. 3.50 moles of C_2H_2 is produced.   **b.** 75.0 g of C_2H_4 reacts.

2. $2H_2(g) + O_2(g) \longrightarrow 2H_2O(g) + 582 \text{ kJ}$

    a. 0.820 mole of H_2 reacts.   **b.** 2.25 g of H_2O is produced.

Answers   **1. a.** 616 kJ absorbed **b.** 471 kJ absorbed   **2. a.** 239 kJ released   **b.** 36.4 kJ released

Checklist for Chapter 7

You are ready to take the Practice Test for Chapter 7. Be sure you have accomplished the following learning goals for this chapter. If not, review the Section listed at the end of the goal. Then apply your new skills and understanding to the Practice Test.

After studying Chapter 7, I can successfully:

_____ State a chemical equation in words, and calculate the total number of atoms of each element in the reactants and products. (7.1)

_____ Write a balanced equation for a chemical reaction from the formulas of the reactants and products. (7.1)

_____ Identify a reaction as a combination, decomposition, single replacement, double replacement, or combustion. (7.2)

_____ Identify an oxidation and reduction reaction. (7.3)

_____ Calculate the number of particles in a mole of a substance. (7.4)

_____ Calculate the molar mass when given the formula of a substance. (7.5)

_____ Convert the grams of a substance to moles; convert the moles of a substance to grams. (7.6)

_____ Use mole–mole factors for the mole relationships in an equation to calculate the moles of another substance in an equation for a chemical reaction. (7.7)

_____ Calculate the mass of a substance in an equation using mole–mole factors and molar masses. (7.8)

_____ Given the masses of reactants, find the limiting reactant and calculate the amount of product formed. (7.9)

_____ Calculate the percent yield, given the actual yield of a product. (7.9)

_____ Given the heat of reaction, describe a reaction as exothermic or endothermic. (7.10)

_____ Calculate the heat, in kilocalories or kilojoules, released or absorbed by a chemical reaction. (7.10)

Practice Test for Chapter 7

The chapter Sections to review are shown in parentheses at the end of each question.

For each of the unbalanced equations in questions 1 through 5, balance the equation and indicate the correct coefficient for the component in the equation written in **boldface type**: (7.1)

A. 1	**B.** 2	**C.** 3	**D.** 4	**E.** 5

1. _____ $Sn(s) + \textbf{Cl}_2(g) \longrightarrow SnCl_4(s)$

2. _____ $Al(s) + H_2O(l) \longrightarrow Al_2O_3(s) + \textbf{H}_2(g)$

3. _____ $C_3H_8(g) + \textbf{O}_2(g) \longrightarrow CO_2(g) + H_2O(g)$

4. _____ $\textbf{NH}_3(g) + O_2(g) \longrightarrow N_2(g) + H_2O(g)$

5. _____ $N_2O(g) \longrightarrow N_2(g) + \textbf{O}_2(g)$

For questions 6 through 10, classify each reaction as one of the following: (7.2)

A. combination	**B.** decomposition	**C.** single replacement
D. double replacement	**E.** combustion	

6. _____ $S(s) + O_2(g) \longrightarrow SO_2(g)$

7. _____ $C_3H_8(g) + 5O_2(g) \longrightarrow 3CO_2(g) + 4H_2O(g)$

8. _____ $CaCO_3(s) \longrightarrow CaO(s) + CO_2(g)$

9. _____ $Mg(s) + 2AgNO_3(aq) \longrightarrow Mg(NO_3)_2(aq) + 2Ag(s)$

10. _____ $Na_2S(aq) + Pb(NO_3)_2(aq) \longrightarrow PbS(s) + 2NaNO_3(aq)$

For questions 11 through 15, identify each reaction as an (O) oxidation or a (R) reduction: (7.3)

11. $Ca \longrightarrow Ca^{2+} + 2\,e^-$ _____   12. $Fe^{3+} + 3\,e^- \longrightarrow Fe$ _____

13. $Al^{3+} + 3\,e^- \longrightarrow Al$ _____   14. $Br_2 + 2\,e^- \longrightarrow 2Br^-$ _____

15. $Sn^{2+} \longrightarrow Sn^{4+} + 2\,e^-$ _____

16. The number of atoms of Na in 0.0500 mole of Na is (7.4)
    A. 0.500 atom   **B.** 6.02×10^{23} atoms   **C.** 3.01×10^{23} atoms
    D. 3.01×10^{22} atoms   **E.** 3.01×10^{25} atoms

17. The number of moles of oxygen (O) in 2.0 moles of $Al(OH)_3$ is (7.4)
    A. 1.0   **B.** 2.0   **C.** 3.0   **D.** 4.0   **E.** 6.0

For questions 18 through 21, calculate the moles of each of the elements in the formula: (7.4)

Tagamet is used to inhibit the production of acid in the stomach. Tagamet has the formula $C_{10}H_{16}N_6S$. Select the correct answer for the number of moles of each of the following in 0.25 mole of Tagamet.

Answers   **A.** 0.25   **B.** 0.50   **C.** 1.5   **D.** 2.5   **E.** 4.0

18. moles of C   **19.** moles of H   **20.** moles of N   **21.** moles of S

22. What is the molar mass of Li_2SO_4? (7.5)
 A. 55.01 g   **B.** 62.10 g   **C.** 103.01 g   **D.** 109.95 g   **E.** 103.11 g

23. What is the molar mass of $NaNO_3$? (7.5)
 A. 34.00 g   **B.** 37.00 g   **C.** 53.00 g   **D.** 75.00 g   **E.** 85.00 g

24. The number of grams in 0.600 mole of Cl_2 is (7.6)
 A. 71.0 g   **B.** 21.3 g   **C.** 42.5 g   **D.** 84.5 g   **E.** 4.30 g

25. How many grams are in 4.00 moles of NH_3? (7.6)
 A. 4.00 g   **B.** 17.1 g   **C.** 34.2 g   **D.** 68.4 g   **E.** 0.240 g

26. The number of moles of water in 3.60 g of H_2O is (7.6)
 A. 0.0500 mole   **B.** 0.100 mole   **C.** 0.200 mole
 D. 0.300 mole   **E.** 0.400 mole

27. 0.200 g of H_2 = _____ mole of H_2 (7.6)
 A. 0.100 mole   **B.** 0.200 mole   **C.** 0.400 mole
 D. 0.500 mole   **E.** 0.0100 mole

For questions 28 through 32, use the reaction: (7.7, 7.8)

$$C_2H_6O(l) + 3O_2(g) \longrightarrow 2CO_2(g) + 3H_2O(l)$$
 Ethanol

28. How many moles of O_2 are needed to react with 1.00 mole of ethanol?
 A. 1.00 mole   **B.** 2.00 moles   **C.** 3.00 moles   **D.** 4.00 moles   **E.** 5.00 moles

29. How many moles of H_2O are produced when 12.0 moles of O_2 reacts?
 A. 3.00 moles   **B.** 6.00 moles   **C.** 8.00 moles   **D.** 12.0 moles   **E.** 36.0 moles

30. How many moles of O_2 would be needed to produce 1.00 mole of CO_2?
 A. 0.670 mole   **B.** 1.00 mole   **C.** 1.50 moles   **D.** 2.00 moles   **E.** 3.00 moles

31. How many grams of CO_2 are produced when 92.0 g of ethanol reacts?
 A. 21.9 g   **B.** 43.8 g   **C.** 87.6 g   **D.** 92.6 g   **E.** 175 g

32. How many grams of water will be produced if 23.0 g of ethanol reacts?
 A. 54.0 g   **B.** 26.8 g   **C.** 18.0 g   **D.** 9.00 g   **E.** 6.00 g

For questions 33 and 34, use the reaction: $C_3H_8(g) + 5O_2(g) \longrightarrow 3CO_2(g) + 4H_2O(g)$ (7.9)

33. If 50.0 g of C_3H_8 and 150. g of O_2 react, the limiting reactant is (7.9)
 A. C_3H_8 **B.** O_2 **C.** both   **D.** neither   **E.** CO_2

34. If 50.0 g of C_3H_8 and 150. g of O_2 react, the mass of CO_2 formed is (7.9)
 A. 150. g   **B.** 206 g   **C.** 200. g   **D.** 124 g   **E.** 132 g

35. In the reaction $N_2(g) + 3H_2(g) \longrightarrow 2NH_3(g)$, a 25.0-g sample of N_2 is reacted. If the actual yield of NH_3 is 20.8 g, the percent yield of NH_3 is (7.9)
 A. 68.2%   **B.** 20.8%   **C.** 25.0%   **D.** 73.4%   **E.** 100.%

Use the following equation and heat of reaction for questions 36 through 39: (7.10)

$$C_2H_6O(l) + 3O_2(g) \longrightarrow 2CO_2(g) + 3H_2O(l) + 1420 \text{ kJ}$$

36. The reaction is
 A. thermic   **B.** endothermic   **C.** exothermic   **D.** nonthermic   **E.** subthermic

37. How much heat, in kilojoules, is released when 0.500 mole of C_2H_6O reacts?
 A. 710. kJ   **B.** 1420 kJ   **C.** 2130 kJ   **D.** 2840 kJ   **E.** 4260 kJ

38. How many grams of CO_2 are produced if 355 kJ of heat are released?
    A. 11.0 g   **B.** 22.0 g   **C.** 33.0 g   **D.** 44.0 g   **E.** 88.0 g

39. How much heat, in kilojoules, is released when 16.0 g of O_2 reacts?
    A. 4260 kJ   **B.** 1420 kJ   **C.** 710. kJ   **D.** 237 kJ   **E.** 118 kJ

For questions 40 through 42, indicate if the reactions are exothermic (**EX**) *or endothermic* (**EN**): (7.10)

40. $N_2(g) + 3H_2(g) \longrightarrow 2NH_3(g) + 84$ kJ

41. $2HCl(g) + 44$ kcal $\longrightarrow H_2(g) + Cl_2(g)$

42. $C_3H_8(g) + 5O_2(g) \longrightarrow 3CO_2(g) + 4H_2O(g) + 531$ kcal

Answers to the Practice Test

1. B	**2.** C	**3.** E	**4.** D	**5.** A
6. A	**7.** E	**8.** B	**9.** C	**10.** D
11. O	**12.** R	**13.** R	**14.** R	**15.** O
16. D	**17.** E	**18.** D	**19.** E	**20.** C
21. A	**22.** D	**23.** E	**24.** C	**25.** D
26. C	**27.** A	**28.** C	**29.** D	**30.** C
31. E	**32.** B	**33.** B	**34.** D	**35.** A
36. C	**37.** A	**38.** B	**39.** D	**40.** EX
41. EN	**42.** EX			

Selected Answers and Solutions to Text Problems

7.1 An equation is balanced when there are equal numbers of atoms of each element on the reactant side and on the product side.
   a. not balanced (2 O ≠ 3 O)   **b.** balanced
   c. not balanced (2 O ≠ 1 O)   **d.** balanced

7.3 Place coefficients in front of formulas until you make the atoms of each element equal on each side of the equation.
   a. $N_2(g) + O_2(g) \longrightarrow 2NO(g)$ **b.** $2HgO(s) \longrightarrow 2Hg(l) + O_2(g)$
   c. $4Fe(s) + 3O_2(g) \longrightarrow 2Fe_2O_3(s)$ **d.** $2Na(s) + Cl_2(g) \longrightarrow 2NaCl(s)$

7.5 **a.** $Mg(s) + 2AgNO_3(aq) \longrightarrow Mg(NO_3)_2(aq) + 2Ag(s)$
   b. $2Al(s) + 3CuSO_4(aq) \longrightarrow Al_2(SO_4)_3(aq) + 3Cu(s)$
   c. $Pb(NO_3)_2(aq) + 2NaCl(aq) \longrightarrow PbCl_2(s) + 2NaNO_3(aq)$
   d. $6HCl(aq) + 2Al(s) \longrightarrow 3H_2(g) + 2AlCl_3(aq)$

7.7 **a.** This is a decomposition reaction where a single reactant splits into two simpler substances.
   b. This is a single replacement reaction where one element in the reacting compound (I in BaI_2) is replaced by the other reactant (Br in Br_2).
   c. This is a combustion reaction where a carbon-containing fuel burns in oxygen to produce carbon dioxide and water.
   d. This is a double replacement reaction where the positive ions in the two reactants exchange places to form two products.
   e. This is a combination reaction where atoms of two different elements bond to form one product.

7.9 **a.** combination   **b.** single replacement   **c.** decomposition
   d. double replacement   **e.** combustion

7.11 **a.** Since this is a combination reaction, the product is a single compound.
   $Mg(s) + Cl_2(g) \longrightarrow MgCl_2(s)$
   b. Since this is a decomposition reaction, the single reactant breaks down into more than one simpler substance.
   $2HBr(g) \longrightarrow H_2(g) + Br_2(l)$
   c. Since this is a single replacement reaction, the uncombined metal element takes the place of the metal element in the compound.
   $Mg(s) + Zn(NO_3)_2(aq) \longrightarrow Zn(s) + Mg(NO_3)_2(aq)$
   d. Since this is a double replacement reaction, the positive ions in the reacting compounds switch places.
   $K_2S(aq) + Pb(NO_3)_2(aq) \longrightarrow PbS(s) + 2KNO_3(aq)$
   e. Since this is a combustion reaction, a fuel burns in oxygen to produce carbon dioxide and water.
   $2C_2H_6(g) + 7O_2(g) \xrightarrow{\Delta} 4CO_2(g) + 6H_2O(g)$

7.13 Oxidation is the loss of electrons; reduction is the gain of electrons.
   a. Na^+ gains an electron to form Na; this is a reduction.
   b. Ni loses electrons to form Ni^{2+}; this is an oxidation.
   c. Cr^{3+} gains electrons to form Cr; this is a reduction.
   d. $2H^+$ gain electrons to form H_2; this is a reduction.

7.15 An oxidized substance has lost electrons; a reduced substance has gained electrons.
   a. Zn loses electrons and is oxidized. Cl_2 gains electrons and is reduced.
   b. Br^- (in NaBr) loses electrons and is oxidized. Cl_2 gains electrons and is reduced.
   c. O^{2-} (in PbO) loses electrons and is oxidized. Pb^{2+} (in PbO) gains electrons and is reduced.
   d. Sn^{2+} loses electrons and oxidized. Fe^{3+} gains electrons and is reduced.

7.17 a. Fe^{3+} gains an electron to form Fe^{2+}; this is a reduction.
   b. Fe^{2+} loses an electron to form Fe^{3+}; this is an oxidation.

7.19 Linoleic acid gains hydrogen atoms and is reduced.

7.21 One mole contains 6.02×10^{23} atoms of an element, molecules of a molecular substance, or formula units of an ionic substance.

7.23 a. $0.500 \text{ mole C} \times \dfrac{6.02 \times 10^{23} \text{ atoms C}}{1 \text{ mole C}} = 3.01 \times 10^{23}$ atoms of C (3 SFs)

b. $1.28 \text{ moles SO}_2 \times \dfrac{6.02 \times 10^{23} \text{ molecules SO}_2}{1 \text{ mole SO}_2} = 7.71 \times 10^{23}$ molecules of SO_2 (3 SFs)

c. $5.22 \times 10^{22} \text{ atoms Fe} \times \dfrac{1 \text{ mole Fe}}{6.02 \times 10^{23} \text{ atoms Fe}} = 0.0867$ mole of Fe (3 SFs)

d. $8.50 \times 10^{24} \text{ molecules C}_2\text{H}_6\text{O} \times \dfrac{1 \text{ mole C}_2\text{H}_6\text{O}}{6.02 \times 10^{23} \text{ molecules C}_2\text{H}_6\text{O}} = 14.1$ moles of C_2H_6O (3 SFs)
(3 SFs)

7.25 1 mole of H_3PO_4 contains 3 moles of H atoms, 1 mole of P atoms, and 4 moles of O atoms.

a. $2.00 \text{ moles H}_3\text{PO}_4 \times \dfrac{3 \text{ moles H}}{1 \text{ mole H}_3\text{PO}_4} = 6.00$ moles of H (3 SFs)

b. $2.00 \text{ moles H}_3\text{PO}_4 \times \dfrac{4 \text{ moles O}}{1 \text{ mole H}_3\text{PO}_4} = 8.00$ moles of O (3 SFs)

c. $2.00 \text{ moles H}_3\text{PO}_4 \times \dfrac{1 \text{ mole P}}{1 \text{ mole H}_3\text{PO}_4} \times \dfrac{6.02 \times 10^{23} \text{ atoms P}}{1 \text{ mole P}} = 1.20 \times 10^{24}$ atoms of P
(3 SFs)

d. $2.00 \text{ moles H}_3\text{PO}_4 \times \dfrac{4 \text{ moles O}}{1 \text{ mole H}_3\text{PO}_4} \times \dfrac{6.02 \times 10^{23} \text{ atoms O}}{1 \text{ mole O}} = 4.82 \times 10^{24}$ atoms of O
(3 SFs)

7.27 The subscripts indicate the moles of each element in one mole of that compound.

a. $1.5 \text{ moles quinine} \times \dfrac{24 \text{ moles H}}{1 \text{ mole quinine}} = 36$ moles of H (2 SFs)

b. $5.0 \text{ moles quinine} \times \dfrac{20 \text{ moles C}}{1 \text{ mole quinine}} = 1.0 \times 10^2$ moles of C (2 SFs)

c. $0.020 \text{ mole quinine} \times \dfrac{2 \text{ moles N}}{1 \text{ mole quinine}} = 0.040$ mole of N (2 SFs)

7.29 a. $2.30 \text{ moles naproxen} \times \dfrac{14 \text{ moles C}}{1 \text{ mole naproxen}} = 32.2$ moles of C (3 SFs)

b. $0.444 \text{ mole naproxen} \times \dfrac{14 \text{ moles H}}{1 \text{ mole naproxen}} = 6.22$ moles of H (3 SFs)

c. $0.0765 \text{ mole naproxen} \times \dfrac{3 \text{ moles O}}{1 \text{ mole naproxen}} = 0.230$ mole of O (3 SFs)

7.31 a. $2 \text{ moles Cl} \times \dfrac{35.45 \text{ g Cl}}{1 \text{ mole Cl}} = 70.90$ g of Cl

$\therefore$ Molar mass of $Cl_2 = 70.90$ g

b. $3 \text{ moles C} \times \dfrac{12.01 \text{ g C}}{1 \text{ mole C}} = 36.03 \text{ g of C}$

$6 \text{ moles H} \times \dfrac{1.008 \text{ g H}}{1 \text{ mole H}} = 6.048 \text{ g of H}$

$3 \text{ moles O} \times \dfrac{16.00 \text{ g O}}{1 \text{ mole O}} = 48.00 \text{ g of O}$

3 moles of C $\qquad = 36.03 \text{ g of C}$

6 moles of H $\qquad = 6.048 \text{ g of H}$

3 moles of O $\qquad = \underline{48.00 \text{ g of O}}$

$\therefore$ Molar mass of $C_3H_6O_3 = 90.08 \text{ g}$

c. $3 \text{ moles Mg} \times \dfrac{24.31 \text{ g Mg}}{1 \text{ mole Mg}} = 72.93 \text{ g of Mg}$

$2 \text{ moles P} \times \dfrac{30.97 \text{ g P}}{1 \text{ mole P}} = 61.94 \text{ g of P}$

$8 \text{ moles O} \times \dfrac{16.00 \text{ g O}}{1 \text{ mole O}} = 128.0 \text{ g of O}$

3 moles of Mg $\qquad = 72.93 \text{ g of Mg}$

2 moles of P $\qquad = 61.94 \text{ g of P}$

8 moles of O $\qquad = \underline{128.0 \text{ g of O}}$

$\therefore$ Molar mass of $Mg_3(PO_4)_2 = 262.9 \text{ g}$

7.33 **a.** $1 \text{ mole Al} \times \dfrac{26.98 \text{ g Al}}{1 \text{ mole Al}} = 26.98 \text{ g of Al}$

$3 \text{ moles F} \times \dfrac{19.00 \text{ g F}}{1 \text{ mole F}} = 57.00 \text{ g of F}$

1 mole of Al $\qquad = 26.98 \text{ g of Al}$

3 moles of F $\qquad = \underline{57.00 \text{ g of F}}$

$\therefore$ Molar mass of $AlF_3 \quad = 83.98 \text{ g}$

b. $2 \text{ moles C} \times \dfrac{12.01 \text{ g C}}{1 \text{ mole C}} = 24.02 \text{ g of C}$

$4 \text{ moles H} \times \dfrac{1.008 \text{ g H}}{1 \text{ mole H}} = 4.032 \text{ g of H}$

$2 \text{ moles Cl} \times \dfrac{35.45 \text{ g Cl}}{1 \text{ mole Cl}} = 70.90 \text{ g of Cl}$

2 moles of C $\qquad = 24.02 \text{ g of C}$

4 moles of H $\qquad = 4.032 \text{ g of H}$

2 moles of Cl $\qquad = \underline{70.90 \text{ g of Cl}}$

$\therefore$ Molar mass of $C_2H_4Cl_2 = 98.95 \text{ g}$

c. $1 \text{ mole Sn} \times \dfrac{118.7 \text{ g Sn}}{1 \text{ mole Sn}} = 118.7 \text{ g of Sn}$

$2 \text{ moles F} \times \dfrac{19.00 \text{ g F}}{1 \text{ mole F}} = 38.00 \text{ g of F}$

1 mole of Sn $\qquad = 118.7 \text{ g of Sn}$

2 moles of F $\qquad = \underline{38.00 \text{ g of F}}$

$\therefore$ Molar mass of $SnF_2 \quad = 156.7 \text{ g}$

7.35 **a.** $1 \text{ mole K} \times \dfrac{39.10 \text{ g K}}{1 \text{ mole K}} = 39.10 \text{ g of K}$

$1 \text{ mole Cl} \times \dfrac{35.45 \text{ g Cl}}{1 \text{ mole Cl}} = 35.45 \text{ g of Cl}$

$\begin{aligned}
1 \text{ mole of K} &= 39.10 \text{ g of K} \\
1 \text{ mole of Cl} &= \underline{35.45 \text{ g of Cl}} \\
\therefore \text{ Molar mass of KCl} &= 74.55 \text{ g}
\end{aligned}$

b. $6 \text{ moles C} \times \dfrac{12.01 \text{ g C}}{1 \text{ mole C}} = 72.06 \text{ g of C}$

$5 \text{ moles H} \times \dfrac{1.008 \text{ g H}}{1 \text{ mole H}} = 5.040 \text{ g of H}$

$1 \text{ mole N} \times \dfrac{14.01 \text{ g N}}{1 \text{ mole N}} = 14.01 \text{ g of N}$

$2 \text{ moles O} \times \dfrac{16.00 \text{ g O}}{1 \text{ mole O}} = 32.00 \text{ g of O}$

$\begin{aligned}
6 \text{ moles of C} &= 72.06 \text{ g of C} \\
5 \text{ moles of H} &= 5.040 \text{ g of H} \\
1 \text{ mole of N} &= 14.01 \text{ g of N} \\
2 \text{ moles of O} &= \underline{32.00 \text{ g of O}} \\
\therefore \text{ Molar mass of } C_6H_5NO_2 &= 123.11 \text{ g}
\end{aligned}$

c. $19 \text{ moles C} \times \dfrac{12.01 \text{ g C}}{1 \text{ mole C}} = 228.2 \text{ g of C}$

$20 \text{ moles H} \times \dfrac{1.008 \text{ g H}}{1 \text{ mole H}} = 20.16 \text{ g of H}$

$1 \text{ mole F} \times \dfrac{19.00 \text{ g F}}{1 \text{ mole F}} = 19.00 \text{ g of F}$

$1 \text{ mole N} \times \dfrac{14.01 \text{ g N}}{1 \text{ mole N}} = 14.01 \text{ g of N}$

$3 \text{ moles O} \times \dfrac{16.00 \text{ g O}}{1 \text{ mole O}} = 48.00 \text{ g of O}$

$\begin{aligned}
19 \text{ moles of C} &= 228.2 \text{ g of C} \\
20 \text{ moles of H} &= 20.16 \text{ g of H} \\
1 \text{ mole of F} &= 19.00 \text{ g of F} \\
1 \text{ mole of N} &= 14.01 \text{ g of N} \\
3 \text{ moles of O} &= \underline{48.00 \text{ g of O}} \\
\therefore \text{ Molar mass of } C_{19}H_{20}FNO_3 &= 329.4 \text{ g}
\end{aligned}$

7.37 **a.** $2 \text{ moles Al} \times \dfrac{26.98 \text{ g Al}}{1 \text{ mole Al}} = 53.96 \text{ g of Al}$

$3 \text{ moles S} \times \dfrac{32.07 \text{ g S}}{1 \text{ mole S}} = 96.21 \text{ g of S}$

$12 \text{ moles O} \times \dfrac{16.00 \text{ g O}}{1 \text{ mole O}} = 192.0 \text{ g of O}$

$\begin{aligned}
2 \text{ moles of Al} &= 53.96 \text{ g of Al} \\
3 \text{ moles of S} &= 96.21 \text{ g of S} \\
12 \text{ moles of O} &= \underline{192.0 \text{ g of O}} \\
\therefore \text{ Molar mass of } Al_2(SO_4)_3 &= 342.2 \text{ g}
\end{aligned}$

b. $1 \text{ mole K} \times \dfrac{39.10 \text{ g K}}{1 \text{ mole K}} = 39.10 \text{ g of K}$

$4 \text{ moles C} \times \dfrac{12.01 \text{ g C}}{1 \text{ mole C}} = 48.04 \text{ g of C}$

$5 \text{ moles H} \times \dfrac{1.008 \text{ g H}}{1 \text{ mole H}} = 5.040 \text{ g of H}$

$6 \text{ moles O} \times \dfrac{16.00 \text{ g O}}{1 \text{ mole O}} = 96.00 \text{ g of O}$

1 mole of K	=	39.10 g of K
4 moles of C	=	48.04 g of C
5 moles of H	=	5.040 g of H
6 moles of O	=	96.00 g of O

$\therefore$ Molar mass of $KC_4H_5O_6$ = 188.18 g

c. $16 \text{ moles C} \times \dfrac{12.01 \text{ g C}}{1 \text{ mole C}} = 192.2 \text{ g of C}$

$19 \text{ moles H} \times \dfrac{1.008 \text{ g H}}{1 \text{ mole H}} = 19.15 \text{ g of H}$

$3 \text{ moles N} \times \dfrac{14.01 \text{ g N}}{1 \text{ mole N}} = 42.03 \text{ g of N}$

$5 \text{ moles O} \times \dfrac{16.00 \text{ g O}}{1 \text{ mole O}} = 80.00 \text{ g of O}$

$1 \text{ mole S} \times \dfrac{32.07 \text{ g S}}{1 \text{ mole S}} = 32.07 \text{ g of S}$

16 moles of C	=	192.2 g of C
19 moles of H	=	19.15 g of H
3 moles of N	=	42.03 g of N
5 moles of O	=	80.00 g of O
1 mole of S	=	32.07 g of S

$\therefore$ Molar mass of $C_{16}H_{19}N_3O_5S$ = 365.5 g

7.39 a. $8 \text{ moles C} \times \dfrac{12.01 \text{ g C}}{1 \text{ mole C}} = 96.08 \text{ g of C}$

$9 \text{ moles H} \times \dfrac{1.008 \text{ g H}}{1 \text{ mole H}} = 9.072 \text{ g of H}$

$1 \text{ mole N} \times \dfrac{14.01 \text{ g N}}{1 \text{ mole N}} = 14.01 \text{ g of N}$

$2 \text{ moles O} \times \dfrac{16.00 \text{ g O}}{1 \text{ mole O}} = 32.00 \text{ g of O}$

8 moles of C	=	96.08 g of C
9 moles of H	=	9.072 g of H
1 mole of N	=	14.01 g of N
2 moles of O	=	32.00 g of O

$\therefore$ Molar mass of $C_8H_9NO_2$ = 151.16 g

b. $3 \text{ moles Ca} \times \dfrac{40.08 \text{ g Ca}}{1 \text{ mole Ca}} = 120.2 \text{ g of Ca}$

$12 \text{ moles C} \times \dfrac{12.01 \text{ g C}}{1 \text{ mole C}} = 144.1 \text{ g of C}$

$$10 \text{ moles H} \times \frac{1.008 \text{ g H}}{1 \text{ mole H}} = 10.08 \text{ g of H}$$

$$14 \text{ moles O} \times \frac{16.00 \text{ g O}}{1 \text{ mole O}} = 224.0 \text{ g of O}$$

3 moles of Ca	= 120.2 g of Ca
12 moles of C	= 144.1 g of C
10 moles of H	= 10.08 g of H
14 moles of O	= 224.0 g of O

$\therefore$ Molar mass of $Ca_3(C_6H_5O_7)_2$ = 498.4 g

c. $17 \text{ moles C} \times \dfrac{12.01 \text{ g C}}{1 \text{ mole C}} = 204.2 \text{ g of C}$

$18 \text{ moles H} \times \dfrac{1.008 \text{ g H}}{1 \text{ mole H}} = 18.14 \text{ g of H}$

$1 \text{ mole F} \times \dfrac{19.00 \text{ g F}}{1 \text{ mole F}} = 19.00 \text{ g of F}$

$3 \text{ moles N} \times \dfrac{14.01 \text{ g N}}{1 \text{ mole N}} = 42.03 \text{ g of N}$

$3 \text{ moles O} \times \dfrac{16.00 \text{ g O}}{1 \text{ mole O}} = 48.00 \text{ g of O}$

17 moles of C	= 204.2 g of C
18 moles of H	= 18.14 g of H
1 mole of F	= 19.00 g of F
3 moles of N	= 42.03 g of N
3 moles of O	= 48.00 g of O

$\therefore$ Molar mass of $C_{17}H_{18}FN_3O_3$ = 331.4 g

7.41 a. $1.50 \text{ moles Na} \times \dfrac{22.99 \text{ g Na}}{1 \text{ mole Na}} = 34.5 \text{ g of Na (3 SFs)}$

b. $2.80 \text{ moles Ca} \times \dfrac{40.08 \text{ g Ca}}{1 \text{ mole Ca}} = 112 \text{ g of Ca (3 SFs)}$

c. Molar mass of CO_2

= 1 mole of C + 2 moles of O = 12.01 g + 2(16.00 g) = 44.01 g

$0.125 \text{ mole CO}_2 \times \dfrac{44.01 \text{ g CO}_2}{1 \text{ mole CO}_2} = 5.50 \text{ g of CO}_2 \text{ (3 SFs)}$

d. Molar mass of Na_2CO_3

= 2 moles of Na + 1 mole of C + 3 moles of O = 2(22.99 g) + 12.01 g + 3(16.00 g)

= 105.99 g

$0.0485 \text{ mole Na}_2\text{CO}_3 \times \dfrac{105.99 \text{ g Na}_2\text{CO}_3}{1 \text{ mole Na}_2\text{CO}_3} = 5.14 \text{ g of Na}_2\text{CO}_3 \text{ (3 SFs)}$

e. Molar mass of PCl_3

= 1 mole of P + 3 moles of Cl = 30.97 g + 3(35.45 g) = 137.32 g

$7.14 \times 10^2 \text{ moles PCl}_3 \times \dfrac{137.32 \text{ g PCl}_3}{1 \text{ mole PCl}_3} = 9.80 \times 10^4 \text{ g of PCl}_3 \text{ (3 SFs)}$

7.43 **a.** $0.150 \text{ mole Ne} \times \dfrac{20.18 \text{ g Ne}}{1 \text{ mole Ne}} = 3.03 \text{ g of Ne}$

b. Molar mass of I_2

$= 2 \text{ moles of I} = 2(126.9 \text{ g}) = 253.8 \text{ g}$

$0.150 \text{ mole } I_2 \times \dfrac{253.8 \text{ g } I_2}{1 \text{ mole } I_2} = 38.1 \text{ g of } I_2 \text{ (3 SFs)}$

c. Molar mass of Na_2O

$= 2 \text{ moles of Na} + 1 \text{ mole of O} = 2(22.99 \text{ g}) + 16.00 \text{ g} = 61.98 \text{ g}$

$0.150 \text{ mole } Na_2O \times \dfrac{61.98 \text{ g } Na_2O}{1 \text{ mole } Na_2O} = 9.30 \text{ g of } Na_2O \text{ (3 SFs)}$

d. Molar mass of $Ca(NO_3)_2$

$= 1 \text{ mole of Ca} + 2 \text{ moles of N} + 6 \text{ moles of O} = 40.08 \text{ g} + 2(14.01 \text{ g}) + 6(16.00 \text{ g}) = 164.10 \text{ g}$

$0.150 \text{ mole } Ca(NO_3)_2 \times \dfrac{164.10 \text{ g } Ca(NO_3)_2}{1 \text{ mole } Ca(NO_3)_2} = 24.6 \text{ g of } Ca(NO_3)_2 \text{ (3 SFs)}$

e. Molar mass of C_6H_{14}

$= 6 \text{ moles of C} + 14 \text{ moles of H} = 6(12.01 \text{ g}) + 14(1.008 \text{ g}) = 86.17 \text{ g}$

$0.150 \text{ mole } C_6H_{14} \times \dfrac{86.17 \text{ g } C_6H_{14}}{1 \text{ mole } C_6H_{14}} = 12.9 \text{ g of } C_6H_{14} \text{ (3 SFs)}$

7.45 **a.** $82.0 \text{ g Ag} \times \dfrac{1 \text{ mole Ag}}{107.9 \text{ g Ag}} = 0.760 \text{ mole of Ag (3 SFs)}$

b. $0.288 \text{ g C} \times \dfrac{1 \text{ mole C}}{12.01 \text{ g C}} = 0.0240 \text{ mole of C (3 SFs)}$

c. Molar mass of NH_3

$= 1 \text{ mole of N} + 3 \text{ moles of H} = 14.01 \text{ g} + 3(1.008 \text{ g}) = 17.03 \text{ g}$

$15.0 \text{ g } NH_3 \times \dfrac{1 \text{ mole } NH_3}{17.03 \text{ g } NH_3} = 0.881 \text{ mole of } NH_3 \text{ (3 SFs)}$

d. Molar mass of CH_4

$= 1 \text{ mole of C} + 4 \text{ moles of H} = 12.01 \text{ g} + 4(1.008 \text{ g}) = 16.04 \text{ g}$

$7.25 \text{ g } CH_4 \times \dfrac{1 \text{ mole } CH_4}{16.04 \text{ g } CH_4} = 0.452 \text{ mole of } CH_4 \text{ (3 SFs)}$

e. Molar mass of Fe_2O_3

$= 2 \text{ moles of Fe} + 3 \text{ moles of O} = 2(55.85 \text{ g}) + 3(16.00 \text{ g}) = 159.7 \text{ g}$

$245 \text{ g } Fe_2O_3 \times \dfrac{1 \text{ mole } Fe_2O_3}{159.7 \text{ g } Fe_2O_3} = 1.53 \text{ moles of } Fe_2O_3 \text{ (3 SFs)}$

7.47 **a.** $25.0 \text{ g He} \times \dfrac{1 \text{ mole He}}{4.003 \text{ g He}} = 6.25 \text{ moles of He (3 SFs)}$

b. Molar mass of O_2

$= 2 \text{ moles of O} = 2(16.00 \text{ g}) = 32.00 \text{ g}$

$25.0 \text{ g } O_2 \times \dfrac{1 \text{ mole } O_2}{32.00 \text{ g } O_2} = 0.781 \text{ mole of } O_2 \text{ (3 SFs)}$

c. Molar mass of $Al(OH)_3$

$= 1 \text{ mole of Al} + 3 \text{ moles of O} + 3 \text{ moles of H} = 26.98 \text{ g} + 3(16.00 \text{ g}) + 3(1.008 \text{ g})$

$= 78.00 \text{ g}$

$25.0 \text{ g } Al(OH)_3 \times \dfrac{1 \text{ mole } Al(OH)_3}{78.00 \text{ g } Al(OH)_3} = 0.321 \text{ mole of } Al(OH)_3 \text{ (3 SFs)}$

d. Molar mass of Ga_2S_3

$= 2$ moles of Ga $+$ 3 moles of S $= 2(69.72$ g$) + 3(32.07$ g$) = 235.7$ g

$25.0 \text{ g } Ga_2S_3 \times \dfrac{1 \text{ mole } Ga_2S_3}{235.7 \text{ g } Ga_2S_3} = 0.106$ mole of Ga_2S_3 (3 SFs)

e. Molar mass of C_4H_{10}

$= 4$ moles of C $+$ 10 moles of H $= 4(12.01$ g$) + 10(1.008$ g$) = 58.12$ g

$25.0 \text{ g } C_4H_{10} \times \dfrac{1 \text{ mole } C_4H_{10}}{58.12 \text{ g } C_4H_{10}} = 0.430$ mole of C_4H_{10} (3 SFs)

7.49 Molar mass of C_2H_5Cl

$= 2$ moles of C $+$ 5 moles of H $+$ 1 mole of Cl $= 2(12.01$ g$) + 5(1.008$ g$) + 35.45$ g $= 64.51$ g

a. $34.0 \text{ g } C_2H_5Cl \times \dfrac{1 \text{ mole } C_2H_5Cl}{64.51 \text{ g } C_2H_5Cl} = 0.527$ mole of C_2H_5Cl (3 SFs)

b. $1.50 \text{ moles } C_2H_5Cl \times \dfrac{64.51 \text{ g } C_2H_5Cl}{1 \text{ mole } C_2H_5Cl} = 96.8$ g of C_2H_5Cl (3 SFs)

7.51 a. Molar mass of $MgSO_4$

$= 1$ mole of Mg $+$ 1 mole of S $+$ 4 moles of O $= 24.31$ g $+$ 32.07 g $+$ 4(16.00 g)

$= 120.38$ g

$5.00 \text{ moles } MgSO_4 \times \dfrac{120.38 \text{ g } MgSO_4}{1 \text{ mole } MgSO_4} = 602$ g of $MgSO_4$ (3 SFs)

b. Molar mass of KI

$= 1$ mole of K $+$ 1 mole of I $= 39.10$ g $+$ 126.9 g $= 166.0$ g

$0.450 \text{ mole KI} \times \dfrac{166.0 \text{ g KI}}{1 \text{ mole KI}} = 74.7$ g of KI (3 SFs)

7.53 Molar mass of N_2O

$= 2$ moles of N $+$ 1 mole of O $= 2(14.01$ g$) + 16.00$ g $= 44.02$ g

a. $1.50 \text{ moles } N_2O \times \dfrac{44.02 \text{ g } N_2O}{1 \text{ mole } N_2O} = 66.0$ g of N_2O (3 SFs)

b. $34.0 \text{ g } N_2O \times \dfrac{1 \text{ mole } N_2O}{44.02 \text{ g } N_2O} = 0.772$ mole of N_2O (3 SFs)

7.55 a. $\dfrac{2 \text{ moles } SO_2}{1 \text{ mole } O_2}$ and $\dfrac{1 \text{ mole } O_2}{2 \text{ moles } SO_2}$; $\dfrac{2 \text{ moles } SO_2}{2 \text{ moles } SO_3}$ and $\dfrac{2 \text{ moles } SO_3}{2 \text{ moles } SO_2}$; $\dfrac{1 \text{ mole } O_2}{2 \text{ moles } SO_3}$ and $\dfrac{2 \text{ moles } SO_3}{1 \text{ mole } O_2}$

b. $\dfrac{4 \text{ moles P}}{5 \text{ moles } O_2}$ and $\dfrac{5 \text{ moles } O_2}{4 \text{ moles P}}$; $\dfrac{4 \text{ moles P}}{2 \text{ moles } P_2O_5}$ and $\dfrac{2 \text{ moles } P_2O_5}{4 \text{ moles P}}$; $\dfrac{5 \text{ moles } O_2}{2 \text{ moles } P_2O_5}$ and $\dfrac{2 \text{ moles } P_2O_5}{5 \text{ moles } O_2}$

7.57 a. $2.6 \text{ moles } H_2 \times \dfrac{1 \text{ mole } O_2}{2 \text{ moles } H_2} = 1.3$ moles of O_2 (2 SFs)

b. $5.0 \text{ moles } O_2 \times \dfrac{2 \text{ moles } H_2}{1 \text{ mole } O_2} = 10.$ moles of H_2 (2 SFs)

c. $2.5 \text{ moles } O_2 \times \dfrac{2 \text{ moles } H_2O}{1 \text{ mole } O_2} = 5.0$ moles of H_2O (2 SFs)

7.59 a. $0.500 \text{ mole } SO_2 \times \dfrac{5 \text{ moles C}}{2 \text{ moles } SO_2} = 1.25$ moles of C (3 SFs)

b. $1.2 \text{ moles C} \times \dfrac{4 \text{ moles CO}}{5 \text{ moles C}} = 0.96$ mole of CO (2 SFs)

c. $0.50 \text{ mole CS}_2 \times \dfrac{2 \text{ moles SO}_2}{1 \text{ mole CS}_2} = 1.0 \text{ mole of SO}_2 \text{ (2 SFs)}$

d. $2.5 \text{ moles C} \times \dfrac{1 \text{ mole CS}_2}{5 \text{ moles C}} = 0.50 \text{ mole of CS}_2 \text{ (2 SFs)}$

7.61 a. $57.5 \text{ g Na} \times \dfrac{1 \text{ mole Na}}{22.99 \text{ g Na}} \times \dfrac{2 \text{ moles Na}_2\text{O}}{4 \text{ moles Na}} \times \dfrac{61.98 \text{ g Na}_2\text{O}}{1 \text{ mole Na}_2\text{O}} = 77.5 \text{ g of Na}_2\text{O} \text{ (3 SFs)}$

b. $18.0 \text{ g Na} \times \dfrac{1 \text{ mole Na}}{22.99 \text{ g Na}} \times \dfrac{1 \text{ mole O}_2}{4 \text{ moles Na}} \times \dfrac{32.00 \text{ g O}_2}{1 \text{ mole O}_2} = 6.26 \text{ g of O}_2 \text{ (3 SFs)}$

c. $75.0 \text{ g Na}_2\text{O} \times \dfrac{1 \text{ mole Na}_2\text{O}}{61.98 \text{ g Na}_2\text{O}} \times \dfrac{1 \text{ mole O}_2}{2 \text{ moles Na}_2\text{O}} \times \dfrac{32.00 \text{ g O}_2}{1 \text{ mole O}_2} = 19.4 \text{ g of O}_2 \text{ (3 SFs)}$

7.63 a. $13.6 \text{ g NH}_3 \times \dfrac{1 \text{ mole NH}_3}{17.03 \text{ g NH}_3} \times \dfrac{3 \text{ moles O}_2}{4 \text{ moles NH}_3} \times \dfrac{32.00 \text{ g O}_2}{1 \text{ mole O}_2} = 19.2 \text{ g of O}_2 \text{ (3 SFs)}$

b. $6.50 \text{ g O}_2 \times \dfrac{1 \text{ mole O}_2}{32.00 \text{ g O}_2} \times \dfrac{2 \text{ moles N}_2}{3 \text{ moles O}_2} \times \dfrac{28.02 \text{ g N}_2}{1 \text{ mole N}_2} = 3.79 \text{ g of N}_2 \text{ (3 SFs)}$

c. $34.0 \text{ g NH}_3 \times \dfrac{1 \text{ mole NH}_3}{17.03 \text{ g NH}_3} \times \dfrac{6 \text{ moles H}_2\text{O}}{4 \text{ moles NH}_3} \times \dfrac{18.02 \text{ g H}_2\text{O}}{1 \text{ mole H}_2\text{O}} = 54.0 \text{ g of H}_2\text{O} \text{ (3 SFs)}$

7.65 a. $28.0 \text{ g NO}_2 \times \dfrac{1 \text{ mole NO}_2}{46.01 \text{ g NO}_2} \times \dfrac{1 \text{ mole H}_2\text{O}}{3 \text{ moles NO}_2} \times \dfrac{18.02 \text{ g H}_2\text{O}}{1 \text{ mole H}_2\text{O}} = 3.66 \text{ g of H}_2\text{O} \text{ (3 SFs)}$

b. $15.8 \text{ g H}_2\text{O} \times \dfrac{1 \text{ mole H}_2\text{O}}{18.02 \text{ g H}_2\text{O}} \times \dfrac{1 \text{ mole NO}}{1 \text{ mole H}_2\text{O}} \times \dfrac{30.01 \text{ g NO}}{1 \text{ mole NO}} = 26.3 \text{ g of NO} \text{ (3 SFs)}$

c. $8.25 \text{ g NO}_2 \times \dfrac{1 \text{ mole NO}_2}{46.01 \text{ g NO}_2} \times \dfrac{2 \text{ moles HNO}_3}{3 \text{ moles NO}_2} \times \dfrac{63.02 \text{ g HNO}_3}{1 \text{ mole HNO}_3} = 7.53 \text{ g of HNO}_3 \text{ (3 SFs)}$

7.67 a. $2\text{PbS}(s) + 3\text{O}_2(g) \longrightarrow 2\text{PbO}(s) + 2\text{SO}_2(g)$

b. $29.9 \text{ g PbS} \times \dfrac{1 \text{ mole PbS}}{239.3 \text{ g PbS}} \times \dfrac{3 \text{ moles O}_2}{2 \text{ moles PbS}} \times \dfrac{32.00 \text{ g O}_2}{1 \text{ mole O}_2} = 6.00 \text{ g of O}_2 \text{ (3 SFs)}$

c. $65.0 \text{ g PbS} \times \dfrac{1 \text{ mole PbS}}{239.3 \text{ g PbS}} \times \dfrac{2 \text{ moles SO}_2}{2 \text{ moles PbS}} \times \dfrac{64.07 \text{ g SO}_2}{1 \text{ mole SO}_2} = 17.4 \text{ g of SO}_2 \text{ (3 SFs)}$

d. $128 \text{ g PbO} \times \dfrac{1 \text{ mole PbO}}{223.2 \text{ g PbO}} \times \dfrac{2 \text{ moles PbS}}{2 \text{ moles PbO}} \times \dfrac{239.3 \text{ g PbS}}{1 \text{ mole PbS}} = 137 \text{ g of PbS} \text{ (3 SFs)}$

7.69 a. $15.0 \text{ g glycine} \times \dfrac{1 \text{ mole glycine}}{75.07 \text{ g glycine}} \times \dfrac{3 \text{ moles O}_2}{2 \text{ moles glycine}} \times \dfrac{32.00 \text{ g O}_2}{1 \text{ mole O}_2} = 9.59 \text{ g of O}_2 \text{ (3 SFs)}$

b. $15.0 \text{ g glycine} \times \dfrac{1 \text{ mole glycine}}{75.07 \text{ g glycine}} \times \dfrac{1 \text{ mole urea}}{2 \text{ moles glycine}} \times \dfrac{60.06 \text{ g urea}}{1 \text{ mole urea}} = 6.00 \text{ g of urea} \text{ (3 SFs)}$

c. $15.0 \text{ g glycine} \times \dfrac{1 \text{ mole glycine}}{75.07 \text{ g glycine}} \times \dfrac{3 \text{ moles CO}_2}{2 \text{ moles glycine}} \times \dfrac{44.01 \text{ g CO}_2}{1 \text{ mole CO}_2} = 13.2 \text{ g of CO}_2 \text{ (3 SFs)}$

7.71 a. The limiting factor is the number of drivers: with only eight drivers available, only eight taxis can be used to pick up passengers.

b. The limiting factor is the number of taxis: only seven taxis are in working condition to be driven.

7.73 a. $3.0 \text{ moles N}_2 \times \dfrac{2 \text{ moles NH}_3}{1 \text{ mole N}_2} = 6.0 \text{ moles of NH}_3$

$5.0 \text{ moles H}_2 \times \dfrac{2 \text{ moles NH}_3}{3 \text{ moles H}_2} = 3.3 \text{ moles of NH}_3 \text{ (smaller amount of product)}$

The limiting reactant is 5.0 moles of H_2. (2 SFs)

b. $8.0 \text{ moles N}_2 \times \dfrac{2 \text{ moles NH}_3}{1 \text{ mole N}_2} = 16 \text{ moles of NH}_3$

$4.0 \text{ moles H}_2 \times \dfrac{2 \text{ moles NH}_3}{3 \text{ moles H}_2} = 2.7 \text{ moles of NH}_3 \text{ (smaller amount of product)}$

The limiting reactant is 4.0 moles of H_2. (2 SFs)

c. $3.0 \text{ moles N}_2 \times \dfrac{2 \text{ moles NH}_3}{1 \text{ mole N}_2} = 6.0 \text{ moles of NH}_3 \text{ (smaller amount of product)}$

$12.0 \text{ moles H}_2 \times \dfrac{2 \text{ moles NH}_3}{3 \text{ moles H}_2} = 8.0 \text{ moles of NH}_3$

The limiting reactant is 3.0 moles of N_2. (2 SFs)

7.75 **a.** $20.0 \text{ g Al} \times \dfrac{1 \text{ mole Al}}{26.98 \text{ g Al}} \times \dfrac{2 \text{ moles AlCl}_3}{2 \text{ moles Al}} \times \dfrac{133.33 \text{ g AlCl}_3}{1 \text{ mole AlCl}_3} = 98.8 \text{ g of AlCl}_3$

$20.0 \text{ g Cl}_2 \times \dfrac{1 \text{ mole Cl}_2}{70.90 \text{ g Cl}_2} \times \dfrac{2 \text{ moles AlCl}_3}{3 \text{ moles Cl}_2} \times \dfrac{133.33 \text{ g AlCl}_3}{1 \text{ mole AlCl}_3} = 25.1 \text{ g of AlCl}_3$
(smaller amount of product)

∴ Cl_2 is the limiting reactant, and 25.1 g of $AlCl_3$ (3 SFs) can be produced.

b. $20.0 \text{ g NH}_3 \times \dfrac{1 \text{ mole NH}_3}{17.03 \text{ g NH}_3} \times \dfrac{6 \text{ moles H}_2O}{4 \text{ moles NH}_3} \times \dfrac{18.02 \text{ g H}_2O}{1 \text{ mole H}_2O} = 31.7 \text{ g of H}_2O$

$20.0 \text{ g O}_2 \times \dfrac{1 \text{ mole O}_2}{32.00 \text{ g O}_2} \times \dfrac{6 \text{ moles H}_2O}{5 \text{ moles O}_2} \times \dfrac{18.02 \text{ g H}_2O}{1 \text{ mole H}_2O} = 13.5 \text{ g of H}_2O$
(smaller amount of product)

∴ O_2 is the limiting reactant, and 13.5 g of H_2O (3 SFs) can be produced.

c. $20.0 \text{ g CS}_2 \times \dfrac{1 \text{ mole CS}_2}{76.15 \text{ g CS}_2} \times \dfrac{2 \text{ moles SO}_2}{1 \text{ mole CS}_2} \times \dfrac{64.07 \text{ g SO}_2}{1 \text{ mole SO}_2} = 33.7 \text{ g of SO}_2$

$20.0 \text{ g O}_2 \times \dfrac{1 \text{ mole O}_2}{32.00 \text{ g O}_2} \times \dfrac{2 \text{ moles SO}_2}{3 \text{ moles O}_2} \times \dfrac{64.07 \text{ g SO}_2}{1 \text{ mole SO}_2} = 26.7 \text{ g of SO}_2$
(smaller amount of product)

∴ O_2 is the limiting reactant, and 26.7 g of SO_2 (3 SFs) can be produced.

7.77 **a.** $25.0 \text{ g SO}_2 \times \dfrac{1 \text{ mole SO}_2}{64.07 \text{ g SO}_2} \times \dfrac{2 \text{ moles SO}_3}{2 \text{ moles SO}_2} \times \dfrac{80.07 \text{ g SO}_3}{1 \text{ mole SO}_3} = 31.2 \text{ g of SO}_3$
(smaller amount of product)

$40.0 \text{ g O}_2 \times \dfrac{1 \text{ mole O}_2}{32.00 \text{ g O}_2} \times \dfrac{2 \text{ moles SO}_3}{1 \text{ mole O}_2} \times \dfrac{80.07 \text{ g SO}_3}{1 \text{ mole SO}_3} = 200. \text{ g of SO}_3$

∴ SO_2 is the limiting reactant, and 31.2 g of SO_3 (3 SFs) can be produced.

b. $25.0 \text{ g Fe} \times \dfrac{1 \text{ mole Fe}}{55.85 \text{ g Fe}} \times \dfrac{1 \text{ mole Fe}_3O_4}{3 \text{ moles Fe}} \times \dfrac{231.6 \text{ g Fe}_3O_4}{1 \text{ mole Fe}_3O_4} = 34.6 \text{ g of Fe}_3O_4$
(smaller amount of product)

$40.0 \text{ g H}_2O \times \dfrac{1 \text{ mole H}_2O}{18.02 \text{ g H}_2O} \times \dfrac{1 \text{ mole Fe}_3O_4}{4 \text{ moles H}_2O} \times \dfrac{231.6 \text{ g Fe}_3O_4}{1 \text{ mole Fe}_3O_4} = 129 \text{ g of Fe}_3O_4$

∴ Fe is the limiting reactant, and 34.6 g of Fe_3O_4 (3 SFs) can be produced.

c. $25.0 \text{ g C}_7H_{16} \times \dfrac{1 \text{ mole C}_7H_{16}}{100.20 \text{ g C}_7H_{16}} \times \dfrac{7 \text{ moles CO}_2}{1 \text{ mole C}_7H_{16}} \times \dfrac{44.01 \text{ g CO}_2}{1 \text{ mole CO}_2} = 76.9 \text{ g of CO}_2$

$40.0 \text{ g O}_2 \times \dfrac{1 \text{ mole O}_2}{32.00 \text{ g O}_2} \times \dfrac{7 \text{ moles CO}_2}{11 \text{ moles O}_2} \times \dfrac{44.01 \text{ g CO}_2}{1 \text{ mole CO}_2} = 35.0 \text{ g of CO}_2$
(smaller amount of product)

∴ O_2 is the limiting reactant, and 35.0 g of CO_2 (3 SFs) can be produced.

7.79 **a.** Theoretical yield of CS_2:

$$40.0 \text{ g C} \times \frac{1 \text{ mole C}}{12.01 \text{ g C}} \times \frac{1 \text{ mole CS}_2}{5 \text{ moles C}} \times \frac{76.15 \text{ g CS}_2}{1 \text{ mole CS}_2} = 50.7 \text{ g of CS}_2$$

Percent yield: $\dfrac{36.0 \text{ g CS}_2 \text{ (actual)}}{50.7 \text{ g CS}_2 \text{ (theoretical)}} \times 100\% = 71.0\%$ (3 SFs)

b. Theoretical yield of CS_2:

$$32.0 \text{ g SO}_2 \times \frac{1 \text{ mole SO}_2}{64.07 \text{ g SO}_2} \times \frac{1 \text{ mole CS}_2}{2 \text{ moles SO}_2} \times \frac{76.15 \text{ g CS}_2}{1 \text{ mole CS}_2} = 19.0 \text{ g of CS}_2$$

Percent yield: $\dfrac{12.0 \text{ g CS}_2 \text{ (actual)}}{19.0 \text{ g CS}_2 \text{ (theoretical)}} \times 100\% = 63.2\%$ (3 SFs)

7.81 Theoretical yield of Al_2O_3:

$$50.0 \text{ g Al} \times \frac{1 \text{ mole Al}}{26.98 \text{ g Al}} \times \frac{2 \text{ moles Al}_2O_3}{4 \text{ moles Al}} \times \frac{101.96 \text{ g Al}_2O_3}{1 \text{ mole Al}_2O_3} = 94.5 \text{ g of Al}_2O_3$$

Use the percent yield to convert theoretical to actual:

$$94.5 \text{ g Al}_2O_3 \times \frac{75.0 \text{ g Al}_2O_3}{100 \text{ g Al}_2O_3} = 70.9 \text{ g of Al}_2O_3 \text{ (actual)} \text{ (3 SFs)}$$

7.83 Theoretical yield of CO_2:

$$30.0 \text{ g C} \times \frac{1 \text{ mole C}}{12.01 \text{ g C}} \times \frac{2 \text{ moles CO}}{3 \text{ moles C}} \times \frac{28.01 \text{ g CO}}{1 \text{ mole CO}} = 46.6 \text{ g of CO}$$

Percent yield: $\dfrac{28.2 \text{ g CO (actual)}}{46.6 \text{ g CO (theoretical)}} \times 100\% = 60.5\%$ (3 SFs)

7.85 In exothermic reactions, the energy of the products is lower than that of the reactants.

7.87 **a.** An exothermic reaction releases energy.
   b. An endothermic reaction has a higher energy level for the products than the reactants.
   c. Metabolism is an exothermic reaction, releasing energy for the body.

7.89 **a.** Heat is released, which makes the reaction exothermic with $\Delta H = -802$ kJ.
   b. Heat is absorbed, which makes the reaction endothermic with $\Delta H = +65.3$ kJ.
   c. Heat is released, which makes the reaction exothermic with $\Delta H = -850$ kJ.

7.91 **a.** $125 \text{ g Cl}_2 \times \dfrac{1 \text{ mole Cl}_2}{70.90 \text{ g Cl}_2} \times \dfrac{-657 \text{ kJ}}{2 \text{ moles Cl}_2} = -579 \text{ kJ}$ (3 SFs)

$\therefore$ 579 kJ are released.

b. $278 \text{ g PCl}_5 \times \dfrac{1 \text{ mole PCl}_5}{208.2 \text{ g PCl}_5} \times \dfrac{+67 \text{ kJ}}{1 \text{ mole PCl}_5} = +89 \text{ kJ}$ (2 SFs)

$\therefore$ 89 kJ are absorbed.

7.93 **a.** Since heat is absorbed, the reaction is endothermic.

b. $18.0 \text{ g CO}_2 \times \dfrac{1 \text{ mole CO}_2}{44.01 \text{ g CO}_2} \times \dfrac{1 \text{ mole glucose}}{6 \text{ moles CO}_2} \times \dfrac{180.2 \text{ g glucose}}{1 \text{ mole glucose}} = 12.3 \text{ g of glucose}$ (3 SFs)

c. $25.0 \text{ g glucose} \times \dfrac{1 \text{ mole glucose}}{180.2 \text{ g glucose}} \times \dfrac{680 \text{ kcal}}{1 \text{ mole glucose}} \times \dfrac{1000 \text{ cal}}{1 \text{ kcal}} \times \dfrac{4.184 \text{ J}}{1 \text{ cal}} \times \dfrac{1 \text{ kJ}}{1000 \text{ J}}$

$= 390 \text{ kJ}$ (2 SFs)

7.95 **a.** $C_6H_{12}O_6(aq) + 6O_2(g) \longrightarrow 6CO_2(g) + 6H_2O(l)$
   b. $6CO_2(g) + 6H_2O(l) \longrightarrow C_6H_{12}O_6(aq) + 6O_2(g)$

7.97 a. 1, 1, 2 combination $\qquad$ **b.** 2, 2, 1 decomposition

7.99 a. reactants NO and O_2; product NO_2 $\qquad$ **b.** $2NO(g) + O_2(g) \longrightarrow 2NO_2(g)$
c. combination

7.101 a. reactant NI_3; products N_2 and I_2 $\qquad$ **b.** $2NI_3(s) \longrightarrow N_2(g) + 3I_2(g)$
c. decomposition

7.103 a. reactants Cl_2 and O_2; product OCl_2 $\qquad$ **b.** $2Cl_2(g) + O_2(g) \longrightarrow 2OCl_2(g)$
c. combination

7.105 1. a. S_2Cl_2

b. Molar mass of S_2Cl_2 = 2(32.07 g) + 2(35.45 g) = 135.04 g (5 SFs)

c. $10.0 \text{ g } S_2Cl_2 \times \dfrac{1 \text{ mole } S_2Cl_2}{135.04 \text{ g } S_2Cl_2}$ = 0.0741 mole of S_2Cl_2 (3 SFs)

2. a. C_6H_6

b. Molar mass of C_6H_6 = 6(12.01 g) + 6(1.008 g) = 78.11 g (4 SFs)

c. $10.0 \text{ g } C_6H_6 \times \dfrac{1 \text{ mole } C_6H_6}{78.11 \text{ g } C_6H_6}$ = 0.128 mole of C_6H_6 (3 SFs)

7.107 a. Molar mass of dipyrithione ($C_{10}H_8N_2O_2S_2$)
= 10(12.01 g) + 8(1.008 g) + 2(14.01 g) + 2(16.00 g) + 2(32.07 g) = 252.3 g (4 SFs)

b. $25.0 \text{ g } C_{10}H_8N_2O_2S_2 \times \dfrac{1 \text{ mole } C_{10}H_8N_2O_2S_2}{252.3 \text{ g } C_{10}H_8N_2O_2S_2}$ = 0.0991 mole of $C_{10}H_8N_2O_2S_2$ (3 SFs)

c. $25.0 \text{ g } C_{10}H_8N_2O_2S_2 \times \dfrac{1 \text{ mole } C_{10}H_8N_2O_2S_2}{252.3 \text{ g } C_{10}H_8N_2O_2S_2} \times \dfrac{10 \text{ moles C}}{1 \text{ mole } C_{10}H_8N_2O_2S_2}$
= 0.991 mole of C (3 SFs)

d. $8.2 \times 10^{24} \text{ N atoms} \times \dfrac{1 \text{ mole N}}{6.02 \times 10^{23} \text{ N atoms}} \times \dfrac{1 \text{ mole } C_{10}H_8O_2N_2S_2}{2 \text{ moles N}}$
= 6.8 moles of $C_{10}H_8O_2N_2S_2$ (2 SFs)

7.109 a. $2NO(g) + O_2(g) \longrightarrow 2NO_2(g)$ $\qquad$ **b.** NO is the limiting reactant.

7.111 a. $N_2(g) + 3H_2(g) \longrightarrow 2NH_3(g)$ $\qquad$ **b.** diagram A; N_2 is the excess reactant.

7.113 a. $2NI_3(s) \longrightarrow N_2(g) + 3I_2(g)$

b. Theoretical yield: from the $6NI_3$ we could obtain

$6NI_3 \times \dfrac{1N_2}{2NI_3} = 3N_2 \qquad 6NI_3 \times \dfrac{3I_2}{2NI_3} = 9I_2$

Actual yield: in the actual products, we obtain $2N_2$ and $6I_2$.
Percent yield based on N_2: $\qquad\qquad$ Percent yield based on I_2:

$\dfrac{2N_2 \text{ (actual)}}{3N_2 \text{ (theoretical)}} \times 100\% = 67\% \qquad \dfrac{6I_2 \text{ (actual)}}{9I_2 \text{ (theoretical)}} \times 100\% = 67\%$

Therefore, the percent yield for the reaction is 67%.

7.115 a. Atoms of a metal and a nonmetal forming an ionic compound is a combination reaction.
b. When a compound of hydrogen and carbon reacts with oxygen, it is a combustion reaction.
c. When magnesium carbonate is heated to produce magnesium oxide and carbon dioxide, it is a decomposition reaction.
d. Zinc replacing copper in $Cu(NO_3)_2$ is a single replacement reaction.

7.117 a. $NH_3(g) + HCl(g) \longrightarrow NH_4Cl(s)$ combination

    b. $C_4H_8(g) + 6O_2(g) \xrightarrow{\Delta} 4CO_2(g) + 4H_2O(g)$ combustion

    c. $2Sb(s) + 3Cl_2(g) \longrightarrow 2SbCl_3(s)$ combination

    d. $2NI_3(s) \longrightarrow N_2(g) + 3I_2(g)$ decomposition

    e. $2KBr(aq) + Cl_2(aq) \longrightarrow 2KCl(aq) + Br_2(l)$ single replacement

    f. $2Fe(s) + 3H_2SO_4(aq) \longrightarrow 3H_2(g) + Fe_2(SO_4)_3(aq)$ single replacement

    g. $Al_2(SO_4)_3(aq) + 6NaOH(aq) \longrightarrow 2Al(OH)_3(s) + 3Na_2SO_4(aq)$ double replacement

7.119 a. Since this is a single replacement reaction, the uncombined metal element takes the place of the metal element in the compound.

    $2HCl(aq) + Zn(s) \longrightarrow H_2(g) + ZnCl_2(aq)$

    b. Since this is a decomposition reaction, the single reactant breaks down into more than one simpler substance.

    $BaCO_3(s) \xrightarrow{\Delta} BaO(s) + CO_2(g)$

    c. Since this is a double replacement reaction, the positive ions in the reacting compounds switch places.

    $HCl(aq) + NaOH(aq) \longrightarrow NaCl(aq) + H_2O(l)$

    d. Since this is a combination reaction, the product is a single compound.

    $2Al(s) + 3F_2(g) \longrightarrow 2AlF_3(s)$

7.121 a. $4Na(s) + O_2(g) \longrightarrow 2Na_2O(s)$ combination

    b. $NaCl(aq) + AgNO_3(aq) \longrightarrow AgCl(s) + NaNO_3(aq)$ double replacement

    c. $C_2H_6O(l) + 3O_2(g) \xrightarrow{\Delta} 2CO_2(g) + 3H_2O(g)$ combustion

7.123 a. N_2 is oxidized; O_2 is reduced.   **b.** H_2 is oxidized; the C (in CO) is reduced.

    c. Mg is oxidized; Br_2 is reduced.

7.125 a. $1 \text{ mole Zn} \times \dfrac{65.41 \text{ g Zn}}{1 \text{ mole Zn}} = 65.41 \text{ g of Zn}$

    $1 \text{ mole S} \times \dfrac{32.07 \text{ g S}}{1 \text{ mole S}} = 32.07 \text{ g of S}$

    $4 \text{ moles O} \times \dfrac{16.00 \text{ g O}}{1 \text{ mole O}} = 64.00 \text{ g of O}$

    1 mole of Zn   =   65.41 g of Zn

    1 mole of S   =   32.07 g of S

    4 moles of O   =   64.00 g of O

    $\therefore$ Molar mass of $ZnSO_4$ = 161.48 g

    b. $1 \text{ mole Ca} \times \dfrac{40.08 \text{ g Ca}}{1 \text{ mole Ca}} = 40.08 \text{ g of Ca}$

    $2 \text{ moles I} \times \dfrac{126.9 \text{ g I}}{1 \text{ mole I}} = 253.8 \text{ g of I}$

    $6 \text{ moles O} \times \dfrac{16.00 \text{ g O}}{1 \text{ mole O}} = 96.00 \text{ g of O}$

    1 mole of Ca   =   40.08 g of Ca

    2 moles of I   =   253.8 g of I

    6 moles of O   =   96.00 g of O

    $\therefore$ Molar mass of $Ca(IO_3)_2$ = 389.9 g

    c. $5 \text{ moles C} \times \dfrac{12.01 \text{ g C}}{1 \text{ mole C}} = 60.05 \text{ g of C}$

    $8 \text{ moles H} \times \dfrac{1.008 \text{ g H}}{1 \text{ mole H}} = 8.064 \text{ g of H}$

$$1 \text{ mole N} \times \frac{14.01 \text{ g N}}{1 \text{ mole N}} = 14.01 \text{ g of N}$$

$$1 \text{ mole Na} \times \frac{22.99 \text{ g Na}}{1 \text{ mole Na}} = 22.99 \text{ g of Na}$$

$$4 \text{ moles O} \times \frac{16.00 \text{ g O}}{1 \text{ mole O}} = 64.00 \text{ g of O}$$

5 moles of C	= 60.05 g of C
8 moles of H	= 8.064 g of H
1 mole of N	= 14.01 g of N
1 mole of Na	= 22.99 g of Na
4 moles of O	= 64.00 g of O

$$\therefore \text{ Molar mass of } C_5H_8NNaO_4 = 169.11 \text{ g}$$

d. $6 \text{ moles C} \times \dfrac{12.01 \text{ g C}}{1 \text{ mole C}} = 72.06 \text{ g of C}$

$$12 \text{ moles H} \times \frac{1.008 \text{ g H}}{1 \text{ mole H}} = 12.10 \text{ g of H}$$

$$2 \text{ moles O} \times \frac{16.00 \text{ g O}}{1 \text{ mole O}} = 32.00 \text{ g of O}$$

6 moles of C	= 72.06 g of C
12 moles of H	= 12.10 g of H
2 moles of O	= 32.00 g of O

$$\therefore \text{ Molar mass of } C_6H_{12}O_2 = 116.16 \text{ g}$$

7.127 a. $0.150 \text{ mole K} \times \dfrac{39.10 \text{ g K}}{1 \text{ mole K}} = 5.87 \text{ g of K (3 SFs)}$

b. $0.150 \text{ mole Cl}_2 \times \dfrac{70.90 \text{ g Cl}_2}{1 \text{ mole Cl}_2} = 10.6 \text{ g of Cl}_2 \text{ (3 SFs)}$

c. $0.150 \text{ mole Na}_2CO_3 \times \dfrac{105.99 \text{ g Na}_2CO_3}{1 \text{ mole Na}_2CO_3} = 15.9 \text{ g of Na}_2CO_3 \text{ (3 SFs)}$

7.129 a. $25.0 \text{ g CO}_2 \times \dfrac{1 \text{ mole CO}_2}{44.01 \text{ g CO}_2} = 0.568 \text{ mole of CO}_2 \text{ (3 SFs)}$

b. $25.0 \text{ g Al}_2O_3 \times \dfrac{1 \text{ mole Al}_2O_3}{101.96 \text{ g Al}_2O_3} = 0.245 \text{ mole of Al}_2O_3 \text{ (3 SFs)}$

c. $25.0 \text{ g MgCl}_2 \times \dfrac{1 \text{ mole MgCl}_2}{95.21 \text{ g MgCl}_2} = 0.263 \text{ mole of MgCl}_2 \text{ (3 SFs)}$

7.131 a. $2NH_3(g) + 5F_2(g) \longrightarrow N_2F_4(g) + 6HF(g)$

b. $4.00 \text{ moles HF} \times \dfrac{2 \text{ moles NH}_3}{6 \text{ moles HF}} = 1.33 \text{ moles of NH}_3 \text{ (3 SFs)}$

$$4.00 \text{ moles HF} \times \frac{5 \text{ moles F}_2}{6 \text{ moles HF}} = 3.33 \text{ moles of F}_2 \text{ (3 SFs)}$$

c. $25.5 \text{ g NH}_3 \times \dfrac{1 \text{ mole NH}_3}{17.03 \text{ g NH}_3} \times \dfrac{5 \text{ moles F}_2}{2 \text{ moles NH}_3} \times \dfrac{38.00 \text{ g F}_2}{1 \text{ mole F}_2} = 142 \text{ g of F}_2 \text{ (3 SFs)}$

d. $3.40 \text{ g NH}_3 \times \dfrac{1 \text{ mole NH}_3}{17.03 \text{ g NH}_3} \times \dfrac{1 \text{ mole N}_2F_4}{2 \text{ moles NH}_3} \times \dfrac{104.02 \text{ g N}_2F_4}{1 \text{ mole N}_2F_4} = 10.4 \text{ g of N}_2F_4 \text{ (3 SFs)}$

7.133 a. $3.00 \text{ moles } H_2O \times \dfrac{2 \text{ moles } H_2O_2}{2 \text{ moles } H_2O} = 3.00$ moles of H_2O_2 (3 SFs)

b. $36.5 \text{ g } O_2 \times \dfrac{1 \text{ mole } O_2}{32.00 \text{ g } O_2} \times \dfrac{2 \text{ moles } H_2O_2}{1 \text{ mole } O_2} \times \dfrac{34.02 \text{ g } H_2O_2}{1 \text{ mole } H_2O_2} = 77.6$ g of H_2O_2 (3 SFs)

c. $12.2 \text{ g } H_2O_2 \times \dfrac{1 \text{ mole } H_2O_2}{34.02 \text{ g } H_2O_2} \times \dfrac{2 \text{ moles } H_2O}{2 \text{ moles } H_2O_2} \times \dfrac{18.02 \text{ g } H_2O}{1 \text{ mole } H_2O} = 6.46$ g of H_2O (3 SFs)

7.135 $12.8 \text{ g } Na \times \dfrac{1 \text{ mole } Na}{22.99 \text{ g } Na} \times \dfrac{2 \text{ moles } NaCl}{2 \text{ moles } Na} \times \dfrac{58.44 \text{ g } NaCl}{1 \text{ mole } NaCl} = 32.5$ g of NaCl

$10.2 \text{ g } Cl_2 \times \dfrac{1 \text{ mole } Cl_2}{70.90 \text{ g } Cl_2} \times \dfrac{2 \text{ moles } NaCl}{1 \text{ mole } Cl_2} \times \dfrac{58.44 \text{ g } NaCl}{1 \text{ mole } NaCl} = 16.8$ g of NaCl

(smaller amount of product)

$\therefore Cl_2$ is the limiting reactant, and 16.8 g of NaCl (3 SFs) can be produced.

7.137 a. $4.00 \text{ moles } H_2O \times \dfrac{1 \text{ mole } C_5H_{12}}{6 \text{ moles } H_2O} = 0.667$ mole of C_5H_{12} (3 SFs)

b. $32.0 \text{ g } O_2 \times \dfrac{1 \text{ mole } O_2}{32.00 \text{ g } O_2} \times \dfrac{5 \text{ moles } CO_2}{8 \text{ moles } O_2} \times \dfrac{44.01 \text{ g } CO_2}{1 \text{ mole } CO_2} = 27.5$ g of CO_2 (3 SFs)

c. $44.5 \text{ g } C_5H_{12} \times \dfrac{1 \text{ mole } C_5H_{12}}{72.15 \text{ g } C_5H_{12}} \times \dfrac{5 \text{ moles } CO_2}{1 \text{ mole } C_5H_{12}} \times \dfrac{44.01 \text{ g } CO_2}{1 \text{ mole } CO_2} = 136$ g of CO_2

$108 \text{ g } O_2 \times \dfrac{1 \text{ mole } O_2}{32.00 \text{ g } O_2} \times \dfrac{5 \text{ moles } CO_2}{8 \text{ moles } O_2} \times \dfrac{44.01 \text{ g } CO_2}{1 \text{ mole } CO_2} = 92.8$ g of CO_2

(smaller amount of product)

$\therefore O_2$ is the limiting reactant, and 92.8 g of CO_2 (3 SFs) can be produced.

7.139 a. Theoretical yield of CO_2:

$22.0 \text{ g } C_2H_2 \times \dfrac{1 \text{ mole } C_2H_2}{26.04 \text{ g } C_2H_2} \times \dfrac{4 \text{ moles } CO_2}{2 \text{ moles } C_2H_2} \times \dfrac{44.01 \text{ g } CO_2}{1 \text{ mole } CO_2} = 74.4$ g of CO_2 (3 SFs)

b. Percent yield: $\dfrac{64.0 \text{ g } CO_2 \text{ (actual)}}{74.4 \text{ g } CO_2 \text{ (theoretical)}} \times 100\% = 86.0\%$ (3 SFs)

7.141 Theoretical yield of C_2H_6:

$28.0 \text{ g } C_2H_2 \times \dfrac{1 \text{ mole } C_2H_2}{26.04 \text{ g } C_2H_2} \times \dfrac{1 \text{ mole } C_2H_6}{1 \text{ mole } C_2H_2} \times \dfrac{30.07 \text{ g } C_2H_6}{1 \text{ mole } C_2H_6} = 32.3$ g of C_2H_6

Percent yield: $\dfrac{24.5 \text{ g } C_2H_6 \text{ (actual)}}{32.3 \text{ g } C_2H_6 \text{ (theoretical)}} \times 100\% = 75.9\%$ (3 SFs)

7.143 a. $3.00 \text{ g } NO \times \dfrac{1 \text{ mole } NO}{30.01 \text{ g } NO} \times \dfrac{+90.2 \text{ kJ}}{2 \text{ moles } NO} = +4.51$ kJ (3 SFs)

$\therefore 4.51$ kJ are required.

b. The equation for the decomposition of NO will be the reverse of the equation for the formation of NO.

$2NO(g) \longrightarrow N_2(g) + O_2(g) + 90.2 \text{ kJ}$

c. $5.00 \text{ g } NO \times \dfrac{1 \text{ mole } NO}{30.01 \text{ g } NO} \times \dfrac{-90.2 \text{ kJ}}{2 \text{ moles } NO} = -7.51$ kJ (3 SFs)

$\therefore 7.51$ kJ are released.

7.145 a. Heat is released, which makes the reaction exothermic.

b. Heat is absorbed, which makes the reaction endothermic.

7.147 a. $K_2O(s) + H_2O(g) \longrightarrow 2KOH(aq)$ combination

b. $2C_8H_{18}(l) + 25O_2(g) \overset{\Delta}{\longrightarrow} 16CO_2(g) + 18H_2O(g)$ combustion

c. $2Fe(OH)_3(s) \longrightarrow Fe_2O_3(s) + 3H_2O(g)$ decomposition

d. $CuS(s) + 2HCl(aq) \longrightarrow CuCl_2(aq) + H_2S(g)$ double replacement

7.149 a. $Fe_3O_4(s) + 4H_2(g) \longrightarrow 3Fe(s) + 4H_2O(g)$

b. $2C_4H_{10}(g) + 13O_2(g) \overset{\Delta}{\longrightarrow} 8CO_2(g) + 10H_2O(g)$

c. $4Al(s) + 3O_2(g) \longrightarrow 2Al_2O_3(s)$

d. $2NaOH(aq) + ZnSO_4(aq) \longrightarrow Zn(OH)_2(s) + Na_2SO_4(aq)$

7.151 a. $3Pb(NO_3)_2(aq) + 2Na_3PO_4(aq) \longrightarrow Pb_3(PO_4)_2(s) + 6NaNO_3(aq)$ double replacement

b. $4Ga(s) + 3O_2(g) \overset{\Delta}{\longrightarrow} 2Ga_2O_3(s)$ combination

c. $2NaNO_3(s) \overset{\Delta}{\longrightarrow} 2NaNO_2(s) + O_2(g)$ decomposition

7.153 a. $124 \text{ g } C_2H_6O \times \dfrac{1 \text{ mole } C_2H_6O}{46.07 \text{ g } C_2H_6O} \times \dfrac{1 \text{ mole } C_6H_{12}O_6}{2 \text{ moles } C_2H_6O} \times \dfrac{180.16 \text{ g } C_6H_{12}O_6}{1 \text{ mole } C_6H_{12}O_6}$

$= 242 \text{ g of } C_6H_{12}O_6 \text{ (3 SFs)}$

b. $0.240 \text{ kg } C_6H_{12}O_6 \times \dfrac{1000 \text{ g}}{1 \text{ kg}} \times \dfrac{1 \text{ mole } C_6H_{12}O_6}{180.16 \text{ g } C_6H_{12}O_6} \times \dfrac{2 \text{ moles } C_2H_6O}{1 \text{ mole } C_6H_{12}O_6} \times \dfrac{46.07 \text{ g } C_2H_6O}{1 \text{ mole } C_2H_6O}$

$= 123 \text{ g of } C_2H_6O \text{ (3 SFs)}$

7.155 a. $4Al(s) + 3O_2(g) \longrightarrow 2Al_2O_3(s)$

b. This is a combination reaction.

c. $4.50 \text{ moles Al} \times \dfrac{3 \text{ moles } O_2}{4 \text{ moles Al}} = 3.38 \text{ moles of } O_2 \text{ (3 SFs)}$

d. $50.2 \text{ g Al} \times \dfrac{1 \text{ mole Al}}{26.98 \text{ g Al}} \times \dfrac{2 \text{ moles Al}_2O_3}{4 \text{ moles Al}} \times \dfrac{101.96 \text{ g Al}_2O_3}{1 \text{ mole Al}_2O_3} = 94.9 \text{ g of Al}_2O_3 \text{ (3 SFs)}$

e. $8.00 \text{ g } O_2 \times \dfrac{1 \text{ mole } O_2}{32.00 \text{ g } O_2} \times \dfrac{2 \text{ moles Al}_2O_3}{3 \text{ moles } O_2} \times \dfrac{101.96 \text{ g Al}_2O_3}{1 \text{ mole Al}_2O_3} = 17.0 \text{ g of Al}_2O_3 \text{ (3 SFs)}$

7.157 a. $4.50 \text{ moles Cr} \times \dfrac{3 \text{ moles } O_2}{4 \text{ moles Cr}} = 3.38 \text{ moles of } O_2 \text{ (3 SFs)}$

b. $24.8 \text{ g Cr} \times \dfrac{1 \text{ mole Cr}}{52.00 \text{ g Cr}} \times \dfrac{2 \text{ moles Cr}_2O_3}{4 \text{ moles Cr}} \times \dfrac{152.0 \text{ g Cr}_2O_3}{1 \text{ mole Cr}_2O_3} = 36.2 \text{ g of Cr}_2O_3 \text{ (3 SFs)}$

c. $26.0 \text{ g Cr} \times \dfrac{1 \text{ mole Cr}}{52.00 \text{ g Cr}} \times \dfrac{2 \text{ moles Cr}_2O_3}{4 \text{ moles Cr}} \times \dfrac{152.0 \text{ g Cr}_2O_3}{1 \text{ mole Cr}_2O_3} = 38.0 \text{ g of Cr}_2O_3$

$8.00 \text{ g } O_2 \times \dfrac{1 \text{ mole } O_2}{32.00 \text{ g } O_2} \times \dfrac{2 \text{ moles Cr}_2O_3}{3 \text{ moles } O_2} \times \dfrac{152.0 \text{ g Cr}_2O_3}{1 \text{ mole Cr}_2O_3} = 25.3 \text{ g of Cr}_2O_3$ (smaller amount of product)

$\therefore O_2$ is the limiting reactant, and $25.3 \text{ g of Cr}_2O_3$ (3 SFs) can be produced.

d. $74.0 \text{ g Cr} \times \dfrac{1 \text{ mole Cr}}{52.00 \text{ g Cr}} \times \dfrac{2 \text{ moles Cr}_2O_3}{4 \text{ moles Cr}} \times \dfrac{152.0 \text{ g Cr}_2O_3}{1 \text{ mole Cr}_2O_3} = 108 \text{ g of Cr}_2O_3$ (smaller amount of product)

$62.0 \text{ g } O_2 \times \dfrac{1 \text{ mole } O_2}{32.00 \text{ g } O_2} \times \dfrac{2 \text{ moles Cr}_2O_3}{3 \text{ moles } O_2} \times \dfrac{152.0 \text{ g Cr}_2O_3}{1 \text{ mole Cr}_2O_3} = 196 \text{ g of Cr}_2O_3$

$\therefore$ Cr is the limiting reactant, and 108 g of Cr_2O_3 (3 SFs) can be produced (theoretical yield).

Percent yield: $\dfrac{87.3 \text{ g } Cr_2O_3 \text{ (actual)}}{108 \text{ g } Cr_2O_3 \text{ (theoretical)}} \times 100\% = 80.8\% \text{ (3 SFs)}$

7.159 a. $0.225 \text{ mole } C_3H_4 \times \dfrac{4 \text{ moles O}_2}{1 \text{ mole } C_3H_4} = 0.900 \text{ mole of O}_2 \text{ (3 SFs)}$

b. $64.0 \text{ g O}_2 \times \dfrac{1 \text{ mole O}_2}{32.00 \text{ g O}_2} \times \dfrac{2 \text{ moles H}_2O}{4 \text{ moles O}_2} \times \dfrac{18.02 \text{ g H}_2O}{1 \text{ mole H}_2O} = 18.0 \text{ g of H}_2O \text{ (3 SFs)}$

c. $78.0 \text{ g } C_3H_4 \times \dfrac{1 \text{ mole } C_3H_4}{40.06 \text{ g } C_3H_4} \times \dfrac{3 \text{ moles CO}_2}{1 \text{ mole } C_3H_4} \times \dfrac{44.01 \text{ g CO}_2}{1 \text{ mole CO}_2} = 257 \text{ g of CO}_2 \text{ (3 SFs)}$

d. Percent yield: $\dfrac{186 \text{ g CO}_2 \text{ (actual)}}{257 \text{ g CO}_2 \text{ (theoretical)}} \times 100\% = 72.4\% \text{ (3 SFs)}$

7.161 a. The heat of reaction (ΔH) is positive, so the reaction is endothermic.

b. $1.5 \text{ moles SO}_3 \times \dfrac{+790 \text{ kJ}}{2 \text{ moles SO}_3} = +590 \text{ kJ (2 SFs)}$

∴ 590 kJ are required.

c. $150 \text{ g O}_2 \times \dfrac{1 \text{ mole O}_2}{32.00 \text{ g O}_2} \times \dfrac{+790 \text{ kJ}}{3 \text{ moles O}_2} = +1200 \text{ kJ (2 SFs)}$

∴ 1200 kJ are required.

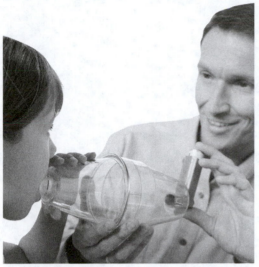

After Whitney's asthma attack, she and her parents have learned to identify and treat her asthma. Because her asthma is mild and intermittent, they follow a management plan to control the symptoms. Whitney participates in physical activity and exercise to maintain lung function and to prevent recurrent attacks. If her symptoms worsen, she uses a bronchodilator to help her breathing. If Whitney has a lung capacity of 3.2 L at 37 °C and 756 mmHg, what is her lung capacity, in liters, at STP?

Credit: Adam Gault/Alamy

LOOKING AHEAD

 The Health icon indicates a question that is related to health and medicine.

8.1 Properties of Gases

Learning Goal: Describe the kinetic molecular theory of gases and the units of measurement used for gases.

REVIEW
Using Significant Figures in Calculations (2.3)
Writing Conversion Factors from Equalities (2.5)
Using Conversion Factors (2.6)

- In a gas, small particles (atoms or molecules) are far apart and moving randomly with high speeds.
- The attractive forces between the particles of a gas are usually very small.
- The actual volume occupied by gas molecules is extremely small compared with the volume that the gas occupies.
- Gas particles are in constant motion, moving rapidly in straight paths.
- The average kinetic energy of gas molecules is proportional to the Kelvin temperature.
- A gas is described by the physical properties of pressure (P), volume (V), temperature (T), and amount (n).
- A gas exerts pressure, which is the force that gas particles exert on the walls of a container.

- Units of gas pressure include torr (Torr), millimeters of mercury (mmHg), pascals (Pa), kilopascals (kPa), lb/in.^{2} (psi), and atmospheres (atm).
- In calculations involving gases, temperatures must use the Kelvin scale.

♦ **Learning Exercise 8.1A**

Answer *true* (T) or *false* (F) for each of the following:

a. _____ Gases are composed of small particles.

b. _____ Gas molecules are usually close together.

c. _____ Gas molecules move rapidly because they are strongly attracted.

d. _____ Distances between gas molecules are large.

e. _____ Gas molecules travel in straight lines until they collide.

f. _____ Pressure is the force of gas particles striking the walls of a container.

g. _____ Kinetic energy decreases with increasing temperature.

h. _____ 450 K is a measure of temperature.

i. _____ 3.56 g is a measure of pressure.

Answers	**a.** T	**b.** F	**c.** F	**d.** T	**e.** T
	f. T	**g.** F	**h.** T	**i.** F	

Units for Measuring Pressure

Unit	Abbreviation	Unit Equivalent to 1 atm
atmosphere	atm	1 atm (exact)
millimeters of Hg	mmHg	760 mmHg (exact)
torr	Torr	760 Torr (exact)
inches of Hg	inHg	29.9 inHg
pounds per square inch	lb/in.^{2} (psi)	14.7 lb/in.^{2}
pascal	Pa	101 325 Pa
kilopascal	kPa	101.325 kPa

♦ **Learning Exercise 8.1B**

Complete the following:

a. 1.50 atm = _____ Torr

b. 550 mmHg = _____ atm

c. 0.725 atm = _____ kPa

d. 1520 Torr = _____ atm

e. 30.5 psi = _____ mmHg

f. 87.6 kPa = _____ Torr

g. During the weather report on TV, the pressure is given as 29.4 inHg. What is this pressure in millimeters of mercury and in atmospheres?

Answers	**a.** 1140 Torr	**b.** 0.72 atm	**c.** 73.5 kPa	**d.** 2.00 atm
	e. 1580 mmHg	**f.** 657 Torr	**g.** 747 mmHg, 0.983 atm	

8.2　Pressure and Volume (Boyle's Law)

Learning Goal: Use the pressure–volume relationship (Boyle's law) to calculate the unknown pressure or volume when the temperature and amount of gas do not change.

- According to Boyle's law, pressure increases if volume decreases and pressure decreases if volume increases, when the temperature and amount of gas do not change.
- If two properties change in opposite directions, the properties have an inverse relationship.
- The volume (V) of a gas changes inversely with the pressure (P) of the gas when T and n do not change:

$$P_1V_1 = P_2V_2$$　The subscripts 1 and 2 represent initial and final conditions.

Piston—

$V_1 = 4$ L
$P_1 = 1$ atm

$V_2 = 2$ L
$P_2 = 2$ atm

Key Terms for Sections 8.1 and 8.2

Match each of the following key terms with the correct description:

   a. kinetic molecular theory   **b.** inverse relationship   **c.** Boyle's law
   d. pressure   **e.** atmospheric pressure

1. _____ the pressure exerted by the atmosphere

2. _____ a force exerted by gas particles when they collide with the sides of a container

3. _____ a relationship in which two properties change in opposite directions

4. _____ the volume of a gas varies inversely with the pressure of a gas, if there is no change in temperature and amount of gas

5. _____ the model that explains the behavior of gas particles

Answers　　**1.** e　　　**2.** d　　　**3.** b　　　**4.** c　　　**5.** a

SAMPLE PROBLEM　Using Boyle's Law

A balloon has a volume of 3.00 L and a pressure of 755 mmHg. What is the volume, in liters, of the balloon at a final pressure of 638 mmHg, if the temperature and amount of gas do not change?

Solution:

STEP 1　State the given and needed quantities.

	Given	Need	Connect
Analyze the Problem	$P_1 = 755$ mmHg $P_2 = 638$ mmHg $V_1 = 3.00$ L	V_2	Boyle's law $P_1V_1 = P_2V_2$
	Factors that do not change: T and n		**Predict:** V increases

STEP 2　Rearrange the gas law equation to solve for the unknown quantity.

$$V_2 = V_1 \times \frac{P_1}{P_2}$$

STEP 3 Substitute values into the gas law equation and calculate.

$$V_2 = 3.00 \text{ L} \times \frac{755 \text{ mmHg}}{638 \text{ mmHg}} = 3.55 \text{ L} \quad \text{As predicted, the volume of the gas increases.}$$

♦ **Learning Exercise 8.2**

Solve each of the following problems using Boyle's law:

a. A sample of 4.0 L of He has a pressure of 800. mmHg. What is the pressure, in millimeters of mercury, when the volume is reduced to 1.0 L, if there is no change in temperature and amount of gas?

b. The maximum volume of air that can fill a person's lungs is 6250 mL at a pressure of 728 mmHg. Inhalation occurs as the pressure in the lungs drops to 721 mmHg with no change in temperature and amount of gas. To what volume, in milliliters, do the lungs expand, if there is no change in temperature and amount of gas?

c. A sample of O_2 at a pressure of 5.00 atm has a volume of 3.00 L. If the gas pressure is changed to 760. mmHg, what volume, in liters, will the O_2 occupy, if there is no change in temperature and amount of gas?

d. A sample of 250. mL of N_2 is initially at a pressure of 2.50 atm. If the pressure changes to 825 mmHg, what is the volume, in milliliters, if there is no change in temperature and amount of gas?

Answers   **a.** 3200 mmHg   **b.** 6310 mL   **c.** 15.0 L   **d.** 576 mL

8.3   Temperature and Volume (Charles's Law)

Learning Goal: Use the temperature–volume relationship (Charles's law) to calculate the unknown temperature or volume when the pressure and amount of gas do not change.

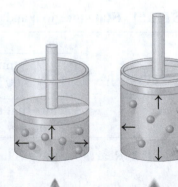

- According to Charles's law, the volume of a gas increases when the temperature of the gas increases, and the volume decreases when the temperature decreases, when the pressure and amount of gas do not change.

- In a direct relationship, two properties increase or decrease together.

- The volume (V) of a gas is directly related to its Kelvin temperature (T) when there is no change in the pressure or amount of gas:

$$\frac{V_1}{T_1} = \frac{V_2}{T_2} \quad \text{The subscripts 1 and 2 represent initial and final conditions.}$$

$T_1 = 200 \text{ K}$
$V_1 = 1 \text{ L}$

$T_2 = 400 \text{ K}$
$V_2 = 2 \text{ L}$

♦ **Learning Exercise 8.3**

Solve each of the following problems using Charles's law:

a. A balloon has a volume of 2.5 L at a temperature of 0 °C. What is the volume, in liters, of the balloon when the temperature rises to 120 °C, if there is no change in pressure and amount of gas?

b. A balloon is filled with He to a volume of 6600 L at a temperature of 223 °C. To what temperature, in degrees Celsius, must the gas be cooled to decrease the volume to 4800 L, if there is no change in pressure and amount of gas?

c. A sample of 750. mL of Ne is heated from 120. °C to 350. °C. What is the volume, in milliliters, if there is no change in pressure and amount of gas?

d. What is the temperature, in degrees Celsius, of 350. mL of O_2 at 22 °C when its volume increases to 0.800 L, if there is no change in pressure and amount of gas?

Answers   **a.** 3.6 L   **b.** 88 °C   **c.** 1190 mL   **d.** 401 °C

8.4   Temperature and Pressure (Gay-Lussac's Law)

Learning Goal: Use the temperature–pressure relationship (Gay-Lussac's law) to calculate the unknown temperature or pressure when the volume and amount of gas do not change.

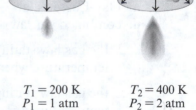

- According to Gay-Lussac's law, the gas pressure increases when the temperature of the gas increases, and the pressure decreases when the temperature decreases, when the volume and amount of gas do not change.

- The pressure (P) of a gas is directly related to its Kelvin temperature (T), when there is no change in the volume or amount of gas:

$$\frac{P_1}{T_1} = \frac{P_2}{T_2}$$ The subscripts 1 and 2 represent initial and final conditions.

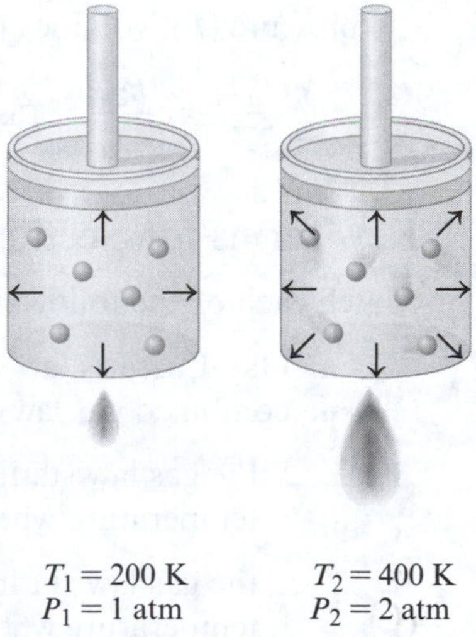

$T_1 = 200$ K $T_2 = 400$ K
$P_1 = 1$ atm $P_2 = 2$ atm

- In a closed container, some of the liquid evaporates and becomes vapor and it creates *vapor pressure*.

♦ **Learning Exercise 8.4**

Solve each of the following problems using Gay-Lussac's law:

a. A sample of He has a pressure of 860. mmHg at a temperature of 225 K. What is the pressure, in millimeters of mercury, when the He reaches a temperature of 675 K, if there is no change in volume or amount of gas?

b. A balloon contains a gas with a pressure of 580. mmHg and a temperature of 227 °C. What is the pressure, in millimeters of mercury, of the gas when the temperature drops to 27 °C, if there is no change in volume or amount of gas?

c. A spray can contains a gas with a pressure of 3.0 atm at a temperature of 17 °C. What is the pressure, in atmospheres, inside the container if the temperature rises to 110 °C, if there is no change in volume or amount of gas?

d. A gas has a pressure of 1200. mmHg at 300. °C. What will the temperature, in degrees Celsius, be when the pressure falls to 1.10 atm, if there is no change in volume or amount of gas?

Answers   **a.** 2580 mmHg   **b.** 348 mmHg   **c.** 4.0 atm   **d.** 126 °C

8.5   The Combined Gas Law

Learning Goal: Use the combined gas law to calculate the unknown pressure, volume, or temperature of a gas when changes in two of these properties are given and the amount of gas does not change.

- When the amount of a gas does not change, the gas laws can be combined into a relationship of pressure (P), volume (V), and temperature (T):

$$\frac{P_1 V_1}{T_1} = \frac{P_2 V_2}{T_2}$$ The subscripts 1 and 2 represent initial and final conditions.

Key Terms for Sections 8.3 to 8.5

Match each of the following key terms with the correct description:

   a. Gay-Lussac's law   **b.** direct relationship   **c.** Charles's law
   d. combined gas law   **e.** vapor pressure

1. _____ the gas law stating that the volume of a gas changes directly with a change in Kelvin temperature when pressure and amount (moles) of the gas do not change

2. _____ the gas law stating that the pressure of a gas changes directly with a change in Kelvin temperature when the number of moles of a gas and its volume do not change

3. _____ a relationship in which two properties change in the same direction

4. _____ a relationship that combines several gas laws relating pressure, volume, and temperature when the amount of gas does not change

5. _____ the pressure exerted by particles of vapor above a liquid

Answers   **1.** c   **2.** a   **3.** b   **4.** d   **5.** e

♦ **Learning Exercise 8.5**

Solve each of the following problems using the combined gas law:

a. A 4.0-L sample of N_2 has a pressure of 1200 mmHg at 220 K. What is the pressure, in millimeters of mercury, of the sample when the volume increases to 20. L at 440 K, if there is no change in the amount of gas?

b. A 25.0-mL bubble forms at the ocean depths when the pressure is 10.0 atm and the temperature is 5.0 °C. What is the volume, in milliliters, of that bubble at the ocean surface when the pressure is 760. mmHg and the temperature is 25 °C, if there is no change in the amount of gas?

c. A 35.0-mL sample of Ar has a pressure of 1.0 atm and a temperature of 15 °C. What is the volume, in milliliters, if the pressure goes to 2.0 atm and the temperature to 45 °C, if there is no change in the amount of gas?

d. A weather balloon is launched from the Earth's surface with a volume of 315 L, a temperature of 12 °C, and pressure of 0.930 atm. What is the volume, in liters, of the balloon in the upper atmosphere when the pressure is 116 mmHg and the temperature is −35 °C, if there is no change in the amount of gas?

e. A 10.0-L sample of gas is emitted from a volcano with a pressure of 1.20 atm and a temperature of 150. °C. What is the volume, in liters, of the gas when its pressure is 0.900 atm and the temperature is −40. °C, if there is no change in the amount of gas?

Answers   **a.** 480 mmHg   **b.** 268 mL   **c.** 19 mL   **d.** 1.60×10^3 L   **e.** 7.34 L

8.6 Volume and Moles (Avogadro's Law)

Learning Goal: Use Avogadro's law to calculate the unknown amount or volume of a gas when the pressure and temperature do not change.

REVIEW
Using Molar Mass as a Conversion Factor (7.3)

- If the number of moles of gas increases, the volume increases; if the number of moles of gas decreases, the volume decreases.

- Avogadro's law states that equal volumes of gases at the same temperature and pressure contain the same number of moles. The volume (V) of a gas is directly related to the number of moles of the gas when the pressure and temperature of the gas do not change:

$$\frac{V_1}{n_1} = \frac{V_2}{n_2}$$ The subscripts 1 and 2 represent initial and final conditions.

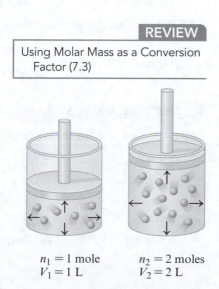

$n_1 = 1$ mole
$V_1 = 1$ L

$n_2 = 2$ moles
$V_2 = 2$ L

- At STP conditions, standard temperature (0 °C) and pressure (1 atm), 1.00 mole of a gas occupies a volume of 22.4 L, the *molar volume*.

♦ **Learning Exercise 8.6A**

Solve each of the following problems using Avogadro's law:

a. A balloon containing 0.50 mole of He has a volume of 4.0 L. What is the volume, in liters, when 1.0 mole of N_2 is added to the balloon, if there is no change in pressure and temperature?

b. A balloon containing 1.0 mole of O_2 has a volume of 15 L. What is the volume, in liters, of the balloon when 3.0 moles of He is added, if there is no change in pressure and temperature?

Answers   **a.** 12 L   **b.** 60. L

SAMPLE PROBLEM **Using Molar Volume**

How many liters would 2.00 moles of N_2 occupy at STP?

Solution:

STEP 1 State the given and needed quantities.

Analyze the Problem	Given	Need	Connect
	2.00 moles of N_2 at STP	liters of N_2 at STP	molar volume (STP)

STEP 2 Write a plan to calculate the needed quantity.

moles of N_2 $\xrightarrow{\text{Molar volume}}$ liters of N_2 (STP)

STEP 3 Write the equalities and conversion factors including 22.4 L/mole at STP.

1 mole of N_2 = 22.4 L of N_2 (STP)

$$\frac{22.4 \text{ L } N_2 \text{ (STP)}}{1 \text{ mole } N_2} \quad \text{and} \quad \frac{1 \text{ mole } N_2}{22.4 \text{ L } N_2 \text{ (STP)}}$$

STEP 4 Set up the problem with factors to cancel units.

$$2.00 \text{ moles } N_2 \times \frac{22.4 \text{ L } N_2 \text{ (STP)}}{1 \text{ mole } N_2} = 44.8 \text{ L of } N_2 \text{ (STP)}$$

♦ **Learning Exercise 8.6B**

Solve each of the following problems using molar volume:

a. What is the volume, in liters, occupied by 18.5 g of N_2 at STP?

b. What is the mass, in grams, of 4.48 L of O_2 at STP?

Answers   **a.** 14.8 L   **b.** 6.40 g

8.7 The Ideal Gas Law

Learning Goal: Use the ideal gas law equation to solve for *P*, *V*, *T*, or *n* of a gas when given three of the four values in the ideal gas law equation. Calculate mass or volume of a gas in a chemical reaction.

*   The ideal gas law $PV = nRT$ gives the relationship between the four properties: pressure, volume, moles, and temperature. When any three properties are given, the fourth can be calculated.
*   *R* is the universal gas constant: 0.0821 L·atm/mole·K or 62.4 L·mmHg/mole·K.
*   The ideal gas law can be used to determine the mass or volume of a gas in a chemical reaction.

Ideal gas constant (*R*)	$\dfrac{0.0821 \text{ L} \cdot \text{atm}}{\text{mole} \cdot \text{K}}$	$\dfrac{62.4 \text{ L} \cdot \text{mmHg}}{\text{mole} \cdot \text{K}}$
Pressure (*P*)	atm	mmHg
Volume (*V*)	L	L
Amount (*n*)	mole	mole
Temperature (*T*)	K	K

SAMPLE PROBLEM   Using the Ideal Gas Law

What is the volume, in liters, of a 8.50-g sample of helium gas at a pressure of 845 mmHg and a temperature of 15.0 °C?

Solution:

STEP 1   State the given and needed quantities.

	Given	Need	Connect
Analyze the Problem	8.50 g of He gas, $P = 845$ mmHg, $T = 15.0\,°C + 273 = 288$ K	*V*	ideal gas law, $PV = nRT$, molar mass

STEP 2   Rearrange the ideal gas law equation to solve for the needed quantity. By dividing both sides of the ideal gas law equation by *P*, we solve for volume, *V*.

$$V = \frac{nRT}{P}$$

STEP 3   Substitute the gas data into the equation, and calculate the needed quantity. To find the moles of He, we use the molar mass 4.003 g/mole.

$$8.50 \text{ g He} \times \frac{1 \text{ mole He}}{4.003 \text{ g He}} = 2.12 \text{ moles of He}$$

$$V = \frac{2.12 \text{ moles} \times 62.4 \text{ (L} \cdot \text{mmHg/mole} \cdot \text{K)} \times 388 \text{ K}}{845 \text{ mmHg}} = 60.7 \text{ L of He}$$

♦ **Learning Exercise 8.7A** 🏃

Use the ideal gas law to solve for the unknown in each of the following:

    a. What volume, in liters, is occupied by 0.250 mole of N_2 at 0 °C and 1.50 atm?

    b. What is the temperature, in degrees Celsius, of 0.500 mole of He that occupies a volume of 15.0 L at a pressure of 1210 mmHg?

    c. What is the pressure, in atmospheres, of 1.50 moles of Ne in a 5.00-L steel container at a temperature of 125 °C?

    d. What is the pressure, in atmospheres, of 8.0 g of O_2 that has a volume of 245 mL at a temperature of 22 °C?

    e. A single-patient hyperbaric chamber has a volume of 650. L. At a temperature of 22 °C, how many grams of O_2 are needed to give a pressure of 1.80 atm?

Answers   **a.** 3.74 L   **b.** 309 °C   **c.** 9.80 atm   **d.** 25 atm   **e.** 1550 g of O_2

Using the Ideal Gas Law for Reactions	
STEP 1	State the given and needed quantities.
STEP 2	Write a plan to convert the given quantity to the needed moles.
STEP 3	Write the equalities and conversion factors for molar mass and mole–mole factors.
STEP 4	Set up the problem to calculate moles of needed quantity.
STEP 5	Convert the moles of needed quantity to mass or volume using the ideal gas law equation.

♦ **Learning Exercise 8.7B**

CORE CHEMISTRY SKILL

Calculating Mass or Volume of a Gas in a Chemical Reaction

Use the ideal gas law equation to determine the quantity of a reactant or product in each of the following:

a. How many grams of KNO_3 must decompose to produce 35.8 L of O_2 at 28 °C and 745 mmHg?

$$2KNO_3(s) \longrightarrow 2KNO_2(s) + O_2(g)$$

b. At a temperature of 325 °C and a pressure of 1.20 atm, how many liters of CO_2 can be produced when 50.0 g of propane (C_3H_8) reacts with O_2?

$$C_3H_8(g) + 5O_2(g) \longrightarrow 3CO_2(g) + 4H_2O(g)$$
Propane

Answers   **a.** 287 g of KNO_3 **b.** 139 L of CO_2

8.8   Partial Pressures (Dalton's Law)

Learning Goal: Use Dalton's law of partial pressures to calculate the total pressure of a mixture of gases.

• In a mixture of two or more gases, the total pressure is the sum of the partial pressures (subscripts 1, 2, 3 . . . represent the partial pressures of the individual gases).

$$P_{total} = P_1 + P_2 + P_3 + \cdots$$

• The partial pressure of a gas in a mixture is the pressure it would exert if it were the only gas in the container.

Key Terms for Sections 8.6 to 8.8

Match each of the following key terms with the correct description:

    a. molar volume   **b.** STP   **c.** Dalton's law
    d. partial pressure   **e.** Avogadro's law **f.** ideal gas law

1. _____ the gas law stating that the total pressure exerted by a mixture of gases in a container is the sum of the pressures that each gas would exert alone

2. _____ the pressure exerted by a single gas in a gas mixture

Oxygen therapy increases the amount of oxygen available to the tissues of the body.

Credit: Levent Konuk/Shutterstock

3. ____ the volume occupied by 1 mole of a gas at STP

4. ____ the law that combines the four measured properties of a gas in the equation $PV = nRT$

5. ____ the gas law stating that the volume of a gas is directly related to the number of moles of gas when pressure and temperature do not change

6. ____ the standard conditions of 0 °C (273 K) and 1 atm, used for the comparison of gases

Answers   **1.** c   **2.** d   **3.** a   **4.** f   **5.** e   **6.** b

Calculating Partial Pressure of a Gas in a Mixture	
STEP 1	Write the equation for the sum of the partial pressures.
STEP 2	Rearrange the equation to solve for the unknown pressure.
STEP 3	Substitute known pressures into the equation, and calculate the unknown pressure.

♦ **Learning Exercise 8.8A**

Solve each of the following problems using Dalton's law:

CORE CHEMISTRY SKILL
Calculating Partial Pressure

a. What is the total pressure, in millimeters of mercury, of a gas sample that contains O_2 at 0.500 atm, N_2 at 132 Torr, and He at 224 mmHg?

b. What is the total pressure, in atmospheres, of a gas sample that contains He at 285 mmHg and O_2 at 1.20 atm?

c. A gas sample containing N_2 and O_2 has a pressure of 1500. mmHg. If the partial pressure of the N_2 is 0.900 atm, what is the partial pressure, in millimeters of mercury, of the O_2 in the mixture?

Answers   **a.** 736 mmHg   **b.** 1.58 atm   **c.** 816 mmHg

♦ **Learning Exercise 8.8B**

Complete the following table for typical blood gas values for partial pressures:

Gas	Alveoli	Oxygenated blood	Deoxygenated blood	Tissues
O_2	_____	_____	_____	_____
CO_2	_____	_____	_____	_____

Answers

Gas	Alveoli	Oxygenated blood	Deoxygenated blood	Tissues
O_2	100 mmHg	100 mmHg	40 mmHg or less	30 mmHg or less
CO_2	40 mmHg	40 mmHg	46 mmHg	50 mmHg or greater

Checklist for Chapter 8

You are ready to take the Practice Test for Chapter 8. Be sure you have accomplished the following learning goals for this chapter. If not, review the Section listed at the end of the goal. Then apply your new skills and understanding to the Practice Test.

After studying Chapter 8, I can successfully:

_____ Describe the kinetic molecular theory of gases. (8.1)

_____ Change the units of pressure from one to another. (8.1)

_____ Use the pressure–volume relationship (Boyle's law) to calculate the unknown pressure or volume when the temperature and amount of gas do not change. (8.2)

_____ Use the temperature–volume relationship (Charles's law) to calculate the unknown temperature or volume when the pressure and amount of gas do not change. (8.3)

_____ Use the temperature–pressure relationship (Gay-Lussac's law) to calculate the unknown temperature or pressure when the volume and amount of gas do not change. (8.4)

_____ Use the combined gas law to calculate the unknown pressure, volume, or temperature of a fixed amount of gas when changes in two of these properties are given when the amount of gas does not change. (8.5)

_____ Describe the relationship between the amount of a gas and its volume (Avogadro's law) when the pressure and temperature do not change and use this relationship in calculations. (8.6)

_____ Use the ideal gas law to solve for pressure, volume, temperature, or amount of a gas. (8.7)

_____ Calculate the total pressure of a gas mixture from the partial pressures. (8.8)

Practice Test for Chapter 8

The chapter Sections to review are shown in parentheses at the end of each question.

For questions 1 through 5, use the kinetic molecular theory of gases to answer true (T) *or false* (F): (8.1)

1. _____ A gas does not have its own volume or shape.

2. _____ The molecules of a gas are moving extremely fast.

3. _____ The collisions of gas molecules with the walls of their container create pressure.

4. _____ Gas molecules are close together and move in straight lines.

5. _____ The attractive forces between molecules are very small.

6. The pressure of a gas will increase when (8.1, 8.2)
   A. the volume increases
   B. the temperature decreases
   C. more molecules of gas are added
   D. molecules of gas are removed
   E. none of these

7. What is the pressure, in atmospheres, of a gas with a pressure of 1200 mmHg? (8.1)
   A. 0.63 atm   **B.** 0.79 atm   **C.** 1.2 atm
   D. 1.6 atm   **E.** 2.0 atm

8. The relationship that the volume of a gas is inversely related to its pressure (when temperature and amount of gas do not change) is known as (8.2)
   A. Boyle's law   **B.** Charles's law   **C.** Gay-Lussac's law
   D. Dalton's law   **E.** Avogadro's law

9. A 6.00-L sample of O_2 has a pressure of 660. mmHg. When the volume is reduced to 2.00 L with no change in temperature and amount of gas, it will have a pressure of (8.2)
   A. 1980 mmHg
   B. 1320 mmHg
   C. 330. mmHg
   D. 220. mmHg
   E. 110. mmHg

10. If the temperature and amount of a gas do not change, but its volume doubles, its pressure will (8.2)
    A. double
    B. triple
    C. decrease to one-half the original pressure
    D. decrease to one-fourth the original pressure
    E. not change

11. If the temperature of a gas in a flexible container is increased (8.3, 8.4)
    A. the pressure will decrease
    B. the volume will increase
    C. the volume will decrease
    D. the number of molecules will increase
    E. none of these

12. When a gas is heated in a closed metal container, the (8.4)
    A. pressure increases
    B. pressure decreases
    C. volume increases
    D. volume decreases
    E. number of molecules increases

13. A sample of N_2 at 110 K has a pressure of 1.0 atm. When the temperature is increased to 360 K when the volume and amount of gas do not change, what is the pressure? (8.4)
    A. 0.50 atm
    B. 1.0 atm
    C. 1.5 atm
    D. 3.3 atm
    E. 4.0 atm

14. A gas sample with a volume of 4.0 L has a pressure of 750 mmHg and a temperature of 77 °C. What is its volume at 277 °C and 250 mmHg, when the amount of gas does not change? (8.5)
    A. 7.6 L
    B. 19 L
    C. 2.1 L
    D. 0.000 56 L
    E. 3.3 L

15. If two gases have the same volume, temperature, and pressure, they also have the same (8.6)
    A. density
    B. number of particles
    C. molar mass
    D. speed
    E. size molecules

16. A tank of O_2 used at home for a patient with pulmonary disease has a pressure of 220 mmHg, 2.0 moles of gas, and a volume of 4.0 L. What will the pressure be when the volume expands to 5.0 L and 3.0 moles of O_2 is added, when the temperature does not change? (8.5, 8.6)
    A. 110 mmHg
    B. 220 mmHg
    C. 440 mmHg
    D. 550 mmHg
    E. 700. mmHg

17. A sample of 2.00 moles of gas initially at STP is converted to a volume of 5.0 L and a temperature of 27 °C. What is the pressure, in atmospheres, if there is no change in the amount of gas? (8.5, 8.6)
    A. 0.12 atm
    B. 5.5 atm
    C. 7.5 atm
    D. 9.8 atm
    E. 10. atm

18. The conditions for standard temperature and pressure (STP) are (8.6)
    A. 0 K, 1 atm
    B. 0 °C, 10 atm
    C. 25 °C, 1 atm
    D. 273 K, 1 atm
    E. 273 K, 0.5 atm

19. The volume occupied by 1.50 moles of CH_4 at STP is (8.6)
    A. 44.1 L
    B. 33.6 L
    C. 22.4 L
    D. 11.2 L
    E. 5.60 L

20. How many grams of O_2 are present in 44.1 L of O_2 at STP? (8.6)
    A. 10.0 g   **B.** 16.0 g   **C.** 32.0 g
    D. 410.0 g   **E.** 63.0 g

21. What is the volume, in liters, of 0.50 mole of N_2 at 25 °C and 2.0 atm? (8.7)
    A. 0.51 L   **B.** 1.0 L   **C.** 4.2 L
    D. 6.1 L   **E.** 24 L

22. 2.50 g of $KClO_3$ decomposes by the equation: $2KClO_3(s) \longrightarrow 2KCl(s) + 3O_2(g)$. What volume, in milliliters, of O_2 is produced at 25 °C and a pressure of 750. mmHg? (8.7)
    A. 9.98 mL   **B.** 759 mL   **C.** 63.6 mL
    D. 506 mL   **E.** 1320 mL

23. A gas mixture contains He with a partial pressure of 0.100 atm, O_2 with a partial pressure of 445 mmHg, and N_2 with a partial pressure of 235 Torr. What is the total pressure, in atmospheres, of the gas mixture? (8.8)
    A. 0.995 atm   **B.** 1.39 atm   **C.** 1.69 atm
    D. 2.00 atm   **E.** 10.0 atm

24. A mixture of O_2 and N_2 has a total pressure of 1040 mmHg. If O_2 has a partial pressure of 510 mmHg, what is the partial pressure of the N_2? (8.8)
    A. 240 mmHg   **B.** 530 mmHg   **C.** 770 mmHg
    D. 1040 mmHg   **E.** 1350 mmHg

25. 3.00 moles of He in a steel container has a pressure of 12.0 atm. What is the pressure after 4.00 moles of Ne is added, if there is no change in temperature? (8.8)
    A. 5.14 atm   **B.** 16.0 atm   **C.** 28.0 atm
    D. 32.0 atm   **E.** 45.0 atm

26. The exchange of gases between the alveoli, blood, and tissues of the body is a result of (8.8)
    A. pressure gradients   **B.** different molecular masses   **C.** shapes of molecules
    D. altitude   **E.** all of these

27. Oxygen moves into the tissues from the blood because its partial pressure (8.8)
    A. in arterial blood is higher than in the tissues
    B. in venous blood is higher than in the tissues
    C. in arterial blood is lower than in the tissues
    D. in venous blood is lower than in the tissues
    E. is equal in the blood and in the tissues

Answers to the Practice Test

1. T	**2.** T	**3.** T	**4.** F	**5.** T
6. C	**7.** D	**8.** A	**9.** A	**10.** C
11. B	**12.** A	**13.** D	**14.** B	**15.** B
16. C	**17.** D	**18.** D	**19.** B	**20.** E
21. D	**22.** B	**23.** A	**24.** B	**25.** C
26. A	**27.** A			

Selected Answers and Solutions to Text Problems

8.1   **a.** At a higher temperature, gas particles have greater kinetic energy, which makes them move faster.

       b. Because there are great distances between the particles of a gas, they can be pushed closer together and still remain a gas.

       c. Gas particles are very far apart, meaning that the mass of a gas in a certain volume is very small, resulting in a low density.

8.3   **a.** The temperature of a gas can be expressed in kelvins.

       b. The volume of a gas can be expressed in milliliters.

       c. The amount of a gas can be expressed in grams.

       d. Pressure can be expressed in millimeters of mercury (mmHg).

8.5   Statements **a**, **d**, and **e** describe the pressure of a gas.

8.7   **a.** $2.00 \ \cancel{\text{atm}} \times \dfrac{760 \text{ Torr}}{1 \ \cancel{\text{atm}}} = 1520 \text{ Torr (3 SFs)}$

       b. $2.00 \ \cancel{\text{atm}} \times \dfrac{14.7 \text{ lb/in.}^2}{1 \ \cancel{\text{atm}}} = 29.4 \text{ lb/in.}^2 \text{ (3 SFs)}$

       c. $2.00 \ \cancel{\text{atm}} \times \dfrac{760 \text{ mmHg}}{1 \ \cancel{\text{atm}}} = 1520 \text{ mmHg (3 SFs)}$

       d. $2.00 \ \cancel{\text{atm}} \times \dfrac{101.325 \text{ kPa}}{1 \ \cancel{\text{atm}}} = 203 \text{ kPa (3 SFs)}$

8.9   As the scuba diver ascends to the surface, external pressure decreases. If the air in the lungs, which is at a higher pressure, were not exhaled, its volume would expand and severely damage the lungs. The pressure of the gas in the lungs must adjust to changes in the external pressure.

8.11   **a.** The pressure is greater in cylinder A. According to Boyle's law, a decrease in volume pushes the gas particles closer together, which will cause an increase in the pressure.

       b. From Boyle's law, we know that pressure is inversely related to volume. The mmHg units must be converted to atm for unit cancellation in the calculation, and because the pressure increases, volume must decrease.

       According to Boyle's law, $P_1 V_1 = P_2 V_2$, then

$$V_2 = V_1 \times \frac{P_1}{P_2} = 220 \text{ mL} \times \frac{650 \ \cancel{\text{mmHg}}}{1.2 \ \cancel{\text{atm}}} \times \frac{1 \ \cancel{\text{atm}}}{760 \ \cancel{\text{mmHg}}} = 160 \text{ mL (2 SFs)}$$

8.13   **a.** The pressure of the gas *increases* to two times the original pressure when the volume is halved.

       b. The pressure *decreases* to one-third the original pressure when the volume expands to three times its initial value.

       c. The pressure *increases* to 10 times the original pressure when the volume decreases to one-tenth of the initial volume.

8.15   From Boyle's law, we know that pressure is inversely related to volume (for example, pressure increases when volume decreases).

       a. Volume increases; pressure must decrease.

$$P_2 = P_1 \times \frac{V_1}{V_2} = 655 \text{ mmHg} \times \frac{10.0 \ \cancel{L}}{20.0 \ \cancel{L}} = 328 \text{ mmHg (3 SFs)}$$

       b. Volume decreases; pressure must increase.

$$P_2 = P_1 \times \frac{V_1}{V_2} = 655 \text{ mmHg} \times \frac{10.0 \ \cancel{L}}{2.50 \ \cancel{L}} = 2620 \text{ mmHg (3 SFs)}$$

c. The mL units must be converted to L for unit cancellation in the calculation, and because the volume increases, pressure must decrease.

$$P_2 = P_1 \times \frac{V_1}{V_2} = 655 \text{ mmHg} \times \frac{10.0 \text{ L}}{13\,800 \text{ mL}} \times \frac{1000 \text{ mL}}{1 \text{ L}} = 475 \text{ mmHg (3 SFs)}$$

d. The mL units must be converted to L for unit cancellation in the calculation, and because the volume decreases, pressure must increase.

$$P_2 = P_1 \times \frac{V_1}{V_2} = 655 \text{ mmHg} \times \frac{10.0 \text{ L}}{1250 \text{ mL}} \times \frac{1000 \text{ mL}}{1 \text{ L}} = 5240 \text{ mmHg (3 SFs)}$$

8.17 From Boyle's law, we know that pressure is inversely related to volume.

a. Pressure decreases; volume must increase.

$$V_2 = V_1 \times \frac{P_1}{P_2} = 50.0 \text{ L} \times \frac{760. \text{ mmHg}}{725 \text{ mmHg}} = 52.4 \text{ L (3 SFs)}$$

b. Pressure increases; volume must decrease.

$$V_2 = V_1 \times \frac{P_1}{P_2} = 50.0 \text{ L} \times \frac{760. \text{ mmHg}}{1520 \text{ mmHg}} = 25.0 \text{ L (3 SFs)}$$

c. The mmHg units must be converted to atm for unit cancellation in the calculation, and because the pressure decreases, volume must increase.

$$P_1 = 760. \text{ mmHg} \times \frac{1 \text{ atm}}{760 \text{ mmHg}} = 1.00 \text{ atm}$$

$$V_2 = V_1 \times \frac{P_1}{P_2} = 50.0 \text{ L} \times \frac{1.00 \text{ atm}}{0.500 \text{ atm}} = 100. \text{ L (3 SFs)}$$

d. The mmHg units must be converted to torr for unit cancellation in the calculation, and because the pressure increases, volume must decrease.

$$P_1 = 760. \text{ mmHg} \times \frac{760 \text{ Torr}}{760 \text{ mmHg}} = 760. \text{ Torr}$$

$$V_2 = V_1 \times \frac{P_1}{P_2} = 50.0 \text{ L} \times \frac{760. \text{ Torr}}{850 \text{ Torr}} = 45 \text{ L (2 SFs)}$$

8.19 Volume increases; initial pressure must have been higher.

$$P_1 = P_2 \times \frac{V_2}{V_1} = 3.62 \text{ atm} \times \frac{9.73 \text{ L}}{5.40 \text{ L}} = 6.52 \text{ atm (3 SFs)}$$

8.21 Pressure decreases; volume must increase.

$$V_2 = V_1 \times \frac{P_1}{P_2} = 5.0 \text{ L} \times \frac{5.0 \text{ atm}}{1.0 \text{ atm}} = 25 \text{ L (2 SFs)}$$

8.23 a. *Inspiration* begins when the diaphragm contracts, causing the lungs to expand. The increased volume of the thoracic cavity reduces the pressure in the lungs such that air flows into the lungs.

b. *Expiration* occurs as the diaphragm relaxes, causing a decrease in the volume of the lungs. The pressure of the air in the lungs increases, and air flows out of the lungs.

c. *Inspiration* occurs when the pressure within the lungs is less than that of the atmosphere.

8.25 According to Charles's law, there is a direct relationship between Kelvin temperature and volume (for example, volume increases when temperature increases, if the pressure and amount of gas do not change).

a. Diagram C shows an increased volume corresponding to an increase in temperature.

b. Diagram A shows a decreased volume corresponding to a decrease in temperature.

c. Diagram B shows no change in volume, which corresponds to no net change in temperature.

8.27 According to Charles's law, the volume of a gas is directly related to the Kelvin temperature. In all gas law computations, temperatures must be in kelvins. (Temperatures in °C are converted to K by the addition of 273.) The initial temperature for all cases here is $T_1 = 15\,°C + 273 = 288\,K$.

a. Volume increases; temperature must have increased.

$$T_2 = T_1 \times \frac{V_2}{V_1} = 288\,K \times \frac{5.00\,L}{2.50\,L} = 576\,K \quad 576\,K - 273 = 303\,°C\ (3\ SFs)$$

b. Volume decreases; temperature must have decreased.

$$T_2 = T_1 \times \frac{V_2}{V_1} = 288\,K \times \frac{1250\,mL}{2.50\,L} \times \frac{1\,L}{1000\,mL} = 144\,K \quad 144\,K - 273 = -129\,°C\ (3\ SFs)$$

c. Volume increases; temperature must have increased.

$$T_2 = T_1 \times \frac{V_2}{V_1} = 288\,K \times \frac{7.50\,L}{2.50\,L} = 864\,K \quad 864\,K - 273 = 591\,°C\ (3\ SFs)$$

d. Volume increases; temperature must have increased.

$$T_2 = T_1 \times \frac{V_2}{V_1} = 288\,K \times \frac{3550\,mL}{2.50\,L} \times \frac{1\,L}{1000\,mL} = 409\,K \quad 409\,K - 273 = 136\,°C\ (3\ SFs)$$

8.29 According to Charles's law, the volume of a gas is directly related to the Kelvin temperature. In all gas law computations, temperatures must be in kelvins. (Temperatures in °C are converted to K by the addition of 273.) The initial temperature for all cases here is $T_1 = 75\,°C + 273 = 348\,K$.

a. When temperature decreases, volume must also decrease.

$$T_2 = 55\,°C + 273 = 328\,K$$

$$V_2 = V_1 \times \frac{T_2}{T_1} = 2500\,mL \times \frac{328\,K}{348\,K} = 2400\,mL\ (2\ SFs)$$

b. When temperature increases, volume must also increase.

$$V_2 = V_1 \times \frac{T_2}{T_1} = 2500\,mL \times \frac{680.\,K}{348\,K} = 4900\,mL\ (2\ SFs)$$

c. When temperature decreases, volume must also decrease.

$$T_2 = -25\,°C + 273 = 248\,K$$

$$V_2 = V_1 \times \frac{T_2}{T_1} = 2500\,mL \times \frac{248\,K}{348\,K} = 1800\,mL\ (2\ SFs)$$

d. When temperature decreases, volume must also decrease.

$$V_2 = V_1 \times \frac{T_2}{T_1} = 2500\,mL \times \frac{240.\,K}{348\,K} = 1700\,mL\ (2\ SFs)$$

8.31 Volume decreases; initial temperature must have been higher.

$$T_2 = 32\,°C + 273 = 305\,K$$

$$T_1 = T_2 \times \frac{V_1}{V_2} = 305\,K \times \frac{0.256\,L}{0.198\,L} = 394\,K \quad 394\,K - 273 = 121\,°C\ (3\ SFs)$$

8.33 According to Gay-Lussac's law, temperature is directly related to pressure. In all gas law computations, temperatures must be in kelvins. (Temperatures in °C are converted to K by the addition of 273.)

a. When temperature decreases, pressure must also decrease.

$$T_1 = 155\,°C + 273 = 428\,K \qquad T_2 = 0\,°C + 273 = 273\,K$$

$$P_2 = P_1 \times \frac{T_2}{T_1} = 1200\,Torr \times \frac{1\,mmHg}{1\,Torr} \times \frac{273\,K}{428\,K} = 770\,mmHg\ (2\ SFs)$$

b. When temperature increases, pressure must also increase.

$T_1 = 12\,°C + 273 = 285\,K$ $T_2 = 35\,°C + 273 = 308\,K$

$$P_2 = P_1 \times \frac{T_2}{T_1} = 1.40\,\text{atm} \times \frac{760\,\text{mmHg}}{1\,\text{atm}} \times \frac{308\,K}{285\,K} = 1150\,\text{mmHg (3 SFs)}$$

8.35 According to Gay-Lussac's law, pressure is directly related to temperature. In all gas law computations, temperatures must be in kelvins. (Temperatures in °C are converted to K by the addition of 273.)

a. Pressure decreases; temperature must have decreased.

$T_1 = 25\,°C + 273 = 298\,K$

$$T_2 = T_1 \times \frac{P_2}{P_1} = 298\,K \times \frac{620.\,\text{mmHg}}{740.\,\text{mmHg}} = 250.\,K \quad 250.\,K - 273 = -23\,°C\ (2\ \text{SFs})$$

b. Pressure increases; temperature must have increased.

$T_1 = -18\,°C + 273 = 255\,K$

$$T_2 = T_1 \times \frac{P_2}{P_1} = 255\,K \times \frac{1250\,\text{Torr}}{0.950\,\text{atm}} \times \frac{1\,\text{atm}}{760\,\text{Torr}} = 441\,K \quad 441\,K - 273 = 168\,°C\ (3\ \text{SFs})$$

8.37 Pressure increases; temperature must have increased.

$T_1 = 22\,°C + 273 = 295\,K$

$$T_2 = \frac{T_1}{P_1} \times P_2 = 295\,K \times \frac{766\,\text{mmHg}}{744\,\text{mmHg}} = 304\,K \quad 304\,K - 273 = 31\,°C\ (2\ \text{SFs})$$

8.39 According to Gay-Lussac's law, pressure is directly related to temperature. In all gas law computations, temperatures must be in kelvins. (Temperatures in °C are converted to K by the addition of 273.)

When temperature increases, pressure must also increase.

$T_1 = 5\,°C + 273 = 278\,K$ $T_2 = 22\,°C + 273 = 295\,K$

$$P_2 = P_1 \times \frac{T_2}{T_1} = 1.8\,\text{atm} \times \frac{295\,K}{278\,K} = 1.9\,\text{atm}\ (2\ \text{SFs})$$

8.41 $\dfrac{P_1 V_1}{T_1} = \dfrac{P_2 V_2}{T_2}$ $\dfrac{P_1 V_1}{T_1} \times \dfrac{T_2 T_1}{P_1 V_1} = \dfrac{P_2 V_2}{T_2} \times \dfrac{T_2 T_1}{P_1 V_1}$ $\therefore\ T_2 = T_1 \times \dfrac{P_2}{P_1} \times \dfrac{V_2}{V_1}$

8.43 $T_1 = 25\,°C + 273 = 298\,K;$ $V_1 = 6.50\,L;$ $P_1 = 845\,\text{mmHg}$

a. $T_2 = 325\,K;$ $V_2 = 1850\,mL = 1.85\,L;$ $P_2 = ?\,\text{atm}$

$$P_2 = P_1 \times \frac{V_1}{V_2} \times \frac{T_2}{T_1} = 845\,\text{mmHg} \times \frac{6.50\,L}{1.85\,L} \times \frac{325\,K}{298\,K} \times \frac{1\,\text{atm}}{760\,\text{mmHg}} = 4.26\,\text{atm (3 SFs)}$$

b. $T_2 = 12\,°C + 273 = 285\,K;$ $V_2 = 2.25\,L;$ $P_2 = ?\,\text{atm}$

$$P_2 = P_1 \times \frac{V_1}{V_2} \times \frac{T_2}{T_1} = 845\,\text{mmHg} \times \frac{6.50\,L}{2.25\,L} \times \frac{285\,K}{298\,K} \times \frac{1\,\text{atm}}{760\,\text{mmHg}} = 3.07\,\text{atm (3 SFs)}$$

c. $T_2 = 47\,°C + 273 = 320.\,K;$ $V_2 = 12.8\,L;$ $P_2 = ?\,\text{atm}$

$$P_2 = P_1 \times \frac{V_1}{V_2} \times \frac{T_2}{T_1} = 845\,\text{mmHg} \times \frac{6.50\,L}{12.8\,L} \times \frac{320.\,K}{298\,K} \times \frac{1\,\text{atm}}{760\,\text{mmHg}} = 0.606\,\text{atm (3 SFs)}$$

8.45 $T_1 = 212\,°C + 273 = 485\,K;$ $V_1 = 124\,mL;$ $P_1 = 1.80\,\text{atm}$

$T_2 = ?\,°C;$ $V_2 = 138\,mL;$ $P_2 = 0.800\,\text{atm}$

$$T_2 = T_1 \times \frac{P_2}{P_1} \times \frac{V_2}{V_1} = 485\,K \times \frac{0.800\,\text{atm}}{1.80\,\text{atm}} \times \frac{138\,mL}{124\,mL} = 240.\,K$$

$240.\,K - 273 = -33\,°C\ (2\ \text{SFs})$

8.47 The volume increases because the number of gas particles in the tire or basketball is increased.

8.49 According to Avogadro's law, the volume of a gas is directly related to the number of moles of gas.
$n_1 = 1.50$ moles Ne; $V_1 = 8.00$ L

a. $V_2 = V_1 \times \dfrac{n_2}{n_1} = 8.00 \text{ L} \times \dfrac{\frac{1}{2}(1.50) \text{ moles Ne}}{1.50 \text{ moles Ne}} = 4.00$ L (3 SFs)

b. $n_2 = 1.50$ moles Ne $+ 3.50$ moles Ne $= 5.00$ moles of Ne

$V_2 = V_1 \times \dfrac{n_2}{n_1} = 8.00 \text{ L} \times \dfrac{5.00 \text{ moles Ne}}{1.50 \text{ moles Ne}} = 26.7$ L (3 SFs)

c. $25.0 \text{ g Ne} \times \dfrac{1 \text{ mole Ne}}{20.18 \text{ g Ne}} = 1.24$ moles of Ne added

$n_2 = 1.50$ moles Ne $+ 1.24$ moles Ne $= 2.74$ moles of Ne

$V_2 = V_1 \times \dfrac{n_2}{n_1} = 8.00 \text{ L} \times \dfrac{2.74 \text{ moles Ne}}{1.50 \text{ moles Ne}} = 14.6$ L (3 SFs)

8.51 At STP, 1 mole of any gas occupies a volume of 22.4 L.

a. $2.24 \text{ L } O_2 \text{ (STP)} \times \dfrac{1 \text{ mole } O_2}{22.4 \text{ L } O_2 \text{ (STP)}} = 0.100$ moles of O_2 (3 SFs)

b. $2.50 \text{ moles } N_2 \times \dfrac{22.4 \text{ L } N_2 \text{ (STP)}}{1 \text{ mole } N_2} = 56.0$ L at STP (3 SFs)

c. $50.0 \text{ g Ar} \times \dfrac{1 \text{ mole Ar}}{39.95 \text{ g Ar}} \times \dfrac{22.4 \text{ L Ar (STP)}}{1 \text{ mole Ar}} = 28.0$ L at STP (3 SFs)

d. $1620 \text{ mL } H_2 \text{ (STP)} \times \dfrac{1 \text{ L } H_2}{1000 \text{ mL } H_2} \times \dfrac{1 \text{ mole } H_2}{22.4 \text{ L } H_2 \text{ (STP)}} \times \dfrac{2.016 \text{ g } H_2}{1 \text{ mole } H_2}$

$= 0.146$ g of H_2 (3 SFs)

8.53 **Analyze the Problem: Using the Ideal Gas Law to Calculate Pressure**

Property	P	V	n	R	T
Given		10.0 L	2.00 moles	$\dfrac{0.0821 \text{ L} \cdot \text{atm}}{\text{mole} \cdot \text{K}}$	27 °C 27 °C + 273 = 300. K
Need	? atm				

$$P = \dfrac{nRT}{V} = \dfrac{(2.00 \text{ moles})\left(\dfrac{0.0821 \text{ L} \cdot \text{atm}}{\text{mole} \cdot \text{K}}\right)(300. \text{ K})}{(10.0 \text{ L})} = 4.93 \text{ atm (3 SFs)}$$

8.55 **Analyze the Problem: Using the Ideal Gas Law to Calculate Moles**

Property	P	V	n	R	T
Given	845 mmHg	20.0 L		$\dfrac{62.4 \text{ L} \cdot \text{mmHg}}{\text{mole} \cdot \text{K}}$	22 °C 22 °C + 273 = 295 K
Need			? moles (? g)		

$$n = \dfrac{PV}{RT} = \dfrac{(845 \text{ mmHg})(20.0 \text{ L})}{\left(\dfrac{62.4 \text{ L} \cdot \text{mmHg}}{\text{mole} \cdot \text{K}}\right)(295 \text{ K})} = 0.918 \text{ mole of } O_2$$

$0.918 \text{ mole } O_2 \times \dfrac{32.00 \text{ g } O_2}{1 \text{ mole } O_2} = 29.4$ g of O_2 (3 SFs)

8.57 $n = 25.0 \text{ g N}_2 \times \dfrac{1 \text{ mole N}_2}{28.02 \text{ g N}_2} = 0.892 \text{ mole of N}_2$

Analyze the Problem: Using the Ideal Gas Law to Calculate Temperature

Property	P	V	n	R	T
Given	630. mmHg	50.0 L	25.0 g (0.892 mole)	$\dfrac{62.4 \text{ L} \cdot \text{mmHg}}{\text{mole} \cdot \text{K}}$	
Need					? K (? °C)

$$T = \frac{PV}{nR} = \frac{(630. \text{ mmHg})(50.0 \text{ L})}{(0.892 \text{ mole})\left(\dfrac{62.4 \text{ L} \cdot \text{mmHg}}{\text{mole} \cdot \text{K}}\right)} = 566 \text{ K} - 273 = 293 \text{ °C (3 SFs)}$$

8.59 **a.** $8.25 \text{ g Mg} \times \dfrac{1 \text{ mole Mg}}{24.31 \text{ g Mg}} \times \dfrac{1 \text{ mole H}_2}{1 \text{ mole Mg}} \times \dfrac{22.4 \text{ L (STP)}}{1 \text{ mole H}_2} = 7.60 \text{ L of H}_2$

released at STP (3 SFs)

b. **Analyze the Problem: Using the Ideal Gas Law to Calculate Moles**

Property	P	V	n	R	T
Given	735 mmHg	5.00 L		$\dfrac{62.4 \text{ L} \cdot \text{mmHg}}{\text{mole} \cdot \text{K}}$	18 °C 18 °C + 273 = 291 K
Need			? moles H₂		

$$n = \frac{PV}{RT} = \frac{(735 \text{ mmHg})(5.00 \text{ L})}{\left(\dfrac{62.4 \text{ L} \cdot \text{mmHg}}{\text{mole} \cdot \text{K}}\right)(291 \text{ K})} = 0.202 \text{ mole of H}_2$$

Analyze the Problem: Reactant and Product

	Reactant	Product
Mass	? g Mg	
Molar Mass	1 mole of Mg = 24.31 g of Mg	
Moles		0.202 mole H₂
Equation	2HCl(aq) + Mg(s) ⟶ H₂(g) + MgCl₂(aq)	

$$0.202 \text{ mole H}_2 \times \frac{1 \text{ mole Mg}}{1 \text{ mole H}_2} \times \frac{24.31 \text{ g Mg}}{1 \text{ mole Mg}} = 4.91 \text{ g of Mg (3 SFs)}$$

8.61 **Analyze the Problem: Reactant and Reactant**

	Reactant 1	Reactant 2
Mass	55.2 g C₄H₁₀	
Molar Mass	1 mole of C₄H₁₀ = 58.12 g of C₄H₁₀	
Moles		? moles O₂
Equation	2C₄H₁₀(g) + 13O₂(g) $\xrightarrow{\Delta}$ 8CO₂(g) + 10H₂O(g)	

$$55.2 \text{ g C}_4\text{H}_{10} \times \frac{1 \text{ mole C}_4\text{H}_{10}}{58.12 \text{ g C}_4\text{H}_{10}} \times \frac{13 \text{ moles O}_2}{2 \text{ moles C}_4\text{H}_{10}} = 6.17 \text{ moles of O}_2$$

Analyze the Problem: Using the Ideal Gas Law to Calculate Volume

Property	P	V	n	R	T
Given	0.850 atm		6.17 moles O_2	$\dfrac{0.0821 \text{ L} \cdot \text{atm}}{\text{mole} \cdot \text{K}}$	25 °C 25 °C + 273 = 298 K
Need		? L			

$$V = \frac{nRT}{P} = \frac{(6.17 \text{ moles})\left(\dfrac{0.0821 \text{ L} \cdot \text{atm}}{\text{mole} \cdot \text{K}}\right)(298 \text{ K})}{(0.850 \text{ atm})} = 178 \text{ L of } O_2 \text{ (3 SFs)}$$

8.63 **Analyze the Problem: Reactant and Reactant**

	Reactant 1	Reactant 2
Mass	5.4 g Al	
Molar Mass	1 mole of Al = 26.98 g of Al	
Moles		? moles O_2 (? L O_2)
Molar Volume		1 mole of O_2 = 22.4 L of O_2 (STP)
Equation	$4Al(s) + 3O_2(g) \xrightarrow{\Delta} 2Al_2O_3(s)$	

$$5.4 \text{ g Al} \times \frac{1 \text{ mole Al}}{26.98 \text{ g Al}} \times \frac{3 \text{ moles } O_2}{4 \text{ moles Al}} \times \frac{22.4 \text{ L } O_2 \text{ (STP)}}{1 \text{ mole } O_2} = 3.4 \text{ L of } O_2 \text{ at STP (2 SFs)}$$

8.65 **Analyze the Problem: Using the Ideal Gas Law to Calculate Moles**

Property	P	V	n	R	T
Given	1.6 atm	640 L		$\dfrac{0.0821 \text{ L} \cdot \text{atm}}{\text{mole} \cdot \text{K}}$	24 °C 24 °C + 273 = 297 K
Need			? moles O_2		

$$n = \frac{PV}{RT} = \frac{(1.6 \text{ atm})(640 \text{ L})}{\left(\dfrac{0.0821 \text{ L} \cdot \text{atm}}{\text{mole} \cdot \text{K}}\right)(297 \text{ K})} = 42 \text{ moles of } O_2$$

$$42 \text{ moles } O_2 \times \frac{32.00 \text{ g } O_2}{1 \text{ mole } O_2} = 1300 \text{ g of } O_2 \text{ (2 SFs)}$$

8.67 To obtain the total pressure of a gas mixture, add up all of the partial pressures using the same pressure unit.

$$P_{total} = P_{nitrogen} + P_{oxygen} + P_{helium} = 425 \text{ Torr} + 115 \text{ Torr} + 225 \text{ Torr} = 765 \text{ Torr (3 SFs)}$$

8.69 Because the total pressure of a gas mixture is the sum of the partial pressures using the same pressure unit, addition and subtraction are used to obtain the "missing" partial pressure.

$$P_{nitrogen} = P_{total} - (P_{oxygen} + P_{helium}) = 925 \text{ Torr} - (425 \text{ Torr} + 75 \text{ Torr}) = 425 \text{ Torr (3 SFs)}$$

8.71 **a.** If oxygen cannot readily cross from the lungs into the bloodstream, then the partial pressure of oxygen will be lower in the blood of an emphysema patient.

b. Breathing a higher concentration of oxygen will help to increase the supply of oxygen in the lungs and blood and raise the partial pressure of oxygen in the blood.

8.73 $P_{total} = P_{oxygen} + P_{nitrogen} + P_{carbon\ dioxide} + P_{water\ vapor}$

$\quad\quad = 93 \text{ mmHg} + 565 \text{ mmHg} + 38 \text{ mmHg} + 47 \text{ mmHg} = 743 \text{ mmHg}$

$$743 \text{ mmHg} \times \frac{1.00 \text{ atm}}{760 \text{ mmHg}} = 0.978 \text{ atm (3 SFs)}$$

8.75 $T_1 = 37\,°C + 273 = 310\,K$; $\quad P_1 = 745\,mmHg$; $\quad V_1 = 3.2\,L$

$T_2 = 0\,°C + 273 = 273\,K$; $\quad P_2 = 760\,mmHg$; $\quad V_2 = ?\,L$

$$V_2 = V_1 \times \frac{P_1}{P_2} \times \frac{T_2}{T_1} = 3.2\,L \times \frac{745\,\cancel{mmHg}}{760\,\cancel{mmHg}} \times \frac{273\,\cancel{K}}{310\,\cancel{K}} = 2.8\,L\ (2\ SFs)$$

8.77 a. True. The flask containing helium has more moles of helium and thus more atoms of helium to collide with the walls of the container to give a higher pressure.

b. True. The mass and volume of each are the same, meaning that the mass/volume ratio or density is the same in both flasks.

8.79 a. 2; The fewest number of gas particles will exert the lowest pressure.

b. 1; The greatest number of gas particles will exert the highest pressure.

8.81 a. A; Volume decreases when temperature decreases (when P and n do not change).

b. C; Volume increases when pressure decreases (when n and T do not change).

c. A; Volume decreases when the number of moles of gas decreases (when T and P do not change).

d. B; Doubling the Kelvin temperature would double the volume, but when half of the gas escapes, the volume would decrease by half. These two opposing effects cancel each other, and there is no overall change in the volume (when P does not change).

e. C; Increasing the moles of gas causes an increase in the volume (when T and P do not change)

8.83 a. The volume of the chest and lungs will decrease when compressed during the Heimlich maneuver.

b. A decrease in volume causes the pressure to increase. A piece of food would be dislodged with a sufficiently high pressure.

8.85 $31\,000\,L\ \cancel{H_2\,(STP)} \times \dfrac{1\,mole\ \cancel{H_2}}{22.4\,L\ \cancel{H_2\,(STP)}} \times \dfrac{2.016\,g\ \cancel{H_2}}{1\,mole\ \cancel{H_2}} \times \dfrac{1\,kg\ H_2}{1000\,g\ \cancel{H_2}} = 2.8\,kg\ of\ H_2\,(2\ SFs)$

8.87 $T_1 = 127\,°C + 273 = 400.\,K$; $\quad P_1 = 2.00\,atm$

$T_2 = ?\,°C$; $\quad P_2 = 0.25\,atm$

$$T_2 = T_1 \times \frac{P_2}{P_1} = 400.\,K \times \frac{0.25\,\cancel{atm}}{2.00\,\cancel{atm}} = 50.\,K \qquad 50.\,K - 273 = -223\,°C$$

8.89 $T_1 = 8\,°C + 273 = 281\,K$; $\quad P_1 = 380\,\cancel{Torr} \times \dfrac{1\,atm}{760\,\cancel{Torr}} = 0.50\,atm$; $\quad V_1 = 750\,L$

$T_2 = -45\,°C + 273 = 228\,K$; $\quad P_2 = 0.20\,atm$; $\quad V_2 = ?\,L$

$$V_2 = V_1 \times \frac{P_1}{P_2} \times \frac{T_2}{T_1} = 750\,L \times \frac{0.50\,\cancel{atm}}{0.20\,\cancel{atm}} \times \frac{228\,\cancel{K}}{281\,\cancel{K}} = 1500\,L\ (2\ SFs)$$

8.91 Analyze the Problem: Using the Ideal Gas Law to Calculate Moles

Property	P	V	n	R	T
Given	2500. mmHg	2.00 L		$\dfrac{62.4\,L \cdot mmHg}{mole \cdot K}$	18 °C 18 °C + 273 = 291 K
Need			? moles		

$$n = \frac{PV}{RT} = \frac{(2500.\,\cancel{mmHg})(2.00\,\cancel{L})}{\left(\dfrac{62.4\,\cancel{L} \cdot \cancel{mmHg}}{mole \cdot \cancel{K}}\right)(291\,\cancel{K})} = 0.275\,mole\ of\ CH_4$$

$$0.275\,\cancel{mole\ CH_4} \times \frac{16.04\,g\ CH_4}{1\,\cancel{mole\ CH_4}} = 4.41\,g\ of\ CH_4\ (3\ SFs)$$

8.93   Analyze the Problem: Using the Ideal Gas Law to Calculate Moles

Property	P	V	n	R	T
Given	1.20 atm	35.0 L		$\dfrac{0.0821 \text{ L} \cdot \text{atm}}{\text{mole} \cdot \text{K}}$	5 °C 5 °C + 273 = 278 K
Need			? moles		

$$n = \frac{PV}{RT} = \frac{(1.20 \text{ atm})(35.0 \text{ L})}{\left(\dfrac{0.0821 \text{ L} \cdot \text{atm}}{\text{mole} \cdot \text{K}}\right)(278 \text{ K})} = 1.84 \text{ moles of } CO_2$$

$$1.84 \text{ moles } CO_2 \times \frac{44.01 \text{ g } CO_2}{1 \text{ mole } CO_2} = 81.0 \text{ g of } CO_2 \text{ (3 SFs)}$$

8.95 $25.0 \text{ g } Zn \times \dfrac{1 \text{ mole } Zn}{65.41 \text{ g } Zn} \times \dfrac{1 \text{ mole } H_2}{1 \text{ mole } Zn} \times \dfrac{22.4 \text{ L (STP)}}{1 \text{ mole } H_2} = 8.56 \text{ L of } H_2 \text{ released at STP (3 SFs)}$

8.97   Analyze the Problem: Using the Ideal Gas Law to Calculate Moles

Property	P	V	n	R	T
Given	725 mmHg	5.00 L		$\dfrac{62.4 \text{ L} \cdot \text{mmHg}}{\text{mole} \cdot \text{K}}$	375 °C 375 °C + 273 = 648 K
Need			? moles H_2O		

$$n = \frac{PV}{RT} = \frac{(725 \text{ mmHg})(5.00 \text{ L})}{\left(\dfrac{62.4 \text{ L} \cdot \text{mmHg}}{\text{mole} \cdot \text{K}}\right)(648 \text{ K})} = 0.0896 \text{ mole of } H_2O$$

Analyze the Problem: Reactant and Product

	Reactant	Product
Mass		? g NH_3
Molar Mass		1 mole of NH_3 = 17.03 g of NH_3
Moles	0.0896 mole H_2O	
Equation	$4NO_2(g) + 6H_2O(g) \longrightarrow 7O_2(g) + 4NH_3(g)$	

$$0.0896 \text{ mole } H_2O \times \frac{4 \text{ moles } NH_3}{6 \text{ moles } H_2O} \times \frac{17.03 \text{ g } NH_3}{1 \text{ mole } NH_3} = 1.02 \text{ g of } NH_3 \text{ (3 SFs)}$$

8.99   Because the partial pressure of nitrogen is to be reported in torr, the atm and mmHg units (for oxygen and argon, respectively) must be converted to torr, as follows:

$$P_{\text{oxygen}} = 0.60 \text{ atm} \times \frac{760 \text{ Torr}}{1 \text{ atm}} = 460 \text{ Torr} \quad \text{and}$$

$$P_{\text{argon}} = 425 \text{ mmHg} \times \frac{1 \text{ Torr}}{1 \text{ mmHg}} = 425 \text{ Torr}$$

$$\therefore P_{\text{nitrogen}} = P_{\text{total}} - (P_{\text{oxygen}} + P_{\text{argon}})$$

$$= 1250 \text{ Torr} - (460 \text{ Torr} + 425 \text{ Torr}) = 370 \text{ Torr (2 SFs)}$$

8.101 $T_1 = 24 \text{ °C} + 273 = 297 \text{ K}; \qquad P_1 = 745 \text{ mmHg}; \qquad V_1 = 425 \text{ mL}$

$T_2 = -95 \text{ °C} + 273 = 178 \text{ K}; \quad P_2 = 0.115 \text{ atm} \times \dfrac{760 \text{ mmHg}}{1 \text{ atm}} = 87.4 \text{ mmHg}; \quad V_2 = ? \text{ mL}$

$$V_2 = V_1 \times \frac{P_1}{P_2} \times \frac{T_2}{T_1} = 425 \text{ mL} \times \frac{745 \text{ mmHg}}{87.4 \text{ mmHg}} \times \frac{178 \text{ K}}{297 \text{ K}} = 2170 \text{ mL (3 SFs)}$$

8.103 $T_1 = 15\,°C + 273 = 288\,K;$ $P_1 = 745\,mmHg;$ $V_1 = 4250\,mL$

$$T_2 = ?\,°C;\quad P_2 = 1.20\,atm \times \frac{760\,mmHg}{1\,atm} = 912\,mmHg;$$

$$V_2 = 2.50\,L \times \frac{1000\,mL}{1\,L} = 2.50 \times 10^3\,mL$$

$$T_2 = T_1 \times \frac{V_2}{V_1} \times \frac{P_2}{P_1} = 288\,K \times \frac{2.50 \times 10^3\,mL}{4250\,mL} \times \frac{912\,mmHg}{745\,mmHg} = 207\,K - 273$$

$$= -66\,°C \text{ (2 SFs)}$$

8.105 Analyze the Problem: Reactant and Product

	Reactant	Product
Mass	881 g C_3H_8	
Molar Mass	1 mole of C_3H_8 = 44.09 g of C_3H_8	
Moles		? moles CO_2 (? L CO_2)
Molar Volume		1 mole of CO_2 = 22.4 L of CO_2 (STP)
Equation	$C_3H_8(g) + 5O_2(g) \xrightarrow{\Delta} 3CO_2(g) + 4H_2O(g)$	

$$881\,g\,C_3H_8 \times \frac{1\,mole\,C_3H_8}{44.09\,g\,C_3H_8} \times \frac{3\,moles\,CO_2}{1\,mole\,C_3H_8} \times \frac{22.4\,L\,CO_2\,(STP)}{1\,mole\,CO_2}$$

$$= 1340\,L\,CO_2 \text{ at STP (3 SFs)}$$

8.107 Analyze the Problem: Using the Ideal Gas Law to Calculate Moles

Property	P	V	n	R	T
Given	1.00 atm	7.50 L		$\dfrac{0.0821\,L \cdot atm}{mole \cdot K}$	37 °C 37 °C + 273 = 310 K
Need			? moles O_2		

$$n = \frac{PV}{RT} = \frac{(1.00\,atm)(7.50\,L)}{\left(\dfrac{0.0821\,L \cdot atm}{mole \cdot K}\right)(310\,K)} = 0.295 \text{ mole of } O_2$$

$$0.295\,mole\,O_2 \times \frac{6\,moles\,H_2O}{6\,moles\,O_2} \times \frac{18.02\,g\,H_2O}{1\,mole\,H_2O} = 5.32\,g \text{ of } H_2O \text{ (smaller amount of product)}$$

$$18.0\,g\,C_6H_{12}O_6 \times \frac{1\,mole\,C_6H_{12}O_6}{180.16\,g\,C_6H_{12}O_6} \times \frac{6\,moles\,H_2O}{1\,mole\,C_6H_{12}O_6} \times \frac{18.02\,g\,H_2O}{1\,mole\,H_2O} = 10.8\,g \text{ of } H_2O$$

Thus, O_2 is the limiting reactant, and 5.32 g of H_2O (3 SFs) can be produced.

8.109 $2400\,g\,O_2 \times \dfrac{1\,mole\,O_2}{32.00\,g\,O_2} = 75 \text{ moles of } O_2$

Analyze the Problem: Using the Ideal Gas Law to Calculate Volume

Property	P	V	n	R	T
Given	3.0 atm		75 moles	$\dfrac{0.0821\,L \cdot atm}{mole \cdot K}$	28 °C 28 °C + 273 = 301 K
Need		? L			

$$V = \frac{nRT}{P} = \frac{(75\,moles)\left(\dfrac{0.0821\,L \cdot atm}{mole \cdot K}\right)(301\,K)}{(3.0\,atm)} = 620\,L \text{ (2 SFs)}$$

8.111 Analyze the Problem: Using the Ideal Gas Law to Calculate Moles

Property	P	V	n	R	T
Given	20. mmHg	4.00 L		$\dfrac{62.4 \text{ L} \cdot \text{mmHg}}{\text{mole} \cdot \text{K}}$	32 °C 32 °C + 273 = 305 K
Need			? moles		

$$n = \frac{PV}{RT} = \frac{(20.\ \text{mmHg})(4.00\ \text{L})}{\left(\dfrac{62.4\ \text{L} \cdot \text{mmHg}}{\text{mole} \cdot \text{K}}\right)(305\ \text{K})} = 0.0042 \text{ mole of } CO_2$$

$$0.0042 \text{ mole } CO_2 \times \frac{44.01 \text{ g } CO_2}{1 \text{ mole } CO_2} = 0.18 \text{ g of } CO_2 \text{ (2 SFs)}$$

8.113 The partial pressure of each gas is proportional to the number of particles of each type of gas that is present. Thus, a ratio of partial pressure to total pressure is equal to the ratio of moles of that gas to the total number of moles of gases that are present:

$$\frac{P_{helium}}{P_{total}} = \frac{n_{helium}}{n_{total}}$$

Solving the equation for the partial pressure of helium yields:

$$P_{helium} = P_{total} \times \frac{n_{helium}}{n_{total}} = 2400 \text{ Torr} \times \frac{2.0 \text{ moles}}{8.0 \text{ moles}} = 600 \text{ Torr } (6.0 \times 10^2 \text{ Torr}) \text{ (2 SFs)}$$

$$P_{oxygen} = P_{total} \times \frac{n_{oxygen}}{n_{total}} = 2400 \text{ Torr} \times \frac{6.0 \text{ moles}}{8.0 \text{ moles}} = 1800 \text{ Torr } (1.8 \times 10^3 \text{ Torr}) \text{ (2 SFs)}$$

Selected Answers to Combining Ideas from Chapters 7 and 8

CI.13 a. Reactants: A and B_2; Products: AB_3

b. $2A + 3B_2 \longrightarrow 2AB_3$

c. combination

CI.15 a. The formula of shikimic acid is $C_7H_{10}O_5$.

b. Molar mass of shikimic acid $(C_7H_{10}O_5)$
$= 7(12.01 \text{ g}) + 10(1.008 \text{ g}) + 5(16.00 \text{ g}) = 174.15 \text{ g (5 SFs)}$

c. $130 \text{ g shikimic acid} \times \dfrac{1 \text{ mole shikimic acid}}{174.15 \text{ g shikimic acid}} = 0.75 \text{ mole of shikimic acid (2 SFs)}$

d. $155 \text{ g star anise} \times \dfrac{0.13 \text{ g shikimic acid}}{2.6 \text{ g star anise}} \times \dfrac{1 \text{ capsule Tamiflu}}{0.13 \text{ g shikimic acid}}$
$= 59 \text{ capsules of Tamiflu (2 SFs)}$

e. Molar mass of Tamiflu $(C_{16}H_{28}N_2O_4)$
$=16(12.01 \text{ g}) + 28(1.008 \text{ g}) + 2(14.01 \text{ g}) + 4(16.00 \text{ g}) = 312.4 \text{ g (4 SFs)}$

f. $500\,000 \text{ people} \times \dfrac{2 \text{ capsules}}{1 \text{ day 1 person}} \times 5 \text{ days} \times \dfrac{75 \text{ mg Tamiflu}}{1 \text{ capsule}} \times \dfrac{1 \text{ g Tamiflu}}{1000 \text{ mg Tamiflu}}$

$\times \dfrac{1 \text{ kg Tamiflu}}{1000 \text{ g Tamiflu}} = 400 \text{ kg of Tamiflu (1 SF)}$

CI.17 a. CH_4 H:C:H or H—C—H (with H above and below C)

b. $7.0 \times 10^6 \text{ gal} \times \dfrac{4 \text{ qt}}{1 \text{ gal}} \times \dfrac{946 \text{ mL}}{1 \text{ qt}} \times \dfrac{0.45 \text{ g}}{1 \text{ mL}} \times \dfrac{1 \text{ kg}}{1000 \text{ g}}$
$= 1.2 \times 10^7 \text{ kg of LNG (methane) (2 SFs)}$

c. Molar mass of methane $(CH_4) = 12.01 \text{ g} + 4(1.008 \text{ g}) = 16.04 \text{ g}$
$7.0 \times 10^6 \text{ gal} \times \dfrac{4 \text{ qt}}{1 \text{ gal}} \times \dfrac{946 \text{ mL}}{1 \text{ qt}} \times \dfrac{0.45 \text{ g}}{1 \text{ mL}} \times \dfrac{1 \text{ mole } CH_4}{16.04 \text{ g}} \times \dfrac{22.4 \text{ L } CH_4 \text{ (STP)}}{1 \text{ mole } CH_4}$
$= 1.7 \times 10^{10} \text{ L of LNG (methane) at STP (2 SFs)}$

d. $CH_4(g) + 2O_2(g) \xrightarrow{\Delta} CO_2(g) + 2H_2O(g) + 883 \text{ kJ}$

e. $7.0 \times 10^6 \text{ gal} \times \dfrac{4 \text{ qt}}{1 \text{ gal}} \times \dfrac{946 \text{ mL}}{1 \text{ qt}} \times \dfrac{0.45 \text{ g}}{1 \text{ mL}} \times \dfrac{1 \text{ mole } CH_4}{16.04 \text{ g}} \times \dfrac{2 \text{ moles } O_2}{1 \text{ mole } CH_4}$
$\times \dfrac{32.00 \text{ g } O_2}{1 \text{ mole } O_2} \times \dfrac{1 \text{ kg } O_2}{1000 \text{ g } O_2}$
$= 4.8 \times 10^7 \text{ kg of } O_2 \text{ (2 SFs)}$

f. $7.0 \times 10^6 \text{ gal} \times \dfrac{4 \text{ qt}}{1 \text{ gal}} \times \dfrac{946 \text{ mL}}{1 \text{ qt}} \times \dfrac{0.45 \text{ g}}{1 \text{ mL}} \times \dfrac{1 \text{ mole } CH_4}{16.04 \text{ g}} \times \dfrac{883 \text{ kJ}}{1 \text{ mole } CH_4}$
$= 6.6 \times 10^{11} \text{ kJ (2 SFs)}$

9

Solutions

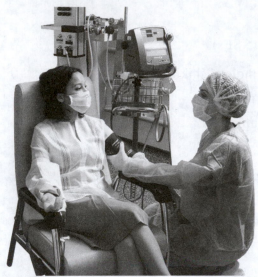

Michelle has hemodialysis three times a week for about 4 h each visit. The treatment removes harmful waste products and extra fluid. Her blood pressure is maintained because the electrolytes potassium and sodium are adjusted each time. Michelle measures the amount of fluid she takes in each day. Her diet is low in protein and includes fruits and vegetables that are low in potassium. She and her husband are now learning about performing hemodialysis at home. Soon she will have her own hemodialysis machine. A typical potassium level in blood plasma is 0.091 g in 500. mL of plasma. What is the molarity of the potassium ion in plasma?

Credit: AJPhoto/Science Source

LOOKING AHEAD

9.1 Solutions
9.2 Electrolytes and Nonelectrolytes

9.3 Solubility
9.4 Solution Concentrations and Reactions

9.5 Dilution of Solutions
9.6 Properties of Solutions

 The Health icon indicates a question that is related to health and medicine.

9.1 Solutions

Learning Goal: Identify the solute and solvent in a solution; describe the formation of a solution.

REVIEW
Identifying Polarity of Molecules (6.8)
Identifying Intermolecular Forces (6.9)

- A solution is a homogeneous mixture that forms when a solute dissolves in a solvent. The solvent is the substance that is present in a greater amount.

- Solvents and solutes can be solids, liquids, or gases. The solution is the same physical state as the solvent.

- A polar solute is soluble in a polar solvent; a nonpolar solute is soluble in a nonpolar solvent.

- Water molecules form hydrogen bonds because the partial positive charge of the hydrogen in one water molecule is attracted to the partial negative charge of oxygen in another water molecule.

- An ionic solute dissolves in water, a polar solvent, because the polar water molecules attract and pull the positive and negative ions into solution. In solution, water molecules surround the ions in a process called hydration.

♦ **Learning Exercise 9.1A**

Indicate the solute and solvent in each of the following:

	Solute	Solvent
a. 10 g of KCl dissolved in 100 g of water	*KCl*	*water*
b. soda water: $CO_2(g)$ dissolved in water	*CO₂*	*water*
c. an alloy composed of 80% Zn and 20% Cu	*Cu*	*Zn*
d. a mixture of O_2 (200 mmHg) and He (500 mmHg)	*O₂*	*He*
e. a solution of 40 mL of CCl_4 and 2 mL of Br_2	*Br₂*	*CCl₄*

Answers   **a.** KCl; water   **b.** CO_2; water   **c.** Cu; Zn
            d. O_2; He   **e.** Br_2; CCl_4

Solute: The substance present in lesser amount

Salt

Water

Solvent: The substance present in greater amount

A solution has at least one solute dispersed in a solvent.

Credit: Pearson Education/Pearson Science

♦ **Learning Exercise 9.1B**

Water is polar, and hexane is nonpolar. In which solvent is each of the following solutes soluble?

a. bromine, Br_2, nonpolar *hexane*   **b.** HCl, polar *water*

c. cholesterol, nonpolar *hexane*   **d.** vitamin D, nonpolar *hexane*

e. vitamin C, polar *water*

Answers   **a.** hexane   **b.** water   **c.** hexane
            d. hexane   **e.** water

9.2   Electrolytes and Nonelectrolytes

Learning Goal: Identify solutes as electrolytes or nonelectrolytes.

- Electrolytes conduct an electrical current because they produce ions in aqueous solutions.
- Strong electrolytes are completely dissociated into ions, whereas weak electrolytes are partially dissociated into ions.
- Nonelectrolytes do not form ions in solution but dissolve as molecules.
- Ions are measured as equivalents (Eq), which is the amount of the ion that is equal to 1 mole of positive or negative electrical charge.

REVIEW

Writing Conversion Factors from Equalities (2.5)
Using Conversion Factors (2.6)
Writing Positive and Negative Ions (6.1)

Key Terms for Sections 9.1 and 9.2

Match each of the following key terms with the correct description:

    a. solution   **b.** nonelectrolyte   **c.** equivalent
    d. solute   **e.** strong electrolyte   **f.** solvent

1. _____ a substance that dissociates as molecules, not ions, when it dissolves in water

2. _____ the substance that comprises the lesser amount in a solution

3. _____ a solute that dissociates 100% into ions in solution

4. ____ the substance that comprises the greater amount in a solution

5. ____ a homogeneous mixture of at least two components called a solute and a solvent

6. ____ the amount of positive or negative ion that supplies 1 mole of electrical charge

Answers   1. b   2. d   3. e   4. f   5. a   6. c

♦ **Learning Exercise 9.2A**

Write an equation for the formation of an aqueous solution of each of the following solid strong electrolytes:

a. $LiCl$

b. $Zn(NO_3)_2$

c. Na_3PO_4

d. K_2SO_4

e. $MgCl_2$

Answers   a. $LiCl(s) \xrightarrow{H_2O} Li^+(aq) + Cl^-(aq)$
b. $Zn(NO_3)_2(s) \xrightarrow{H_2O} Zn^{2+}(aq) + 2NO_3^-(aq)$
c. $Na_3PO_4(s) \xrightarrow{H_2O} 3Na^+(aq) + PO_4^{3-}(aq)$
d. $K_2SO_4(s) \xrightarrow{H_2O} 2K^+(aq) + SO_4^{2-}(aq)$
e. $MgCl_2(s) \xrightarrow{H_2O} Mg^{2+}(aq) + 2Cl^-(aq)$

♦ **Learning Exercise 9.2B**

Indicate whether an aqueous solution of each of the following contains only ions, only molecules, or mostly molecules and a few ions. Write an equation for the formation of the solution.

a. glucose, $C_6H_{12}O_6$, a nonelectrolyte   b. NaOH, a strong electrolyte

c. $CaCl_2$, a strong electrolyte   d. HF, a weak electrolyte

Answers   a. $C_6H_{12}O_6(s) \xrightarrow{H_2O} C_6H_{12}O_6(aq)$ only molecules
b. $NaOH(s) \xrightarrow{H_2O} Na^+(aq) + OH^-(aq)$ only ions
c. $CaCl_2(s) \xrightarrow{H_2O} Ca^{2+}(aq) + 2Cl^-(aq)$ only ions
d. $HF(aq) \overset{H_2O}{\rightleftharpoons} H^+(aq) + F^-(aq)$ mostly molecules and a few ions

♦ **Learning Exercise 9.2C**

Calculate the following:

  a. the number of equivalents in 1 mole of Zn^{2+}

  b. the number of equivalents in 2.5 moles of Cl^-

  c. the number of equivalents in 2.0 moles of Ca^{2+}

Answers   **a.** 2 Eq   **b.** 2.5 Eq   **c.** 4.0 Eq

♦ **Learning Exercise 9.2D**

A maintenance IV solution contains 40 mEq Na^+, 35 mEq K^+, 40 mEq Cl^-, 20 mEq lactate$^-$, and HPO_4^{2-}.

  a. What is the total number of milliequivalents of cations?

  b. What is the total number of milliequivalents of anions?

  c. What is the number of milliequivalents of HPO_4^{2-}?

Answers   **a.** 75 mEq   **b.** 75 mEq   **c.** 15 mEq

♦ **Learning Exercise 9.2E**

A replacement IV solution contains 140 mEq Na^+, 10 mEq K^+, 5 mEq Ca^{2+}, 3 mEq Mg^{2+}, 47 mEq acetate$^-$, 8 mEq citrate^{3-}, and Cl^-.

  a. What is the total number of milliequivalents of cations?

  b. What is the total number of milliequivalents of anions?

  c. What is the number of milliequivalents of Cl^-?

Answers   **a.** 158 mEq   **b.** 158 mEq   **c.** 103 mEq

9.3 Solubility

Learning Goal: Define solubility; distinguish between an unsaturated and a saturated solution. Identify an ionic compound as soluble or insoluble.

- The amount of solute that dissolves depends on the nature of the solute and solvent.
- Solubility describes the maximum amount of a solute that dissolves in 100. g of solvent at a given temperature.
- A saturated solution contains the maximum amount of dissolved solute at a certain temperature, whereas an unsaturated solution contains less than this amount.
- An increase in temperature increases the solubility of most solids, but decreases the solubility of gases in water.
- Henry's law states that the solubility of gases in liquids is directly related to the partial pressure of that gas above the liquid.
- The solubility rules describe the kinds of ionic combinations that are soluble or insoluble in water.

Solubility Rules for Ionic Compounds in Water

An ionic compound is soluble in water if it contains one of the following:

Positive Ions	Li^+, Na^+, K^+, Rb^+, Cs^+, NH_4^+
Negative Ions	NO_3^-, $C_2H_3O_2^-$
	Cl^-, Br^-, I^- except when combined with Ag^+, Pb^{2+}, or Hg_2^{2+}
	SO_4^{2-} except when combined with Ba^{2+}, Pb^{2+}, Ca^{2+}, Sr^{2+}, or Hg_2^{2+}

Ionic compounds that do not contain at least one of these ions are usually insoluble.

CdS

PbI$_2$

If an ionic compound contains a combination of a cation and an anion that is not soluble, that ionic compound is insoluble.

◆ **Learning Exercise 9.3A**

Identify each of the following solutions as saturated or unsaturated:

a. _____ A sugar cube dissolves when added to a cup of hot tea.

b. _____ A KCl crystal added to a KCl solution does not change in size.

c. _____ A layer of sugar forms in the bottom of a glass of iced tea.

d. _____ The rate of crystal formation is equal to the rate of dissolving.

e. _____ Upon heating, all the sugar in a solution dissolves.

Answers	**a.** unsaturated	**b.** saturated	**c.** saturated
	d. saturated	**e.** unsaturated	

♦ **Learning Exercise 9.3B**

Use the solubility data for $NaNO_3$ listed below to answer each of the following problems:

Temperature (°C)	Solubility (g of $NaNO_3$/100. g of H_2O)
40	100. g
60	120. g
80	150. g
100	200. g

a. How many grams of $NaNO_3$ will dissolve in 100. g of water at 40 °C?

b. How many grams of $NaNO_3$ will dissolve in 300. g of water at 60 °C?

c. A solution is prepared using 200. g of water and 350. g of $NaNO_3$ at 80 °C. Will any solute remain undissolved? If so, how much?

d. Will 250. g of $NaNO_3$ completely dissolve when added to 150. g of water at 100 °C?

Answers   **a.** 100. g   **b.** 360. g
c. 300. g of $NaNO_3$ will dissolve leaving 50. g of $NaNO_3$ that will not dissolve.
d. Yes, 250. g of $NaNO_3$ will completely dissolve.

♦ **Learning Exercise 9.3C**

> **CORE CHEMISTRY SKILL**
> Using Solubility Rules

Predict whether the following ionic compounds are soluble or insoluble in water:

a. _____ NaCl   **b.** _____ $LiNO_3$ **c.** _____ $SrSO_4$

d. _____ Ag_2S **e.** _____ $BaSO_4$ **f.** _____ Na_2CO_3

g. _____ Na_2S **h.** _____ $MgCl_2$ **i.** _____ BaS

Answers   **a.** soluble   **b.** soluble   **c.** insoluble   **d.** insoluble   **e.** insoluble
f. soluble   **g.** soluble   **h.** soluble   **i.** insoluble

♦ **Learning Exercise 9.3D**

Will an insoluble compound form (Yes or No) when the following solutions are mixed? If an insoluble compound forms, write its formula.

a. $NaCl(aq)$ and $Pb(NO_3)_2(aq)$ _____

b. $BaCl_2(aq)$ and $Na_2SO_4(aq)$ _____

c. $K_3PO_4(aq)$ and $NaNO_3(aq)$ _____

d. $Na_2S(aq)$ and $AgNO_3(aq)$ _____

Answers   **a.** Yes; $PbCl_2$ **b.** Yes; $BaSO_4$ **c.** No   **d.** Yes; Ag_2S

	Writing Equations for the Formation of an Insoluble Ionic Compound
STEP 1	Write the ions of the reactants.
STEP 2	Write the combinations of ions, and determine if any are insoluble.
STEP 3	Write the ionic equation including any solid.
STEP 4	Write the net ionic equation.

♦ **Learning Exercise 9.3E**

For the mixtures that form ionic compounds in Learning Exercise 9.3D, write the ionic equation and the net ionic equation.

a.

b.

c.

d.

Answers

a. $2Na^+(aq) + 2Cl^-(aq) + Pb^{2+}(aq) + 2NO_3^-(aq) \longrightarrow 2Na^+(aq) + PbCl_2(s) + 2NO_3^-(aq)$

$2Cl^-(aq) + Pb^{2+}(aq) \longrightarrow PbCl_2(s)$

b. $Ba^{2+}(aq) + 2Cl^-(aq) + 2Na^+(aq) + SO_4^{2-}(aq) \longrightarrow 2Na^+(aq) + BaSO_4(s) + 2Cl^-(aq)$

$Ba^{2+}(aq) + SO_4^{2-}(aq) \longrightarrow BaSO_4(s)$

c. No insoluble ionic compound forms.

d. $2Na^+(aq) + S^{2-}(aq) + 2Ag^+(aq) + 2NO_3^-(aq) \longrightarrow 2Na^+(aq) + Ag_2S(s) + 2NO_3^-(aq)$

$S^{2-}(aq) + 2Ag^+(aq) \longrightarrow Ag_2S(s)$

9.4 Solution Concentrations and Reactions

REVIEW

Learning Goal: Calculate the concentration of a solute in a solution; use concentration units to calculate the amount of solute or solution. Given the volume and concentration of a solution, calculate the amount of another reactant or product in a reaction.

Calculating Percentages (1.4)
Using Molar Mass as a Conversion Factor (7.6)
Using Mole–Mole Factors (7.7)

- The concentration of a solution is the relationship between the amount of solute, in grams, moles, or milliliters, and the amount of solution, in grams, milliliters, or liters.
- A mass percent (m/m) expresses the ratio of the mass of solute to the mass of solution multiplied by 100%. Mass units must be the same such as grams of solute/grams of solution × 100%.

$$\text{Mass percent (m/m)} = \frac{\text{mass of solute}}{\text{mass of solution}} \times 100\%$$

- A volume percent (v/v) expresses the ratio of the volume of solute to volume of solution multiplied by 100%. Volume units must be the same such as milliliters of solute/milliliters of solution × 100%.

$$\text{Volume percent (v/v)} = \frac{\text{volume of solute}}{\text{volume of solution}} \times 100\%$$

- A mass/volume percent (m/v) expresses the ratio of the mass of solute to the volume of solution multiplied by 100%. Mass units for the solute must be in grams and the volume units for the solution must be in milliliters.

$$\text{Mass/volume percent (m/v)} = \frac{\text{grams of solute}}{\text{milliliters of solution}} \times 100\%$$

- Molarity is a concentration term that indicates the number of moles of solute dissolved in 1 L (1000 mL) of solution.

$$\text{Molarity (M)} = \frac{\text{moles of solute}}{\text{liters of solution}}$$

- Concentrations can be used as conversion factors to relate the amount of solute and the volume or mass of the solution.

$$\frac{\text{amount of solute}}{\text{amount of solution}} \quad \text{and} \quad \frac{\text{amount of solution}}{\text{amount of solute}}$$

Key Terms for Sections 9.3 and 9.4

Match each of the following key terms with the correct description:

a. mass percent (m/m)	**b.** concentration	**c.** solubility
d. unsaturated	**e.** molarity	**f.** saturated

1. __e__ the number of moles of solute in 1 L of solution

2. __b__ the amount of solute dissolved in a certain amount of solution

3. __f__ a solution containing the maximum amount of solute that can dissolve at a given temperature

4. __d__ a solution with less than the maximum amount of solute that can dissolve

5. __a__ the concentration of a solution in terms of mass of solute in 100. g of solution

6. __c__ the maximum amount of solute that can dissolve in 100. g of solvent at a given temperature

Answers   **1.** e   **2.** b   **3.** f   **4.** d   **5.** a   **6.** c

SAMPLE PROBLEM   Calculating Mass Percent (m/m) Concentration

What is the mass percent (m/m) when 2.4 g of $NaHCO_3$ dissolves in water to make 120. g of a solution of $NaHCO_3$?

Solution:

STEP 1 State the given and needed quantities.

	Given	Need	Connect
Analyze the Problem	2.4 g of $NaHCO_3$ solute, 120. g of $NaHCO_3$ solution	mass percent (m/m)	$\dfrac{\text{mass of solute}}{\text{mass of solution}} \times 100\%$

STEP 2 Write the concentration expression.

$$\text{Mass percent (m/m)} = \frac{\text{grams of solute}}{\text{grams of solution}} \times 100\%$$

STEP 3 Substitute solute and solution quantities into the expression and calculate.

$$\text{Mass percent (m/m)} = \frac{2.4 \text{ g } NaHCO_3}{120. \text{ g } NaHCO_3 \text{ solution}} \times 100\% = 2.0\% \text{ (m/m)}$$

♦ Learning Exercise 9.4A

CORE CHEMISTRY SKILL
Calculating Concentration

Calculate the percent concentration for each of the following solutions:

a. mass percent (m/m) for 18.0 g of NaCl in 90.0 g of a solution of NaCl

$m/m = \frac{18.0\text{g}}{90.0\text{g}} \times 100 = 20.0\%$

b. mass/volume percent (m/v) for 4.0 g of KOH in 50.0 mL of a solution of KOH

$\frac{4.0\text{g}}{50.0\text{mL}} \times 100\% = 8.0\%$

c. mass/volume percent (m/v) for 5.0 g of KCl in 2.0 L of a solution of KCl

$\frac{5.0\text{g}}{2000\text{mL}} \times 100 = 0.25\%$

d. volume percent (v/v) for 18 mL of ethanol in 350 mL of a mouthwash solution

$\frac{18\text{ mL}}{350\text{ mL}} \times 100 = 5.1\%$

Answers **a.** 20.0% (m/m) **b.** 8.0% (m/v) **c.** 0.25% (m/v) **d.** 5.1% (v/v)

SAMPLE PROBLEM Using Concentration to Calculate Mass of Solute

How many grams of KI are needed to prepare 225 g of a 4.0% (m/m) KI solution?

Solution:

STEP 1 State the given and needed quantities.

	Given	Need	Connect
Analyze the Problem	225 g of 4.0% (m/m) KI solution	grams of KI	mass percent factor $\frac{\text{g of solute}}{100. \text{ g of solution}}$

STEP 2 Write a plan to calculate the mass.

$$\text{grams of KI solution} \xrightarrow{\%\text{(m/m) factor}} \text{grams of KI}$$

STEP 3 Write equalities and conversion factors.

4.0 g of KI = 100. g of KI solution

$$\frac{4.0 \text{ g KI}}{100. \text{ g KI solution}} \quad \text{and} \quad \frac{100. \text{ g KI solution}}{4.0 \text{ g KI}}$$

STEP 4 Set up the problem to calculate the mass.

$$225 \text{ g } \cancel{\text{KI solution}} \times \frac{4.0 \text{ g KI}}{100. \text{ g } \cancel{\text{KI solution}}} = 9.0 \text{ g of KI}$$

♦ Learning Exercise 9.4B

CORE CHEMISTRY SKILL
Using Concentration as a Conversion Factor

Calculate the number of grams of solute in each of the following solutions:

a. grams of glucose ($C_6H_{12}O_6$) in 480 g of a 5.0% (m/m) intravenous (IV) glucose solution

$480g \times \frac{5.0g}{100g} = 24g$

b. grams of lidocaine hydrochloride in 50.0 g of a 2.0% (m/m) lidocaine hydrochloride solution

$50.0g \times \frac{2.0g}{100g} = 1.0g$

c. grams of KCl in 1.2 L of a 4.0% (m/v) KCl solution

$1200mL \times \frac{4.0g}{100g} = 48g$

d. grams of NaCl in 1.50 L of a 2.00% (m/v) NaCl solution

$1500mL \times \frac{2.00g}{100g} = 30.0g$

Answers **a.** 24 g **b.** 1.0 g **c.** 48 g **d.** 30.0 g

♦ Learning Exercise 9.4C

Use percent concentration factors to calculate each of the following:

a. How many milliliters of a 1.00% (m/v) NaCl solution contain 2.00 g of NaCl?

$2.00g \times \frac{100}{1.00g} = 200mL$

b. How many milliliters of a 5.0% (m/v) glucose ($C_6H_{12}O_6$) solution contain 24 g of glucose?

$24g \times \frac{100}{5.0g} = 480mL$

c. How many milliliters of a 2.5% (v/v) methanol (CH_4O) solution contain 7.0 mL of methanol?

$7.0mL \times \frac{100g}{2.5g} = 280mL$

d. How many grams of a 15% (m/m) NaOH solution contain 7.5 g of NaOH?

$7.5g \times \frac{100g}{15.0g} = 50g$

Answers **a.** 200. mL **b.** 480 mL **c.** 280 mL **d.** 50. g

SAMPLE PROBLEM Calculating Molarity

A physiological glucose solution contains 25 g of glucose ($C_6H_{12}O_6$) in 0.50 L of solution. What is the molarity (M) of the glucose solution?

Solution:

STEP 1 State the given and needed quantities.

	Given	Need	Connect
Analyze the Problem	25 g of glucose, 0.50 L of solution	molarity (mole/L)	molar mass of glucose, $\dfrac{\text{moles of solute}}{\text{liters of solution}}$

To calculate the moles of glucose, we need to write the equality and conversion factors for the molar mass of glucose.

1 mole of glucose = 180.16 g of glucose

$$\frac{180.16 \text{ g glucose}}{1 \text{ mole glucose}} \quad \text{and} \quad \frac{1 \text{ mole glucose}}{180.16 \text{ g glucose}}$$

$$\text{moles of glucose} = 25 \text{ g glucose} \times \frac{1 \text{ mole glucose}}{180.16 \text{ g glucose}} = 0.14 \text{ mole of glucose}$$

STEP 2 Write the concentration expression.

$$\text{Molarity (M)} = \frac{\text{moles of solute}}{\text{liters of solution}}$$

STEP 3 Substitute solute and solution quantities into the expression and calculate.

$$\text{Molarity (M)} = \frac{0.14 \text{ mole glucose}}{0.50 \text{ L solution}} = 0.28 \text{ M glucose solution}$$

◆ **Learning Exercise 9.4D**

Calculate the molarity (M) of the following solutions:

a. 1.45 moles of HCl in 0.250 L of a solution of HCl

$$M = \frac{1.45 \text{ mol HCl}}{0.250 L} = 5.80 \text{ M}$$

b. 10.0 moles of glucose ($C_6H_{12}O_6$) in 2.50 L of a glucose solution

$$M = \frac{10.00 \text{ mol}}{2.50 L} = 4.00 \text{ M}$$

c. 80.0 g of NaOH in 1.60 L of a solution of NaOH

$$80.0 g \times \frac{1 \text{ mol}}{39.997 g} = 2.00 \text{ mol} \qquad M = \frac{2.00 \text{ mol}}{1.60 L} = 1.25 \text{ M}$$

d. 38.8 g of NaBr in 175 mL of a solution of NaBr

$$38.8 g \times \frac{1 \text{ mol}}{102.89} = 0.38 \text{ mol} \qquad M = \frac{0.38 \text{ mol}}{175 mL} = 2.17 \times 10^{-03}$$

Answers   **a.** 5.80 M   **b.** 4.00 M   **c.** 1.25 M   **d.** 2.15 M

SAMPLE PROBLEM   Using Molarity

How many grams of NaOH are in 0.250 L of a 5.00 M NaOH solution?

Solution:

STEP 1 State the given and needed quantities.

Analyze the Problem	Given	Need	Connect
	0.250 L of a 5.00 M NaOH solution	grams of NaOH	molar mass of NaOH, molarity

STEP 2 Write a plan to calculate the mass.

liters of solution $\xrightarrow{\text{Molarity}}$ moles of solute $\xrightarrow{\text{Molar mass}}$ grams of solute

STEP 3 Write equalities and conversion factors.

$$1 \text{ L of solution} = 5.00 \text{ moles of NaOH} \qquad 1 \text{ mole of NaOH} = 40.00 \text{ g of NaOH}$$

$$\frac{5.00 \text{ moles NaOH}}{1 \text{ L solution}} \text{ and } \frac{1 \text{ L solution}}{5.00 \text{ moles NaOH}} \qquad \frac{40.00 \text{ g NaOH}}{1 \text{ mole NaOH}} \text{ and } \frac{1 \text{ mole NaOH}}{40.00 \text{ g NaOH}}$$

STEP 4 Set up the problem to calculate the mass.

$$0.250 \text{ L solution} \times \frac{5.00 \text{ moles NaOH}}{1 \text{ L solution}} \times \frac{40.00 \text{ g NaOH}}{1 \text{ mole NaOH}} = 50.0 \text{ g of NaOH}$$

♦ **Learning Exercise 9.4E**

Calculate the quantity of solute in each of the following solutions:

a. moles of HCl in 1.50 L of a 6.00 M HCl solution

$$1.50 \text{ L} \times \frac{1 \text{ L}}{6.00 \text{ mol}} \times \frac{1 \text{ mol}}{36.458} = 9.11 \text{ moles}$$

b. moles of KOH in 0.750 L of a 10.0 M KOH solution

$$0.750 \text{ L} \times \frac{1 \text{ L}}{10.0 \text{ mol}} \times \frac{1 \text{ mol}}{56.105} =$$

c. grams of NaOH in 0.500 L of a 4.40 M NaOH solution

$$0.500 \text{ L} \times \frac{4.40 \text{ mol}}{1 \text{ L}} \times \frac{39.998 \text{ s}}{1 \text{ mol}} = 88.0 \text{ g}$$

d. grams of NaCl in 285 mL of a 1.75 M NaCl solution

$$0.285 \text{ L} \times \frac{1.75 \text{ mol}}{1 \text{ L}} \times \frac{58.44 \text{ g}}{1 \text{ mol}} = 29.1 \text{ g}$$

Answers   **a.** 9.00 moles   **b.** 7.50 moles   **c.** 88.0 g   **d.** 29.1 g

♦ **Learning Exercise 9.4F**

Calculate the volume, in milliliters, needed of each solution to obtain the given quantity of solute:

a. 1.50 moles of $Mg(OH)_2$ from a 2.00 M $Mg(OH)_2$ solution

$$1.50 \text{ mol} \times \frac{1 \text{ mol}}{41.312 g} \times \frac{1000 mL}{2.00 \text{ mol}}$$

b. 0.150 mole of glucose from a 2.20 M glucose solution

c. 18.5 g of KI from a 3.00 M KI solution

d. 18.0 g of NaOH from a 6.00 M NaOH solution

Answers　　a. 750 mL　　b. 68.2 mL　　c. 37.1 mL　　d. 75.0 mL

♦ **Learning Exercise 9.4G**

a. A patient receives 150 mL of a 10.% (m/v) dextrose solution each hour.

　i. How many grams of dextrose are given in 1 h?

　ii. How many grams of dextrose are given in 8 h?

b. A patient receives 125 mL of half-normal saline, which is a 0.45% (m/v) NaCl solution, each hour.

　i. How many grams of NaCl are given in 1 h?

　ii. How many grams of NaCl are given in 12 h?

Answers　　a. i 15 g　　a. ii 120 g　　b. i 0.56 g　　b. ii 6.8 g

Calculations Involving Solutions in Chemical Reactions	
STEP 1	State the given and needed quantities.
STEP 2	Write a plan to calculate the needed quantity.
STEP 3	Write equalities and conversion factors including mole–mole and concentration factors.
STEP 4	Set up the problem to calculate the needed quantity.

♦ **Learning Exercise 9.4H**

For the following reaction:

$2AgNO_3(aq) + H_2SO_4(aq) \longrightarrow Ag_2SO_4(s) + 2HNO_3(aq)$

a. How many milliliters of a 1.5 M $AgNO_3$ solution will react with 40.0 mL of a 1.0 M H_2SO_4 solution?

b. How many grams of Ag_2SO_4 will be produced?

Answers   **a.** 53 mL   **b.** 12 g of Ag_2SO_4

> **CORE CHEMISTRY SKILL**
> Calculating the Quantity of a Reactant or Product for a Chemical Reaction in Solution

♦ **Learning Exercise 9.4I**

Calculate the milliliters of a 1.80 M KOH solution that react with 18.5 mL of a 2.20 M HCl solution.
$HCl(aq) + KOH(aq) \longrightarrow H_2O(l) + KCl(aq)$

Answer   22.6 mL of KOH solution

9.5 Dilution of Solutions

Learning Goal: Describe the dilution of a solution; calculate the unknown concentration or volume when a solution is diluted.

> **REVIEW**
> Solving Equations (1.4)

- Dilution is the process of mixing a solution with solvent to obtain a solution with a lower concentration.
- For dilutions, use the expression $C_1V_1 = C_2V_2$ and solve for the unknown value.

SAMPLE PROBLEM Concentration of a Diluted Solution

What is the molarity when 150. mL of a 2.00 M NaCl solution is diluted to a volume of 400. mL?

Solution:

STEP 1 Prepare a table of the concentrations and volumes of the solutions.

	Given	Need	Connect
Analyze the Problem	$C_1 = 2.00$ M $V_1 = 150.$ mL $V_2 = 400.$ mL	C_2	$C_1V_1 = C_2V_2$ **Predict:** C_2 decreases

STEP 2 Rearrange the dilution expression to solve for the unknown quantity.

$$C_1V_1 = C_2V_2 \qquad C_2 = C_1 \times \frac{V_1}{V_2}$$

STEP 3 Substitute the known quantities into the dilution expression and calculate.

$$C_2 = 2.00 \text{ M} \times \frac{150. \text{ mL}}{400. \text{ mL}} = 0.750 \text{ M}$$

♦ **Learning Exercise 9.5**

Solve each of the following dilution problems:

a. What is the molarity when 100. mL of a 5.0 M KCl solution is diluted with water to give a final volume of 200. mL?

b. What is the molarity of the diluted solution when 5.0 mL of a 1.5 M KCl solution is diluted to a total volume of 25 mL?

c. What is the molarity when 250 mL of an 8.0 M NaOH solution is diluted with 750 mL of water? Assume that the volumes add.

d. What is the final volume when 42 mL of a 1.0 M NaCl solution is diluted to give a 0.20 M NaCl solution?

e. What volume of a 6.0 M HCl solution is needed to prepare 300. mL of a 1.0 M HCl solution? How much water must be added? Assume that the volumes add.

Answers   **a.** 2.5 M   **b.** 0.30 M   **c.** 2.0 M   **d.** 210 mL
            e. $V_1 = 50.$ mL; add 250. mL of water

9.6   Properties of Solutions

Learning Goal: Identify a mixture as a solution, a colloid, or a suspension. Describe how the number of particles in a solution affects the freezing point, the boiling point, and the osmotic pressure of a solution.

- Colloids contain particles that do not settle out and pass through filters but not through semipermeable membranes.
- Suspensions are composed of large particles that settle out of solution.
- A solute added to water decreases the vapor pressure, increases the boiling point, decreases the freezing point, and increases osmotic pressure.
- In the process of osmosis, water (solvent) moves through a semipermeable membrane from the solution that has a lower solute concentration to a solution where the solute concentration is higher.
- Osmotic pressure is the pressure that prevents the flow of water into a more concentrated solution.
- Isotonic solutions have osmotic pressures equal to that of body fluids. A hypotonic solution has a lower osmotic pressure than body fluids; a hypertonic solution has a higher osmotic pressure.
- A red blood cell maintains its volume in an isotonic solution, but it swells (hemolysis) in a hypotonic solution and shrinks (crenation) in a hypertonic solution.
- In dialysis, water and small solute particles can pass through a dialyzing membrane, while larger particles such as blood cells cannot.

Key Terms for Sections 9.5 and 9.6

Match each of the following key terms with the correct description:

a. hypertonic	**b.** dilution	**c.** hemolysis
d. osmosis	**e.** semipermeable membrane	**f.** colloid

1. _____ the particles that pass through filters but are too large to pass through semipermeable membranes

2. _____ the swelling and bursting of red blood cells when placed in a hypotonic solution

3. _____ a solution that has a higher osmotic pressure than the red blood cells of the body

4. _____ a process by which water is added to a solution and decreases its concentration

5. _____ a membrane that permits the passage of small particles while blocking larger particles

6. _____ the flow of solvent through a semipermeable membrane into a solution of higher solute concentration

Answers   **1.** f   **2.** c   **3.** a   **4.** b   **5.** e   **6.** d

◆ **Learning Exercise 9.6A**

Identify each of the following as a solution, colloid, or suspension:

a. _____ the solute consists of single atoms, ions, or small molecules

b. _____ settles out with gravity

c. _____ retained by filters

d. _____ cannot pass through a membrane

e. _____ contains large particles that are visible

Answers   **a.** solution   **b.** suspension   **c.** suspension
             d. colloid, suspension   **e.** suspension

Freezing Point Lowering/Boiling Point Elevation	
STEP 1	State the given and needed quantities.
STEP 2	Determine the number of moles of solute particles.
STEP 3	Determine the temperature change using the moles of solute particles and the degrees Celsius change per mole of particles.
STEP 4	Subtract the temperature change from the freezing point or add the temperature change to the boiling point.

♦ **Learning Exercise 9.6B**

For the following solutes, each in 1.00 kg of water, calculate:

(1) the number of moles of particles present in the solution

(2) the freezing point change and freezing point of the solution

(3) the boiling point change and boiling point of the solution

CORE CHEMISTRY SKILL

Calculating the Boiling Point/
Freezing Point of a Solution

Solute	Moles of Solute Particles	Freezing Point	Boiling Point
a. 1.00 mole of fructose (nonelectrolyte)			
b. 1.50 moles of KCl (strong electrolyte)			
c. 1.25 moles of Ca(NO₃)₂ (strong electrolyte)			

Answers

Solute	Moles of Solute Particles	Freezing Point	Boiling Point
a. 1.00 mole of fructose (nonelectrolyte)	1.00 mole	$-1.86\,°C$	$100.52\,°C$
b. 1.50 moles of KCl (strong electrolyte)	3.00 moles	$-5.58\,°C$	$101.56\,°C$
c. 1.25 moles of Ca(NO₃)₂ (strong electrolyte)	3.75 moles	$-6.98\,°C$	$101.95\,°C$

♦ **Learning Exercise 9.6C**

Fill in the blanks:

In osmosis, the direction of solvent flow is from the **(a.)** [higher/lower] solute concentration to the **(b.)** [higher/lower] solute concentration. A semipermeable membrane separates 5% (m/v) and 10% (m/v) sucrose solutions. The **(c.)** _____ % (m/v) solution has the greater osmotic pressure. Water will move from the **(d.)** _____ % (m/v) solution into the **(e.)** _____ % (m/v) solution. The compartment that contains the **(f.)** _____ % (m/v) solution increases in volume.

Answers **a.** lower **b.** higher **c.** 10 **d.** 5 **e.** 10 **f.** 10

♦ **Learning Exercise 9.6D**

A semipermeable membrane separates a 2% starch solution from a 10% starch solution. Complete each of the following with 2% or 10%:

a. Water will flow from the _____ starch solution to the _____ starch solution.

b. The volume of the _____ starch solution will increase and the volume of the _____ starch solution will decrease.

Answers **a.** 2%, 10% **b.** 10%, 2%

♦ **Learning Exercise 9.6E**

Fill in the blanks:

A **(a.)** _____ % (m/v) NaCl solution and a **(b.)** _____ % (m/v) glucose solution are isotonic to the body fluids. A red blood cell placed in these solutions does not change in volume because these solutions are **(c.)** _____ tonic. When a red blood cell is placed in water, it undergoes **(d.)** _____ because water is **(e.)** _____ tonic. A 20% (m/v) glucose solution will cause a red blood cell to undergo **(f.)** _____ because the 20% (m/v) glucose solution is **(g.)** _____ tonic.

Answers **a.** 0.9   **b.** 5   **c.** iso   **d.** hemolysis
       e. hypo   **f.** crenation **g.** hyper

♦ **Learning Exercise 9.6F**

Are the following solutions **a.** hypotonic, **b.** hypertonic, or **c.** isotonic, compared with a red blood cell?

1. ____ 5% (m/v) glucose **2.** ____ 3% (m/v) NaCl **3.** ____ 2% (m/v) glucose

4. ____ water   **5.** ____ 0.9% (m/v) NaCl **6.** ____ 10% (m/v) glucose

Answers **1.** c **2.** b **3.** a **4.** a **5.** c **6.** b

♦ **Learning Exercise 9.6G**

Indicate whether the following will cause a red blood cell to undergo

  a. crenation   **b.** hemolysis   **c.** no change

1. ____ 10% (m/v) NaCl **2.** ____ 1% (m/v) glucose **3.** ____ 5% (m/v) glucose

4. ____ 0.5% (m/v) NaCl **5.** ____ 10% (m/v) glucose **6.** ____ water

Answers **1.** a **2.** b **3.** c **4.** b **5.** a **6.** b

♦ **Learning Exercise 9.6H** 🏃

A dialysis bag contains starch, glucose, NaCl, protein, and urea.

a. When the dialysis bag is placed in water, what components would you expect to dialyze through the bag? Why?

b. Which components will stay inside the dialysis bag? Why?

Answers   **a.** Glucose, Na^+, Cl^- ions, and urea are solution particles. Solution particles will pass through semipermeable membranes.
b. Starch and protein form colloidal particles; because of their size, colloidal particles are retained by semipermeable membranes.

Checklist for Chapter 9

You are ready to take the Practice Test for Chapter 9. Be sure you have accomplished the following learning goals for this chapter. If not, review the Section listed at the end of the goal. Then apply your new skills and understanding to the Practice Test.

After studying Chapter 9, I can successfully:

_____ Identify the solute and solvent in a solution and describe the process of dissolving an ionic solute in water. (9.1)

_____ Identify the components in a solution of an electrolyte or a nonelectrolyte. (9.2)

_____ Calculate the number of equivalents in a given number of moles. (9.2)

_____ Identify a saturated and an unsaturated solution. (9.3)

_____ Describe the effects of temperature and nature of a solute on its solubility in a solvent. (9.3)

_____ Determine the solubility of an ionic compound in water. (9.3)

_____ Determine if an insoluble compound forms after two aqueous solutions of ionic compounds are mixed. (9.3)

_____ Write an ionic and a net ionic equation for the formation of an insoluble compound. (9.3)

_____ Calculate the percent concentration, m/m, v/v, and m/v, of a solute in a solution. (9.4)

_____ Use percent concentration to calculate the amount of solute or solution. (9.4)

_____ Calculate the molarity of a solution. (9.4)

_____ Use molarity as a conversion factor to calculate the moles (or grams) of a solute or the volume of the solution. (9.4)

_____ Use the molarity of a solution in a chemical reaction to calculate the volume or quantity of a reactant or product. (9.4)

_____ Calculate the unknown concentration or volume when a solution is diluted. (9.5)

_____ Identify a mixture as a solution, a colloid, or a suspension. (9.6)

_____ Use the concentration of particles in a solution to calculate the freezing point and boiling point of the solution. (9.6)

_____ Identify a solution as isotonic, hypotonic, or hypertonic. (9.6)

_____ Explain the processes of osmosis and dialysis. (9.6)

Practice Test for Chapter 9

The chapter Sections to review are shown in parentheses at the end of each question.

For questions 1 through 4, indicate if each of the following is more soluble in (W) *water, a polar solvent, or* (B) *benzene, a nonpolar solvent:* (9.1)

1. $I_2(s)$, nonpolar

2. $NaBr(s)$, polar

3. $KI(s)$, polar

4. $C_6H_{12}(l)$, nonpolar

5. When dissolved in water, $Ca(NO_3)_2(s)$ dissociates into (9.2)
 A. $Ca^{2+}(aq) + (NO_3)_2{}^{2-}(aq)$
 B. $Ca^+(aq) + NO_3{}^-(aq)$
 C. $Ca^{2+}(aq) + 2NO_3{}^-(aq)$
 D. $Ca^{2+}(aq) + 2N^{5+}(aq) + 2O_3{}^{6-}(aq)$
 E. $CaNO_3{}^+(aq) + NO_3{}^-(aq)$

6. What is the number of equivalents in 2 moles of Mg^{2+}? (9.2)
 A. 0.50 Eq   **B.** 1 Eq   **C.** 1.5 Eq   **D.** 2 Eq   **E.** 4 Eq

7. The solubility of NH_4Cl is 46 g in 100. g of water at 40 °C. How many grams of NH_4Cl can dissolve in 500. g of water at 40 °C? (9.3)
 A. 9.2 g   **B.** 46 g   **C.** 100 g   **D.** 184 g   **E.** 230 g

For questions 8 through 11, indicate if each of the following is soluble (S) *or insoluble* (I) *in water:* (9.3)

8. NaCl   **9.** AgCl   **10.** $BaSO_4$ **11.** FeO

12. Which of the following ionic compounds is soluble in water? (9.3)
 A. FeS   **B.** $BaCO_3$ **C.** K_2SO_4 **D.** PbS   **E.** MgO

13. The insoluble ionic compound that forms when a solution of NaCl mixes with a $Pb(NO_3)_2$ solution is (9.3)
 A. Na_2Pb **B.** $ClNO_3$ **C.** $NaNO_3$ **D.** $PbCl_2$ **E.** no insoluble ionic compound forms

14. Which of the following ionic compounds is insoluble in water? (9.3)
 A. $CuCl_2$ **B.** $Pb(NO_3)_2$ **C.** K_2CO_3 **D.** $(NH_4)_2SO_4$ **E.** $CaCO_3$

15. A solution made by dissolving 48 g of KCl in 1200 mL of solution has a percent (m/m) concentration of (9.4)
 A. 0.040%   **B.** 0.40%   **C.** 4.0%   **D.** 25%   **E.** 40.%

16. A solution containing 6.0 g of NaCl in 1500 g of solution has a percent (m/m) concentration of (9.4)
 A. 0.40%   **B.** 0.25%   **C.** 4.0%   **D.** 0.90%   **E.** 2.5%

17. A solution containing 1.20 g of sucrose in 50.0 g of solution has a percent (m/m) concentration of (9.4)
 A. 0.600%   **B.** 1.20%   **C.** 2.40%   **D.** 30.0%   **E.** 41.6%

18. The amount of lactose in 250 g of a 3.0% (m/m) lactose solution for infant formula is (9.4)
 A. 0.15 g   **B.** 1.2 g   **C.** 6.0 g   **D.** 7.5 g   **E.** 30 g

19. The mass of solution needed to obtain 0.40 g of glucose from a 5.0% (m/m) glucose solution is (9.4)
    A. 1.0 g   **B.** 2.0 g   **C.** 4.0 g   **D.** 5.0 g   **E.** 8.0 g

20. The amount of NaCl needed to prepare 50.0 g of a 4.00% (m/m) NaCl solution is (9.4)
    A. 20.0 g   **B.** 15.0 g   **C.** 10.0 g   **D.** 4.00 g   **E.** 2.00 g

21. The number of grams of NaOH needed to prepare 7.5 mL of a 5.0 M NaOH solution is (9.4)
    A. 1.5 g   **B.** 3.8 g   **C.** 6.7 g   **D.** 15 g   **E.** 38 g

For questions 22 through 24, consider a 20.0-g sample of a solution that contains 2.0 g of NaOH: (9.4)

22. The percent (m/m) concentration of the solution is
    A. 1.0%   **B.** 4.0%   **C.** 5.0%   **D.** 10.%   **E.** 20.%

23. The number of moles of NaOH in the sample is
    A. 0.050 mole   **B.** 0.40 mole   **C.** 1.0 mole   **D.** 2.5 moles   **E.** 4.0 moles

24. If the sample has a volume of 0.025 L, what is the molarity of the solution?
    A. 0.10 M   **B.** 0.5 M   **C.** 1.0 M   **D.** 1.5 M   **E.** 2.0 M

25. What mass of Ag_2SO_4 forms when 25.0 mL of a 0.111 M $AgNO_3$ solution reacts? (9.4)
$$2AgNO_3(aq) + H_2SO_4(aq) \longrightarrow Ag_2SO_4(s) + 2HNO_3(aq)$$
    A. 0.866 g   **B.** 1.74 g   **C.** 866 g   **D.** 0.433 g   **E.** 2.78 g

26. A 20.-mL sample of a 5.0 M HCl solution is diluted with water to give 100. mL of solution. The final concentration of the HCl solution is (9.5)
    A. 10 M   **B.** 5.0 M   **C.** 2.0 M   **D.** 1.0 M   **E.** 0.50 M

27. Water is added to 200. mL of a 4.00 M KNO_3 solution to give 400. mL of solution. The final concentration of the KNO_3 solution is (9.5)
    A. 1.00 M   **B.** 2.00 M   **C.** 4.00 M   **D.** 0.500 M   **E.** 0.100 M

28. 5.0 mL of a 2.0 M KOH solution is diluted with water to give a 0.40 M solution. The final volume of the KOH solution is (9.5)
    A. 10. mL   **B.** 25 mL   **C.** 40. mL   **D.** 250 mL   **E.** 400 mL

For questions 29 through 33, indicate whether each statement describes a (9.6)

    A. solution   **B.** colloid   **C.** suspension

29. _____ can be separated by filtering

30. _____ can be separated by semipermeable membranes

31. _____ passes through semipermeable membranes

32. _____ contains single atoms, ions, or small molecules of solute

33. _____ settles out upon standing

34. A solution contains 0.50 mole of $CaCl_2$ in 1.0 kg of water. The freezing point of the solution is _____ the freezing point of water. (9.6)
    A. higher than   **B.** lower than   **C.** the same as

35. If a solution contains 0.50 mole of the strong electrolyte $CaCl_2$ in 1.0 kg of water, what is the freezing point of the solution? (9.6)
    A. 0 °C   **B.** 2.8 °C   **C.** −2.8 °C   **D.** 0.93 °C   **E.** −0.93 °C

36. In osmosis, the net flow of water is (9.6)
    A. between solutions of equal concentrations
    B. from higher solute concentration to lower solute concentration
    C. from lower solute concentration to higher solute concentration
    D. from a colloid to a solution of equal concentration
    E. from lower solvent concentration to higher solvent concentration

37. A red blood cell undergoes hemolysis when placed in a solution that is (9.6)

    A. isotonic   **B.** hypotonic   **C.** hypertonic   **D.** colloidal   **E.** semitonic

38. A red blood cell undergoes crenation when placed in a solution that is (9.6)

    A. isotonic   **B.** hypotonic   **C.** hypertonic   **D.** colloidal   **E.** semitonic

39. A solution that has the same osmotic pressure as body fluids is (9.6)

    A. 0.1% (m/v) NaCl   **B.** 0.9% (m/v) NaCl   **C.** 5% (m/v) NaCl

    D. 10% (m/v) glucose   **E.** 15% (m/v) glucose

Answers to the Practice Test

1. B	**2.** W	**3.** W	**4.** B	**5.** C
6. E	**7.** E	**8.** S	**9.** I	**10.** I
11. I	**12.** C	**13.** D	**14.** E	**15.** C
16. A	**17.** C	**18.** D	**19.** E	**20.** E
21. A	**22.** D	**23.** A	**24.** E	**25.** D
26. D	**27.** B	**28.** B	**29.** C	**30.** B, C
31. A	**32.** A	**33.** C	**34.** B	**35.** C
36. C	**37.** B	**38.** C	**39.** B	

Selected Answers and Solutions to Text Problems

9.1 The component present in the smaller amount is the solute; the component present in the larger amount is the solvent.
   a. NaCl, solute; water, solvent
   b. water, solute; ethanol, solvent
   c. oxygen, solute; nitrogen, solvent

9.3 The K^+ and I^- ions at the surface of the solid are pulled into solution by the polar water molecules, where the hydration process surrounds separate ions with water molecules.

9.5 **a.** $CaCO_3$ (an ionic solute) would be soluble in water (a polar solvent).
   b. Retinol (a nonpolar solute) would be soluble in CCl_4 (a nonpolar solvent).
   c. Sucrose (a polar solute) would be soluble in water (a polar solvent).
   d. Cholesterol (a nonpolar solute) would be soluble in CCl_4 (a nonpolar solvent).

9.7 The strong electrolyte KF completely dissociates into K^+ and F^- ions when it dissolves in water. When the weak electrolyte HF dissolves in water, there are a few ions of H^+ and F^- present, but mostly dissolved HF molecules.

9.9 Strong electrolytes dissociate completely into ions.
   a. $KCl(s) \xrightarrow{H_2O} K^+(aq) + Cl^-(aq)$
   b. $CaCl_2(s) \xrightarrow{H_2O} Ca^{2+}(aq) + 2Cl^-(aq)$
   c. $K_3PO_4(s) \xrightarrow{H_2O} 3K^+(aq) + PO_4^{3-}(aq)$
   d. $Fe(NO_3)_3(s) \xrightarrow{H_2O} Fe^{3+}(aq) + 3NO_3^-(aq)$

9.11 **a.** An aqueous solution of a weak electrolyte like acetic acid will contain mostly $HC_2H_3O_2$ molecules, with a few H^+ ions and a few $C_2H_3O_2^-$ ions.
   b. An aqueous solution of a strong electrolyte like NaBr will contain only the ions Na^+ and Br^-.
   c. An aqueous solution of a nonelectrolyte like fructose will contain only $C_6H_{12}O_6$ molecules.

9.13 **a.** K_2SO_4 is a strong electrolyte because only ions are present in the K_2SO_4 solution.
   b. NH_3 is a weak electrolyte because only a few NH_4^+ and OH^- ions are present in the solution.
   c. $C_6H_{12}O_6$ is a nonelectrolyte because only $C_6H_{12}O_6$ molecules are present in the solution.

9.15 **a.** $1 \text{ mole } K^+ \times \dfrac{1 \text{ Eq } K^+}{1 \text{ mole } K^+} = 1 \text{ Eq of } K^+$

   b. $2 \text{ moles } OH^- \times \dfrac{1 \text{ Eq } OH^-}{1 \text{ mole } OH^-} = 2 \text{ Eq of } OH^-$

   c. $1 \text{ mole } Ca^{2+} \times \dfrac{2 \text{ Eq } Ca^{2+}}{1 \text{ mole } Ca^{2+}} = 2 \text{ Eq of } Ca^{2+}$

   d. $3 \text{ moles } CO_3^{2-} \times \dfrac{2 \text{ Eq } CO_3^{2-}}{1 \text{ mole } CO_3^{2-}} = 6 \text{ Eq of } CO_3^{2-}$

9.17 $1.00 \text{ L} \times \dfrac{154 \text{ mEq } Na^+}{1 \text{ L}} \times \dfrac{1 \text{ Eq } Na^+}{1000 \text{ mEq } Na^+} \times \dfrac{1 \text{ mole } Na^+}{1 \text{ Eq } Na^+} = 0.154 \text{ mole of } Na^+ \text{ (3 SFs)}$

   $1.00 \text{ L} \times \dfrac{154 \text{ mEq } Cl^-}{1 \text{ L}} \times \dfrac{1 \text{ Eq } Cl^-}{1000 \text{ mEq } Cl^-} \times \dfrac{1 \text{ mole } Cl^-}{1 \text{ Eq } Cl^-} = 0.154 \text{ mole of } Cl^- \text{ (3 SFs)}$

9.19 In any solution, the total equivalents of anions must be equal to the equivalents of cations. mEq of anions = 40. mEq/L Cl^- + 15 mEq/L HPO_4^{2-} = 55 mEq/L of anions mEq of Na^+ = mEq of anions = 55 mEq/L of Na^+

9.21 a. The solution must be saturated because no additional solute dissolves.
  b. The solution was unsaturated because the sugar cube dissolves completely.
  c. The solution is unsaturated because the uric acid concentration does not exceed its solubility of 7 mg/100 mL at 37 °C.

9.23 a. At 20 °C, KCl has a solubility of 34 g of KCl in 100. g of H_2O. Because 25 g of KCl is less than the maximum amount that can dissolve in 100. g of H_2O at 20 °C, the KCl solution is unsaturated.
  b. At 20 °C, $NaNO_3$ has a solubility of 88 g of $NaNO_3$ in 100. g of H_2O. Using the solubility as a conversion factor, we can calculate the maximum amount of $NaNO_3$ that can dissolve in 25 g of H_2O:

$$25 \text{ g } H_2O \times \frac{88 \text{ g } NaNO_3}{100. \text{ g } H_2O} = 22 \text{ g of } NaNO_3 \text{ (2 SFs)}$$

Because 11 g of $NaNO_3$ is less than the maximum amount that can dissolve in 25 g of H_2O at 20 °C, the $NaNO_3$ solution is unsaturated.

  c. At 20 °C, sugar has a solubility of 204 g of $C_{12}H_{22}O_{11}$ in 100. g of H_2O. Using the solubility as a conversion factor, we can calculate the maximum amount of sugar that can dissolve in 125 g of H_2O:

$$125 \text{ g } H_2O \times \frac{204 \text{ g sugar}}{100. \text{ g } H_2O} = 255 \text{ g of sugar (3 SFs)}$$

Because 400. g of $C_{12}H_{22}O_{11}$ exceeds the maximum amount that can dissolve in 125 g of H_2O at 20 °C, the sugar solution is saturated, and excess undissolved sugar will be present on the bottom of the container.

9.25 a. At 20 °C, KCl has a solubility of 34 g of KCl in 100. g of H_2O.
  ∴ 200. g of H_2O will dissolve:

$$200. \text{ g } H_2O \times \frac{34 \text{ g KCl}}{100. \text{ g } H_2O} = 68 \text{ g of KCl (2 SFs)}$$

At 20 °C, 68 g of KCl will remain in solution.

  b. Since 80. g of KCl dissolves at 50 °C and 68 g remains in solution at 20 °C, the mass of solid KCl that crystallizes after cooling is (80. g KCl − 68 g KCl =) 12 g of KCl. (2 SFs)

9.27 a. In general, the solubility of solid solutes (like sugar) in water increases as temperature is increased.
  b. The solubility of a gaseous solute (CO_2) in water is less at a higher temperature.
  c. The solubility of a gaseous solute in water is less at a higher temperature, and the CO_2 pressure in the can is increased. When the can of warm soda is opened, more CO_2 is released, producing more spray.

9.29 a. Salts containing Li^+ ions are soluble.
  b. Salts containing S^{2-} ions are usually insoluble.
  c. Salts containing CO_3^{2-} ions are usually insoluble.
  d. Salts containing K^+ ions are soluble.
  e. Salts containing NO_3^- ions are soluble.

9.31 a. No solid forms; salts containing K^+ and Na^+ are soluble.
  b. Solid silver sulfide (Ag_2S) forms:

$$2AgNO_3(aq) + K_2S(aq) \longrightarrow Ag_2S(s) + 2KNO_3(aq)$$
$$2Ag^+(aq) + 2NO_3^-(aq) + 2K^+(aq) + S^{2-}(aq) \longrightarrow Ag_2S(s) + 2K^+(aq) + 2NO_3^-(aq)$$
$$2Ag^+(aq) + S^{2-}(aq) \longrightarrow Ag_2S(s) \qquad \text{Net ionic equation}$$

c. Solid calcium sulfate ($CaSO_4$) forms:

$$CaCl_2(aq) + Na_2SO_4(aq) \longrightarrow CaSO_4(s) + 2NaCl(aq)$$

$$Ca^{2+}(aq) + 2Cl^-(aq) + 2Na^+(aq) + SO_4^{2-}(aq) \longrightarrow CaSO_4(s) + 2Na^+(aq) + 2Cl^-(aq)$$

$$Ca^{2+}(aq) + SO_4^{2-}(aq) \longrightarrow CaSO_4(s) \quad \text{Net ionic equation}$$

d. Solid copper phosphate ($Cu_3(PO_4)_2$) forms:

$$3CuCl_2(aq) + 2Li_3PO_4(aq) \longrightarrow Cu_3(PO_4)_2(s) + 6LiCl(aq)$$

$$3Cu^{2+}(aq) + 6Cl^-(aq) + 6Li^+(aq) + 2PO_4^{3-}(aq) \longrightarrow Cu_3(PO_4)_2(s) + 6Li^+(aq) + 6Cl^-(aq)$$

$$3Cu^{2+}(aq) + 2PO_4^{3-}(aq) \longrightarrow Cu_3(PO_4)_2(s) \quad \text{Net ionic equation}$$

9.33 Mass percent (m/m) $= \dfrac{\text{mass of solute (g)}}{\text{mass of solution (g)}} \times 100\%$

a. mass of solution $= 25$ g KCl $+ 125$ g $H_2O = 150.$ g of solution

$$\dfrac{25 \text{ g KCl}}{150. \text{ g solution}} \times 100\% = 17\% \text{ (m/m) KCl solution (2 SFs)}$$

b. $\dfrac{12 \text{ g sucrose}}{225 \text{ g solution}} \times 100\% = 5.3\% \text{ (m/m) sucrose solution (2 SFs)}$

c. $\dfrac{8.0 \text{ g } CaCl_2}{80.0 \text{ g solution}} \times 100\% = 10.\% \text{ (m/m) } CaCl_2 \text{ solution (2 SFs)}$

9.35 $355 \text{ mL solution} \times \dfrac{22.5 \text{ mL alcohol}}{100. \text{ mL solution}} = 79.9 \text{ mL of alcohol (3 SFs)}$

9.37 Mass/volume percent (m/v) $= \dfrac{\text{mass of solute (g)}}{\text{volume of solution (mL)}} \times 100\%$

a. $\dfrac{75 \text{ g } Na_2SO_4}{250 \text{ mL solution}} \times 100\% = 30.\% \text{ (m/v) } Na_2SO_4 \text{ solution (2 SFs)}$

b. $\dfrac{39 \text{ g sucrose}}{355 \text{ mL solution}} \times 100\% = 11\% \text{ (m/v) sucrose solution (2 SFs)}$

9.39 a. $50. \text{ g solution} \times \dfrac{5.0 \text{ g KCl}}{100. \text{ g solution}} = 2.5 \text{ g of KCl (2 SFs)}$

b. $1250 \text{ mL solution} \times \dfrac{4.0 \text{ g } NH_4Cl}{100. \text{ mL solution}} = 50. \text{ g of } NH_4Cl \text{ (2 SFs)}$

c. $250. \text{ mL solution} \times \dfrac{10.0 \text{ mL acetic acid}}{100. \text{ mL solution}} = 25.0 \text{ mL of acetic acid (3 SFs)}$

9.41 a. $5.0 \text{ g } LiNO_3 \times \dfrac{100. \text{ g solution}}{25 \text{ g } LiNO_3} = 20. \text{ g of } LiNO_3 \text{ solution (2 SFs)}$

b. $40.0 \text{ g KOH} \times \dfrac{100. \text{ mL solution}}{10.0 \text{ g KOH}} = 400. \text{ mL of KOH solution (3 SFs)}$

c. $2.0 \text{ mL formic acid} \times \dfrac{100. \text{ mL solution}}{10.0 \text{ mL formic acid}} = 20. \text{ mL of formic acid solution (2 SFs)}$

9.43 Molarity (M) $= \dfrac{\text{moles of solute}}{\text{liters of solution}}$

a. $\dfrac{2.00 \text{ moles glucose}}{4.00 \text{ L solution}} = 0.500 \text{ M glucose solution (3 SFs)}$

b. $\dfrac{4.00 \text{ g } \cancel{\text{KOH}}}{2.00 \text{ L solution}} \times \dfrac{1 \text{ mole KOH}}{56.11 \text{ g } \cancel{\text{KOH}}} = 0.0356 \text{ M KOH solution (3 SFs)}$

c. $\dfrac{5.85 \text{ g } \cancel{\text{NaCl}}}{400. \text{ mL } \cancel{\text{solution}}} \times \dfrac{1 \text{ mole NaCl}}{58.44 \text{ g } \cancel{\text{NaCl}}} \times \dfrac{1000 \text{ mL } \cancel{\text{solution}}}{1 \text{ L solution}} = 0.250 \text{ M NaCl solution (3 SFs)}$

9.45 a. $2.00 \text{ L } \cancel{\text{solution}} \times \dfrac{1.50 \text{ moles } \cancel{\text{NaOH}}}{1 \text{ L } \cancel{\text{solution}}} \times \dfrac{40.00 \text{ g NaOH}}{1 \text{ mole } \cancel{\text{NaOH}}} = 120. \text{ g of NaOH (3 SFs)}$

b. $4.00 \text{ L } \cancel{\text{solution}} \times \dfrac{0.200 \text{ mole } \cancel{\text{KCl}}}{1 \text{ L } \cancel{\text{solution}}} \times \dfrac{74.55 \text{ g KCl}}{1 \text{ mole } \cancel{\text{KCl}}} = 59.6 \text{ g of KCl (3 SFs)}$

c. $25.0 \text{ mL } \cancel{\text{solution}} \times \dfrac{1 \text{ L } \cancel{\text{solution}}}{1000 \text{ mL } \cancel{\text{solution}}} \times \dfrac{6.00 \text{ moles } \cancel{\text{HCl}}}{1 \text{ L } \cancel{\text{solution}}} \times \dfrac{36.46 \text{ g HCl}}{1 \text{ mole } \cancel{\text{HCl}}}$
$= 5.47 \text{ g of HCl (3 SFs)}$

9.47 a. $3.00 \text{ moles } \cancel{\text{KBr}} \times \dfrac{1 \text{ L solution}}{2.00 \text{ moles } \cancel{\text{KBr}}} = 1.50 \text{ L of KBr solution (3 SFs)}$

b. $15.0 \text{ moles } \cancel{\text{NaCl}} \times \dfrac{1 \text{ L solution}}{1.50 \text{ moles } \cancel{\text{NaCl}}} = 10.0 \text{ L of NaCl solution (3 SFs)}$

c. $0.0500 \text{ mole } \cancel{\text{Ca(NO}_3)_2} \times \dfrac{1 \text{ L } \cancel{\text{solution}}}{0.800 \text{ mole } \cancel{\text{Ca(NO}_3)_2}} \times \dfrac{1000 \text{ mL solution}}{1 \text{ L } \cancel{\text{solution}}}$
$= 62.5 \text{ mL of Ca(NO}_3)_2 \text{ solution (3 SFs)}$

9.49 a. $12.5 \text{ g } \cancel{\text{Na}_2\text{CO}_3} \times \dfrac{1 \text{ mole } \cancel{\text{Na}_2\text{CO}_3}}{105.99 \text{ g } \cancel{\text{Na}_2\text{CO}_3}} \times \dfrac{1 \text{ L } \cancel{\text{solution}}}{0.120 \text{ mole } \cancel{\text{Na}_2\text{CO}_3}} \times \dfrac{1000 \text{ mL solution}}{1 \text{ L } \cancel{\text{solution}}}$
$= 983 \text{ mL of Na}_2\text{CO}_3 \text{ solution (3 SFs)}$

b. $0.850 \text{ mole } \cancel{\text{NaNO}_3} \times \dfrac{1 \text{ L } \cancel{\text{solution}}}{0.500 \text{ mole } \cancel{\text{NaNO}_3}} \times \dfrac{1000 \text{ mL solution}}{1 \text{ L } \cancel{\text{solution}}}$
$= 1700 \text{ mL } (1.70 \times 10^3 \text{ mL}) \text{ of NaNO}_3 \text{ solution (3 SFs)}$

c. $30.0 \text{ g } \cancel{\text{LiOH}} \times \dfrac{1 \text{ mole } \cancel{\text{LiOH}}}{23.95 \text{ g } \cancel{\text{LiOH}}} \times \dfrac{1 \text{ L } \cancel{\text{solution}}}{2.70 \text{ moles } \cancel{\text{LiOH}}} \times \dfrac{1000 \text{ mL solution}}{1 \text{ L } \cancel{\text{solution}}}$
$= 464 \text{ mL of LiOH solution (3 SFs)}$

9.51 a. $50.0 \text{ mL } \cancel{\text{solution}} \times \dfrac{1 \text{ L } \cancel{\text{solution}}}{1000 \text{ mL } \cancel{\text{solution}}} \times \dfrac{1.50 \text{ moles KCl}}{1 \text{ L } \cancel{\text{solution}}} = 0.0750 \text{ mole of KCl}$

$0.0750 \text{ mole } \cancel{\text{KCl}} \times \dfrac{1 \text{ mole } \cancel{\text{PbCl}_2}}{2 \text{ moles } \cancel{\text{KCl}}} \times \dfrac{278.1 \text{ g PbCl}_2}{1 \text{ mole } \cancel{\text{PbCl}_2}} = 10.4 \text{ g of PbCl}_2 \text{ (3 SFs)}$

b. $50.0 \text{ mL } \cancel{\text{solution}} \times \dfrac{1 \text{ L } \cancel{\text{solution}}}{1000 \text{ mL } \cancel{\text{solution}}} \times \dfrac{1.50 \text{ moles KCl}}{1 \text{ L } \cancel{\text{solution}}} = 0.0750 \text{ mole of KCl}$

$0.0750 \text{ mole } \cancel{\text{KCl}} \times \dfrac{1 \text{ mole } \cancel{\text{Pb(NO}_3)_2}}{2 \text{ moles } \cancel{\text{KCl}}} \times \dfrac{1 \text{ L } \cancel{\text{solution}}}{2.00 \text{ moles } \cancel{\text{Pb(NO}_3)_2}} \times \dfrac{1000 \text{ mL solution}}{1 \text{ L } \cancel{\text{solution}}}$
$= 18.8 \text{ mL of Pb(NO}_3)_2 \text{ solution (3 SFs)}$

c. $30.0 \text{ mL } \cancel{\text{solution}} \times \dfrac{1 \text{ L } \cancel{\text{solution}}}{1000 \text{ mL } \cancel{\text{solution}}} \times \dfrac{0.400 \text{ mole } \cancel{\text{Pb(NO}_3)_2}}{1 \text{ L } \cancel{\text{solution}}} \times \dfrac{2 \text{ moles KCl}}{1 \text{ mole } \cancel{\text{Pb(NO}_3)_2}}$
$= 0.0240 \text{ mole of KCl}$

$20.0 \text{ mL } \cancel{\text{solution}} \times \dfrac{1 \text{ L solution}}{1000 \text{ mL } \cancel{\text{solution}}} = 0.0200 \text{ L of solution}$

$\text{molarity (M)} = \dfrac{\text{moles of solute}}{\text{liters of solution}} = \dfrac{0.0240 \text{ mole KCl}}{0.0200 \text{ L solution}} = 1.20 \text{ M KCl solution (3 SFs)}$

9.53 **a.** $15.0 \text{ g Mg} \times \dfrac{1 \text{ mole Mg}}{24.31 \text{ g Mg}} \times \dfrac{2 \text{ moles HCl}}{1 \text{ mole Mg}} \times \dfrac{1 \text{ L solution}}{6.00 \text{ moles HCl}} \times \dfrac{1000 \text{ mL solution}}{1 \text{ L solution}}$

$= 206 \text{ mL of HCl solution (3 SFs)}$

b. $0.500 \text{ L solution} \times \dfrac{2.00 \text{ moles HCl}}{1 \text{ L solution}} \times \dfrac{1 \text{ mole H}_2}{2 \text{ moles HCl}} \times \dfrac{22.4 \text{ L H}_2 \text{ (STP)}}{1 \text{ mole H}_2}$

$= 11.2 \text{ L of H}_2 \text{ at STP (3 SFs)}$

c. $n = \dfrac{PV}{RT} = \dfrac{(735 \text{ mmHg})(5.20 \text{ L})}{\left(\dfrac{62.4 \text{ L} \cdot \text{mmHg}}{\text{mole} \cdot \text{K}}\right)(298 \text{ K})} = 0.206 \text{ mole of H}_2 \text{ gas}$

$0.206 \text{ mole H}_2 \times \dfrac{2 \text{ moles HCl}}{1 \text{ mole H}_2} = 0.412 \text{ mole of HCl}$

$45.2 \text{ mL solution} \times \dfrac{1 \text{ L solution}}{1000 \text{ mL solution}} = 0.0452 \text{ L of solution}$

$\text{molarity (M)} = \dfrac{\text{moles of solute}}{\text{liters of solution}} = \dfrac{0.412 \text{ mole HCl}}{0.0452 \text{ L solution}} = 9.12 \text{ M HCl solution (3 SFs)}$

9.55 **a.** $1 \text{ K} \times \dfrac{100. \text{ mL solution}}{1 \text{ K}} \times \dfrac{20. \text{ g mannitol}}{100. \text{ mL solution}} = 20. \text{ g of mannitol (2 SFs)}$

b. $12 \text{ K} \times \dfrac{100. \text{ mL solution}}{1 \text{ K}} \times \dfrac{20. \text{ g mannitol}}{100. \text{ mL solution}} = 240 \text{ g of mannitol (2 SFs)}$

9.57 **a.** $100. \text{ g glucose} \times \dfrac{100. \text{ mL solution}}{5 \text{ g glucose}} \times \dfrac{1 \text{ L}}{1000 \text{ mL}} = 2 \text{ L of glucose solution (1 SF)}$

b. $2.0 \text{ g NaCl} \times \dfrac{100. \text{ mL solution}}{0.90 \text{ g NaCl}} = 220 \text{ mL of NaCl solution (2 SFs)}$

9.59 $5.0 \text{ mL solution} \times \dfrac{10. \text{ g CaCl}_2}{100. \text{ mL solution}} = 0.50 \text{ g of CaCl}_2 \text{ (2 SFs)}$

9.61 Adding water (solvent) to the soup increases the volume and decreases the tomato soup concentration.

9.63 $C_1 V_1 = C_2 V_2$

a. $C_2 = C_1 \times \dfrac{V_1}{V_2} = 6.0 \text{ M} \times \dfrac{2.0 \text{ L}}{6.0 \text{ L}} = 2.0 \text{ M HCl solution (2 SFs)}$

b. $C_2 = C_1 \times \dfrac{V_1}{V_2} = 12 \text{ M} \times \dfrac{0.50 \text{ L}}{3.0 \text{ L}} = 2.0 \text{ M NaOH solution (2 SFs)}$

c. $C_2 = C_1 \times \dfrac{V_1}{V_2} = 25\% \times \dfrac{10.0 \text{ mL}}{100.0 \text{ mL}} = 2.5\% \text{ (m/v) KOH solution (2 SFs)}$

d. $C_2 = C_1 \times \dfrac{V_1}{V_2} = 15\% \times \dfrac{50.0 \text{ mL}}{250 \text{ mL}} = 3.0\% \text{ (m/v) H}_2\text{SO}_4 \text{ solution (2 SFs)}$

9.65 $C_1 V_1 = C_2 V_2$

a. $V_2 = V_1 \times \dfrac{C_1}{C_2} = 20.0 \text{ mL} \times \dfrac{6.0 \text{ M}}{1.5 \text{ M}} = 80. \text{ mL of diluted HCl solution (2 SFs)}$

b. $V_2 = V_1 \times \dfrac{C_1}{C_2} = 50.0 \text{ mL} \times \dfrac{10.0 \%}{2.0 \%} = 250 \text{ mL of diluted LiCl solution (2 SFs)}$

c. $V_2 = V_1 \times \dfrac{C_1}{C_2} = 50.0 \text{ mL} \times \dfrac{6.00 \text{ M}}{0.500 \text{ M}} = 600. \text{ mL of diluted } H_3PO_4 \text{ solution (3 SFs)}$

d. $V_2 = V_1 \times \dfrac{C_1}{C_2} = 75 \text{ mL} \times \dfrac{12 \text{ \%}}{5.0 \text{ \%}} = 180 \text{ mL of diluted glucose solution (2 SFs)}$

9.67 $C_1V_1 = C_2V_2$

a. $V_1 = V_2 \times \dfrac{C_2}{C_1} = 255 \text{ mL} \times \dfrac{0.200 \text{ M}}{4.00 \text{ M}} = 12.8 \text{ mL of the } HNO_3 \text{ solution (3 SFs)}$

b. $V_1 = V_2 \times \dfrac{C_2}{C_1} = 715 \text{ mL} \times \dfrac{0.100 \text{ M}}{6.00 \text{ M}} = 11.9 \text{ mL of the } MgCl_2 \text{ solution (3 SFs)}$

c. $V_2 = 0.100 \text{ L} \times \dfrac{1000 \text{ mL}}{1 \text{ L}} = 100. \text{ mL}$

$V_1 = V_2 \times \dfrac{C_2}{C_1} = 100. \text{ mL} \times \dfrac{0.150 \text{ M}}{8.00 \text{ M}} = 1.88 \text{ mL of the KCl solution (3 SFs)}$

9.69 $C_1V_1 = C_2V_2$

$V_1 = V_2 \times \dfrac{C_2}{C_1} = 500. \text{ mL} \times \dfrac{5.0\%}{25\%} = 1.0 \times 10^2 \text{ mL of the glucose solution (2 SFs)}$

9.71 a. A solution cannot be separated by a semipermeable membrane.

b. A suspension settles out upon standing.

9.73 a. When 1.0 mole of glycerol (a nonelectrolyte) dissolves in water, it does not dissociate into ions and so will only produce 1.0 mole of particles. Similarly, 2.0 moles of ethylene glycol (also a non-electrolyte) dissolves as molecules to produce only 2.0 moles of particles in water. The ethylene glycol solution has more particles in 1.0 L of water and will thus have a lower freezing point.

b. When 0.50 mole of the strong electrolyte KCl dissolves in water, it will produce 1.0 mole of particles because each formula unit of KCl dissociates to give two particles, K^+ and Cl^-. When 0.50 mole of the strong electrolyte $MgCl_2$ dissolves in water, it will produce 1.5 moles of particles because each formula unit of $MgCl_2$ dissociates to give three particles, Mg^{2+} and $2Cl^-$. Thus, a solution of 0.50 mole of $MgCl_2$ in 2.0 L of water will have the lower freezing point.

9.75 a. A solution of 1.36 moles of the nonelectrolyte CH_4O in 1.00 kg of water contains 1.36 moles of particles in 1.00 kg of water and has a freezing point change of

$$1.36 \text{ moles particles} \times \dfrac{1.86 \, ^\circ\text{C}}{1 \text{ mole particles}} = 2.53 \, ^\circ\text{C}$$

The freezing point would be $0.00 \, ^\circ\text{C} - 2.53 \, ^\circ\text{C} = -2.53 \, ^\circ\text{C}$

b. $640. \text{ g } C_3H_8O_2 \times \dfrac{1 \text{ mole } C_3H_8O_2}{76.09 \text{ g } C_3H_8O_2} = 8.41 \text{ moles of } C_3H_8O_2 \text{ (3 SFs)}$

A solution of 8.41 moles of the nonelectrolyte $C_3H_8O_2$ in 1.00 kg of water contains 8.41 moles of particles in 1.00 kg of water and has a freezing point change of

$$8.41 \text{ moles particles} \times \dfrac{1.86 \, ^\circ\text{C}}{1 \text{ mole particles}} = 15.6 \, ^\circ\text{C}$$

The freezing point would be $0.00 \, ^\circ\text{C} - 15.6 \, ^\circ\text{C} = -15.6 \, ^\circ\text{C}$

c. $111 \text{ g } KCl \times \dfrac{1 \text{ mole } KCl}{74.55 \text{ g } KCl} = 1.49 \text{ moles of KCl (3 SFs)}$

A solution of 1.49 moles of the strong electrolyte KCl in 1.00 kg of water contains 2.98 moles of particles (1.49 moles of K^+ and 1.49 moles of Cl^-) in 1.00 kg of water and has a freezing point change of

$$2.98 \text{ moles particles} \times \dfrac{1.86 \, ^\circ\text{C}}{1 \text{ mole particles}} = 5.54 \, ^\circ\text{C}$$

The freezing point would be $0.00 \, ^\circ\text{C} - 5.54 \, ^\circ\text{C} = -5.54 \, ^\circ\text{C}$

9.77 a. The 10% (m/v) starch solution has the higher solute concentration, more solute particles, and therefore the higher osmotic pressure.
 b. Initially, water will flow out of the 1% (m/v) starch solution into the more concentrated 10% (m/v) starch solution.
 c. The volume of the 10% (m/v) starch solution will increase due to inflow of water.

9.79 Water will flow from a region of higher solvent concentration (which corresponds to a lower solute concentration) to a region of lower solvent concentration (which corresponds to a higher solute concentration).
 a. B; The volume level will rise as water flows into compartment B, which contains the 10% (m/v) sucrose solution.
 b. A; The volume level will rise as water flows into compartment A, which contains the 8% (m/v) albumin solution.
 c. B; The volume level will rise as water flows into compartment B, which contains the 10% (m/v) starch solution.

9.81 a. 1 mole NaCl(s) $\longrightarrow$ 1 mole Na$^+$(aq) + 1 mole Cl$^-$(aq) = 2 moles of particles

$$\frac{0.45 \text{ g NaCl}}{100. \text{ mL solution}} \times \frac{1 \text{ mole NaCl}}{58.44 \text{ g NaCl}} \times \frac{2 \text{ moles particles}}{1 \text{ mole NaCl}} \times \frac{1000 \text{ mL solution}}{1 \text{ L solution}}$$

= 0.15 mole of particles (2 SFs)
 b. 1 mole C$_6$H$_{12}$O$_6$(s) $\longrightarrow$ 1 mole C$_6$H$_{12}$O$_6$(aq) = 1 mole of particles

$$\frac{50. \text{ g C}_6\text{H}_{12}\text{O}_6}{100. \text{ mL solution}} \times \frac{1 \text{ mole C}_6\text{H}_{12}\text{O}_6}{180.16 \text{ g C}_6\text{H}_{12}\text{O}_6} \times \frac{1 \text{ mole particles}}{1 \text{ mole C}_6\text{H}_{12}\text{O}_6} \times \frac{1000 \text{ mL solution}}{1 \text{ L solution}}$$

= 2.8 moles of particles (2 SFs)

9.83 A red blood cell has the same osmotic pressure as a 5% (m/v) glucose solution or a 0.9% (m/v) NaCl solution.
 a. Distilled water is a hypotonic solution when compared with a red blood cell's contents.
 b. A 1% (m/v) glucose solution is a hypotonic solution.
 c. A 0.9% (m/v) NaCl solution is isotonic with a red blood cell's contents.
 d. A 15% (m/v) glucose solution is a hypertonic solution.

9.85 Colloids cannot pass through the semipermeable dialysis membrane; water and solutions freely pass through semipermeable membranes.
 a. Sodium and chloride ions will both pass through the membrane into the distilled water.
 b. The amino acid alanine can pass through a dialysis membrane; the colloid starch will not.
 c. Sodium and chloride ions will both be present in the water surrounding the dialysis bag; the colloid starch will not.
 d. Urea will diffuse through the dialysis bag into the surrounding water.

9.87 a. Given 0.45 g/dL Cl$^-$ **Need** mEq/L

 Plan g $\rightarrow$ mole $\rightarrow$ Eq $\rightarrow$ mEq; dL $\rightarrow$ L $\dfrac{1 \text{ mole Cl}^-}{35.45 \text{ g Cl}^-}$ $\dfrac{1 \text{ Eq}}{1 \text{ mole Cl}^-}$ $\dfrac{1000 \text{ mEq}}{1 \text{ Eq}}$ $\dfrac{10 \text{ dL}}{1 \text{ L}}$

 Set-up $\dfrac{0.45 \text{ g Cl}^-}{1 \text{ dL}} \times \dfrac{1 \text{ mole Cl}^-}{35.45 \text{ g Cl}^-} \times \dfrac{1 \text{ Eq}}{1 \text{ mole Cl}^-} \times \dfrac{1000 \text{ mEq}}{1 \text{ Eq}} \times \dfrac{10 \text{ dL}}{1 \text{ L}}$

 = 130 mEq/L (2 SFs)
 b. The normal range for Cl$^-$ is 95 to 105 mEq/L of blood plasma, so a concentration of 130 mEq/L is above the normal range.

9.89 a. The levels of Ca^{2+}, Mg^{2+}, and HCO$_3^-$ in Michelle's blood serum need to be increased by dialysis.
 b. The levels of K$^+$, Na$^+$, and Cl$^-$ in Michelle's blood serum need to be decreased by dialysis.

9.91 **a.** 1; A solution will form because both the solute and the solvent are polar.
 b. 2; Two layers will form because one component is nonpolar and the other is polar.
 c. 1; A solution will form because both the solute and the solvent are nonpolar.

9.93 **a.** 3; A nonelectrolyte will show no dissociation.
 b. 1; A weak electrolyte will show some dissociation, producing a few ions, but mostly remaining as molecules.
 c. 2; A strong electrolyte will be completely dissociated into ions.

9.95 **a.** Beaker 3; Solid silver chloride (AgCl) will precipitate when the two solutions are mixed.
 b. $NaCl(aq) + AgNO_3(aq) \longrightarrow AgCl(s) + NaNO_3(aq)$
 $Na^+(aq) + Cl^-(aq) + Ag^+(aq) + NO_3^-(aq) \longrightarrow AgCl(s) + Na^+(aq) + NO_3^-(aq)$
 c. $Ag^+(aq) + Cl^-(aq) \longrightarrow AgCl(s)$ Net ionic equation

9.97 A "brine" saltwater solution has a high concentration of Na^+ and Cl^- ions, which is hypertonic to the cucumber. The skin of the cucumber acts like a semipermeable membrane; therefore, water flows from the more dilute solution inside the cucumber into the more concentrated brine solution that surrounds it. The loss of water causes the cucumber to become a wrinkled pickle.

9.99 **a.** 2; Water will flow into the B (8% starch solution) side.
 b. 1; Water will continue to flow equally in both directions; no change in volumes.
 c. 3; Water will flow into the A (5% sucrose solution) side.
 d. 2; Water will flow into the B (1% sucrose solution) side.

9.101 Because iodine is a nonpolar molecule, it will dissolve in hexane, a nonpolar solvent. Iodine does not dissolve in water because water is a polar solvent.

9.103 At 20 °C, KNO_3 has a solubility of 32 g of KNO_3 in 100. g of H_2O.
 a. 200. g of H_2O will dissolve:

$$200. \text{ g } H_2O \times \frac{32 \text{ g } KNO_3}{100. \text{ g } H_2O} = 64 \text{ g of } KNO_3 \text{ (2 SFs)}$$

Because 32 g of KNO_3 is less than the maximum amount that can dissolve in 200. g of H_2O at 20 °C, the KNO_3 solution is unsaturated.
 b. 50. g of H_2O will dissolve:

$$50. \text{ g } H_2O \times \frac{32 \text{ g } KNO_3}{100. \text{ g } H_2O} = 16 \text{ g of } KNO_3 \text{ (2 SFs)}$$

Because 19 g of KNO_3 exceeds the maximum amount that can dissolve in 50. g of H_2O at 20 °C, the KNO_3 solution is saturated, and excess undissolved KNO_3 will be present on the bottom of the container.
 c. 150. g of H_2O will dissolve:

$$150. \text{ g } H_2O \times \frac{32 \text{ g } KNO_3}{100. \text{ g } H_2O} = 48 \text{ g of } KNO_3 \text{ (2 SFs)}$$

Because 68 g of KNO_3 exceeds the maximum amount that can dissolve in 150. g of H_2O at 20 °C, the KNO_3 solution is saturated, and excess undissolved KNO_3 will be present on the bottom of the container.

9.105 **a.** K^+ salts are soluble.
 b. Most salts containing SO_4^{2-} are soluble.
 c. Salts containing S^{2-} are usually insoluble.
 d. Salts containing NO_3^- ions are soluble.
 e. Salts containing OH^- are usually insoluble.

9.107 a. Solid silver bromide (AgBr) forms:

$AgNO_3(aq) + LiBr(aq) \longrightarrow AgBr(s) + LiNO_3(aq)$

$Ag^+(aq) + NO_3^-(aq) + Li^+(aq) + Br^-(aq) \longrightarrow AgBr(s) + Li^+(aq) + NO_3^-(aq)$

$Ag^+(aq) + Br^-(aq) \longrightarrow AgBr(s)$ Net ionic equation

b. none; no solid forms; salts containing K^+ and Na^+ are soluble.

c. Solid barium sulfate ($BaSO_4$) forms:

$Na_2SO_4(aq) + BaCl_2(aq) \longrightarrow BaSO_4(s) + 2NaCl(aq)$

$2Na^+(aq) + SO_4^{2-}(aq) + Ba^{2+}(aq) + 2Cl^-(aq) \longrightarrow BaSO_4(s) + 2Na^+(aq) + 2Cl^-(aq)$

$Ba^{2+}(aq) + SO_4^{2-}(aq) \longrightarrow BaSO_4(s)$ Net ionic equation

9.109 mass of solution $= 15.5$ g $Na_2SO_4 + 75.5$ g $H_2O = 91.0$ g of solution

$$\frac{15.5 \text{ g Na}_2\text{SO}_4}{91.0 \text{ g solution}} \times 100\% = 17.0\% \text{ (m/m) Na}_2\text{SO}_4 \text{ solution (3 SFs)}$$

9.111 $4.5 \text{ mL propyl alcohol} \times \dfrac{100. \text{ mL solution}}{12 \text{ mL propyl alcohol}} = 38 \text{ mL of propyl alcohol solution (2 SFs)}$

9.113 $86.0 \text{ g KOH} \times \dfrac{100. \text{ mL solution}}{12 \text{ g KOH}} \times \dfrac{1 \text{ L solution}}{1000 \text{ mL solution}} = 0.72 \text{ L of KOH solution (2 SFs)}$

9.115 $\dfrac{8.0 \text{ g NaOH}}{400. \text{ mL solution}} \times \dfrac{1 \text{ mole NaOH}}{40.00 \text{ g NaOH}} \times \dfrac{1000 \text{ mL solution}}{1 \text{ L solution}} = 0.50 \text{ M NaOH solution (2 SFs)}$

9.117 a. $2.5 \text{ L solution} \times \dfrac{3.0 \text{ moles Al(NO}_3)_3}{1 \text{ L solution}} \times \dfrac{213.0 \text{ g Al(NO}_3)_3}{1 \text{ mole Al(NO}_3)_3}$

$= 1600 \text{ g of Al(NO}_3)_3 \text{ (2 SFs)}$

b. $75 \text{ mL solution} \times \dfrac{1 \text{ L solution}}{1000 \text{ mL solution}} \times \dfrac{0.50 \text{ mole C}_6\text{H}_{12}\text{O}_6}{1 \text{ L solution}} \times \dfrac{180.16 \text{ g C}_6\text{H}_{12}\text{O}_6}{1 \text{ mole C}_6\text{H}_{12}\text{O}_6}$

$= 6.8 \text{ g of C}_6\text{H}_{12}\text{O}_6 \text{ (2 SFs)}$

c. $235 \text{ mL solution} \times \dfrac{1 \text{ L solution}}{1000 \text{ mL solution}} \times \dfrac{1.80 \text{ moles LiCl}}{1 \text{ L solution}} \times \dfrac{42.39 \text{ g LiCl}}{1 \text{ mole LiCl}}$

$= 17.9 \text{ g of LiCl (3 SFs)}$

9.119 $C_1V_1 = C_2V_2$

a. $C_2 = C_1 \times \dfrac{V_1}{V_2} = 0.200 \text{ M} \times \dfrac{25.0 \text{ mL}}{50.0 \text{ mL}} = 0.100 \text{ M NaBr solution (3 SFs)}$

b. $C_2 = C_1 \times \dfrac{V_1}{V_2} = 12.0\% \times \dfrac{15.0 \text{ mL}}{40.0 \text{ mL}} = 4.50\% \text{ (m/v) K}_2\text{SO}_4 \text{ solution (3 SFs)}$

c. $C_2 = C_1 \times \dfrac{V_1}{V_2} = 6.00 \text{ M} \times \dfrac{75.0 \text{ mL}}{255 \text{ mL}} = 1.76 \text{ M NaOH solution (3 SFs)}$

9.121 $C_1V_1 = C_2V_2$

a. $V_1 = V_2 \times \dfrac{C_2}{C_1} = 250 \text{ mL} \times \dfrac{3.0 \%}{10.0 \%} = 75 \text{ mL of the HCl solution (2 SFs)}$

b. $V_1 = V_2 \times \dfrac{C_2}{C_1} = 500. \text{ mL} \times \dfrac{0.90 \%}{5.0 \%} = 90. \text{ mL of the NaCl solution (2 SFs)}$

c. $V_1 = V_2 \times \dfrac{C_2}{C_1} = 350. \text{ mL} \times \dfrac{2.00 \text{ M}}{6.00 \text{ M}} = 117 \text{ mL of the NaOH solution (3 SFs)}$

9.123 a. A solution of 0.580 mole of the nonelectrolyte lactose in 1.00 kg of water contains 0.580 mole of particles in 1.00 kg of water and has a freezing point change of

$$0.580 \text{ mole particles} \times \frac{1.86\,°\text{C}}{1 \text{ mole particles}} = 1.08\,°\text{C}$$

The freezing point would be $0.00\,°\text{C} - 1.08\,°\text{C} = -1.08\,°\text{C}$

b. $45.0 \text{ g KCl} \times \dfrac{1 \text{ mole KCl}}{74.55 \text{ g KCl}} = 0.604$ mole of KCl (3 SFs)

A solution of 0.604 mole of the strong electrolyte KCl in 1.00 kg of water contains 1.21 moles of particles (0.604 mole of K^+ and 0.604 mole of Cl^-) in 1.00 kg of water and has a freezing point change of

$$1.21 \text{ moles particles} \times \frac{1.86\,°\text{C}}{1 \text{ mole particles}} = 2.25\,°\text{C}$$

The freezing point would be $0.00\,°\text{C} - 2.25\,°\text{C} = -2.25\,°\text{C}$

c. A solution of 1.5 moles of the strong electrolyte K_3PO_4 in 1.00 kg of water contains 6.0 moles of particles (4.5 moles of K^+ and 1.5 moles of PO_4^{3-}) in 1.00 kg of water and has a freezing point change of

$$6.0 \text{ moles particles} \times \frac{1.86\,°\text{C}}{1 \text{ mole particles}} = 11\,°\text{C}$$

The freezing point would be $0.00\,°\text{C} - 11\,°\text{C} = -11\,°\text{C}$

9.125 $450 \text{ mg Al(OH)}_3 \times \dfrac{1 \text{ g Al(OH)}_3}{1000 \text{ mg Al(OH)}_3} \times \dfrac{1 \text{ mole Al(OH)}_3}{78.00 \text{ g Al(OH)}_3} \times \dfrac{3 \text{ moles HCl}}{1 \text{ mole Al(OH)}_3}$

$\times \dfrac{1 \text{ L solution}}{0.20 \text{ mole HCl}} \times \dfrac{1000 \text{ mL solution}}{1 \text{ L solution}} = 87 \text{ mL}$ of the HCl solution (2 SFs)

9.127 A dialysis solution contains no sodium or urea so that these substances will flow out of the blood into the dialyzing solution. High levels of potassium are maintained in the dialyzing solution so that the potassium will dialyze into the blood to increase the K^+ level in the bloodstream back to a more normal range.

9.129 a. mass of NaCl = 25.50 g − 24.10 g = 1.40 g of NaCl

mass of solution = 36.15 g − 24.10 g = 12.05 g of solution

$$\text{mass percent (m/m)} = \frac{1.40 \text{ g NaCl}}{12.05 \text{ g solution}} \times 100\% = 11.6\% \text{ (m/m) NaCl solution (3 SFs)}$$

b. $\text{molarity (M)} = \dfrac{1.40 \text{ g NaCl}}{10.0 \text{ mL solution}} \times \dfrac{1 \text{ mole NaCl}}{58.44 \text{ g NaCl}} \times \dfrac{1000 \text{ mL solution}}{1 \text{ L solution}}$

$= 2.40$ M NaCl solution (3 SFs)

c. $C_1V_1 = C_2V_2 \quad C_2 = C_1 \times \dfrac{V_1}{V_2} = 2.40 \text{ M} \times \dfrac{10.0 \text{ mL}}{60.0 \text{ mL}} = 0.400$ M NaCl solution (3 SFs)

9.131 At 18 °C, KF has a solubility of 92 g of KF in 100. g of H_2O.

a. 25 g of H_2O will dissolve:

$$25 \text{ g H}_2\text{O} \times \frac{92 \text{ g KF}}{100. \text{ g H}_2\text{O}} = 23 \text{ g of KF (2 SFs)}$$

Because 35 g of KF exceeds the maximum amount that can dissolve in 25 g of H_2O at 18 °C, the KF solution is saturated, and excess undissolved KF will be present on the bottom of the container.

b. 50. g of H_2O will dissolve:

$$50.\ \cancel{g\ H_2O} \times \frac{92\ g\ KF}{100.\ \cancel{g\ H_2O}} = 46\ g\ of\ KF\ (2\ SFs)$$

Because 42 g of KF is less than the maximum amount that can dissolve in 50. g of H_2O at 18 °C, the solution is unsaturated.

c. 150. g of H_2O will dissolve:

$$150.\ \cancel{g\ H_2O} \times \frac{92\ g\ KF}{100.\ \cancel{g\ H_2O}} = 140\ g\ of\ KF\ (2\ SFs)$$

Because 145 g of KF exceeds the maximum amount that can dissolve in 150. g of H_2O at 18 °C, the KF solution is saturated, and excess undissolved KF will be present on the bottom of the container.

9.133 mass of solution: 70.0 g HNO_3 + 130.0 g H_2O = 200.0 g of solution

a. $\dfrac{70.0\ g\ HNO_3}{200.0\ g\ solution} \times 100\% = 35.0\%\ (m/m)\ HNO_3$ solution (3 SFs)

b. $200.0\ \cancel{g\ solution} \times \dfrac{1\ mL\ solution}{1.21\ \cancel{g\ solution}} = 165\ mL$ of solution (3 SFs)

c. $\dfrac{70.0\ g\ HNO_3}{165\ mL\ solution} \times 100\% = 42.4\%\ (m/v)\ HNO_3$ solution (3 SFs)

d. $\dfrac{70.0\ \cancel{g\ HNO_3}}{165\ \cancel{mL\ solution}} \times \dfrac{1\ mole\ HNO_3}{63.02\ \cancel{g\ HNO_3}} \times \dfrac{1000\ \cancel{mL\ solution}}{1\ L\ solution}$

$= 6.73\ M\ HNO_3$ solution (3 SFs)

9.135 a. total positive charge $= \dfrac{45\ mEq\ Na^+}{1\ L\ solution} + \dfrac{20.\ mEq\ K^+}{1\ L\ solution}$

$$= \frac{65\ mEq\ positive\ charge}{1\ L\ solution}\ (2\ SFs)$$

b. total negative charge $= \dfrac{35\ mEq\ Cl^-}{1\ L\ solution} + \dfrac{30.\ mEq\ citrate^{3-}}{1\ L\ solution}$

$$= \frac{65\ mEq\ negative\ charge}{1\ L\ solution}\ (2\ SFs)$$

Reaction Rates and Chemical Equilibrium

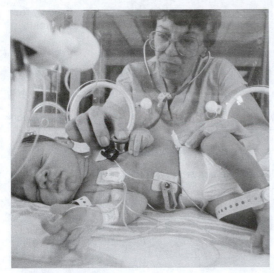

Credit: Fuse/Getty Images

Andrew is born 12 weeks early with infant respiratory distress syndrome (IRDS). After treatment with surfactants and an air–oxygen mixture for several days, he could breathe on his own. As a two-year-old, blood tests show that Andrew's red-blood cell count and blood oxygen levels are below normal, and he is diagnosed with children's iron-deficiency anemia. The hemoglobin (Hb) in red blood cells contains iron, which carries oxygen from the lungs to the muscles and tissues of the body. There is an equilibrium between the gain of oxygen by the hemoglobin in the alveoli and the loss of oxygen at the tissues.

$$Hb(aq) + O_2(aq) \rightleftharpoons HbO_2(aq)$$

Treatment for Andrew's anemia includes a diet that includes foods with high levels of iron such as chicken, fish, spinach, iron-fortified cereal, eggs, peas, beans, peanut butter, and whole-grain bread. After eight months, his hemoglobin increased to the normal range. In anemia, how does the low concentration of hemoglobin affect the concentration of O_2 transported to the tissues?

LOOKING AHEAD

 The Health icon indicates a question that is related to health and medicine.

10.1 Rates of Reactions

Learning Goal: Describe how temperature, concentration, and catalysts affect the rate of a reaction.

- The rate of a reaction is the speed at which reactants are consumed or products are formed.
- At higher temperatures, reaction rates increase because reactants move faster, collide more often, and produce more collisions, which have the required energy of activation.
- Increasing the concentrations of reactants or lowering the energy of activation by adding a catalyst increases the rate of a reaction.
- The reaction rate slows when the temperature or the concentration of a reactant is decreased.

♦ **Learning Exercise 10.1A**

Indicate the effect of each of the following on the rate of a chemical reaction as:
          increase (I)   decrease (D)   no change (N)

a. ____ adding a catalyst

b. ____ running the reaction at a lower temperature

c. ____ doubling the concentrations of the reactants

d. ____ removing a catalyst

e. ____ running the experiment under the same conditions in a different laboratory

f. ____ increasing the temperature

g. ____ using lower concentrations of reactants

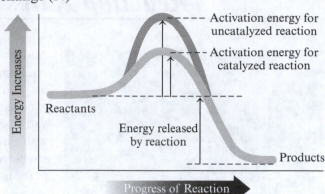

A catalyst lowers the activation energy.

Answers   **a.** I   **b.** D   **c.** I   **d.** D
              e. N   **f.** I   **g.** D

♦ **Learning Exercise 10.1B**

For the following reaction, $NO_2(g) + CO(g) \longrightarrow NO(g) + CO_2(g)$, indicate the effect of each of the following on the reaction rate as:
          increase (I)   decrease (D)   no change (N)

a. ____ adding CO

b. ____ running the experiment under the same conditions two days later

c. ____ removing NO_2

d. ____ adding a catalyst

e. ____ adding NO_2

Answers   **a.** I   **b.** N   **c.** D   **d.** I   **e.** I

10.2   Chemical Equilibrium

Learning Goal: Use the concept of reversible reactions to explain chemical equilibrium.

- A reversible reaction proceeds in both the forward and reverse directions.
- Chemical equilibrium is achieved when the rate of the forward reaction becomes equal to the rate of the reverse reaction.
- In a system at equilibrium, there is no change in the concentrations of reactants and products.
- At equilibrium, the concentrations of reactants are typically different from the concentrations of products.

♦ **Learning Exercise 10.2**

Indicate whether each of the following indicates a system at equilibrium (E) or not (NE):

a. _____ The rate of the forward reaction is faster than the rate of the reverse reaction.

b. _____ There is no change in the concentrations of reactants and products.

c. _____ The rate of the forward reaction is equal to the rate of the reverse reaction.

d. _____ The concentrations of reactants are decreasing.

e. _____ The concentrations of products are increasing.

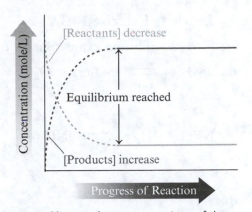

At equilibrium, the concentrations of the reactants and products are constant.

Answers   **a.** NE   **b.** E   **c.** E   **d.** NE   **e.** NE

10.3 Equilibrium Constants

Learning Goal: Calculate the equilibrium constant for a reversible reaction given the concentrations of reactants and products at equilibrium.

REVIEW

Using Significant Figures in Calculations (2.3)

Balancing a Chemical Equation (7.1)

Calculating Concentration (9.4)

- The equilibrium expression for a system at equilibrium is the ratio of the concentrations of the products to the concentrations of the reactants with the concentration of each substance raised to an exponent that is equal to its coefficient in the balanced chemical equation.

- For the general equation, $aA + bB \rightleftharpoons cC + dD$, the equilibrium expression is written as:

$$K_c = \frac{[\text{Products}]}{[\text{Reactants}]} = \frac{[C]^c [D]^d}{[A]^a [B]^b}$$

- The square brackets around each substance indicate that the concentrations are given in moles/L (M).

SAMPLE PROBLEM   Writing Equilibrium Constant Expressions

Write the equilibrium constant expressions for the following reaction:

$$N_2(g) + O_2(g) \rightleftharpoons 2NO(g)$$

Solution:

	Given	Need	Connect
Analyze the Problem	equation	equilibrium constant expressions	$\frac{[\text{products}]}{[\text{reactants}]}$

STEP 1 Write the balanced chemical equation.

$$N_2(g) + O_2(g) \rightleftharpoons 2NO(g)$$

STEP 2 Write the concentrations of the products as the numerator and the reactants as the denominator.

$$\frac{[\text{Products}]}{[\text{Reactants}]} \longrightarrow \frac{[NO]}{[N_2][O_2]}$$

STEP 3 Write any coefficient in the equation as an exponent.

$$K_c = \frac{[NO]^2}{[N_2][O_2]}$$

♦ Learning Exercise 10.3A

CORE CHEMISTRY SKILL
Writing the Equilibrium Expression

Write the equilibrium expression for each of the following reactions:

a. $2SO_3(g) \rightleftarrows 2SO_2(g) + O_2(g)$

b. $2NO(g) + Br_2(g) \rightleftarrows 2NOBr(g)$

c. $N_2(g) + 3H_2(g) \rightleftarrows 2NH_3(g)$

d. $2NO_2(g) \rightleftarrows N_2O_4(g)$

Answers

a. $K_c = \frac{[SO_2]^2[O_2]}{[SO_3]^2}$ **b.** $K_c = \frac{[NOBr]^2}{[NO]^2[Br_2]}$ **c.** $K_c = \frac{[NH_3]^2}{[N_2][H_2]^3}$ **d.** $K_c = \frac{[N_2O_4]}{[NO_2]^2}$

Calculating the K_c Value	
STEP 1	State the given and needed quantities.
STEP 2	Write the equilibrium expression.
STEP 3	Substitute equilibrium (molar) concentrations and calculate K_c.

♦ Learning Exercise 10.3B

CORE CHEMISTRY SKILL
Calculating an Equilibrium Constant

Calculate the numerical value of K_c using the following equilibrium concentrations:

a. $H_2(g) + I_2(g) \rightleftarrows 2HI(g)$ $[H_2] = 0.28$ M $[I_2] = 0.28$ M $[HI] = 2.0$ M

b. $2NO_2(g) \rightleftarrows N_2(g) + 2O_2(g)$ $[NO_2] = 0.60$ M $[N_2] = 0.010$ M $[O_2] = 0.020$ M

c. $N_2(g) + 3H_2(g) \rightleftarrows 2NH_3(g)$ $[N_2] = 0.50$ M $[H_2] = 0.20$ M $[NH_3] = 0.80$ M

Answers

a. $K_c = \frac{[HI]^2}{[H_2][I_2]} = \frac{[2.0]^2}{[0.28][0.28]} = 51$

b. $K_c = \frac{[N_2][O_2]^2}{[NO_2]^2} = \frac{[0.010][0.020]^2}{[0.60]^2} = 1.1 \times 10^{-5}$

c. $K_c = \frac{[NH_3]^2}{[N_2][H_2]^3} = \frac{[0.80]^2}{[0.50][0.20]^3} = 1.6 \times 10^2$

10.4 Using Equilibrium Constants

Learning Goal: Use an equilibrium constant to predict the extent of reaction and to calculate equilibrium concentrations.

- A large K_c indicates that a reaction at equilibrium has more products than reactants; a small K_c indicates that a reaction at equilibrium has more reactants than products.
- The concentration of a component in an equilibrium mixture is calculated from the K_c and the concentrations of all the other components.

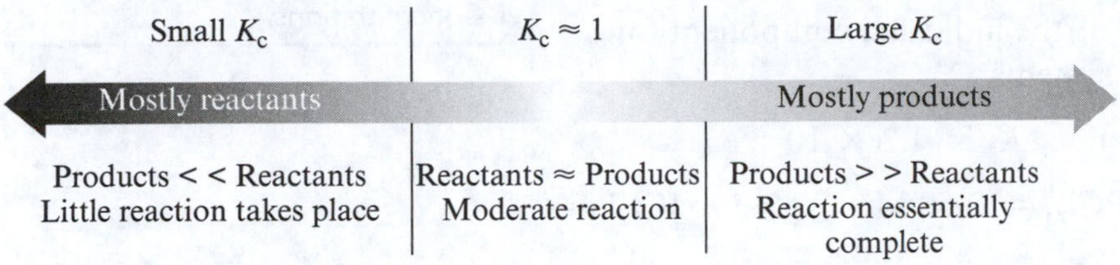

Small K_c	$K_c \approx 1$	Large K_c	
Mostly reactants		Mostly products	
Products < < Reactants Little reaction takes place	Reactants ≈ Products Moderate reaction	Products > > Reactants Reaction essentially complete	The values of equilibrium constants can be less than 1, about equal to 1, or greater than 1.

♦ Learning Exercise 10.4A

Consider the reaction $2NOBr(g) \rightleftharpoons 2NO(g) + Br_2(g)$

a. Write the equilibrium expression for the reaction.

b. If the equilibrium constant, K_c, for the reaction is 2×10^3, does the equilibrium mixture contain mostly reactants, mostly products, or both reactants and products? Explain.

Answers

a. $K_c = \dfrac{[NO]^2[Br_2]}{[NOBr]^2}$

b. A large K_c ($>>1$) means that the equilibrium mixture contains mostly products.

♦ Learning Exercise 10.4B

Consider the reaction $2HI(g) \rightleftharpoons H_2(g) + I_2(g)$

a. Write the equilibrium expression for the reaction.

b. If the K_c for the reaction is 1.6×10^{-2}, does the equilibrium mixture contain mostly reactants, mostly products, or both reactants and products? Explain.

Answers

a. $K_c = \dfrac{[H_2][I_2]}{[HI]^2}$

b. A small K_c ($<<1$) means that the equilibrium mixture contains mostly reactants.

	Using the Equilibrium Constant
STEP 1	State the given and needed quantities.
STEP 2	Write the equilibrium expression and solve for the needed concentration.
STEP 3	Substitute the equilibrium (molar) concentrations and calculate the needed concentration.

♦ **Learning Exercise 10.4C**

Write the equilibrium constant expression for the reaction and calculate the molar concentration of the indicated component for each of the following equilibrium systems:

<table>
<tr><td>CORE CHEMISTRY SKILL</td></tr>
<tr><td>Calculating Equilibrium Concentrations</td></tr>
</table>

a. $PCl_5(g) \rightleftharpoons PCl_3(g) + Cl_2(g)$ $K_c = 1.2 \times 10^{-2}$

$[PCl_5] = 2.50$ M $[PCl_3] = 0.50$ M $[Cl_2] = ?$

b. $CO(g) + H_2O(g) \rightleftharpoons CO_2(g) + H_2(g)$ $K_c = 1.6$

$[CO] = 1.0$ M $[H_2O] = 0.80$ M $[CO_2] = ?$ $[H_2] = 1.2$ M

Answers

a. $K_c = \dfrac{[PCl_3][Cl_2]}{[PCl_5]}$ $[Cl_2] = 6.0 \times 10^{-2}$ M

b. $K_c = \dfrac{[CO_2][H_2]}{[CO][H_2O]}$ $[CO_2] = 1.1$ M

10.5 Changing Equilibrium Conditions: Le Châtelier's Principle

Learning Goal: Use Le Châtelier's principle to describe the changes made in equilibrium concentrations when reaction conditions change.

- Le Châtelier's principle states that a change in the concentration of a reactant or product, the temperature, or the volume of gases will cause the reaction to shift in the direction that relieves the stress.
- The addition of a catalyst does not affect the equilibrium position of a reaction.

Key Terms for Sections 10.1 to 10.5

Match each of the following key terms with the correct description:

a. activation energy   b. equilibrium   c. catalyst
d. equilibrium expression   e. Le Châtelier's principle   f. reversible reaction

1. _____ a substance that lowers the activation energy and increases the rate of reaction

2. _____ the ratio of the concentrations of products to those of the reactants raised to exponents equal to their coefficients

3. ____ the energy required to convert reactants to products in a chemical reaction

4. ____ states that a stress placed on a reaction at equilibrium causes the equilibrium to shift in the direction that relieves the stress

5. ____ a system in which the rate of the forward reaction is equal to the rate of the reverse reaction

6. ____ occurs when a reaction consists of both a forward reaction and a reverse reaction

Answers **1.** c **2.** d **3.** a **4.** e **5.** b **6.** f

♦ Learning Exercise 10.5A

CORE CHEMISTRY SKILL
Using Le Châtelier's Principle

For each of the following changes at equilibrium, indicate whether the equilibrium shifts in the direction of products (P), reactants (R), or does not change (N):

$N_2(g) + O_2(g) + 180\text{ kJ} \rightleftarrows 2NO(g)$

1. ____ adding more $O_2(g)$

2. ____ removing some $N_2(g)$

3. ____ removing some $NO(g)$

4. ____ increasing the temperature

5. ____ reducing the volume of the container

6. ____ increasing the volume of the container

Answers **1.** P **2.** R **3.** P **4.** P **5.** N **6.** N

♦ Learning Exercise 10.5B

For each of the following changes at equilibrium, indicate whether the equilibrium shifts in the direction of products (P), reactants (R), or does not change (N):

$2NOBr(g) \rightleftarrows 2NO(g) + Br_2(g) + 340\text{ kJ}$

1. ____ adding more $NO(g)$

2. ____ removing some $Br_2(g)$

3. ____ removing some $NOBr(g)$

4. ____ increasing the temperature

5. ____ decreasing the temperature

6. ____ increasing the volume of the container

Answers **1.** R **2.** P **3.** R **4.** R **5.** P **6.** P

♦ Learning Exercise 10.5C

Use the following diagram of a gaseous reaction at equilibrium (the white atoms are nitrogen atoms and the black atoms are oxygen atoms) to answer questions **a** to **f**:

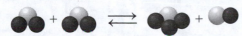

a. Write the balanced chemical equation for the reaction.

b. Write the equilibrium expression for the reaction.

Answer **c** to **f** with shifts in the direction of the products (P), shifts in the direction of the reactants (R), or no change (N).

c. What is the effect on the equilibrium when more NO_3 is added? _____

d. What is the effect on the equilibrium when more NO_2 is added? _____

e. What is the effect on the equilibrium when some NO is removed? _____

f. What is the effect on the equilibrium when the volume of the container is decreased? _____

Answers

a. $2NO_2(g) \rightleftharpoons NO_3(g) + NO(g)$

b. $K_c = \dfrac{[NO_3][NO]}{[NO_2]^2}$

c. R   **d.** P   **e.** P   **f.** N

Checklist for Chapter 10

You are ready to take the Practice Test for Chapter 10. Be sure you have accomplished the following learning goals for this chapter. If not, review the Section listed at the end of the goal. Then apply your new skills and understanding to the Practice Test.

After studying Chapter 10, I can successfully:

_____ Describe the factors that increase or decrease the rate of a reaction. (10.1)

_____ Write the equations for the forward and reverse reactions of a reversible reaction. (10.2)

_____ Explain how equilibrium occurs when the rate of a forward reaction is equal to the rate of a reverse reaction. (10.2)

_____ Write the equilibrium expression for a reaction system at equilibrium. (10.3)

_____ Calculate the equilibrium constant from the equilibrium concentrations. (10.3)

_____ Use the equilibrium constant to determine whether an equilibrium mixture contains mostly reactants, mostly products, or about the same amounts of reactants and products. (10.4)

_____ Use the equilibrium constant to determine the equilibrium concentration of a component in the reaction. (10.4)

_____ Use Le Châtelier's principle to describe the shift in a system at equilibrium when stress is applied to the system. (10.5)

Practice Test for Chapter 10

The chapter Sections to review are shown in parentheses at the end of each question.

1. The number of molecular collisions increases when (10.1)
  A. more reactants are added
  B. products are removed
  C. the energy of collision is below the energy of activation
  D. the reaction temperature is lowered
  E. the reacting molecules have an incorrect orientation upon impact

2. The energy of activation is lowered when (10.1)
  A. more reactants are added
  B. products are removed

    C. a catalyst is used
    D. the reaction temperature is lowered
    E. the reaction temperature is raised

3. Food deteriorates more slowly in a refrigerator because (10.1)
    A. more reactants are added
    B. products are removed
    C. the energy of activation is higher
    D. fewer collisions have the energy of activation
    E. collisions have the wrong orientation upon impact

4. A reaction reaches equilibrium when (10.2)
    A. the rate of the forward reaction is faster than the rate of the reverse reaction
    B. the rate of the reverse reaction is faster than the rate of the forward reaction
    C. the concentrations of reactants and products are changing
    D. fewer collisions have the energy of activation
    E. the rate of the forward reaction is equal to the rate of the reverse reaction

5. The equilibrium expression for the following reaction is (10.3)
$$2NOCl(g) \rightleftharpoons 2NO(g) + Cl_2(g)$$

A. $\dfrac{[NO][Cl_2]}{[NOCl]}$ B. $\dfrac{[NOCl_2]^2}{[NO]^2[Cl_2]}$ C. $\dfrac{[NOCl_2]}{[NO][Cl_2]}$ D. $\dfrac{[NO^2][Cl_2]}{[NOCl]}$ E. $\dfrac{[NO]^2[Cl_2]}{[NOCl]^2}$

6. The equilibrium expression for the following reaction is (10.3)
$$CO(g) + 2H_2(g) \rightleftharpoons CH_4O(g)$$

A. $[CO][2H_2]$ B. $\dfrac{[CO][H_2]}{[CH_4O]}$ C. $\dfrac{[CH_4O]}{[CO][H_2]^2}$ D. $\dfrac{1}{[CH_4O]}$ E. $\dfrac{[CO][H_2]^2}{[CH_4O]}$

7. The equilibrium expression for the following reaction is (10.3)
$$C_3H_8(g) + 5O_2(g) \rightleftharpoons 3CO_2(g) + 4H_2O(g)$$

A. $\dfrac{[CO_2][H_2O]}{[C_3H_8][O_2]}$ B. $\dfrac{[C_3H_8][O_2]^5}{[CO_2]^3[H_2O]^4}$ C. $\dfrac{[CO_2]^3[H_2O]^4}{[C_3H_8][O_2]^5}$ D. $[CO_2]^3[H_2O]^4$ E. $\dfrac{[CO_2]^3}{[C_3H_8][O_2]^5}$

8. The equilibrium equation that has the equilibrium expression $\dfrac{[H_2S]^2}{[H_2]^2[S_2]}$ is (10.3)

    A. $H_2S(g) \rightleftharpoons H_2(g) + S_2(g)$
    B. $2H_2S(g) \rightleftharpoons H_2(g) + S_2(g)$
    C. $2H_2(g) + S_2(g) \rightleftharpoons 2H_2S(g)$
    D. $2H_2S(g) \rightleftharpoons 2H_2(g)$
    E. $2H_2(g) \rightleftharpoons 2H_2S(g)$

9. Use the given equilibrium concentrations to calculate the numerical value of the equilibrium constant for (10.3)
$$COBr_2(g) \rightleftharpoons CO(g) + Br_2(g)$$

$[COBr_2] = 0.93$ M $[CO] = 0.42$ M $[Br_2] = 0.42$ M
A. 0.19   B. 0.39   C. 0.42   D. 2.2   E. 5.3

10. Use the given equilibrium concentrations to calculate the numerical value of the equilibrium constant for (10.3)
$$2NO(g) + O_2(g) \rightleftharpoons 2NO_2(g)$$

$[NO] = 2.7$ M $[O_2] = 1.0$ M $[NO_2] = 3.0$ M
A. 0.81   B. 1.1   C. 1.2   D. 8.1   E. 9.0

11. Use the given equilibrium concentrations to calculate $[PCl_5]$ for the decomposition of PCl_5 that has a $K_c = 0.050$. (10.4)

$$PCl_5(g) \rightleftharpoons PCl_3(g) + Cl_2(g)$$

$[PCl_3] = 0.20 \text{ M}$ $[Cl_2] = 0.20 \text{ M}$

  A. 0.01 M   **B.** 0.050 M   **C.** 0.20 M

  D. 0.40 M   **E.** 0.80 M

12. The reaction that has a much greater concentration of products than reactants at equilibrium has a K_c value of (10.4)

  A. 1.6×10^{-15} **B.** 2×10^{-11} **C.** 1.2×10^{-5}

  D. 3×10^{-3} **E.** 1.4×10^{5}

13. The reaction that has a much greater concentration of products than reactants at equilibrium has a K_c value of (10.4)

  A. 1.1×10^{-4} **B.** 2×10^{-1} **C.** 1.2×10^{2}

  D. 2×10^{4} **E.** 1.3×10^{12}

14. The reaction that has about the same concentration of reactants and products at equilibrium has a K_c value of (10.4)

  A. 1.4×10^{-12} **B.** 2×10^{-8} **C.** 1.2

  D. 3×10^{2} **E.** 1.3×10^{7}

*For questions 15 through 19, answer **A**, **B**, or **C** for the correct change in equilibrium caused by each of the following:* (10.5)

$$PCl_5(g) + heat \rightleftharpoons PCl_3(g) + Cl_2(g)$$

  A. shifts in the direction of the products   **B.** shifts in the direction of the reactants

  C. no change

15. _____ adding more $Cl_2(g)$ **16.** _____ increasing the temperature

17. _____ removing some $PCl_3(g)$ **18.** _____ decreasing the volume

19. _____ removing some $PCl_5(g)$

*For questions 20 through 24, answer **A**, **B**, or **C** for the correct change in equilibrium caused by each of the following:* (10.5)

$$2NO(g) + O_2(g) \rightleftharpoons 2NO_2(g) + heat$$

  A. shifts in the direction of the products   **B.** shifts in the direction of the reactants

  C. no change

20. _____ adding more $NO(g)$ **21.** _____ increasing the temperature

22. _____ adding a catalyst   **23.** _____ decreasing the volume

24. _____ removing some $NO_2(g)$

Answers to the Practice Test

1. A	**2.** C	**3.** D	**4.** E	**5.** E
6. C	**7.** C	**8.** C	**9.** A	**10.** C
11. E	**12.** E	**13.** A	**14.** C	**15.** B
16. A	**17.** A	**18.** B	**19.** B	**20.** A
21. B	**22.** C	**23.** A	**24.** A	

Selected Answers and Solutions to Text Problems

10.1 **a.** The rate of a reaction indicates how fast the products form or how fast the reactants are used up.
b. At room temperature, more of the reactants will have the energy necessary to proceed to products (the activation energy) than at the lower temperature of the refrigerator, so the rate of formation of bread mold will be faster.

10.3 Adding $Br_2(g)$ molecules increases the concentration of reactants, which increases the number of collisions that take place between the reactants.

10.5 **a.** Adding more reactant increases the number of collisions that take place between the reactants, which increases the reaction rate.
b. Increasing the temperature increases the kinetic energy of the reactant molecules, which increases the number of collisions and makes the collisions more forceful. The rate of reaction will be increased.
c. Adding a catalyst lowers the energy of activation, which increases the reaction rate.
d. Removing a reactant decreases the number of collisions that take place between the reactants, which decreases the reaction rate.

10.7 A reversible reaction is one in which a forward reaction converts reactants to products while a reverse reaction converts products to reactants.

10.9 **a.** When the rate of the forward reaction is faster than the rate of the reverse reaction, the process is not at equilibrium.
b. When the concentrations of the reactants and the products do not change, the process is at equilibrium.
c. When the rate of either the forward or reverse reaction does not change, the process is at equilibrium.

10.11 The reaction has reached equilibrium because the number of reactants and products does not change; the concentrations of A and B have not changed between 3 h and 4 h.

10.13 In the expression for K_c, the products are divided by the reactants, with each concentration raised to a power equal to its coefficient in the balanced chemical equation.

a. $K_c = \dfrac{[CS_2][H_2]^4}{[CH_4][H_2S]^2}$

b. $K_c = \dfrac{[N_2][O_2]}{[NO]^2}$

c. $K_c = \dfrac{[CS_2][O_2]^4}{[SO_3]^2[CO_2]}$

d. $K_c = \dfrac{[H_2]^3[CO]}{[CH_4][H_2O]}$

10.15 $K_c = \dfrac{[XY]^2}{[X_2][Y_2]} = \dfrac{[6]^2}{[1][1]} = 36$

10.17 $K_c = \dfrac{[NO_2]^2}{[N_2O_4]} = \dfrac{[0.21]^2}{[0.030]} = 1.5\,(2\,SFs)$

10.19 $K_c = \dfrac{[CH_4][H_2O]}{[CO][H_2]^3} = \dfrac{[1.8][2.0]}{[0.51][0.30]^3} = 260\,(2\,SFs)$

10.21 Calculate the value of K_c for each diagram and compare to the given value ($K_c = 4$).

for **A**: $K_c = \dfrac{[XY]^2}{[X_2][Y_2]} = \dfrac{[6]^2}{[1][1]} = 36$

for **B**: $K_c = \dfrac{[XY]^2}{[X_2][Y_2]} = \dfrac{[4]^2}{[2][2]} = 4$

for **C**: $K_c = \dfrac{[XY]^2}{[X_2][Y_2]} = \dfrac{[2]^2}{[3][3]} = 0.44$

Therefore, diagram **B** represents the equilibrium mixture.

10.23 a. A large K_c value indicates that the equilibrium mixture contains mostly products.
   b. A large K_c value indicates that the equilibrium mixture contains mostly products.
   c. A small K_c value indicates that the equilibrium mixture contains mostly reactants.

10.25 $K_c = \dfrac{[HI]^2}{[H_2][I_2]} = 54$

Rearrange the K_c expression to solve for $[H_2]$ and substitute in known values.

$[H_2] = \dfrac{[HI]^2}{K_c[I_2]} = \dfrac{[0.030]^2}{54[0.015]} = 1.1 \times 10^{-3}\text{ M (2 SFs)}$

10.27 $K_c = \dfrac{[NO]^2[Br_2]}{[NOBr]^2} = 2.0$

Rearrange the K_c expression to solve for $[NOBr]$ and substitute in known values.

$[NOBr]^2 = \dfrac{[NO]^2[Br_2]}{K_c} = \dfrac{[2.0]^2[1.0]}{2.0} = 2.0$

Take the square root of both sides of the equation.

$[NOBr] = \sqrt{2.0} = 1.4\text{ M (2 SFs)}$

10.29 a. Adding more reactant shifts equilibrium in the direction of the product.
   b. Adding more product shifts equilibrium in the direction of the reactant.
   c. Increasing the temperature of an endothermic reaction shifts equilibrium in the direction of the product.
   d. Increasing the volume of the container shifts the equilibrium in the direction of the reactant, which has more moles of gas.
   e. No shift in equilibrium occurs when a catalyst is added.

10.31 a. Adding more reactant shifts equilibrium in the direction of the product.
   b. Increasing the temperature of an endothermic reaction shifts equilibrium in the direction of the product to remove heat.
   c. Removing product shifts equilibrium in the direction of the product.
   d. No shift in equilibrium occurs when a catalyst is added.
   e. Removing reactant shifts equilibrium in the direction of the reactants.

10.33 a. When an athlete first arrives at high altitude, the oxygen concentration is decreased.
   b. Removing reactant shifts equilibrium in the direction of the reactants.

10.35 a. Adding more reactant shifts equilibrium in the direction of the product.
   b. Adding more product shifts equilibrium in the direction of the reactants.
   c. Removing reactant shifts equilibrium in the direction of the reactants.

10.37 Without enough iron, Andrew will not have sufficient hemoglobin. When Hb is decreased, the equilibrium shifts in the direction of the reactants, Hb and O_2.

10.39 a. $K_c = \dfrac{[CO_2][H_2O]^2}{[CH_4][O_2]^2}$

   b. $K_c = \dfrac{[N_2]^2[H_2O]^6}{[NH_3]^4[O_2]^3}$

10.41 At equilibrium, the diagram shows mostly reactants and a few products, so the equilibrium constant K_c for the reaction would have a small value.

10.43 T_2 is lower than T_1. This would cause the exothermic reaction shown to shift in the direction of the products to add heat; more product is seen in the T_2 diagram.

10.45 a. Increasing the temperature of an exothermic reaction shifts equilibrium in the direction of the reactants.
  b. Decreasing volume favors the side of the reaction with fewer moles of gas, so there is a shift in the direction of the product.
  c. Adding a catalyst does not shift equilibrium.
  d. Adding more reactant shifts equilibrium in the direction of the product.

10.47 a. A large K_c value indicates that the equilibrium mixture contains mostly product.
  b. A K_c value close to 1 indicates that the equilibrium mixture contains both reactants and products.
  c. A small K_c value indicates that the equilibrium mixture contains mostly reactants.

10.49 a. $K_c = \dfrac{[N_2][H_2]^3}{[NH_3]^2}$

  b. $K_c = \dfrac{[3.0][0.50]^3}{[0.20]^2} = 9.4 \ (2 \ SFs)$

10.51 $K_c = \dfrac{[N_2O_4]}{[NO_2]^2} = 5.0$

Rearrange the K_c expression to solve for $[N_2O_4]$ and substitute in known values.
$[N_2O_4] = K_c \times [NO_2]^2 = 5.0 \times [0.50]^2 = 1.3 \ M \ (2 \ SFs)$

10.53 a. When the reactant $[O_2]$ increases, the rate of the forward reaction increases to shift the equilibrium in the direction of the product.
  b. When the product $[O_2]$ increases, the rate of the reverse reaction increases to shift the equilibrium in the direction of the reactant.
  c. When the reactant $[O_2]$ increases, the rate of the forward reaction increases to shift the equilibrium in the direction of the product.
  d. When the product $[O_2]$ increases, the rate of the reverse reaction increases to shift the equilibrium in the direction of the reactants.

10.55 Decreasing the volume of an equilibrium mixture shifts the equilibrium in the direction of the side of the reaction that has the fewer number of moles of gas. No shift occurs when there are an equal number of moles of gas on both sides of the equation.
  a. With 3 moles of gas on the reactant side and 2 moles of gas on the product side, decreasing the volume will shift equilibrium in the direction of the product.
  b. With 2 moles of gas on the reactant side and 3 moles of gas on the product side, decreasing the volume will shift equilibrium in the direction of the reactant.
  c. With 4 moles of gas on the reactant side and 5 moles of gas on the product side, decreasing the volume will shift equilibrium in the direction of the reactants.

10.57 $K_c = \dfrac{[CO][Cl_2]}{[COCl_2]} = 0.68$

Rearrange the K_c expression to solve for $[COCl_2]$ and substitute in known values.
$[COCl_2] = \dfrac{[CO][Cl_2]}{K_c} = \dfrac{[0.40][0.74]}{0.68} = 0.44 \ M \ (2 \ SFs)$

10.59 a. $K_c = \dfrac{[NO]^2[Br_2]}{[NOBr]^2}$

b. When the concentrations are substituted into the equilibrium expression, the result is 1.0, which is not equal to K_c (2.0). Therefore, the system is not at equilibrium.

$$K_c = \frac{[NO]^2[Br_2]}{[NOBr]^2} = \frac{[1.0]^2[1.0]}{[1.0]^2} = 1.0 \ (2 \ SFs)$$

c. Since the calculated value in part **b** is less than K_c, the rate of the forward reaction will initially increase.

d. When the system has reestablished equilibrium, the $[Br_2]$ and $[NO]$ will have increased, and the $[NOBr]$ will have decreased.

10.61 a. When more product molecules are added to an equilibrium mixture, the system shifts in the direction of the reactants. This will cause a decrease in the equilibrium concentration of the product H_2O.

b. When the temperature is increased for an endothermic reaction, the system shifts in the direction of the products to remove heat. This will cause an increase in the equilibrium concentration of the product H_2O.

c. Increasing the volume of the reaction container favors the side of the reaction with the greater number of moles of gas, so this system shifts in the direction of the reactants. This will cause a decrease in the equilibrium concentration of the product H_2O.

d. Decreasing the volume of the reaction container favors the side of the reaction with the fewer moles of gas, so this system shifts in the direction of the products. This will cause an increase in the equilibrium concentration of the product H_2O.

e. When a catalyst is added, the rates of both forward and reverse reactions increase; the equilibrium position does not change. No change will be observed in the equilibrium concentration of the product H_2O.

10.63 Increasing the volume of an equilibrium mixture shifts the system toward the side of the reaction that has the greater number of moles of gas; decreasing the volume shifts the equilibrium toward the side of the reaction that has the fewer moles of gas.

a. Since the product side has the greater number of moles of gas, increasing the volume of the container will increase the yield of product.

b. Since the product side has the greater number of moles of gas, increasing the volume of the container will increase the yield of products.

c. Since the product side has the fewer moles of gas, decreasing the volume of the container will increase the yield of product.

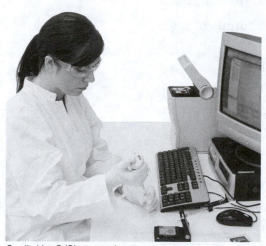

Credit: Lisa S./Shutterstock

When Larry is in the emergency room after an automobile accident, a blood sample is analyzed by Brianna, a clinical laboratory technician. The results show that Larry's blood pH is 7.30 and the partial pressure of CO_2 gas is above the desired level. Blood pH is typically in the range of 7.35 to 7.45, and a value less than 7.35 indicates a state of acidosis. Larry has respiratory acidosis because an increase in the partial pressure of CO_2 gas in the bloodstream decreases the pH. In the emergency room, Larry is given an IV containing bicarbonate to increase his blood pH.

$$HCO_3^-(aq) + H^+(aq) \rightleftharpoons H_2CO_3(aq) \rightleftharpoons CO_2(g) + H_2O(l)$$

Soon Larry's blood pH and partial pressure of CO_2 gas return to normal. What is the pH of a sample of blood that has $[H_3O^+] = 3.5 \times 10^{-8}$ M?

LOOKING AHEAD

 The Health icon indicates a question that is related to health and medicine.

11.1 Acids and Bases

Learning Goal: Describe and name acids and bases.

REVIEW

Writing Ionic Formulas (6.2)

- In water, an Arrhenius acid produces hydrogen ions, H^+, and an Arrhenius base produces hydroxide ions, OH^-.
- An acid with a simple nonmetal anion is named by placing the prefix *hydro* in front of the name of the anion, and changing its *ide* ending to *ic acid*.
- An acid with a polyatomic anion is named as an *ic acid* when its anion ends in *ate* and as an *ous acid* when its anion ends in *ite*.
- Typical Arrhenius bases are named as hydroxides.

♦ **Learning Exercise 11.1A**

Indicate if each of the following characteristics describes an acid or a base:

a. _A_ turns litmus red

b. _A_ tastes sour

c. _B_ contains more OH^- than H_3O^+

d. _A_ neutralizes bases

e. _b_ tastes bitter

f. _B_ turns litmus blue

g. _A_ contains more H_3O^+ than OH^-

h. _B_ neutralizes acids

Answers   **a.** acid   **b.** acid   **c.** base   **d.** acid
e. base   **f.** base   **g.** acid   **h.** base

♦ **Learning Exercise 11.1B**

Fill in the blanks with the formula or name of an acid or base.

a. KOH _____

b. _____ sodium hydroxide

c. _____ sulfurous acid

d. _____ chlorous acid

e. $Ba(OH)_2$ _____

f. H_2CO_3 _____

g. $Zn(OH)_2$ _____

h. _____ lithium hydroxide

i. $HBrO_3$ _____

j. H_3PO_3 _____

Answers   **a.** potassium hydroxide   **b.** NaOH   **c.** H_2SO_3
d. $HClO_2$ **e.** barium hydroxide   **f.** carbonic acid
g. zinc hydroxide   **h.** LiOH   **i.** bromic acid
j. phosphorous acid

11.2   Brønsted–Lowry Acids and Bases

Learning Goal: Identify conjugate acid–base pairs for Brønsted–Lowry acids and bases.

- According to the Brønsted–Lowry theory, acids are hydrogen ion, H^+, donors, and bases are hydrogen ion, H^+, acceptors.
- In solution, hydrogen ions, H^+, from acids bond to polar water molecules to form hydronium ions, H_3O^+.
- Conjugate acid–base pairs are molecules or ions linked by the loss and gain of one hydrogen ion, H^+.
- Every hydrogen ion transfer reaction involves two acid–base conjugate pairs.

Writing Conjugate Acid–Base Pairs	
STEP 1	Identify the reactant that loses H^+ as the acid.
STEP 2	Identify the reactant that gains H^+ as the base.
STEP 3	Write the conjugate acid–base pairs.

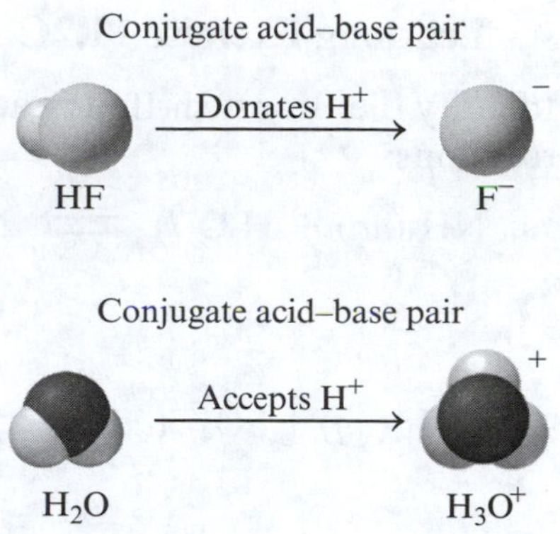

◆ Learning Exercise 11.2A

Complete the following:

Conjugate acid–base pair

$$\text{HF} \xrightarrow{\text{Donates } H^+} F^-$$

Acid	Conjugate Base
a. H_2O	OH^-
b. H_2SO_4	HSO_4
c. HCl	Cl^-
d. HCO_3^-	CO_3^{2-}
e. HNO_3	NO_3^-
f. NH_4^+	NH_3^-
g. H_2S	HS^-
h. H_3PO_4	$H_2PO_4^-$

Conjugate acid–base pair

$$H_2O \xrightarrow{\text{Accepts } H^+} H_3O^+$$

Answers   **a.** OH^- **b.** HSO_4^- **c.** HCl **d.** CO_3^{2-}

              e. HNO_3 **f.** NH_3 **g.** H_2S **h.** H_3PO_4

◆ Learning Exercise 11.2B

Identify the reactant that is a Brønsted–Lowry acid and the reactant that is a Brønsted–Lowry base in each of the following:

a. $HBr(aq) + CO_3^{2-}(aq) \longrightarrow Br^-(aq) + HCO_3^-(aq)$

    A   B

b. $HSO_4^-(aq) + OH^-(aq) \rightleftharpoons SO_4^{2-}(aq) + H_2O(l)$

    A   B

c. $NH_4^+(aq) + H_2O(l) \rightleftharpoons NH_3(aq) + H_3O^+(aq)$

    A   B

d. $SO_4^{2-}(aq) + HCl(aq) \longrightarrow HSO_4^-(aq) + Cl^-(aq)$

    B   A

Answers   **a.** Brønsted–Lowry acid: HBr; Brønsted–Lowry base CO_3^{2-}

              b. Brønsted–Lowry acid: HSO_4^-; Brønsted–Lowry base OH^-

              c. Brønsted–Lowry acid: NH_4^+; Brønsted–Lowry base H_2O

              d. Brønsted–Lowry acid: HCl; Brønsted–Lowry base SO_4^{2-}

♦ Learning Exercise 11.2C

CORE CHEMISTRY SKILL
Identifying Conjugate Acid–Base Pairs

Identify the Brønsted–Lowry acid–base pairs in each of the following reactions:

a. $NH_3(aq) + H_2O(l) \rightleftarrows NH_4^+(aq) + OH^-(aq)$
Base Acid Conj. Acid conj. base

b. $NH_4^+(aq) + SO_4^{2-}(aq) \rightleftarrows NH_3(aq) + HSO_4^-(aq)$
Acid Base Conj. base conj. Acid

c. $HCO_3^-(aq) + H_2O(l) \rightleftarrows CO_3^{2-}(aq) + H_3O^+(aq)$
Acid Base Conj. Base Conj. Acid

d. $HNO_3(aq) + OH^-(aq) \longrightarrow NO_3^-(aq) + H_2O(l)$
Acid base Conj. Base Conj. Acid

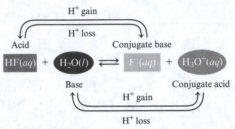

HF, an acid, loses one H^+ to form its conjugate base F^-. Water acts as a base by gaining one H^+ to form its conjugate acid H_3O^+.

Answers
a. H_2O/OH^- and NH_4^+/NH_3
b. NH_4^+/NH_3 and HSO_4^-/SO_4^{2-}
c. HCO_3^-/CO_3^{2-} and H_3O^+/H_2O
d. HNO_3/NO_3^- and H_2O/OH^-

11.3 Strengths of Acids and Bases

Learning Goal: Write equations for the dissociation of strong and weak acids and bases; identify the direction of reaction.

- Strong acids dissociate completely in water, and the H^+ is accepted by H_2O acting as a base.
- A weak acid dissociates slightly in water, producing only small amounts of H_3O^+.
- All hydroxides of Group 1A (1) and most hydroxides of Group 2A (2) are strong bases, which dissociate completely in water.
- In an aqueous ammonia solution, a weak base, NH_3, accepts only a small percentage of hydrogen ions to form the conjugate acid, NH_4^+.
- As the strength of an acid decreases, the strength of the conjugate base increases. By comparing relative strengths, the direction of an acid–base reaction can be predicted.

Study Note

Only six common acids are strong acids; other acids are weak acids.

HI	HBr
$HClO_4$	HCl
H_2SO_4	HNO_3

Example: Is H_2S a strong or a weak acid?

Solution: H_2S is a weak acid because it is not one of the six strong acids.

♦ **Learning Exercise 11.3A**

Identify each of the following as a strong acid, a weak acid, a strong base, or a weak base:

a. HNO_3 *strong acid*

b. H_2CO_3 *weak acid*

c. $H_2PO_4^-$ *weak acid*

d. NH_3 *weak base*

e. LiOH   *strong base*

f. H_3BO_3 *weak acid*

g. $Ca(OH)_2$ *strong base*

h. H_2SO_4 *strong acid.*

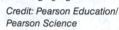

Answers   **a.** strong acid   **b.** weak acid   **c.** weak acid
              d. weak base   **e.** strong base   **f.** weak acid
              g. strong base   **h.** strong acid

Acetic acid ($HC_2H_3O_2$) is a weak acid because it dissociates slightly in water, producing mosly molecules and a few H_3O^+ and $C_2H_3O_2^-$ ions.
Credit: Pearson Education/ Pearson Science

♦ **Learning Exercise 11.3B**

Using Table 11.3 in the text, identify the stronger acid in each of the following pairs of acids:

a. HCl or H_2CO_3 _____

b. HNO_2 or HCN _____

c. H_2S or HBr _____

d. H_2SO_4 or HSO_4^- _____

e. HF or H_3PO_4 _____

Answers   **a.** HCl   **b.** HNO_2 **c.** HBr   **d.** H_2SO_4 **e.** H_3PO_4

Study Note

In an acid–base reaction, the relative strengths of the two acids or two bases indicate whether the equilibrium mixture contains mostly reactants or products:

Example: Does the following equilibrium mixture contain mostly reactants or products?

$$NO_3^-(aq) + H_2O(l) \rightleftharpoons HNO_3(aq) + OH^-(aq)$$

Solution: The reactants contain the weaker base and acid (NO_3^- and H_2O), which means the equilibrium mixture contains mostly reactants.

♦ **Learning Exercise 11.3C**

Using Table 11.3 in the text, indicate whether each of the following equilibrium mixtures contains mostly reactants or products:

a. $H_2SO_4(aq) + H_2O(l) \rightleftharpoons H_3O^+(aq) + HSO_4^-(aq)$ *products*

b. $I^-(aq) + H_3O^+(aq) \rightleftharpoons H_2O(l) + HI(aq)$ *reactants*

c. $NH_3(aq) + H_2O(l) \rightleftharpoons NH_4^+(aq) + OH^-(aq)$ *reactants*

d. $HCl(aq) + CO_3^{2-}(aq) \rightleftharpoons Cl^-(aq) + HCO_3^-(aq)$ *products*

Answers   **a.** products   **b.** reactants   **c.** reactants   **d.** products

11.4 Dissociation of Weak Acids and Bases

REVIEW
Balancing a Chemical
Equation (7.1)

Learning Goal: Write the expression for the dissociation of a weak acid or weak base.

- An acid or base dissociation expression is the the ratio of the concentrations of products to the reactants when the concentration of water is considered a constant and not included.

- For an acid dissociation expression, as with other equilibrium expressions, the molar concentrations of the products are divided by the molar concentrations of the reactants. (Because water is a pure liquid with a constant concentration, it is omitted.)

$$HA(aq) + H_2O(l) \rightleftharpoons H_3O^+(aq) + A^-(aq)$$

$$K_a = \frac{[H_3O^+][A^-]}{[HA]}$$

- The numerical value of the acid dissociation expression is the acid dissociation constant, K_a.

- An acid or a base with a large dissociation constant is more dissociated than an acid or base with a small dissociation constant.

- Equilibrium mixtures of acids or bases with dissociation constants greater than one ($K > 1$) contain mostly products, whereas constants smaller than one ($K < 1$) contain mostly reactants.

SAMPLE PROBLEM   Writing an Acid Dissociation Expression

Write the acid dissociation expression for hydrogen carbonate, H_2CO_3, a weak acid.

Solution:

STEP 1   Write the balanced chemical equation.

$$H_2CO_3(aq) + H_2O(l) \rightleftharpoons H_3O^+(aq) + HCO_3^-(aq)$$

STEP 2   Write the concentrations of the products as the numerator and the reactants as the denominator.

$$K_a = \frac{[H_3O^+][HCO_3^-]}{[H_2CO_3]}$$

♦ Learning Exercise 11.4A

Write the equation for the dissociation and the acid dissociation expression for each of the following weak acids:

a. $HCN(aq) + H_2O(l) \rightleftharpoons H_3O^+(aq) + CN^-(aq)$

$$K_a = [H_3O^+][CN^-]$$

b. $HNO_2(aq) + H_2O(l) \rightleftharpoons H_3O^+(aq) + NO_2^-(aq)$

c. H_2S (first dissociation only)

Answers

a. $HCN(aq) + H_2O(l) \rightleftharpoons H_3O^+(aq) + CN^-(aq)$ $K_a = \dfrac{[H_3O^+][CN^-]}{[HCN]}$

b. $HNO_2(aq) + H_2O(l) \rightleftharpoons H_3O^+(aq) + NO_2^-(aq) \quad K_a = \dfrac{[H_3O^+][NO_2^-]}{[HNO_2]}$

c. $H_2S(aq) + H_2O(l) \rightleftharpoons H_3O^+(aq) + HS^-(aq) \quad K_a = \dfrac{[H_3O^+][HS^-]}{[H_2S]}$

♦ **Learning Exercise 11.4B**

For each of the following pairs of K_a values, indicate which one belongs to the weaker acid:

a. 5.2×10^{-5} or 3.8×10^{-3} _____
b. 3.0×10^{-8} or 1.6×10^{-10} _____
c. 4.5×10^{-2} or 7.2×10^{-6} _____

Answers   **a.** 5.2×10^{-5} **b.** 1.6×10^{-10} **c.** 7.2×10^{-6}

♦ **Learning Exercise 11.4C**

Indicate whether the equilibrium mixture contains mostly reactants or products for each of the following K_a values:

a. 5.2×10^{-5} *reactants*
b. 3.0×10^8 *product*
c. 4.5×10^{-3} *reactant*
d. 7.2×10^{15} *product*

Answers   **a.** reactants   **b.** products   **c.** reactants   **d.** products

11.5   Dissociation of Water

Learning Goal: Use the water dissociation expression to calculate the $[H_3O^+]$ and $[OH^-]$ in an aqueous solution.

- In pure water, a few water molecules transfer hydrogen ions to other water molecules, producing small, but equal, amounts of each ion such that $[H_3O^+]$ and $[OH^-]$ each $= 1.0 \times 10^{-7}$ M at 25 °C.

$$H_2O(l) + H_2O(l) \rightleftharpoons H_3O^+(aq) + OH^-(aq)$$

- The water dissociation expression $[H_3O^+][OH^-] = K_w$, applies to all aqueous solutions.
- The water dissociation constant, $K_w = 1.0 \times 10^{-14}$ at 25 °C.
- In acidic solutions, the $[H_3O^+]$ is greater than the $[OH^-]$. In basic solutions, the $[OH^-]$ is greater than the $[H_3O^+]$.

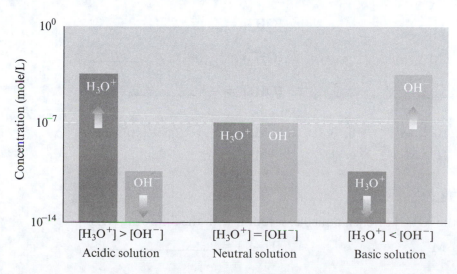

♦ **Learning Exercise 11.5A**

Indicate whether each of the following solutions is acidic, basic, or neutral:

a. $[H_3O^+] = 2.5 \times 10^{-9}$ M ___basic___

b. $[OH^-] = 1.6 \times 10^{-2}$ M ___basic___

c. $[H_3O^+] = 7.9 \times 10^{-3}$ M ___acidic___

d. $[OH^-] = 2.9 \times 10^{-12}$ M ___acidic___

Answers   **a.** basic   **b.** basic   **c.** acidic   **d.** acidic

CORE CHEMISTRY SKILL
Calculating $[H_3O^+]$ and $[OH^-]$ in Solutions

SAMPLE PROBLEM   Calculating $[H_3O^+]$ in Aqueous Solutions

What is the $[H_3O^+]$ in a urine sample that has $[OH^-] = 4.0 \times 10^{-10}$ M?

Solution:

STEP 1 State the given and needed quantities.

Analyze the Problem	Given	Need	Connect
	$[OH^-] = 4.0 \times 10^{-10}$ M	$[H_3O^+]$	$K_w = [H_3O^+][OH^-]$

STEP 2 Write the K_w for water and solve for the unknown $[H_3O^+]$.

$$K_w = [H_3O^+][OH^-] = 1.0 \times 10^{-14}$$

$$[H_3O^+] = \frac{1.0 \times 10^{-14}}{[OH^-]}$$

STEP 3 Substitute the known $[OH^-]$ into the equation and calculate.

$$[H_3O^+] = \frac{1.0 \times 10^{-14}}{[4.0 \times 10^{-10}]} = 2.5 \times 10^{-5} \text{ M}$$

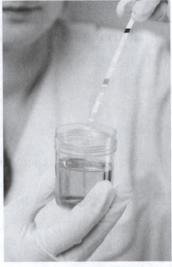

A dipstick is used to measure the acidity of a urine sample.
Credit: Alexander Gospodinov/Fotolia

♦ **Learning Exercise 11.5B**

Use the K_w to calculate the $[OH^-]$ of each aqueous solution with the following $[H_3O^+]$:

a. $[H_3O^+] = 1.0 \times 10^{-3}$ M $[OH^-] =$ ___1.0×10^{-11}___

b. $[H_3O^+] = 3.0 \times 10^{-10}$ M $[OH^-] =$ ___3.3×10^{-5}___

c. $[H_3O^+] = 4.0 \times 10^{-6}$ M $[OH^-] =$ ___2.5×10^{-9}___

d. $[H_3O^+] = 2.8 \times 10^{-13}$ M $[OH^-] =$ ___3.6×10^{-2}___

e. $[H_3O^+] = 8.6 \times 10^{-7}$ M $[OH^-] =$ ___1.2×10^{-8}___

Answers   **a.** 1.0×10^{-11} M   **b.** 3.3×10^{-5} M   **c.** 2.5×10^{-9} M
d. 3.6×10^{-2} M   **e.** 1.2×10^{-8} M

♦ **Learning Exercise 11.5C**

Use the K_w to calculate the $[OH^-]$ of each aqueous solution with the following $[OH^-]$:

a. $[OH^-] = 1.0 \times 10^{-10}$ M $[H_3O^+] =$ ___1.0×10^{-4}___

b. $[OH^-] = 2.0 \times 10^{-5}$ M $[H_3O^+] =$ ___5.0×10^{-10}___

c. $[OH^-] = 4.5 \times 10^{-7}$ M $\qquad$ $[H_3O^+] =$ _2.2×10^{-8}_

d. $[OH^-] = 8.0 \times 10^{-4}$ M $\qquad$ $[H_3O^+] =$ _1.3×10^{-11}_

e. $[OH^-] = 5.5 \times 10^{-8}$ M $\qquad$ $[H_3O^+] =$ _1.8×10^{-7}_

Answers $\qquad$ **a.** 1.0×10^{-4} M $\qquad$ **b.** 5.0×10^{-10} M $\qquad$ **c.** 2.2×10^{-8} M
$\qquad\qquad$ **d.** 1.3×10^{-11} M $\qquad$ **e.** 1.8×10^{-7} M

11.6 The pH Scale

Learning Goal: Calculate pH from $[H_3O^+]$; given the pH, calculate the $[H_3O^+]$ and $[OH^-]$ of a solution.

- The pH scale is a range of numbers from 0 to 14 related to the $[H_3O^+]$ of a solution.
- A neutral solution has a pH of 7.0. In an acidic solution, the pH is below 7.0, and in a basic solution, the pH is above 7.0.
- Mathematically, pH is the negative logarithm of the hydronium ion concentration: $pH = -\log[H_3O^+]$.

♦ **Learning Exercise 11.6A**

State whether each of the following pH values is acidic, basic, or neutral:

a. _basic_ blood plasma, pH = 7.4 $\qquad$ **b.** _acidic_ soft drink, pH = 2.8

c. _acidic_ maple syrup, pH = 6.8 $\qquad$ **d.** _acidic_ beans, pH = 5.0

e. _acidic_ tomatoes, pH = 4.2 $\qquad$ **f.** _acidic_ lemon juice, pH = 2.2

g. _neutral_ saliva, pH = 7.0 $\qquad$ **h.** _basic_ egg white, pH = 7.8

i. _basic_ lye, pH = 12.4 $\qquad$ **j.** _acidic_ strawberries, pH = 3.0

Answers $\qquad$ **a.** basic $\qquad$ **b.** acidic $\qquad$ **c.** acidic $\qquad$ **d.** acidic $\qquad$ **e.** acidic
$\qquad\qquad$ **f.** acidic $\qquad$ **g.** neutral $\qquad$ **h.** basic $\qquad$ **i.** basic $\qquad$ **j.** acidic

SAMPLE PROBLEM $\quad$ **Calculating pH from $[H_3O^+]$**

A solution of bleach has a $[H_3O^+] = 2.5 \times 10^{-11}$ M. What is the pH of the bleach solution?

Solution:

STEP 1 State the given and needed quantities.

Analyze the Problem	Given	Need	Connect
	$[H_3O^+] = 2.5 \times 10^{-11}$ M	pH	pH equation

STEP 2 Enter the $[H_3O^+]$ into the pH equation and calculate.

$$pH = -\log[H_3O^+] = -\log[2.5 \times 10^{-11}]$$

Calculator Procedure $\qquad\qquad\qquad\qquad\qquad$ **Calculator Display**

2.5 [EE or EXP] [+/−] 11 [log] [+/−] $\quad$ or $\quad$ [+/−] [log] 2.5 [EE or EXP] [+/−] 2.5 = $\quad$ 10.60205999

STEP 3 Adjust the number of SFs on the *right* of the decimal point. In a pH value, the number to the *left* of the decimal point is an *exact* number derived from the power of 10. Thus, the two SFs in the coefficient determine that there are two SFs after the decimal point in the pH value.

Coefficient		Power of ten	
2.5	$\times$	10^{-11} M	$pH = -\log[2.5 \times 10^{-11}] = 10.60$
Two SFs		Exact	Exact Two SFs after decimal point

♦ **Learning Exercise 11.6B**

KEY MATH SKILL

Calculating pH from $[H_3O^+]$

Calculate the pH of each of the following solutions:

a. $[H_3O^+] = 5.0 \times 10^{-3}$ M _2.30_   **b.** $[OH^-] = 4.0 \times 10^{-6}$ M _8.60_

c. $[H_3O^+] = 7.5 \times 10^{-8}$ M _7.12_   **d.** $[OH^-] = 2.5 \times 10^{-10}$ M _4.40_

e. $[H_3O^+] = 3.4 \times 10^{-8}$ M _7.47_   **f.** $[OH^-] = 7.8 \times 10^{-2}$ M _12.89_

Answers   **a.** 2.30   **b.** 8.60   **c.** 7.12   **d.** 4.40   **e.** 7.47   **f.** 12.89

SAMPLE PROBLEM   Calculating $[H_3O^+]$ from pH

Calculate the $[H_3O^+]$ for a solution with a pH of 4.60.

Solution:

STEP 1 State the given and needed quantities.

Analyze the Problem	Given	Need	Connect
	pH = 4.60	$[H_3O^+]$	pH equation

STEP 2 Enter the pH value into the inverse log equation and calculate.

$$[H_3O^+] = 10^{-pH} = 10^{-4.60}$$

Calculator Procedure	**Calculator Display**
(2nd) (log) (+/−) 4.60 or 4.60 (+/−) (2nd) (log) =	2.511886432 E-05

STEP 3 Adjust the SFs for the coefficient. Because the pH value 4.60 has two digits to the *right* of the decimal point, the coefficient for $[H_3O^+]$ is written with two SFs.

$$[H_3O^+] = 2.5 \times 10^{-5} \text{ M}$$
Two SFs

♦ **Learning Exercise 11.6C**

Complete the following table:

KEY MATH SKILL

Calculating $[H_3O^+]$ from pH

$[H_3O^+]$	$[OH^-]$	pH
a. _1.0 × 10⁻²_	1×10^{-12} M	_2.0_
b. _1.0 × 10⁻⁸_	_1.0×10⁻⁶_	8.0
c. 5.0×10^{-11} M	_2.0 ×10⁻⁴_	_10.30_
d. _____	_____	7.80
e. _____	_____	4.25
f. 2.0×10^{-10} M	_____	_____

Answers	$[H_3O^+]$	$[OH^-]$	pH
a.	1×10^{-2} M	1×10^{-12} M	2.0
b.	1×10^{-8} M	1×10^{-6} M	8.0
c.	5.0×10^{-11} M	2.0×10^{-4} M	10.30
d.	1.6×10^{-8} M	6.3×10^{-7} M	7.80
e.	5.6×10^{-5} M	1.8×10^{-10} M	4.25
f.	2.0×10^{-10} M	5.0×10^{-5} M	9.70

11.7 Reactions of Acids and Bases

Learning Goal: Write balanced equations for reactions of acids with metals, carbonates or bicarbonates, and bases; calculate the molarity or volume of an acid from titration information.

- Acids react with many metals to yield hydrogen gas, H_2, and a salt.
- Acids react with carbonates and bicarbonates to yield CO_2, H_2O, and a salt.
- Acids neutralize bases in a reaction that produces water and a salt.
- The net ionic equation for any strong acid–strong base neutralization is $H^+(aq) + OH^-(aq) \longrightarrow H_2O(l)$.
- In a balanced neutralization equation, an equal number of moles of H^+ and OH^- must react.
- In a laboratory procedure called titration, an acid or base sample is neutralized.
- A titration is used to determine the volume or concentration of an acid or a base from the laboratory data.

CORE CHEMISTRY SKILL

Writing Equations for Reactions of Acids and Bases

♦ **Learning Exercise 11.7A**

Complete and balance each of the following reactions of acids:

a. ___2___ $HCl(aq) +$ _____ $Fe(s) \longrightarrow$ ___H_2___ $+$ _____ $FeCl_2(aq)$

b. ___2___ $HCl(aq) +$ _____ $Li_2CO_3(aq) \longrightarrow$ ___CO_2___ $+$ ___H_2O___ $+ 2LiCl_2$

c. _____ $HBr(aq) +$ _____ $KHCO_3(s) \longrightarrow$ _____ $CO_2(g) +$ _____ $H_2O(l) +$ _____

d. _____ $H_2SO_4(aq) +$ _____ $Al(s) \longrightarrow$ _____ $+$ _____ $Al_2(SO_4)_3(aq)$

Answers

a. $2HCl(aq) + Fe(s) \longrightarrow H_2(g) + FeCl_2(aq)$

b. $2HCl(aq) + Li_2CO_3(aq) \longrightarrow CO_2(g) + H_2O(l) + 2LiCl(aq)$

c. $HBr(aq) + KHCO_3(s) \longrightarrow CO_2(g) + H_2O(l) + KBr(aq)$

d. $3H_2SO_4(aq) + 2Al(s) \longrightarrow 3H_2(g) + Al_2(SO_4)_3(aq)$

Balancing Equations for Neutralization of Acids and Bases	
STEP 1	Write the reactants and products.
STEP 2	Balance the H^+ in the acid with the OH^- in the base.
STEP 3	Balance the H_2O with the H^+ and the OH^-.
STEP 4	Write the salt from the remaining ions.

◆ **Learning Exercise 11.7B**

Complete and balance each of the following neutralization reactions:

a. ____ $H_2SO_4(aq)$ + ____ $NaOH(aq)$ ⟶ _____ + _____

b. ____ $HCl(aq)$ + ____ $Mg(OH)_2(s)$ ⟶ _____ + _____

c. ____ $HNO_3(aq)$ + ____ $Al(OH)_3(s)$ ⟶ _____ + _____

d. ____ $H_3PO_4(aq)$ + ____ $Ca(OH)_2(s)$ ⟶ _____ + _____

Answers   a.

　　　　　　b.

　　　　　　c.

　　　　　　d.

◆ **Learning Ex**

Complete and ba

a. ____ $H_3PO_4($

b. _____

c. _____

d. _____

Answers

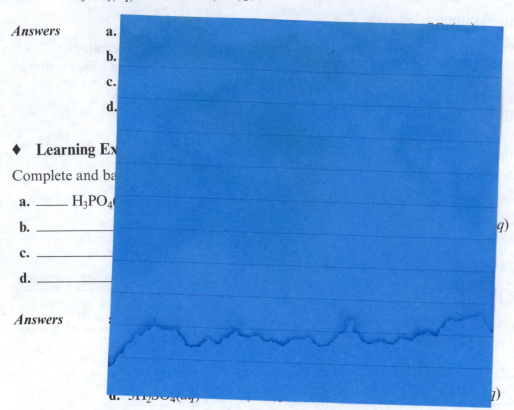

d. $5H_2SO_4(aq)$ ⟶ ____ $)$

SAMPLE PROBLEM **Titration of an Acid for Molarity**

A 32.0-mL (0.0320 L) sample of an HCl solution is placed in a flask
with a few drops of indicator. If 46.2 mL (0.0462 L) of a 0.214 M
NaOH solution is needed to reach the endpoint, what is the molarity
of the HCl solution?

　　　$NaOH(aq) + HCl(aq) \longrightarrow H_2O(l) + NaCl(aq)$

Solution:

STEP 1 State the given and needed quantities and concentrations.

	Given	Need	Connect
Analyze the Problem	32.0 mL (0.0320 L) of HCl solution, 46.2 mL (0.0462 L) of 0.214 M NaOH solution	molarity of HCl solution	molarity, mole–mole factor
	Neutralization Equation		
	$NaOH(aq) + HCl(aq) \longrightarrow H_2O(l) + NaCl(aq)$		

STEP 2 Write a plan to calculate the molarity.

Molarity of   Mole–mole   divide by
NaOH   factor   liters HCl

liters of NaOH solution $\longrightarrow$ moles of NaOH $\longrightarrow$ moles of HCl $\longrightarrow$ molarity of HCl solution

STEP 3 State equalities and conversion factors, including concentrations.

1 L of NaOH solution = 0.214 mole of NaOH   1 mole of NaOH = 1 mole of HCl

$$\frac{0.214 \text{ mole NaOH}}{1 \text{ L NaOH solution}} \text{ and } \frac{1 \text{ L NaOH solution}}{0.214 \text{ mole NaOH}} \qquad \frac{1 \text{ mole NaOH}}{1 \text{ mole HCl}} \text{ and } \frac{1 \text{ mole HCl}}{1 \text{ mole NaOH}}$$

STEP 4 Set up the problem to calculate the needed quantity.

$$0.0462 \text{ L NaOH solution} \times \frac{0.214 \text{ mole NaOH}}{1 \text{ L NaOH solution}} \times \frac{1 \text{ mole HCl}}{1 \text{ mole NaOH}} = 0.009\ 89 \text{ mole of HCl}$$

$$\text{molarity of HCl solution} = \frac{0.009\ 89 \text{ mole HCl}}{0.0320 \text{ L soution}} = 0.309 \text{ M HCl solution}$$

SAMPLE PROBLEM   Titration of an Acid for Volume

A 15.7-mL (0.0157 L) sample of a 0.165 M H_2SO_4 solution reacts completely with a 0.187 M NaOH solution. How many liters of the NaOH solution are needed?

$$H_2SO_4(aq) + 2NaOH(aq) \longrightarrow 2H_2O(l) + Na_2SO_4(aq)$$

Solution:

STEP 1 State the given and needed quantities and concentrations.

	Given	**Need**	**Connect**
Analyze the Problem	0.0157 L of 0.165 M H_2SO_4 solution, 0.187 M NaOH solution	liters of NaOH solution	molarity, mole–mole factor
	Neutralization Equation		
	$H_2SO_4(aq) + 2NaOH(aq) \longrightarrow 2H_2O(l) + Na_2SO_4(aq)$		

STEP 2 Write a plan to calculate the volume.

Molarity of   Mole–mole   Molarity of
H_2SO_4 factor   NaOH

liters of H_2SO_4 solution $\longrightarrow$ moles of H_2SO_4 $\longrightarrow$ moles of NaOH $\longrightarrow$ liters of NaOH solution

STEP 3 State equalities and conversion factors, including concentrations.

1 L of H_2SO_4 solution = 0.165 mole of H_2SO_4 2 moles of NaOH = 1 mole of H_2SO_4

$$\frac{0.165 \text{ mole } H_2SO_4}{1 \text{ L } H_2SO_4 \text{ solution}} \text{ and } \frac{1 \text{ L } H_2SO_4 \text{ solution}}{0.165 \text{ mole } H_2SO_4} \qquad \frac{2 \text{ moles NaOH}}{1 \text{ mole } H_2SO_4} \text{ and } \frac{1 \text{ mole } H_2SO_4}{2 \text{ moles NaOH}}$$

1 L of NaOH solution = 0.187 mole of NaOH

$$\frac{0.187 \text{ mole NaOH}}{1 \text{ L NaOH solution}} \text{ and } \frac{1 \text{ L NaOH solution}}{0.187 \text{ mole NaOH}}$$

STEP 4 Set up the problem to calculate the needed quantity.

$$0.0157 \text{ L } H_2SO_4 \text{ solution} \times \frac{0.165 \text{ mole } H_2SO_4}{1 \text{ L } H_2SO_4 \text{ solution}} \times \frac{2 \text{ moles NaOH}}{1 \text{ mole } H_2SO_4} \times \frac{1 \text{ L NaOH solution}}{0.187 \text{ mole NaOH}}$$

$$= 0.0277 \text{ L of NaOH solution}$$

♦ **Learning Exercise 11.7D**

Solve the following problems using the titration data given:

a. A 5.00-mL sample of HCl solution is placed in a flask. If 15.0 mL of a 0.200 M NaOH solution is required for neutralization, what is the molarity of the HCl solution?

$$HCl(aq) + NaOH(aq) \longrightarrow H_2O(l) + NaCl(aq)$$

b. How many milliliters of a 0.200 M KOH solution are required to neutralize completely 8.50 mL of a 0.500 M H_2SO_4 solution?

$$H_2SO_4(aq) + 2KOH(aq) \longrightarrow 2H_2O(l) + K_2SO_4(aq)$$

c. A 10.0-mL sample of H_3PO_4 solution is placed in a flask. If titration requires 42.0 mL of a 0.100 M NaOH solution for complete neutralization, what is the molarity of the H_3PO_4 solution?

$$H_3PO_4(aq) + 3NaOH(aq) \longrightarrow 3H_2O(l) + Na_3PO_4(aq)$$

d. A 24.6-mL sample of HCl solution reacts completely with 33.0 mL of a 0.222 M NaOH solution. What is the molarity of the HCl solution (see reaction in part **a**)?

Answers	**a.** 0.600 M HCl solution	**b.** 42.5 mL
	c. 0.140 M H_3PO_4 solution	**d.** 0.298 M HCl solution

11.8 Buffers

Learning Goal: Describe the role of buffers in maintaining the pH of a solution; calculate the pH of a buffer.

- A buffer solution resists a change in pH when small amounts of acid or base are added.
- A buffer contains either (1)a weak acid and its salt or (2) a weak base and its salt.
- The weak acid reacts with added OH^-, and the anion of the salt reacts with added H_3O^+, which maintains the pH of the solution.
- The weak base reacts with added H_3O^+, and the cation of the salt reacts with added OH^-, which maintains the pH of the solution.
- The $[H_3O^+]$ is calculated by solving the K_a expression for $[H_3O^+]$.
- The pH of a buffer is calculated using the acid dissociation expression and the concentrations of the weak acid and its conjugate base.

Key Terms for Sections 11.1 to 11.8

Match each of the following key terms with the correct description:

    a. acid   **b.** K_a **c.** pH   **d.** neutralization
    e. buffer   **f.** titration   **g.** dissociation

1. _____ the acid dissociation constant

2. _____ a reaction between an acid and a base to form water and a salt

3. _____ a substance that forms H^+ in water

4. _____ a mixture of a weak acid (or base) and its salt that maintains the pH of a solution

5. _____ a measure of the acidity of a solution

6. _____ the separation of an acid or base into ions in water

7. _____ the addition of base to an acid sample to determine the concentration of the acid

Answers   **1.** b   **2.** d   **3.** a   **4.** e   **5.** c   **6.** g   **7.** f

♦ **Learning Exercise 11.8A**

State whether each of the following represents a buffer system and explain why:

a. $HCl + NaCl$ **b.** K_2SO_4

c. H_2CO_3 **d.** $H_2CO_3 + KHCO_3$

Answers
    a. No. A strong acid and its salt do not make a buffer.
    b. No. A salt alone cannot act as a buffer.
    c. No. A weak acid alone cannot act as a buffer.
    d. Yes. A weak acid and its salt act as a buffer system.

♦ **Learning Exercise 11.8B**

A buffer system that functions in the fluid of the cells contains dihydrogen phosphate, $H_2PO_4^-$, and its conjugate base, hydrogen phosphate, HPO_4^{2-}.

    $H_2PO_4^-(aq) + H_2O(l) \rightleftharpoons H_3O^+(aq) + HPO_4^{2-}(aq)$

a. The purpose of this buffer system is to _____
    1. maintain $H_2PO_4^-$ **2.** maintain HPO_4^{2-} **3.** maintain pH

b. The weak acid is needed to _____
    1. provide the conjugate base   **2.** neutralize added OH^- **3.** provide the conjugate acid

c. If H_3O^+ is added, it is neutralized by _____
    1. HPO_4^{2-} **2.** H_2O **3.** OH^-

d. When OH^- is added, the equilibrium shifts in the direction of _____
    1. the reactants   **2.** the products   **3.** does not change

Answers   **a.** 3   **b.** 2   **c.** 1   **d.** 2

SAMPLE PROBLEM Calculating the pH of a Buffer

CORE CHEMISTRY SKILL
Calculating the pH of a Buffer

The K_a for acetic acid, $HC_2H_3O_2$, is 1.8×10^{-5}. What is the pH of a buffer prepared with 0.050 M $HC_2H_3O_2$ and 0.050 M $C_2H_3O_2^-$?

$$HC_2H_3O_2(aq) + H_2O(l) \rightleftharpoons H_3O^+(aq) + C_2H_3O_2^-(aq)$$

Solution:

STEP 1 State the given and needed quantities.

	Given	Need	Connect
Analyze the Problem	0.050 M $HC_2H_3O_2$, 0.050 M $C_2H_3O_2^-$	pH	K_a expression
	Equation		
	$HC_2H_3O_2(aq) + H_2O(l) \rightleftharpoons H_3O^+(aq) + C_2H_3O_2^-(aq)$		

STEP 2 Write the K_a expression and rearrange for $[H_3O^+]$.

$$K_a = \frac{[H_3O^+][C_2H_3O_2^-]}{[HC_2H_3O_2]}$$

$$[H_3O^+] = K_a \times \frac{[HC_2H_3O_2]}{[C_2H_3O_2^-]}$$

STEP 3 Substitute [HA] and [A⁻] into the K_a expression.

$$[H_3O^+] = 1.8 \times 10^{-5} \times \frac{[0.050]}{[0.050]} = 1.8 \times 10^{-5}\ \text{M}$$

STEP 4 Use $[H_3O^+]$ to calculate pH. Placing the $[H_3O^+]$ into the pH equation gives the pH of the buffer.

$$\text{pH} = -\log[1.8 \times 10^{-5}] = 4.74$$

♦ Learning Exercise 11.8C

The K_a for acetic acid, $HC_2H_3O_2$, is 1.8×10^{-5}.

a. What is the pH of a buffer that contains 1.0 M $HC_2H_3O_2$ and 0.50 M $NaC_2H_3O_2$?

b. What is the pH of a buffer that contains 1.0 M $HC_2H_3O_2$ and 0.10 M $NaC_2H_3O_2$?

c. What is the pH of a buffer that contains 0.10 M $HC_2H_3O_2$ and 1.0 M $NaC_2H_3O_2$?

Answers

a. $[H_3O^+] = 1.8 \times 10^{-5} \times \frac{[1.0]}{[0.50]} = 3.6 \times 10^{-5}\ \text{M}$ $\quad \text{pH} = -\log[3.6 \times 10^{-5}] = 4.44$

b. $[H_3O^+] = 1.8 \times 10^{-5} \times \frac{[1.0]}{[0.1]} = 1.8 \times 10^{-4}\ \text{M}$ $\quad \text{pH} = -\log[1.8 \times 10^{-4}] = 3.74$

c. $[H_3O^+] = 1.8 \times 10^{-5} \times \frac{[0.1]}{[1.0]} = 1.8 \times 10^{-6}\ \text{M}$ $\quad \text{pH} = -\log[1.8 \times 10^{-6}] = 5.74$

Checklist for Chapter 11

You are ready to take the Practice Test for Chapter 11. Be sure you have accomplished the following learning goals for this chapter. If not, review the Section listed at the end of the goal. Then apply your new skills and understanding to the Practice Test.

After studying Chapter 11, I can successfully:

____ Describe the properties of Arrhenius acids and bases and write their names. (11.1)

____ Describe the Brønsted–Lowry concept of acids and bases. (11.2)

____ Write conjugate acid–base pairs for an acid–base reaction. (11.2)

____ Write equations for the dissociation of strong and weak acids and bases. (11.3)

____ Write the expression for the dissociation constant for a weak acid or base. (11.4)

____ Use K_a values to compare the strengths of acids. (11.4)

____ Use the water dissociation expression to calculate $[H_3O^+]$ and $[OH^-]$. (11.5)

____ Calculate pH from the $[H_3O^+]$ or $[OH^-]$ of a solution. (11.6)

____ Write a balanced equation for the reactions of acids with metals, carbonates or bicarbonates, and bases. (11.7)

____ Calculate the molarity or volume of an acid or base from titration information. (11.7)

____ Describe the role of buffers in maintaining the pH of a solution. (11.8)

____ Calculate the pH of a buffer solution. (11.8)

Practice Test for Chapter 11

The chapter Sections to review are shown in parentheses at the end of each question.

1. An acid is a compound that when placed in water yields this characteristic ion. (11.1)
   A. H_3O^+ B. OH^- C. Na^+ D. Cl^- E. CO_3^{2-}

2. $MgCl_2$ would be classified as a(an) (11.1)
   A. acid   B. base   C. salt   D. buffer   E. nonelectrolyte

3. $Mg(OH)_2$ would be classified as a (11.1)
   A. weak acid   B. strong base   C. salt   D. buffer   E. nonelectrolyte

4. Which of the following would turn litmus blue? (11.1)
   A. HCl   B. NH_4Cl C. Na_2SO_4 D. KOH   E. $NaNO_3$

5. What is the name of $HClO_3$? (11.1)
   A. hypochlorous acid   B. chloric acid   C. chlorous acid
   D. chloric trioxide acid   E. perchloric acid

6. What is the name of NH_4OH? (11.1)
   A. ammonium oxide   B. nitrogen tetrahydride hydroxide
   C. ammonium hydroxide   D. perammonium hydroxide
   E. amine oxide hydride

7. Which of the following is a conjugate acid–base pair? (11.2)
   A. HCl/HNO_3 B. HNO_2/NO_2^- C. NaOH/KOH   D. HSO_4^-/HCO_3^- E. H_2S/S^{2-}

8. The conjugate base of HSO_4^- is (11.2)
   A. SO_4^{2-} B. H_2SO_4 C. HS^- D. H_2S E. SO_3^{2-}

9. In which of the following reactions does H_2O act as an acid? (11.2)
   A. $H_3PO_4(aq) + H_2O(l) \longrightarrow H_3O^+(aq) + H_2PO_4^-(aq)$
   B. $H_2SO_4(aq) + H_2O(l) \longrightarrow H_3O^+(aq) + HSO_4^-(aq)$
   C. $H_2O(l) + HS^-(aq) \longrightarrow H_3O^+(aq) + S^{2-}(aq)$
   D. $HCl(aq) + NaOH(aq) \longrightarrow H_2O(l) + NaCl(aq)$
   E. $NH_3(g) + H_2O(l) \longrightarrow NH_4^+(aq) + OH^-(aq)$

10. Which of the following acids has the smallest K_a value? (11.3)
   A. HNO_3 B. H_2SO_4 C. HCl   D. H_2CO_3 E. HBr

11. Using the following K_a values, identify the strongest acid in the group: (11.3)
   A. 7.5×10^{-3} B. 1.8×10^{-5} C. 4.5×10^8 D. 4.9×10^{-10} E. 3.2×10^4

12. Acetic acid is a weak acid because (11.3)
   A. it forms a dilute acid solution   B. it is isotonic
   C. it is slightly dissociated in water   D. it is a nonpolar molecule
   E. it can form a buffer

13. A weak base when added to water (11.3)
   A. makes the solution slightly basic   B. does not affect the pH
   C. dissociates completely   D. does not dissociate
   E. makes the solution slightly acidic

14. The weak acid formic acid dissociates in water.

   $$HCHO_2(aq) + H_2O(l) \rightleftharpoons H_3O^+(aq) + CHO_2^-(aq)$$

   The correctly written acid dissociation constant expression is: (11.4)

   A. $K_a = \dfrac{[HCHO_2]}{[H_3O^+][CHO_2^-]}$ B. $K_a = \dfrac{[HCHO_2]}{[CHO_2^-]}$

   C. $K_a = \dfrac{[H_3O^+][CHO_2^-]}{[HCHO_2][H_2O]}$ D. $K_a = \dfrac{[H_3O^+][CHO_2^-]}{[HCHO_2]}$

   E. $K_a = \dfrac{[HCHO_2][H_2O]}{[H_3O^+][CHO_2^-]}$

15. In the K_w for H_2O at 25 °C, the $[H_3O^+]$ has the value (11.5)
   A. 1.0×10^{-7} M   B. 1.0×10^{-1} M   C. 1.0×10^{-14} M
   D. 1.0×10^{-6} M   E. 1.0×10^{12} M

For questions 16 and 17, consider a solution with $[H_3O^+] = 1 \times 10^{-11}$ M. (11.5)

16. The hydroxide ion concentration is
   A. 1×10^{-1} M   B. 1×10^{-3} M   C. 1×10^{-4} M   D. 1×10^{-7} M   E. 1×10^{-11} M

17. The solution is
   A. acidic   B. basic   C. neutral   D. a buffer   E. neutralized

For questions 18 and 19, consider a solution with $[OH^-] = 1 \times 10^{-5}$ M. (11.5)

18. The hydrogen ion concentration of the solution is
   A. 1×10^{-5} M   B. 1×10^{-7} M   C. 1×10^{-9} M   D. 1×10^{-10} M   E. 1×10^{-14} M

19. The solution is
   A. acidic   B. basic   C. neutral   D. a buffer   E. neutralized

20. What is the $[H_3O^+]$ of a solution that has a pH = 8.7? (11.6)
   A. 7×10^{-8} B. 2×10^{-9} C. 2×10^9 D. 8×10^{-7} E. 2×10^{-8}

21. Of the following pH values, which is the most acidic? (11.6)

    A. 8.0   **B.** 5.5   **C.** 1.5   **D.** 3.2   **E.** 9.0

22. Of the following pH values, which is the most basic? (11.6)

    A. 10.0   **B.** 4.0   **C.** 2.2   **D.** 11.5   **E.** 9.0

23. Which is an equation for neutralization of an acid and a base? (11.7)

    A. $CaCO_3(s) \longrightarrow CaO(s) + CO_2(g)$

    B. $Na_2SO_4(s) \longrightarrow 2Na^+(aq) + SO_4^{2-}(aq)$

    C. $H_2SO_4(aq) + 2NaOH(aq) \longrightarrow 2H_2O(l) + Na_2SO_4(aq)$

    D. $Na_2O(s) + SO_3(g) \longrightarrow Na_2SO_4(aq)$

    E. $H_2CO_3(aq) \longrightarrow CO_2(g) + H_2O(l)$

24. What is the molarity of a 10.0-mL sample of HCl solution that is neutralized by 15.0 mL of a 2.0 M NaOH solution? (11.7)

    A. 0.50 M HCl   **B.** 1.0 M HCl   **C.** 1.5 M HCl   **D.** 2.0 M HCl   **E.** 3.0 M HCl

25. In a titration, 6.0 moles of NaOH will completely neutralize _____ mole(s) of H_2SO_4. (11.7)

    A. 1.0   **B.** 2.0   **C.** 3.0   **D.** 6.0   **E.** 12

26. What is the name given to components in the body that keep blood pH within its normal 7.35 to 7.45 range? (11.8)

    A. nutrients   **B.** buffers   **C.** metabolites   **D.** regufluids   **E.** neutralizers

27. A buffer system (11.8)

    A. maintains a pH of 7.0   **B.** contains only a weak base

    C. contains only a salt   **D.** contains a strong acid and its salt

    E. maintains the pH of a solution

28. Which of the following would act as a buffer system? (11.8)

    A. HCl   **B.** Na_2CO_3 **C.** $NaOH + NaNO_3$

    D. NH_4OH **E.** $NaHCO_3 + H_2CO_3$

29. The K_a of formic acid is 1.8×10^{-4}. What is the pH of a buffer made from 0.50 M formic acid and 0.050 M sodium formate? (11.8)

    A. 3.74   **B.** 4.74   **C.** 5.74   **D.** 6.74   **E.** 7.00

Answers to the Practice Test

1. A	**2.** C	**3.** B	**4.** D	**5.** B
6. C	**7.** B	**8.** A	**9.** E	**10.** C
11. A	**12.** R	**13.** R	**14.** D	**15.** A
16. B	**17.** B	**18.** C	**19.** B	**20.** B
21. C	**22.** D	**23.** C	**24.** E	**25.** C
26. B	**27.** E	**28.** E	**29.** B	

Selected Answers and Solutions to Text Problems

11.1   **a.** Acids taste sour.
    b. Acids neutralize bases.
    c. Acids produce H^+ ions in water.
    d. Barium hydroxide is the name of a base.
    e. Both acids and bases are electrolytes.

11.3   Acids containing a simple nonmetal anion use the prefix *hydro*, followed by the name of the anion with its *ide* ending changed to *ic acid*. When the anion is an oxygen-containing polyatomic ion, the *ate* ending of the polyatomic anion is replaced with *ic acid*. Acids with one oxygen less than the common *ic acid* name are named as *ous acids*. Bases are named as ionic compounds containing hydroxide anions.
    a. hydrochloric acid   **b.** calcium hydroxide   **c.** perchloric acid
    d. strontium hydroxide   **e.** sulfurous acid   **f.** bromous acid

11.5   **a.** RbOH   **b.** HF   **c.** H_3PO_4
    d. LiOH   **e.** NH_4OH **f.** HIO_4

11.7   A Brønsted–Lowry acid donates a hydrogen ion (H^+), whereas a Brønsted–Lowry base accepts a hydrogen ion.
    a. HI is the acid (H^+ donor); H_2O is the base (H^+ acceptor).
    b. H_2O is the acid (H^+ donor); F^- is the base (H^+ acceptor).
    c. H_2S is the acid (H^+ donor); $C_2H_5-NH_2$ is the base (H^+ acceptor).

11.9   To form the conjugate base, remove a hydrogen ion (H^+) from the acid.
    a. F^- **b.** OH^- **c.** HPO_3^{2-}
    d. SO_4^{2-} **e.** ClO_2^-

11.11   To form the conjugate acid, add a hydrogen ion (H^+) to the base.
    a. HCO_3^- **b.** H_3O^+ **c.** H_3PO_4
    d. HBr   **e.** $HClO_4$

11.13   The acid donates an H^+ to form the conjugate base, and the base accepts an H^+ to form the conjugate acid.
    a. In the reaction, the acid H_2CO_3 donates an H^+ to the base H_2O. The conjugate acid–base pairs are H_2CO_3/HCO_3^- and H_3O^+/H_2O.
    b. In the reaction, the acid HCN donates an H^+ to the base NO_2^-. The conjugate acid–base pairs are HCN/CN^- and HNO_2/NO_2^-.
    c. In the reaction, the acid HF donates an H^+ to the base CHO_2^-. The conjugate acid–base pairs are HF/F^- and $HCHO_2/CHO_2^-$.

11.15 $NH_4^+(aq) + H_2O(l) \rightleftharpoons NH_3(aq) + H_3O^+(aq)$

11.17   A strong acid is a good H^+ donor, whereas its conjugate base is a poor H^+ acceptor.

11.19   Use Table 11.3 to answer (the stronger acid will be closer to the top of the table).
    a. HBr is the stronger acid.
    b. HSO_4^- is the stronger acid.
    c. H_2CO_3 is the stronger acid.

11.21   Use Table 11.3 to answer (the weaker acid will be closer to the bottom of the table).
    a. HSO_4^- is the weaker acid.
    b. HF is the weaker acid.
    c. HCO_3^- is the weaker acid.

11.23 a. From Table 11.3, we see that H_2CO_3 is a weaker acid than H_3O^+ and that H_2O is a weaker base than HCO_3^-. Thus, the solution will contain mostly reactants at equilibrium.

b. From Table 11.3, we see that NH_4^+ is a weaker acid than H_3O^+ and that H_2O is a weaker base than NH_3. Thus, the solution will contain mostly reactants at equilibrium.

c. From Table 11.3, we see that NH_4^+ is a weaker acid than HNO_2 and that NO_2^- is a weaker base than NH_3. Thus, the solution will contain mostly products at equilibrium.

11.25 $NH_4^+(aq) + SO_4^{2-}(aq) \rightleftharpoons NH_3(aq) + HSO_4^-(aq)$

This equilibrium contains mostly reactants because NH_4^+ is a weaker acid than HSO_4^-, and SO_4^{2-} is a weaker base than NH_3.

11.27 a. True

b. False; a strong acid has a large value of K_a.

c. False; a strong acid has a weak conjugate base.

d. True

e. False; a strong acid is completely dissociated in aqueous solution.

11.29 The smaller the K_a value, the weaker the acid. The weaker acid has the stronger conjugate base.

a. H_2SO_3, which has a larger K_a value than HS^-, is the stronger acid.

b. The conjugate base HSO_3^- is formed by removing an H^+ from the acid H_2SO_3.

c. The stronger acid, H_2SO_3, has the weaker conjugate base, HSO_3^-.

d. The weaker acid, HS^-, has the stronger conjugate base, S^{2-}.

e. The stronger acid, H_2SO_3, dissociates more and produces more ions.

11.31 $H_3PO_4(aq) + H_2O(l) \rightleftharpoons H_3O^+(aq) + H_2PO_4^-(aq)$

The K_a expression is the ratio of the [products] divided by the [reactants] with [H_2O] considered constant and part of the K_a:

$$K_a = \frac{[H_3O^+][H_2PO_4^-]}{[H_3PO_4]} = 7.5 \times 10^{-3}$$

11.33 In pure water, $[H_3O^+] = [OH^-]$ because one of each is produced every time a hydrogen ion is transferred from one water molecule to another.

11.35 In an acidic solution, the $[H_3O^+]$ is greater than the $[OH^-]$, which means that at 25 °C, the $[H_3O^+]$ is greater than 1.0×10^{-7} M and the $[OH^-]$ is less than 1.0×10^{-7} M.

11.37 The value of $K_w = [H_3O^+][OH^-] = 1.0 \times 10^{-14}$ at 25 °C.

If $[H_3O^+]$ needs to be calculated from $[OH^-]$, then rearranging the K_w for $[H_3O^+]$ gives

$$[H_3O^+] = \frac{1.0 \times 10^{-14}}{[OH^-]}.$$

If $[OH^-]$ needs to be calculated from $[H_3O^+]$, then rearranging the K_w for $[OH^-]$ gives

$$[OH^-] = \frac{1.0 \times 10^{-14}}{[H_3O^+]}.$$

A neutral solution has $[OH^-] = [H_3O^+]$. If the $[OH^-] > [H_3O^+]$, the solution is basic; if the $[H_3O^+] > [OH^-]$, the solution is acidic.

a. $[OH^-] = \dfrac{1.0 \times 10^{-14}}{[H_3O^+]} = \dfrac{1.0 \times 10^{-14}}{[2.0 \times 10^{-5}]} = 5.0 \times 10^{-10}$ M; since $[H_3O^+] > [OH^-]$,

the solution is acidic.

b. $[OH^-] = \dfrac{1.0 \times 10^{-14}}{[H_3O^+]} = \dfrac{1.0 \times 10^{-14}}{[1.4 \times 10^{-9}]} = 7.1 \times 10^{-6}$ M; since $[OH^-] > [H_3O^+]$,

the solution is basic.

c. $[H_3O^+] = \dfrac{1.0 \times 10^{-14}}{[OH^-]} = \dfrac{1.0 \times 10^{-14}}{[8.0 \times 10^{-3}]} = 1.3 \times 10^{-12}$ M; since $[OH^-] > [H_3O^+]$,

the solution is basic.

d. $[H_3O^+] = \dfrac{1.0 \times 10^{-14}}{[OH^-]} = \dfrac{1.0 \times 10^{-14}}{[3.5 \times 10^{-10}]} = 2.9 \times 10^{-5}$ M; since $[H_3O^+] > [OH^-]$,

the solution is acidic.

11.39 The value of $K_w = [H_3O^+][OH^-] = 1.0 \times 10^{-14}$ at 25 °C.

When $[OH^-]$ is known, the $[H_3O^+]$ can be calculated by rearranging the K_w for $[H_3O^+]$:

$$[H_3O^+] = \dfrac{1.0 \times 10^{-14}}{[OH^-]}$$

a. $[H_3O^+] = \dfrac{1.0 \times 10^{-14}}{[OH^-]} = \dfrac{1.0 \times 10^{-14}}{[1.0 \times 10^{-9}]} = 1.0 \times 10^{-5}$ M (2 SFs)

b. $[H_3O^+] = \dfrac{1.0 \times 10^{-14}}{[OH^-]} = \dfrac{1.0 \times 10^{-14}}{[1.0 \times 10^{-6}]} = 1.0 \times 10^{-8}$ M (2 SFs)

c. $[H_3O^+] = \dfrac{1.0 \times 10^{-14}}{[OH^-]} = \dfrac{1.0 \times 10^{-14}}{[2.0 \times 10^{-5}]} = 5.0 \times 10^{-10}$ M (2 SFs)

d. $[H_3O^+] = \dfrac{1.0 \times 10^{-14}}{[OH^-]} = \dfrac{1.0 \times 10^{-14}}{[4.0 \times 10^{-13}]} = 2.5 \times 10^{-2}$ M (2 SFs)

11.41 The value of $K_w = [H_3O^+][OH^-] = 1.0 \times 10^{-14}$ at 25 °C.

When $[H_3O^+]$ is known, the $[OH^-]$ can be calculated by rearranging the K_w for $[OH^-]$:

$$[OH^-] = \dfrac{1.0 \times 10^{-14}}{[H_3O^+]}$$

a. $[OH^-] = \dfrac{1.0 \times 10^{-14}}{[H_3O^+]} = \dfrac{1.0 \times 10^{-14}}{[4.0 \times 10^{-2}]} = 2.5 \times 10^{-13}$ M (2 SFs)

b. $[OH^-] = \dfrac{1.0 \times 10^{-14}}{[H_3O^+]} = \dfrac{1.0 \times 10^{-14}}{[5.0 \times 10^{-6}]} = 2.0 \times 10^{-9}$ M (2 SFs)

c. $[OH^-] = \dfrac{1.0 \times 10^{-14}}{[H_3O^+]} = \dfrac{1.0 \times 10^{-14}}{[2.0 \times 10^{-4}]} = 5.0 \times 10^{-11}$ M (2 SFs)

d. $[OH^-] = \dfrac{1.0 \times 10^{-14}}{[H_3O^+]} = \dfrac{1.0 \times 10^{-14}}{[7.9 \times 10^{-9}]} = 1.3 \times 10^{-6}$ M (2 SFs)

11.43 In a neutral solution, the $[H_3O^+] = 1 \times 10^{-7}$ M at 25 °C.

$pH = -\log[H_3O^+] = -\log[1 \times 10^{-7}] = 7.0$. The pH value contains one *decimal place*, which represents the one significant figure in the coefficient 1.

11.45 An acidic solution has a pH less than 7.0. A basic solution has a pH greater than 7.0. A neutral solution has a pH equal to 7.0. (All at 25 °C.)
    a. basic (pH 7.38 > 7.0)   **b.** acidic (pH 2.8 < 7.0)
    c. acidic (pH 5.52 < 7.0)   **d.** acidic (pH 4.2 < 7.0)
    e. basic (pH 7.6 > 7.0)

11.47 Since pH is a logarithmic scale, an increase or decrease of 1 pH unit changes the $[H_3O^+]$ by a factor of 10. Thus, a pH of 3 ($[H_3O^+] = 10^{-3}$ M, or 0.001 M) is 10 times more acidic than a pH of 4 ($[H_3O^+] = 10^{-4}$ M, or 0.0001 M).

11.49 $pH = -\log[H_3O^+]$

Since the value of $K_w = [H_3O^+][OH^-] = 1.0 \times 10^{-14}$ at 25 °C, if $[H_3O^+]$ needs to be calculated

from $[OH^-]$, rearranging the K_w for $[H_3O^+]$ gives $[H_3O^+] = \dfrac{1.0 \times 10^{-14}}{[OH^-]}$.

a. $pH = -\log[H_3O^+] = -\log[1 \times 10^{-4}] = 4.0$ (1 SF on the right of the decimal point)

b. $pH = -\log[H_3O^+] = -\log[3 \times 10^{-9}] = 8.5$ (1 SF on the right of the decimal point)

c. $[H_3O^+] = \dfrac{1.0 \times 10^{-14}}{[1 \times 10^{-5}]} = 1 \times 10^{-9}$ M

$pH = -\log[1 \times 10^{-9}] = 9.0$ (1 SF on the right of the decimal point)

d. $[H_3O^+] = \dfrac{1.0 \times 10^{-14}}{[2.5 \times 10^{-11}]} = 4.0 \times 10^{-4}$ M

$pH = -\log[4.0 \times 10^{-4}] = 3.40$ (2 SFs on the right of the decimal point)

e. $pH = -\log[H_3O^+] = -\log[6.7 \times 10^{-8}] = 7.17$ (2 SFs on the right of the decimal point)

f. $[H_3O^+] = \dfrac{1.0 \times 10^{-14}}{[8.2 \times 10^{-4}]} = 1.2 \times 10^{-11}$ M

$pH = -\log[1.2 \times 10^{-11}] = 10.92$ (2 SFs on the right of the decimal point)

11.51 On a calculator, pH is calculated by entering $-log$, followed by the coefficient *EE* (*EXP*) key and the power of 10 followed by the change sign (+/−) key. On some calculators, the concentration is entered first (coefficient *EXP* –power) followed by *log* and +/− key.

$[H_3O^+] = \dfrac{1.0 \times 10^{-14}}{[OH^-]}$; $[OH^-] = \dfrac{1.0 \times 10^{-14}}{[H_3O^+]}$; $pH = -\log[H_3O^+]$; $[H_3O^+] = 10^{-pH}$

Food	$[H_3O^+]$	$[OH^-]$	pH	Acidic, Basic, or Neutral?
Rye bread	1.6×10^{-9} M	6.3×10^{-6} M	8.80	Basic
Tomatoes	2.3×10^{-5} M	4.3×10^{-10} M	4.64	Acidic
Peas	6.2×10^{-7} M	1.6×10^{-8} M	6.21	Acidic

$[H_3O^+]$ and $[OH^-]$ all have 2 SFs here, so all pH values have 2 decimal places on the right of the decimal point.

11.53 $[H_3O^+] = 10^{-pH} = 10^{-6.92} = 1.2 \times 10^{-7}$ M (2 SFs)

11.55 Acids react with active metals to form $H_2(g)$ and a salt of the metal. The reaction of acids with carbonates or bicarbonates yields CO_2, H_2O, and a salt. In a neutralization reaction, an acid and a base react to form H_2O and a salt.

a. $2HBr(aq) + ZnCO_3(s) \longrightarrow CO_2(g) + H_2O(l) + ZnBr_2(aq)$

b. $2HCl(aq) + Zn(s) \longrightarrow H_2(g) + ZnCl_2(aq)$

c. $HCl(aq) + NaHCO_3(s) \longrightarrow CO_2(g) + H_2O(l) + NaCl(aq)$

d. $H_2SO_4(aq) + Mg(OH)_2(s) \longrightarrow 2H_2O(l) + MgSO_4(aq)$

11.57 In balancing a neutralization equation, the number of H^+ and OH^- must be equalized by placing coefficients in front of the formulas for the acid and base.

a. $2HCl(aq) + Mg(OH)_2(s) \longrightarrow 2H_2O(l) + MgCl_2(aq)$

b. $H_3PO_4(aq) + 3LiOH(aq) \longrightarrow 3H_2O(l) + Li_3PO_4(aq)$

11.59 The products of a neutralization are water and a salt. In balancing a neutralization equation, the number of H^+ and OH^- must be equalized by placing coefficients in front of the formulas for the acid and base.
  a. $H_2SO_4(aq) + 2NaOH(aq) \longrightarrow 2H_2O(l) + Na_2SO_4(aq)$
  b. $3HCl(aq) + Fe(OH)_3(s) \longrightarrow 3H_2O(l) + FeCl_3(aq)$
  c. $H_2CO_3(aq) + Mg(OH)_2(s) \longrightarrow 2H_2O(l) + MgCO_3(s)$

11.61 In the titration equation, 1 mole of HCl reacts with 1 mole of NaOH.

$$28.6 \text{ mL NaOH solution} \times \frac{1 \text{ L solution}}{1000 \text{ mL solution}} \times \frac{0.145 \text{ mole NaOH}}{1 \text{ L solution}} \times \frac{1 \text{ mole HCl}}{1 \text{ mole NaOH}}$$

$$= 0.004\ 15 \text{ mole of HCl}$$

$$5.00 \text{ mL HCl solution} \times \frac{1 \text{ L solution}}{1000 \text{ mL solution}} = 0.005\ 00 \text{ L of HCl solution}$$

$$\text{molarity (M) of HCl} = \frac{\text{moles of solute}}{\text{liters of solution}} = \frac{0.004\ 15 \text{ mole HCl}}{0.005\ 00 \text{ L solution}}$$

$$= 0.830 \text{ M HCl solution (3 SFs)}$$

11.63 In the titration equation, 1 mole of H_2SO_4 reacts with 2 moles of KOH.

$$38.2 \text{ mL KOH solution} \times \frac{1 \text{ L solution}}{1000 \text{ mL solution}} \times \frac{0.163 \text{ mole KOH}}{1 \text{ L solution}} \times \frac{1 \text{ mole H}_2\text{SO}_4}{2 \text{ moles KOH}}$$

$$= 0.003\ 11 \text{ mole of H}_2\text{SO}_4$$

$$25.0 \text{ mL H}_2\text{SO}_4 \text{ solution} \times \frac{1 \text{ L solution}}{1000 \text{ mL solution}} = 0.0250 \text{ L of H}_2\text{SO}_4 \text{ solution}$$

$$\text{molarity (M) of H}_2\text{SO}_4 = \frac{\text{moles of solute}}{\text{liters of solution}} = \frac{0.003\ 11 \text{ mole H}_2\text{SO}_4}{0.0250 \text{ L solution}}$$

$$= 0.124 \text{ M H}_2\text{SO}_4 \text{ solution (3 SFs)}$$

11.65 In the titration equation, 1 mole of H_3PO_4 reacts with 3 moles of NaOH.

$$50.0 \text{ mL H}_3\text{PO}_4 \text{ solution} \times \frac{1 \text{ L H}_3\text{PO}_4 \text{ solution}}{1000 \text{ mL H}_3\text{PO}_4 \text{ solution}} \times \frac{0.0224 \text{ mole H}_3\text{PO}_4}{1 \text{ L H}_3\text{PO}_4 \text{ solution}} \times \frac{3 \text{ moles NaOH}}{1 \text{ mole H}_3\text{PO}_4}$$

$$\times \frac{1 \text{ L NaOH solution}}{0.204 \text{ mole NaOH}} \times \frac{1000 \text{ mL NaOH solution}}{1 \text{ L NaOH solution}} = 16.5 \text{ mL of NaOH solution (3 SFs)}$$

11.67 A buffer system is a solution that contains a weak acid and a salt containing its conjugate base (or a weak base and a salt containing its conjugate acid).
  a. A solution of the strong base NaOH and the salt NaCl would not be a buffer system.
  b. These two substances can be combined in solution to create a buffer system because one is the weak acid H_2CO_3 and the other is a salt containing its conjugate base HCO_3^-.
  c. These two substances can be combined in solution to create a buffer system because one is the weak acid HF and the other is a salt containing its conjugate base F^-.
  d. A solution of the salts KCl and NaCl would not be a buffer system.

11.69 a. The purpose of this buffer system is to (3) maintain pH.
  b. The salt of the weak acid is needed to (1) provide the conjugate base and (2) neutralize added H_3O^+.
  c. If OH^- is added, it is neutralized by (3) H_3O^+.
  d. When H_3O^+ is added, the equilibrium shifts in the direction of the (1) reactants.

11.71 $HNO_2(aq) + H_2O(l) \rightleftarrows NO_2^-(aq) + H_3O^+(aq)$

Rearrange the K_a for $[H_3O^+]$ and use it to calculate the pH.

$$[H_3O^+] = K_a \times \frac{[HNO_2]}{[NO_2^-]} = 4.5 \times 10^{-4} \times \frac{[0.10 \text{ M}]}{[0.10 \text{ M}]} = 4.5 \times 10^{-4} \text{ M}$$

pH $= -\log[H_3O^+] = -\log[4.5 \times 10^{-4}] = 3.35$ (2 SFs on the right of the decimal point)

11.73 $HF(aq) + H_2O(l) \rightleftarrows F^-(aq) + H_3O^+(aq)$

Rearrange the K_a for $[H_3O^+]$ and use it to calculate the pH.

$$[H_3O^+] = K_a \times \frac{[HF]}{[F^-]} = 3.5 \times 10^{-4} \times \frac{[0.10 \text{ M}]}{[0.10 \text{ M}]} = 3.5 \times 10^{-4} \text{ M}$$

pH $= -\log[3.5 \times 10^{-4}] = 3.46$ (2 SFs on the right of the decimal point)

$$[H_3O^+] = K_a \times \frac{[HF]}{[F^-]} = 3.5 \times 10^{-4} \times \frac{[0.060 \text{ M}]}{[0.120 \text{ M}]} = 1.75 \times 10^{-4} \text{ M}$$

pH $= -\log[1.75 \times 10^{-4}] = 3.76$ (2 SFs on the right of the decimal point)

∴ The solution with 0.10 M HF/0.10 M NaF is more acidic.

11.75 If you breathe fast, CO_2 is expelled and the equilibrium shifts to lower $[H_3O^+]$, which raises the pH.

11.77 If large amounts of HCO_3^- are lost, the equilibrium shifts to higher $[H_3O^+]$, which lowers the pH.

11.79 pH $= -\log[H_3O^+] = -\log[2.0 \times 10^{-4}] = 3.70$ (2 SFs on the right of the decimal point)

11.81 $[H_3O^+] = 10^{-pH} = 10^{-3.60} = 2.5 \times 10^{-4}$ M (2 SFs)

11.83 $2HCl(aq) + CaCO_3(s) \longrightarrow CO_2(g) + H_2O(l) + CaCl_2(aq)$

11.85 From the neutralization equation in problem 11.83, 1 mole of $CaCO_3$ reacts with 2 moles of HCl.

$$100. \text{ mL HCl solution} \times \frac{1 \text{ L HCl solution}}{1000 \text{ mL HCl solution}} \times \frac{0.0400 \text{ mole HCl}}{1 \text{ L HCl solution}} \times \frac{1 \text{ mole CaCO}_3}{2 \text{ moles HCl}}$$

$$\times \frac{100.09 \text{ g CaCO}_3}{1 \text{ mole CaCO}_3} = 0.200 \text{ g of CaCO}_3 \text{ (3 SFs)}$$

11.87 **a.** This diagram represents a weak acid; only a few HX molecules dissociate into H_3O^+ and X^- ions.

b. This diagram represents a strong acid; all of the HX molecules dissociate into H_3O^+ and X^- ions.

11.89 **a.** H_2SO_4 is an acid.   **b.** RbOH is a base.

c. $Ca(OH)_2$ is a base.   **d.** HI is an acid.

11.91

Acid	Conjugate Base
H_2O	OH^-
HCN	CN^-
HNO_2	NO_2^-
H_3PO_4	$H_2PO_4^-$

11.93 **a.** Hyperventilation will lower the CO_2 concentration in the blood, which lowers the $[H_2CO_3]$, which decreases the $[H_3O^+]$ and increases the blood pH.

b. Breathing into a paper bag will increase the CO_2 concentration in the blood, increase the $[H_2CO_3]$, increase $[H_3O^+]$, and lower the blood pH back toward the normal range.

11.95 An acidic solution has a pH less than 7.0. A neutral solution has a pH equal to 7.0. A basic solution has a pH greater than 7.0. (All at 25 °C.)
    a. acidic (pH 5.2 < 7.0)   **b.** basic (pH 7.5 > 7.0)
    c. basic (pH 8.1 > 7.0)   **d.** acidic (pH 2.5 < 7.0)

11.97 **a.** acid; bromous acid   **b.** base; cesium hydroxide
       c. salt; magnesium nitrate   **d.** acid; perchloric acid

11.99

Acid	Conjugate Base
HI	I^-
HCl	Cl^-
NH_4^+	NH_3
H_2S	HS^-

11.101 Use Table 11.3 to answer (the stronger acid will be closer to the top of the table).
    a. HF is the stronger acid.   **b.** H_3O^+ is the stronger acid.
    c. HNO_2 is the stronger acid.   **d.** HCO_3^- is the stronger acid.

11.103 $[H_3O^+] = \dfrac{1.0 \times 10^{-14}}{[OH^-]}$; $pH = -\log[H_3O^+]$

    a. $pH = -\log[H_3O^+] = -\log[2.0 \times 10^{-8}] = 7.70$ (2 SFs on the right of the decimal point)

    b. $pH = -\log[5.0 \times 10^{-2}] = 1.30$ (2 SFs on the right of the decimal point)

    c. $[H_3O^+] = \dfrac{1.0 \times 10^{-14}}{[3.5 \times 10^{-4}]} = 2.9 \times 10^{-11}$ M

        $pH = -\log[2.9 \times 10^{-11}] = 10.54$ (2 SFs on the right of the decimal point)

    d. $[H_3O^+] = \dfrac{1.0 \times 10^{-14}}{[0.0054]} = 1.9 \times 10^{-12}$ M

        $pH = -\log[1.9 \times 10^{-12}] = 11.73$ (2 SFs on the right of the decimal point)

11.105 **a.** basic (pH > 7.0)   **b.** acidic (pH < 7.0)
       c. basic (pH > 7.0)   **d.** basic (pH > 7.0)

11.107 If the pH is given, the $[H_3O^+]$ can be found by using the relationship $[H_3O^+] = 10^{-pH}$.
The $[OH^-]$ can be found by rearranging $K_w = [H_3O^+][OH^-] = 1 \times 10^{-14}$.

    a. $pH = 3.00$; $[H_3O^+] = 10^{-pH} = 10^{-3.00} = 1.0 \times 10^{-3}$ M (2 SFs)

        $[OH^-] = \dfrac{1.0 \times 10^{-14}}{[H_3O^+]} = \dfrac{1.0 \times 10^{-14}}{[1.0 \times 10^{-3}]} = 1.0 \times 10^{-11}$ M (2 SFs)

    b. $pH = 6.2$; $[H_3O^+] = 10^{-pH} = 10^{-6.2} = 6 \times 10^{-7}$ M (1 SF)

        $[OH^-] = \dfrac{1.0 \times 10^{-14}}{[H_3O^+]} = \dfrac{1.0 \times 10^{-14}}{[6.3 \times 10^{-7}]} = 2 \times 10^{-8}$ M (1 SF)

    c. $pH = 8.85$; $[H_3O^+] = 10^{-pH} = 10^{-8.85} = 1.4 \times 10^{-9}$ M (2 SFs)

        $[OH^-] = \dfrac{1.0 \times 10^{-14}}{[H_3O^+]} = \dfrac{1.0 \times 10^{-14}}{[1.4 \times 10^{-9}]} = 7.1 \times 10^{-6}$ M (2 SFs)

    d. $pH = 11.00$; $[H_3O^+] = 10^{-pH} = 10^{-11.00} = 1.0 \times 10^{-11}$ M (2 SFs)

        $[OH^-] = \dfrac{1.0 \times 10^{-14}}{[H_3O^+]} = \dfrac{1.0 \times 10^{-14}}{[1.0 \times 10^{-11}]} = 1.0 \times 10^{-3}$ M (2 SFs)

11.109 a. Solution A, with a pH of 4.5, is more acidic.

    b. In solution A, the $[H_3O^+] = 10^{-pH} = 10^{-4.5} = 3 \times 10^{-5}$ M (1 SF)

       In solution B, the $[H_3O^+] = 10^{-pH} = 10^{-6.7} = 2 \times 10^{-7}$ M (1 SF)

    c. In solution A, the $[OH^-] = \dfrac{1.0 \times 10^{-14}}{[H_3O^+]} = \dfrac{1.0 \times 10^{-14}}{[3 \times 10^{-5}]} = 3 \times 10^{-10}$ M (1 SF)

       In solution B, the $[OH^-] = \dfrac{1.0 \times 10^{-14}}{[H_3O^+]} = \dfrac{1.0 \times 10^{-14}}{[2 \times 10^{-7}]} = 5 \times 10^{-8}$ M (1 SF)

11.111 The $[OH^-]$ can be calculated from the moles of NaOH (each NaOH produces 1 OH^-) and the volume of the solution (in L).

$$0.225 \text{ g NaOH} \times \frac{1 \text{ mole NaOH}}{40.00 \text{ g NaOH}} \times \frac{1 \text{ mole OH}^-}{1 \text{ mole NaOH}} = 0.005\,63 \text{ mole of OH}^-$$

$$[OH^-] = \frac{0.005\,63 \text{ mole OH}^-}{0.250 \text{ L solution}} = 0.0225 \text{ M (3 SFs)}$$

11.113 $2.5 \text{ g HCl} \times \dfrac{1 \text{ mole HCl}}{36.46 \text{ g HCl}} \times \dfrac{1 \text{ mole H}_3O^+}{1 \text{ mole HCl}} = 0.069$ mole of H_3O^+ (2 SFs)

$$[H_3O^+] = \frac{0.069 \text{ mole H}_3O^+}{0.425 \text{ L solution}} = 0.16 \text{ M (2 SFs)}$$

pH $= -\log[H_3O^+] = -\log[0.16] = 0.80$ (2 SFs on the right of the decimal point)

11.115 a. To form the conjugate base, remove a hydrogen ion (H^+) from the acid.

       1. HS^- **2.** $H_2PO_4^-$

    b. 1. $K_a = \dfrac{[H_3O^+][HS^-]}{[H_2S]}$ **2.** $K_a = \dfrac{[H_3O^+][H_2PO_4^-]}{[H_3PO_4]}$

    c. H_2S (see Table 11.3; the weaker acid will be closer to the bottom of the table)

11.117 a. NH_4^+/NH_3 and HNO_3/NO_3^-

       From Table 11.3, we see that NH_4^+ is a weaker acid than HNO_3 and that NO_3^- is a weaker base than NH_3. Thus, the equilibrium mixture will contain mostly products.

    b. H_3O^+/H_2O and HBr/Br^-

       From Table 11.3, we see that H_3O^+ is a weaker acid than HBr and that Br^- is a weaker base than H_2O. Thus, the equilibrium mixture will contain mostly products.

11.119 a. $H_2SO_4(aq) + ZnCO_3(s) \longrightarrow CO_2(g) + H_2O(l) + ZnSO_4(aq)$

    b. $6HNO_3(aq) + 2Al(s) \longrightarrow 3H_2(g) + 2Al(NO_3)_3(aq)$

11.121 KOH (strong base) $\longrightarrow K^+(aq) + OH^-(aq)$ (100% dissociation)

    $[OH^-] = 0.050$ M $= 5.0 \times 10^{-2}$ M

    a. $[H_3O^+] = \dfrac{1.0 \times 10^{-14}}{[OH^-]} = \dfrac{1.0 \times 10^{-14}}{[5.0 \times 10^{-2}]} = 2.0 \times 10^{-13}$ M (2 SFs)

    b. pH $= -\log[H_3O^+] = -\log[2.0 \times 10^{-13}] = 12.70$ (2 SFs on the right of the decimal point)

    c. $H_2SO_4(aq) + 2KOH(aq) \longrightarrow 2H_2O(l) + K_2SO_4(aq)$

    d. In the titration equation, 1 mole of H_2SO_4 reacts with 2 moles of KOH.

$$40.0 \text{ mL H}_2SO_4 \text{ solution} \times \frac{1 \text{ L H}_2SO_4 \text{ solution}}{1000 \text{ mL H}_2SO_4 \text{ solution}} \times \frac{0.035 \text{ mole H}_2SO_4}{1 \text{ L H}_2SO_4 \text{ solution}} \times \frac{2 \text{ moles KOH}}{1 \text{ mole H}_2SO_4}$$

$$\times \frac{1 \text{ L KOH solution}}{0.050 \text{ mole KOH}} \times \frac{1000 \text{ mL KOH solution}}{1 \text{ L KOH solution}} = 56 \text{ mL of KOH solution (2 SFs)}$$

11.123 a. In the titration equation, 1 mole of HCl reacts with 1 mole of NaOH.

$$HCl(aq) + NaOH(aq) \longrightarrow H_2O(l) + NaCl(aq)$$

$$25.0 \text{ mL HCl solution} \times \frac{1 \text{ L HCl solution}}{1000 \text{ mL HCl solution}} \times \frac{0.288 \text{ mole HCl}}{1 \text{ L HCl solution}} \times \frac{1 \text{ mole NaOH}}{1 \text{ mole HCl}}$$

$$\times \frac{1 \text{ L NaOH solution}}{0.150 \text{ mole NaOH}} \times \frac{1000 \text{ mL NaOH solution}}{1 \text{ L NaOH solution}} = 48.0 \text{ mL of NaOH solution (3 SFs)}$$

b. In the titration equation, 1 mole of H_2SO_4 reacts with 2 moles of NaOH.

$$H_2SO_4(aq) + 2NaOH(aq) \longrightarrow 2H_2O(l) + Na_2SO_4(aq)$$

$$10.0 \text{ mL } H_2SO_4 \text{ solution} \times \frac{1 \text{ L } H_2SO_4 \text{ solution}}{1000 \text{ mL } H_2SO_4 \text{ solution}} \times \frac{0.560 \text{ mole } H_2SO_4}{1 \text{ L } H_2SO_4 \text{ solution}} \times \frac{2 \text{ moles NaOH}}{1 \text{ mole } H_2SO_4}$$

$$\times \frac{1 \text{ L NaOH solution}}{0.150 \text{ mole NaOH}} \times \frac{1000 \text{ mL NaOH solution}}{1 \text{ L NaOH solution}} = 74.7 \text{ mL of NaOH solution (3 SFs)}$$

11.125 In the titration equation, 1 mole of H_2SO_4 reacts with 2 moles of NaOH.

$$45.6 \text{ mL NaOH solution} \times \frac{1 \text{ L solution}}{1000 \text{ mL solution}} \times \frac{0.205 \text{ mole NaOH}}{1 \text{ L solution}} \times \frac{1 \text{ mole } H_2SO_4}{2 \text{ moles NaOH}}$$

$$= 0.004 \, 67 \text{ mole of } H_2SO_4$$

$$20.0 \text{ mL } H_2SO_4 \text{ solution} \times \frac{1 \text{ L solution}}{1000 \text{ mL solution}} = 0.0200 \text{ L of } H_2SO_4 \text{ solution}$$

$$\text{molarity (M) of } H_2SO_4 = \frac{\text{moles of solute}}{\text{liters of solution}} = \frac{0.004 \, 67 \text{ mole } H_2SO_4}{0.0200 \text{ L solution}}$$

$$= 0.234 \text{ M } H_2SO_4 \text{ solution (3 SFs)}$$

11.127 This buffer solution is made from the weak acid H_3PO_4 and a salt containing its conjugate base $H_2PO_4^-$.

a. Add acid: $H_2PO_4^-(aq) + H_3O^+(aq) \longrightarrow H_3PO_4(aq) + H_2O(l)$

b. Add base: $H_3PO_4(aq) + OH^-(aq) \longrightarrow H_2PO_4^-(aq) + H_2O(l)$

c. $[H_3O^+] = K_a \times \dfrac{[H_3PO_4]}{[H_2PO_4^-]} = 7.5 \times 10^{-3} \times \dfrac{[0.50 \text{ M}]}{[0.20 \text{ M}]} = 1.9 \times 10^{-2} \text{ M}$

$pH = -\log[1.9 \times 10^{-2}] = 1.72$ (2 SFs on the right of the decimal point)

11.129 a. The $[H_3O^+]$ can be found by using the relationship $[H_3O^+] = 10^{-pH}$.

$$[H_3O^+] = 10^{-pH} = 10^{-4.2} = 6 \times 10^{-5} \text{ M (1 SF)}$$

$$[OH^-] = \frac{1.0 \times 10^{-14}}{[H_3O^+]} = \frac{1.0 \times 10^{-14}}{[6 \times 10^{-5}]} = 2 \times 10^{-10} \text{ M (1 SF)}$$

b. $[H_3O^+] = 10^{-pH} = 10^{-6.5} = 3 \times 10^{-7} \text{ M (1 SF)}$

$$[OH^-] = \frac{1.0 \times 10^{-14}}{[H_3O^+]} = \frac{1.0 \times 10^{-14}}{[3 \times 10^{-7}]} = 3 \times 10^{-8} \text{ M (1 SF)}$$

c. In the titration equation, 1 mole of $CaCO_3$ reacts with 1 mole of H_2SO_4.

$$1.0 \text{ kL solution} \times \frac{1000 \text{ L solution}}{1 \text{ kL solution}} \times \frac{6 \times 10^{-5} \text{ mole } H_3O^+}{1 \text{ L solution}}$$

$$\times \frac{1 \text{ mole } H_2SO_4}{2 \text{ moles } H_3O^+} \times \frac{1 \text{ mole } CaCO_3}{1 \text{ mole } H_2SO_4} \times \frac{100.09 \text{ g } CaCO_3}{1 \text{ mole } CaCO_3} = 3 \text{ g of } CaCO_3 \text{ (1 SF)}$$

Selected Answers to Combining Ideas from Chapters 9 to 11

CI.19 a. $[H_2] = \dfrac{2.02 \text{ g } H_2}{10.0 \text{ L}} \times \dfrac{1 \text{ mole } H_2}{2.016 \text{ g } H_2} = 0.100 \text{ M}$

$[S_2] = \dfrac{10.3 \text{ g } S_2}{10.0 \text{ L}} \times \dfrac{1 \text{ mole } S_2}{64.14 \text{ g } S_2} = 0.0161 \text{ M}$

$[H_2S] = \dfrac{68.2 \text{ g } H_2S}{10.0 \text{ L}} \times \dfrac{1 \text{ mole } H_2S}{34.09 \text{ g } H_2S} = 0.200 \text{ M}$

$K_c = \dfrac{[H_2S]^2}{[H_2]^2[S_2]} = \dfrac{[0.200]^2}{[0.100]^2[0.0161]} = 248 \text{ (3 SFs)}$

b. If more H_2 (a reactant) is added, the equilibrium will shift in the direction of the products.

c. If the volume decreases from 10.0 L to 5.00 L (with no change in temperature), the equilibrium will shift in the direction of the products (fewer moles of gas).

d. $[H_2] = \dfrac{0.300 \text{ mole } H_2}{5.00 \text{ L}} = 0.0600 \text{ M}$

$[H_2S] = \dfrac{2.50 \text{ moles } H_2S}{5.00 \text{ L}} = 0.500 \text{ M}$

$K_c = \dfrac{[H_2S]^2}{[H_2]^2[S_2]} = 248$

Rearrange the expression to solve for $[S_2]$ and substitute in known values.

$[S_2] = \dfrac{[H_2S]^2}{[H_2]^2 K_c} = \dfrac{[0.500]^2}{[0.0600]^2(248)} = 0.280 \text{ M (3 SFs)}$

CI.21 a. $6HCl(aq) + 2M(s) \longrightarrow 3H_2(g) + 2MCl_3(aq)$

b. $34.8 \text{ mL solution} \times \dfrac{1 \text{ L solution}}{1000 \text{ mL solution}} \times \dfrac{0.520 \text{ mole HCl}}{1 \text{ L solution}} \times \dfrac{3 \text{ moles } H_2}{6 \text{ moles HCl}}$

$= 0.009 \, 05 \text{ mole of } H_2 \text{ (3 SFs)}$

$T = 24 \,°C + 273 = 297 \text{ K}$

Rearrange the ideal gas law $PV = nRT$ to solve for V,

$V = \dfrac{nRT}{P} = \dfrac{(0.009 \, 05 \text{ mole})\left(\dfrac{62.4 \text{ L} \cdot \text{mmHg}}{\text{mole} \cdot \text{K}}\right)(297 \text{ K})}{(720. \text{ mmHg})} \times \dfrac{1000 \text{ mL}}{1 \text{ L}} = 233 \text{ mL of } H_2 \text{ (3 SFs)}$

c. $34.8 \text{ mL solution} \times \dfrac{1 \text{ L solution}}{1000 \text{ mL solution}} \times \dfrac{0.520 \text{ mole HCl}}{1 \text{ L solution}} \times \dfrac{2 \text{ moles M}}{6 \text{ moles HCl}}$

$= 6.03 \times 10^{-3} \text{ mole of M (3 SFs)}$

d. $\dfrac{0.420 \text{ g M}}{6.03 \times 10^{-3} \text{ mole M}} = 69.7 \text{ g/mole of M (3 SFs); } \therefore \text{ metal M is gallium}$

e. $6HCl(aq) + 2Ga(s) \longrightarrow 3H_2(g) + 2GaCl_3(aq)$

CI.23 a. $\dfrac{8.57 \text{ g KOH}}{850. \text{ mL solution}} \times \dfrac{1 \text{ mole KOH}}{56.11 \text{ g KOH}} \times \dfrac{1000 \text{ mL solution}}{1 \text{ L solution}} = 0.180 \text{ M KOH solution (3 SFs)}$

b. $[H_3O^+] = \dfrac{K_w}{[OH^-]} = \dfrac{1.0 \times 10^{-14}}{[0.180]} = 5.56 \times 10^{-14} \text{ M}$

$pH = -\log[H_3O^+] = -\log[5.56 \times 10^{-14}] = 13.255 \text{ (3 SFs on the right of the decimal point)}$

c. $H_2SO_4(aq) + 2KOH(aq) \longrightarrow 2H_2O(l) + K_2SO_4(aq)$

d. $10.0 \text{ mL KOH solution} \times \dfrac{1 \text{ L KOH solution}}{1000 \text{ mL KOH solution}} \times \dfrac{0.180 \text{ mole KOH}}{1 \text{ L KOH solution}} \times \dfrac{1 \text{ mole } H_2SO_4}{2 \text{ moles KOH}}$

$\times \dfrac{1 \text{ L } H_2SO_4 \text{ solution}}{0.250 \text{ mole } H_2SO_4} \times \dfrac{1000 \text{ mL } H_2SO_4 \text{ solution}}{1 \text{ L } H_2SO_4 \text{ solution}}$

$= 3.60 \text{ mL of } H_2SO_4 \text{ solution (3 SFs)}$

CI.25 a. $[H_2CO_3] = \dfrac{0.0149 \text{ g } H_2CO_3}{200. \text{ mL solution}} \times \dfrac{1 \text{ mole } H_2CO_3}{62.03 \text{ g } H_2CO_3} \times \dfrac{1000 \text{ mL solution}}{1 \text{ L solution}}$

$= 1.20 \times 10^{-3} \text{ M (3 SFs)}$

b. $[HCO_3^-] = \dfrac{0.403 \text{ g } NaHCO_3}{200. \text{ mL solution}} \times \dfrac{1 \text{ mole } NaHCO_3}{84.01 \text{ g } NaHCO_3} \times \dfrac{1 \text{ mole } HCO_3^-}{1 \text{ mole } NaHCO_3} \times \dfrac{1000 \text{ mL solution}}{1 \text{ L solution}}$

$= 0.0240 \text{ M (3 SFs)}$

c. $[H_3O^+] = K_a \times \dfrac{[H_2CO_3]}{[HCO_3^-]} = 7.9 \times 10^{-7} \times \dfrac{[1.20 \times 10^{-3}]}{[0.0240]} = 4.0 \times 10^{-8} \text{ M (2 SFs)}$

d. $pH = -\log[4.0 \times 10^{-8}] = 7.40$ (2 SFs on the right of the decimal point)

e. $HCO_3^-(aq) + H_3O^+(aq) \longrightarrow H_2CO_3(aq) + H_2O(l)$

f. $H_2CO_3(aq) + OH^-(aq) \longrightarrow HCO_3^-(aq) + H_2O(l)$

Introduction to Organic Chemistry: Hydrocarbons

12

Three months after her house fire, Diane remains in the hospital's burn unit, where she continues to have treatment for burns, including removing damaged tissue, skin grafts, and preventing infection. For large burn areas, some of Diane's own skin cells are removed and grown for three weeks to give a new layer of skin. Dehydration and infection are decreased by covering the burn areas.

The arson investigators determine that gasoline is the primary accelerant used to start the fire at Diane's house along with kerosene and some diesel fuel. About 70% of kerosene consists of alkanes and cycloalkanes; the rest is aromatic hydrocarbons. Write a balanced chemical equation for the combustion of $C_{11}H_{24}$, which is a compound in kerosene.

Credit: Andrew Poertner/AP Images

LOOKING AHEAD

12.1 Organic Compounds
12.2 Alkanes
12.3 Alkanes with Substituents

12.4 Properties of Alkanes
12.5 Alkenes and Alkynes
12.6 Cis–Trans Isomers

12.7 Addition Reactions for Alkenes
12.8 Aromatic Compounds

 The Health icon indicates a question that is related to health and medicine.

12.1 Organic Compounds

Learning Goal: Identify the properties of organic or inorganic compounds.

> **REVIEW**
> Drawing Lewis Structures (6.6)
> Predicting Shape (6.8)

- Organic compounds are compounds of carbon and hydrogen that have covalent bonds, have low melting and boiling points, burn vigorously in air, are mostly nonpolar molecules, and are usually more soluble in nonpolar solvents than in water.
- An expanded structural formula shows all of the atoms and the bonds connected to each atom.
- A condensed structural formula depicts each carbon atom and its attached hydrogen atoms as a group.
- A molecular formula gives the total number of each kind of atom.

Ball-and-Stick Model	Expanded Structural Formula	Condensed Structural Formula	Molecular Formula
	$\begin{matrix} & H & H & H \\ & \vert & \vert & \vert \\ H- & C- & C- & C-H \\ & \vert & \vert & \vert \\ & H & H & H \end{matrix}$	$CH_3—CH_2—CH_3$	C_3H_8

♦ **Learning Exercise 12.1A**

Identify each of the following as characteristic of organic or inorganic compounds:

1. <u>Organic</u> have covalent bonds   2. <u>Organic</u> have low boiling points

3. <u>Organic</u> burn in air   4. <u>Inorganic</u> are soluble in water

5. <u>Inorganic</u> have high melting points   6. <u>Organic</u> are soluble in nonpolar solvents

7. <u>Inorganic</u> have ionic bonds   8. _____ form long carbon chains

9. _____ contain carbon and hydrogen   10. _____ do not burn in air

Answers
    1. organic   2. organic   3. organic   4. inorganic
    5. inorganic   6. organic   7. inorganic   8. organic
    9. organic   10. inorganic

♦ **Learning Exercise 12.1B**

1. What type of three-dimensional representation is shown here?

 a.

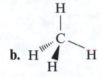 b.

2. Draw the expanded structural formula for methane.

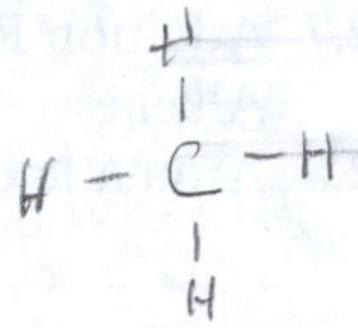

Vegetable oil, an organic compound, is not soluble in water.

Credit: Pearson Education/ Pearson Science

Answers   1. a. ball-and-stick model   b. wedge–dash model

           H
           |
2. 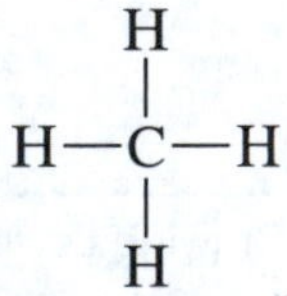   H—C—H
           |
           H

12.2   Alkanes

Learning Goal: Write the IUPAC names and draw the condensed structural and line-angle formulas for alkanes and cycloalkanes.

- Alkanes are hydrocarbons that have only single bonds, C—C and C—H.
- Each carbon in an alkane has four bonds arranged so that the bonded atoms are in the corners of a tetrahedron.
- A line-angle formula shows the carbon atoms as the corners or ends of a zigzag line.
- In an IUPAC (International Union of Pure and Applied Chemistry) name, the prefix indicates the number of carbon atoms, and the suffix describes the family of the compound. For example, in the name *propane*, the prefix *prop* indicates a chain of three carbon atoms and the ending *ane* indicates single bonds (alkane). The names of the first six alkanes follow:

Number of Carbon Atoms	Name	Condensed Structural Formula	Line-Angle Formula
1	Methane	CH_4	
2	Ethane	$CH_3—CH_3$	
3	Propane	$CH_3—CH_2—CH_3$	
4	Butane	$CH_3—CH_2—CH_2—CH_3$	
5	Pentane	$CH_3—CH_2—CH_2—CH_2—CH_3$	
6	Hexane	$CH_3—CH_2—CH_2—CH_2—CH_2—CH_3$	

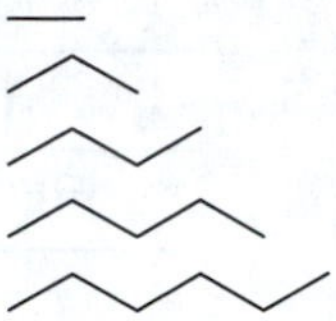

◆ **Learning Exercise 12.2A**

Indicate if each of the following is an expanded structural formula (E), a condensed structural formula (C), a line-angle formula (L), or a molecular formula (M):

CORE CHEMISTRY SKILL
Naming and Drawing Alkanes

1. _C_ $CH_3—CH_3$

2. _M_ C_5H_{12}

3. _C_ $CH_3—CH_2—CH_3$

4. _E_
$$H-\overset{\overset{\displaystyle H}{|}}{\underset{\underset{\displaystyle H}{|}}{C}}-\overset{\overset{\displaystyle H}{|}}{\underset{\underset{\displaystyle H}{|}}{C}}-\overset{\overset{\displaystyle H}{|}}{\underset{\underset{\displaystyle H}{|}}{C}}-H$$

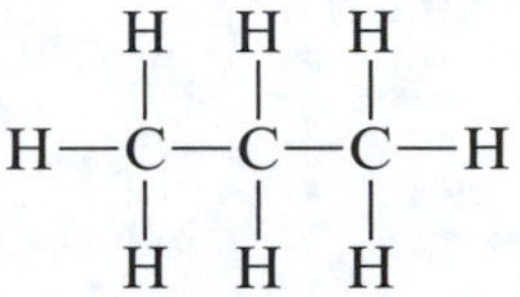

5. _L_

Answers   **1.** C   **2.** M   **3.** C   **4.** E   **5.** L

Drawing Formulas for an Alkane	
STEP 1	Draw the carbon chain.
STEP 2	Draw the expanded structural formula by adding the hydrogen atoms using single bonds to each of the carbon atoms.
STEP 3	Draw the condensed structural formula by combining the H atoms with each C atom.
STEP 4	Draw the line-angle formula as a zigzag line in which the ends and corners represent C atoms.

◆ **Learning Exercise 12.2B**

Draw the condensed structural formula for each of the following expanded structural or line-angle formulas:

1. 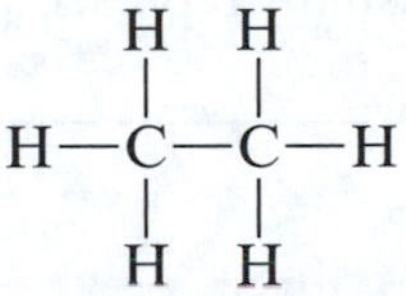 $H-\overset{\overset{\displaystyle H}{|}}{\underset{\underset{\displaystyle H}{|}}{C}}-\overset{\overset{\displaystyle H}{|}}{\underset{\underset{\displaystyle H}{|}}{C}}-H$ $CH_3 CH_3$

2. 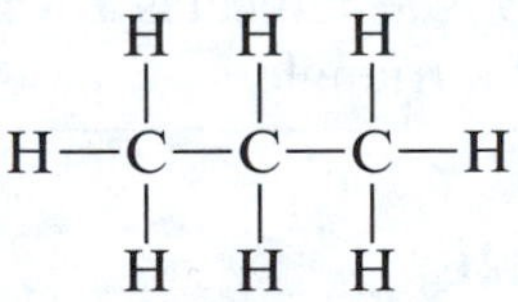 $H-\overset{\overset{\displaystyle H}{|}}{\underset{\underset{\displaystyle H}{|}}{C}}-\overset{\overset{\displaystyle H}{|}}{\underset{\underset{\displaystyle H}{|}}{C}}-\overset{\overset{\displaystyle H}{|}}{\underset{\underset{\displaystyle H}{|}}{C}}-H$ $CH_3 - CH_2 - CH_3$

3. 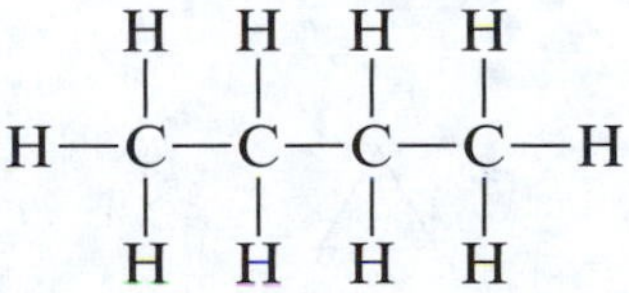 $H-\overset{\overset{\displaystyle H}{|}}{\underset{\underset{\displaystyle H}{|}}{C}}-\overset{\overset{\displaystyle H}{|}}{\underset{\underset{\displaystyle H}{|}}{C}}-\overset{\overset{\displaystyle H}{|}}{\underset{\underset{\displaystyle H}{|}}{C}}-\overset{\overset{\displaystyle H}{|}}{\underset{\underset{\displaystyle H}{|}}{C}}-H$ $CH_3-CH_2 CH_2-CH_3$

4.  $CH_3 - CH_2 - CH_2 - CH_2 - CH_3$

Answers
1. $CH_3—CH_3$
2. $CH_3—CH_2—CH_3$
3. $CH_3—CH_2—CH_2—CH_3$
4. $CH_3—CH_2—CH_2—CH_2—CH_3$

♦ **Learning Exercise 12.2C**

Draw the condensed structural formula and write the name for the continuous-chain alkane with each of the following molecular formulas:

1. C_2H_6 _____CH$_3$ –CH$_3$ ethane_____

2. C_3H_8 _____CH$_3$ –CH$_2$ –CH$_3$ propane_____

3. C_4H_{10} _____

4. C_5H_{12} _____

5. C_6H_{14} _____

Answers   1. $CH_3—CH_3$, ethane
2. $CH_3—CH_2—CH_3$, propane
3. $CH_3—CH_2—CH_2—CH_3$, butane
4. $CH_3—CH_2—CH_2—CH_2—CH_3$, pentane
5. $CH_3—CH_2—CH_2—CH_2—CH_2—CH_3$, hexane

♦ **Learning Exercise 12.2D**

Draw the line-angle formula for each of the following:

1. butane _____   2. pentane _____

3. hexane  _____   4. octane _____

5. cyclohexane _____

Answers   1.    2.    3.

4.    5.

Study Note

A cycloalkane is named by adding the prefix *cyclo* to the name of the corresponding alkane. For example, the name *cyclopropane* indicates a cyclic structure of three carbon atoms, usually represented by the geometric shape of a triangle.

♦ **Learning Exercise 12.2E**

Write the IUPAC name for each of the following cycloalkanes:

1. 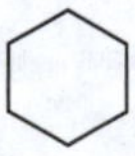 cyclohexane   2. cyclobutane   3.  cyclopentane   4.  cyclopropane

Answers   1. cyclohexane   2. cyclobutane   3. cyclopentane   4. cyclopropane

12.3 Alkanes with Substituents

Learning Goal: Write the IUPAC names for alkanes with substituents and draw their condensed structural and line-angle formulas.

- In the IUPAC system, each substituent is numbered and listed alphabetically in front of the name of the longest chain.
- Carbon groups that are substituents are named as alkyl groups or alkyl substituents. An alkyl group is named by replacing the *ane* of the alkane name with *yl*. For example, CH_3— is methyl (from CH_4 methane), and CH_3—CH_2— is ethyl (from CH_3—CH_3 ethane).
- In a haloalkane, a halogen atom, —F, —Cl, —Br, or —I replaces a hydrogen atom in an alkane.
- A halogen atom is named as a substituent (*fluoro, chloro, bromo, iodo*) attached to the alkane chain.

SAMPLE PROBLEM   Naming Alkanes with Substituents

Write the IUPAC name for the following compound:

$$CH_3-CH_2-\overset{\overset{\displaystyle CH_3}{|}}{CH}-CH_3$$

Solution:

STEP 1 Write the alkane name for the longest chain of carbon atoms. In this compound, the longest chain has four carbon atoms, which is named butane.

STEP 2 Number the carbon atoms from the end nearer a substituent. In this compound, the methyl group (substituent) is on carbon 2.

STEP 3 Give the location and name for each substituent (alphabetical order) as a prefix to the name of the main chain. The compound with a methyl group on carbon 2 of a four-carbon chain butane is named 2-methylbutane.

When there are two or more of the same substituent, a prefix (*di, tri, tetra*) is used in front of the name. Commas are used to separate the numbers for the location of the substituents.

$$CH_3-CH_2-\overset{\overset{\displaystyle CH_3}{|}}{CH}-\overset{\overset{\displaystyle CH_3}{|}}{CH}-CH_3$$

2,3-Dimethylpentane

◆ Learning Exercise 12.3A

Write the IUPAC name for each of the following compounds:

1. $CH_3-\overset{\overset{\displaystyle CH_3}{|}}{CH}-CH_3$

methyl propane

2. $CH_3-\overset{\overset{\displaystyle CH_3}{|}}{CH}-CH_2-\overset{\overset{\displaystyle CH_3}{|}}{CH}-CH_2-CH_3$

2,4-dimethylhexane

3.

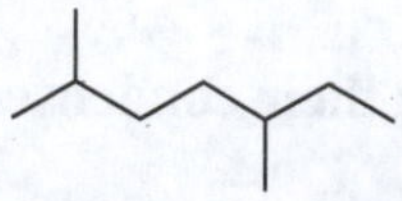

2,5-dimethylheptane

4. $CH_3-\overset{\overset{\displaystyle CH_3}{|}}{\underset{\underset{\displaystyle CH_3}{|}}{C}}-\overset{\overset{\displaystyle CH_3}{|}}{CH}-CH_3$

2,2,3-trimethylbutane

Answers   1. methylpropane   2. 2,4-dimethylhexane   3. 2,5-dimethylheptane
4. 2,2,3-trimethylbutane

♦ **Learning Exercise 12.3B**

Write the IUPAC name for each of the following:

1. CH_3-CH_2-Br

Bromo ethane

2. $CH_3-CH_2-\overset{\overset{\displaystyle Cl}{|}}{\underset{\underset{\displaystyle Cl}{|}}{C}}-CH_2-CH_3$

3,3-dichloropentane

3.

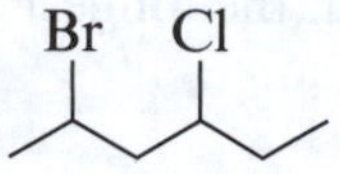

2-bromo,4-chlorohexane

4. $CH_3-CH_2-CH_2-\overset{\overset{\displaystyle F}{|}}{CH}-Cl$

1-chloro,1-florobutane

5.

bromocyclopropane

Answers   1. bromoethane   2. 3,3-dichloropentane
3. 2-bromo-4-chlorohexane   4. 1-chloro-1-fluorobutane
5. bromocyclopropane

Drawing Condensed Structural and Line-Angle Formulas from IUPAC Names	
STEP 1	Draw the main chain of carbon atoms.
STEP 2	Number the chain and place the substituents on the carbons indicated by the numbers.
STEP 3	For the condensed structural formula, add the correct number of hydrogen atoms to give four bonds to each C atom.

♦ **Learning Exercise 12.3C**

Draw the condensed structural formulas for **1**, **2**, and **3** and the line-angle formulas for **4**, **5**, and **6**.

1. 2-methylbutane

$CH_3-\overset{\overset{\displaystyle CH_3}{|}}{CH}-CH_2-CH_3$

2. 2-methylpentane

$CH_3-\overset{\underset{\underset{\displaystyle CH_3}{|}}{}}{CH}-CH_2-CH_3-CH_3$

3. 4-ethyl-2-methylhexane

4. 2,2,4-trimethylheptane

$$CH_3 - CH - CH_2 - CH - CH_2 - CH_3$$
$$\qquad\quad | \qquad\qquad\quad |$$
$$\qquad\quad CH_3 \qquad\qquad C H_2$$
$$\qquad\qquad\qquad\qquad\qquad\quad |$$
$$\qquad\qquad\qquad\qquad\qquad\quad CH_3$$

5. 3-ethylpentane

6. methylcyclohexane

Answers

1.
$$\overset{\displaystyle CH_3}{\underset{|}{}}$$
$$CH_3-CH-CH_2-CH_3$$

2.
$$\overset{\displaystyle CH_3}{\underset{|}{}}$$
$$CH_3-CH-CH_2-CH_2-CH_3$$

3.
$$\overset{\displaystyle CH_3}{\underset{|}{}}\qquad\overset{\displaystyle CH_2-CH_3}{\underset{|}{}}$$
$$CH_3-CH-CH_2-CH-CH_2-CH_3$$

4.

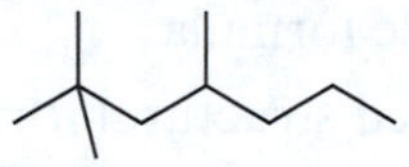

5.

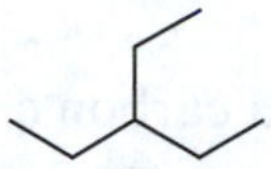

6.

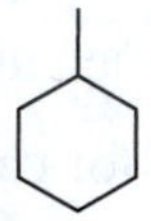

♦ **Learning Exercise 12.3D**

Draw the condensed structural formulas for **1, 2,** and **3** and the line-angle formulas for **4, 5,** and **6.**

1. chloroethane

$$CH_2 - CH_3$$
$$\ \ |$$
$$\ \ Cl$$

2. bromomethane

$$Br - CH_3$$

3. 3-bromo-1-chloropentane

$$CH_2 - CH_2 - CH - CH_2 - CH_3$$
$$\ |\qquad\qquad\quad |$$
$$\ Cl\qquad\qquad\ Br$$

4. 1,1-dichlorohexane

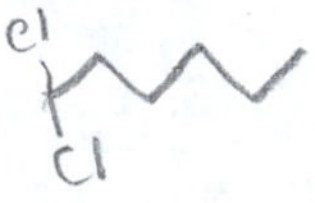

5. 2,2,3-trichlorobutane

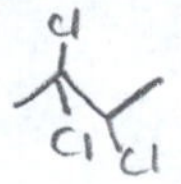

6. 2,4-dibromo-2,4-dichloropentane

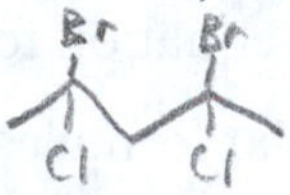

Answers

1. CH_3-CH_2-Cl

2. CH_3-Br

3.
$$\qquad\qquad\qquad\overset{\displaystyle Br}{\underset{|}{}}$$
$$Cl-CH_2-CH_2-CH-CH_2-CH_3$$

4.

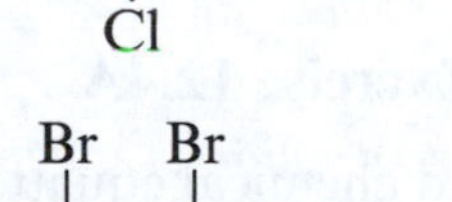

5.

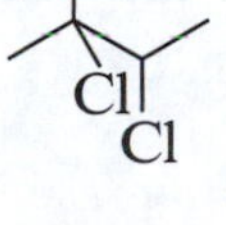

6.

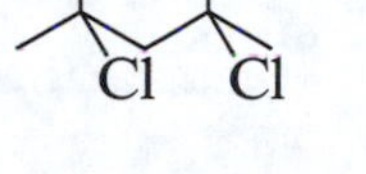

12.4 Properties of Alkanes

Learning Goal: Identify the properties of alkanes and write a balanced chemical equation for combustion.

- Alkanes are less dense than water, and mostly unreactive, except that they burn vigorously in air.
- Alkanes are found in natural gas, gasoline, and diesel fuels.
- Alkanes are nonpolar and insoluble in water, and they have low boiling points.
- In combustion, an alkane reacts with oxygen at a high temperature to produce carbon dioxide, water, and energy.

Key Terms for Sections 12.1 to 12.4

Match each of the following key terms with the correct description:

a. line-angle formula	**b.** isomers	**c.** alkane
d. condensed structural formula	**e.** combustion	**f.** cycloalkane
g. substituent		

1. _____ groups of atoms such as an alkyl group or a halogen bonded to a carbon chain

2. _____ a type of formula for carbon compounds that indicates the carbon atoms as corners or ends of a zigzag line

3. _____ compounds having the same molecular formula but a different arrangement of atoms

4. _____ a straight-chain hydrocarbon that contains only carbon–carbon and carbon–hydrogen single bonds

5. _____ an alkane that exists as a cyclic structure

6. _____ the chemical reaction of an alkane and oxygen that yields CO_2, H_2O, and energy

7. _____ the type of formula that shows the arrangement of the carbon atoms grouped with their attached H atoms

Answers   **1.** g   **2.** a   **3.** b   **4.** c   **5.** f   **6.** e   **7.** d

SAMPLE PROBLEM   Combustion of Alkanes

Write the balanced chemical equation for the combustion of methane.

Solution: Write the molecular formulas for the reactants: methane (CH_4) and oxygen (O_2).
Write the products CO_2 and H_2O and balance the equation.

$$CH_4 + O_2 \xrightarrow{\Delta} CO_2 + H_2O + energy$$

$$CH_4 + 2O_2 \xrightarrow{\Delta} CO_2 + 2H_2O + energy \quad \text{Balanced}$$

♦ **Learning Exercise 12.4A**

Write a balanced chemical equation for the complete combustion of each of the following:

1. propane   _____

2. hexane   _____

3. cyclobutane   _____

4. 2,3-dimethylpentane _____

Answers

1. $C_3H_8 + 5O_2 \xrightarrow{\Delta} 3CO_2 + 4H_2O + energy$

2. $2C_6H_{14} + 19O_2 \xrightarrow{\Delta} 12CO_2 + 14H_2O + energy$

3. $C_4H_8 + 6O_2 \xrightarrow{\Delta} 4CO_2 + 4H_2O + energy$

4. $C_7H_{16} + 11O_2 \xrightarrow{\Delta} 7CO_2 + 8H_2O + energy$

♦ **Learning Exercise 12.4B**

Indicate which property in each of the following pairs is more likely to be a property of hexane:

a. density is 0.66 g/mL or 1.66 g/mL _____

b. solid or liquid at room temperature _____

c. soluble or insoluble in water _____

d. flammable or nonflammable _____

e. melting point 95 °C or −95 °C _____

Answers
    a. 0.66 g/mL   **b.** liquid   **c.** insoluble
    d. flammable   **e.** −95 °C

12.5 Alkenes and Alkynes

Learning Goal: Write the IUPAC names and draw the condensed structural and line-angle formulas for alkenes and alkynes.

- Alkenes are unsaturated hydrocarbons that contain one or more carbon–carbon double bonds.
- In alkenes, each carbon in the double bond is connected to two other groups in addition to the other carbon of the double bond, all being in the same plane and having bond angles of 120°.
- Alkynes are unsaturated hydrocarbons that contain at least one carbon–carbon triple bond.
- The IUPAC names of alkenes are derived by changing the *ane* ending of the parent alkane to *ene*. For example, the IUPAC name of $H_2C{=}CH_2$ is ethene. It has a common name of ethylene. In alkenes, the longest carbon chain containing the double bond is numbered from the end nearer the double bond. In cycloalkenes with substituents, the double-bond carbons are given positions of 1 and 2, and the ring is numbered to give the next lower numbers to the substituents.

Ethene

CH₃—CH=CH₂   H₂C=CH—CH₂—CH₃   CH₃—CH=CH—CH₃
Propene (propylene)   1-Butene   2-Butene

- The IUPAC names of alkynes are derived by changing the *ane* ending of the parent alkane to *yne*. In alkynes, the longest carbon chain containing the triple bond is numbered from the end nearer the triple bond.

HC≡CH   CH₃—C≡CH   CH₃—CH₂—C≡CH
Ethyne   Propyne   1-Butyne

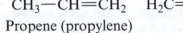

H—C≡C—H

Ethyne

Ball-and-stick models show the double bond of an alkene and the triple bond of an alkyne.

♦ **Learning Exercise 12.5A**

Classify each of the following as an alkane, alkene, cycloalkene, or alkyne:

1. _____ $CH_3-CH_2-CH_3$ 2. _____ (structure)

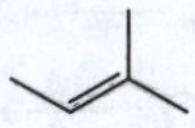

3. _____ $CH_3-C\equiv C-CH_3$ 4. _____ $CH_3-CH_2-CH=\overset{\displaystyle CH_3}{\underset{\displaystyle |}{C}}-CH_2-CH_3$

Answers   **1.** alkane   **2.** alkene   **3.** alkyne   **4.** alkene

Naming Alkenes and Alkynes	
STEP 1	Name the longest carbon chain that contains the double or triple bond.
STEP 2	Number the carbon chain starting from the end nearer the double or triple bond.
STEP 3	Give the location and name for each substituent (alphabetical order) as a prefix to the alkene or alkyne name.

♦ **Learning Exercise 12.5B**

Write the IUPAC name for each of the following alkenes:

1. $CH_3-CH=CH_2$

_____

2. $CH_3-\overset{\displaystyle CH_3}{\underset{\displaystyle |}{C}}=CH-CH_3$

_____

3.  (structure)

_____

4. $H_2C=CH-\overset{\displaystyle Cl}{\underset{\displaystyle |}{CH}}-CH_2-\overset{\displaystyle CH_3}{\underset{\displaystyle |}{CH}}-CH_3$

_____

5. $CH_3-CH=\overset{\displaystyle CH_2-CH_3}{\underset{\displaystyle |}{C}}-CH_2-CH_3$

_____

6.  (structure)

_____

Answers   **1.** propene   **2.** 2-methyl-2-butene   **3.** cyclohexene
             4. 3-chloro-5-methyl-1-hexene   **5.** 3-ethyl-2-pentene   **6.** 1-butene

♦ **Learning Exercise 12.5C**

Write the IUPAC name for each of the following alkynes:

1. $HC \equiv CH$

_____

2. $CH_3 - C \equiv CH$

_____

3. $CH_3 - CH_2 - C \equiv CH$

_____

4. $CH_3 - \overset{\displaystyle CH_3}{\overset{|}{CH}} - C \equiv C - CH_3$

_____

Answers   1. ethyne
               3. 1-butyne

2. propyne
4. 4-methyl-2-pentyne

♦ **Learning Exercise 12.5D**

Draw the condensed structural formulas for **1** and **2** and the line-angle formulas for **3** and **4**.

1. 2-pentyne

2. 2-chloro-2-butene

3. 3-bromo-2-methyl-2-pentene

4. 4-methylcyclohexene

Answers   1. $CH_3 - C \equiv C - CH_2 - CH_3$

2. $CH_3 - \overset{\displaystyle Cl}{\overset{|}{C}} = CH - CH_3$

3.

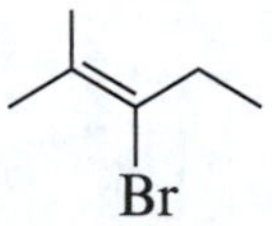

4.

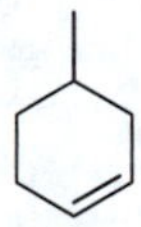

12.6   Cis–Trans Isomers

Learning Goal: Draw the condensed structural formulas and write the names for the cis–trans isomers of alkenes.

- Cis–trans isomers are possible for alkenes because there is no rotation around the rigid double bond.
- In the cis isomer, carbon groups or halogen atoms are attached on the same side of the double bond, whereas in the trans isomer, they are attached on opposite sides of the double bond.

♦ **Learning Exercise 12.6A**

Draw the condensed structural formulas for the cis and trans isomers of 1,2-dibromoethene.

cis-2-Butene

Answers   In the cis isomer, the bromine (halogen) atoms are attached on the same side of the double bond; in the trans isomer, they are on opposite sides.

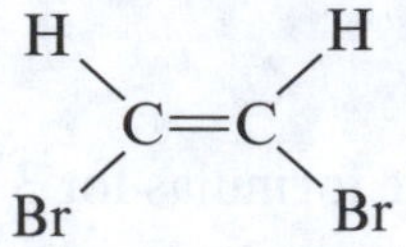

$$H \quad\quad\quad H$$
$$\diagdown \;C{=}C\; \diagup$$
$$Br \quad\quad\quad Br$$
cis-1,2-Dibromoethene

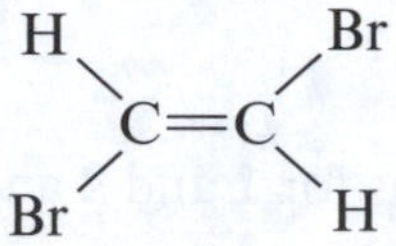

$$H \quad\quad\quad Br$$
$$\diagdown \;C{=}C\; \diagup$$
$$Br \quad\quad\quad H$$
trans-1,2-Dibromoethene

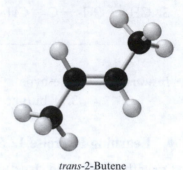

trans-2-Butene

Ball-and-stick models show the cis and trans isomers of 2-butene.

♦ **Learning Exercise 12.6B**

Name the following alkenes using the cis–trans prefix where isomers are possible:

1. 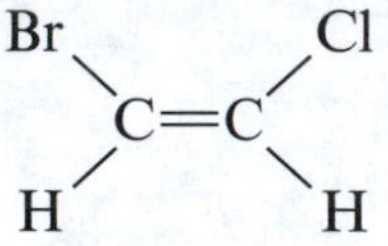
$$Br \quad\quad\quad Cl$$
$$\diagdown \;C{=}C\; \diagup$$
$$H \quad\quad\quad H$$

_____

2. 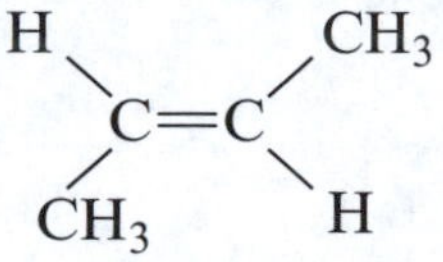
$$H \quad\quad\quad CH_3$$
$$\diagdown \;C{=}C\; \diagup$$
$$CH_3 \quad\quad\quad H$$

_____

3.
$$H \quad\quad\quad H$$
$$\diagdown \;C{=}C\; \diagup$$
$$CH_3 \quad\quad\quad H$$

_____

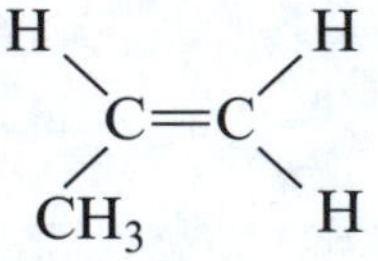

4. 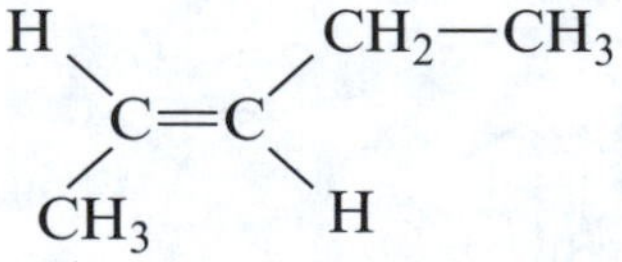
$$H \quad\quad\quad CH_2{-}CH_3$$
$$\diagdown \;C{=}C\; \diagup$$
$$CH_3 \quad\quad\quad H$$

_____

Answers   **1.** *cis*-1-bromo-2-chloroethene   **2.** *trans*-2-butene
            3. propene (not a cis–trans isomer)   **4.** *trans*-2-pentene

12.7   Addition Reactions for Alkenes

Learning Goal: Draw the condensed structural and line-angle formulas and write the names for the organic products of addition reactions of alkenes.

• The addition of small molecules to the double bond is a characteristic reaction of alkenes.
• Hydrogenation adds hydrogen atoms to the double bond of an alkene to yield an alkane.

$$H_2C{=}CH_2 + H_2 \xrightarrow{\;\;Pt\;\;} CH_3{-}CH_3$$

- Hydration, in the presence of a strong acid, adds water to a double bond. For an asymmetrical alkene, the H— from H—OH bonds to the carbon in the double bond that has the greater number of hydrogen atoms.

$$CH_3{-}CH{=}CH_2 + H_2O \xrightarrow{H^+} CH_3{-}\overset{\displaystyle OH}{\overset{|}{CH}}{-}CH_3$$

- Polymers are large molecules prepared from the bonding of many small alkene units called *monomers*.

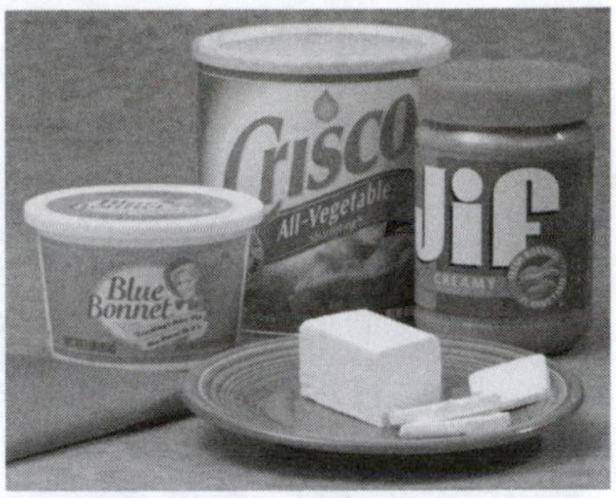

Hydrogenation is used to convert unsaturated fats in vegetable oils to saturated fats that make a more solid product.

Credit: Pearson Education Inc./Pearson Science

Summary of Addition Reactions

Name of Addition Reaction	Reactants	Conditions/ Catalysts	Product
Hydrogenation	Alkene + H_2	Pt, Ni, or Pd	Alkane
Hydration	Alkene + H_2O	H^+ (strong acid)	Alcohol
Polymerization	Alkene monomers	High temperature, high pressure	Polymer

♦ Learning Exercise 12.7A

CORE CHEMISTRY SKILL

Writing Equations for Hydrogenation, Hydration, and Polymerization

Draw the condensed structural or line-angle formulas for the products of the following reactions:

1. $CH_3{-}CH_2{-}CH{=}CH_2 + H_2 \xrightarrow{Pt}$
2. 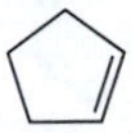 $+ H_2 \xrightarrow{Pt}$
3. $CH_3{-}CH{=}CH_2 + H_2 \xrightarrow{Pt}$
4. 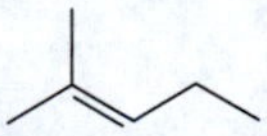 $+ H_2O \xrightarrow{H^+}$
5. $CH_3{-}CH{=}CH_2 + H_2O \xrightarrow{H^+}$
6. 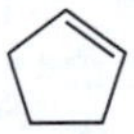 $+ H_2O \xrightarrow{H^+}$

Answers

1. $CH_3{-}CH_2{-}CH_2{-}CH_3$
2. 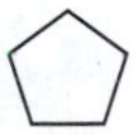
3. $CH_3{-}CH_2{-}CH_3$
4. 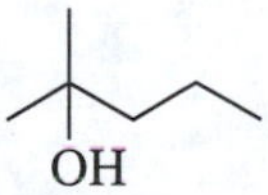
5. $CH_3{-}\overset{\displaystyle OH}{\overset{|}{CH}}{-}CH_3$
6. 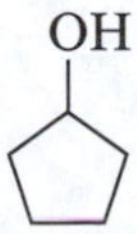

♦ **Learning Exercise 12.7B**

Draw the condensed structural formula for the starting monomer for each of the following polymers:

1. 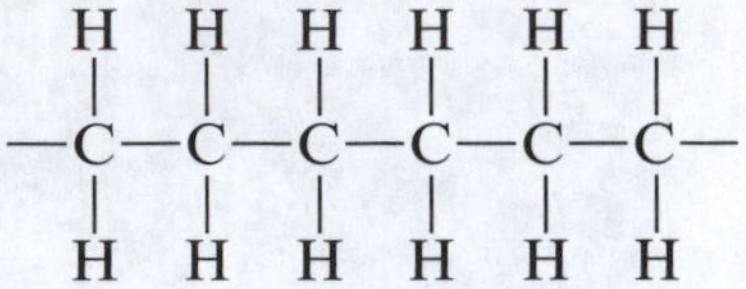
```
    H   H   H   H   H   H
    |   |   |   |   |   |
 —C —C —C —C —C —C—
    |   |   |   |   |   |
    H   H   H   H   H   H
```

2. 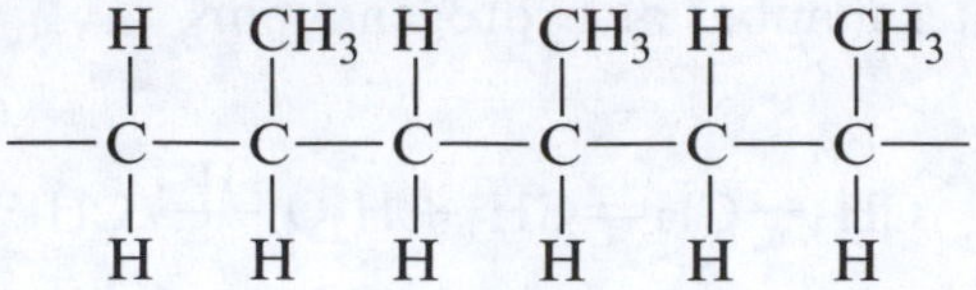
```
    H  CH₃  H  CH₃  H  CH₃
    |   |   |   |   |   |
 —C —C —C —C —C —C—
    |   |   |   |   |   |
    H   H   H   H   H   H
```

3. 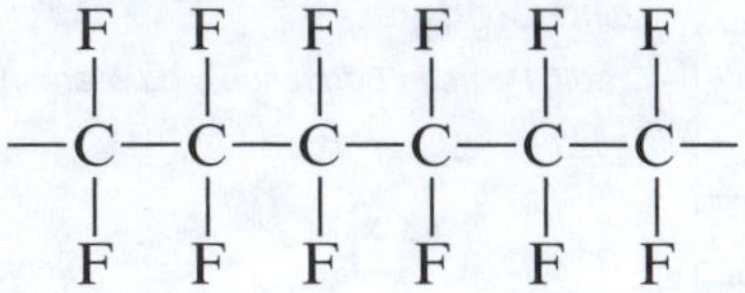
```
    F   F   F   F   F   F
    |   |   |   |   |   |
 —C —C —C —C —C —C—
    |   |   |   |   |   |
    F   F   F   F   F   F
```

Answers   **1.** $H_2C{=}CH_2$ **2.** 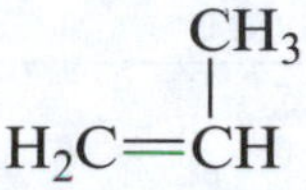 $H_2C{=}CH$ with CH_3 substituent   **3.** $F_2C{=}CF_2$

♦ **Learning Exercise 12.7C**

Draw the structure of the polymer formed from the addition of three monomers of 1,1-difluoroethene.

Answer
```
    F   H   F   H   F   H
    |   |   |   |   |   |
 —C —C —C —C —C —C—
    |   |   |   |   |   |
    F   H   F   H   F   H
```

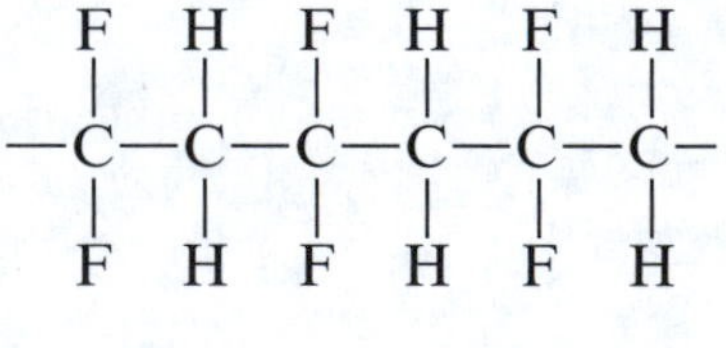

Synthetic polymers are used to replace diseased veins and arteries.

12.8   Aromatic Compounds

Learning Goal: Describe the bonding in benzene; name aromatic compounds and draw their line-angle formulas.

- Most aromatic compounds contain benzene, a cyclic structure containing six CH units. The structure of benzene can be represented as a hexagon with alternating single and double bonds or with a circle in the center.

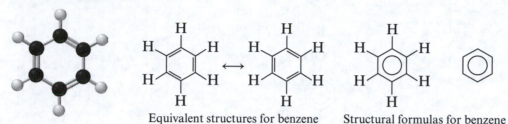

Equivalent structures for benzene   Structural formulas for benzene

- When benzene has only one substituent, the ring is not numbered. The names of many aromatic compounds use the parent name benzene, although many common names were retained as IUPAC names, such as toluene, aniline, and phenol.

  Toluene   Aniline   Phenol

- When there are two or more substituents, the benzene ring is numbered to give the lowest numbers to the substituents. The position of two substituents on the ring is often shown by *o*, *m*, and *p*, representing the prefixes *ortho* (1,2-), *meta* (1,3-), and *para* (1,4-).

Key Terms for Sections 12.5 to 12.8

Match each of the following key terms with the correct description:

    a. alkene   **b.** hydrogenation   **c.** alkyne
    d. hydration   **e.** benzene

1. _____ contains a cyclic structure with six CH units

2. _____ the addition of H_2 to a carbon–carbon double bond

3. _____ a compound that contains a carbon–carbon double bond

4. _____ the addition of H_2O to a carbon–carbon double bond

5. _____ a compound that contains a carbon–carbon triple bond

Answers   **1.** e   **2.** b   **3.** a   **4.** d   **5.** c

♦ **Learning Exercise 12.8**

Write the IUPAC name for each of the following:

1.

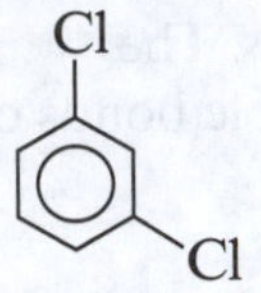

2.

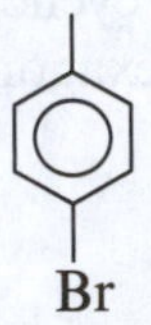

3.

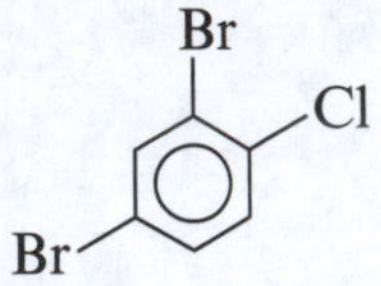

4.

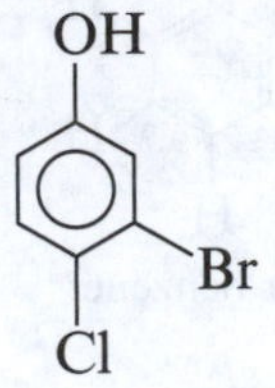

Answers   **1.** 1,3-dichlorobenzene

3. 1,3-dibromo-4-chlorobenzene

2. 4-bromotoluene

4. 3-bromo-4-chlorophenol

Checklist for Chapter 12

You are ready to take the Practice Test for Chapter 12. Be sure you have accomplished the following learning goals for this chapter. If not, review the Section listed at the end of the goal. Then apply your new skills and understanding to the Practice Test.

After studying Chapter 12, I can successfully:

_____ Identify properties as characteristic of organic or inorganic compounds. (12.1)

_____ Identify the number of bonds to carbon. (12.1)

_____ Describe the shape around carbon in organic compounds. (12.1)

_____ Draw the expanded structural, condensed structural, and line-angle formulas for an alkane. (12.2)

_____ Use the IUPAC system to write the names for alkanes and cycloalkanes. (12.2)

_____ Use the IUPAC system to write the names for alkanes with substituents. (12.3)

_____ Draw the condensed structural and line-angle formulas for alkanes with substituents. (12.3)

_____ Describe some physical properties of alkanes. (12.4)

_____ Write and balance equations for the combustion of alkanes. (12.4)

_____ Identify the structural features of alkenes and alkynes. (12.5)

_____ Name alkenes and alkynes using IUPAC rules and draw their condensed structural and line-angle formulas. (12.5)

_____ Identify alkenes that exist as cis–trans isomers; draw their condensed structural and line-angle formulas and write their names. (12.6)

_____ Draw the condensed structural and line-angle formulas for the products of the addition of hydrogen and water to alkenes. (12.7)

_____ Draw a condensed structural formula for a section of a polymer. (12.7)

_____ Write the names and draw the line-angle formulas for compounds that contain a benzene ring. (12.8)

Practice Test for Chapter 12

The chapter Sections to review are shown in parentheses at the end of each question.

For questions 1 through 8, indicate whether the following characteristics are typical of organic (O) or inorganic (I) compounds: (12.1)

1. _____ high melting points

2. _____ fewer compounds

3. _____ covalent bonds

4. _____ soluble in water

5. _____ ionic bonds

6. _____ flammable

7. _____ low boiling points

8. _____ soluble in nonpolar solvents

For questions 9 through 13, match the formula with the correct name: (12.2)

    A. methane   **B.** ethane   **C.** propane   **D.** pentane   **E.** heptane

9. _____ $CH_3-CH_2-CH_3$

10. _____ $CH_3-CH_2-CH_2-CH_2-CH_2-CH_2-CH_3$

11. _____ CH_4

12. _____ ⌒⌄⌒

13. _____ CH_3-CH_3

For questions 14 through 17, match the compounds with one of the following families: (12.2, 12.5)

    A. alkane   **B.** alkene   **C.** alkyne   **D.** cycloalkene

14. $CH_3-CH_2-CH=CH-CH_3$

15. 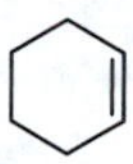 ⬡

16. $CH_3-CH_2-\overset{\overset{\displaystyle CH_3}{|}}{C}H-CH_2-CH_3$

17. $CH_3-CH_2-C\equiv CH$

For questions 18 through 21, match the formula with the correct name: (12.2, 12.3)

    A. butane   **B.** 1,1-dichlorobutane   **C.** 1,2-dichloropropane

    D. 3,5-dimethylhexane   **E.** 2,4-dimethylhexane   **F.** 1,2-dichlorobutane

18. $CH_3-CH_2-CH_2-CH_3$

19. $CH_3-\overset{\overset{\displaystyle CH_3}{|}}{C}H-CH_2-\overset{\overset{\displaystyle CH_3}{|}}{C}H-CH_2-CH_3$

20. 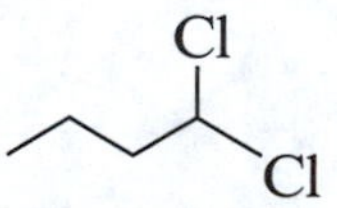 (structure with Cl, Cl)

21. 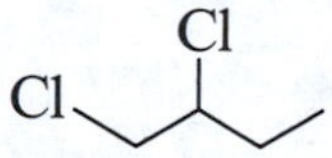 Cl (structure with Cl)

For questions 22 through 24, match the formula with the correct name: (12.2, 12.3)

    A. methylcyclopentane   **B.** cyclobutane
    C. cyclohexane   **D.** ethylcyclopentane

22. _____ ⬡   23. 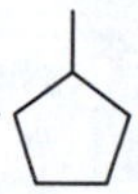 _____ ⬠   24. _____ ⬠

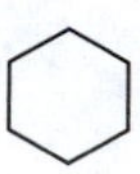

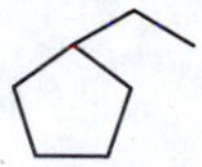

For questions 25 through 27, match the formula with the correct name: (12.3)

A. 2,4-dichloropentane
B. chlorocyclopentane
C. 1,2-dichloropentane
D. 4,5-dichloropentane

25. _____ $CH_3-CH_2-CH_2-\overset{\overset{\displaystyle Cl}{|}}{CH}-CH_2-Cl$

26. _____ (cyclopentane ring with Cl)

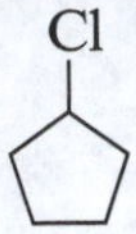

27. _____ $CH_3-\overset{\overset{\displaystyle Cl}{|}}{CH}-CH_2-\overset{\overset{\displaystyle Cl}{|}}{CH}-CH_3$

For questions 28 through 30, refer to the following compounds (A) and (B): (12.2, 12.5)

$H_2C=CH-CH_3$ $H_2\overset{\overset{\displaystyle CH_2}{\diagup\diagdown}}{C-CH_2}$

    (A)   (B)

28. Compound (A) is a(an)

  A. alkane   **B.** alkene   **C.** cycloalkane   **D.** alkyne   **E.** aromatic

29. Compound (B) is named

  A. propane   **B.** propylene   **C.** cyclobutane   **D.** cyclopropane   **E.** cyclopropene

30. Compound (A) is named

  A. propane   **B.** propene   **C.** 2-propene   **D.** propyne   **E.** 1-butene

31. The correctly balanced equation for the complete combustion of ethane is (12.4)

  A. $C_2H_6 + O_2 \xrightarrow{\Delta} 2CO + 3H_2O + \text{energy}$

  B. $C_2H_6 + O_2 \xrightarrow{\Delta} CO_2 + H_2O + \text{energy}$

  C. $C_2H_6 + 2O_2 \xrightarrow{\Delta} 2CO_2 + 3H_2O + \text{energy}$

  D. $2C_2H_6 + 7O_2 \xrightarrow{\Delta} 4CO_2 + 6H_2O + \text{energy}$

  E. $2C_2H_6 + 4O_2 \xrightarrow{\Delta} 4CO_2 + 6H_2O + \text{energy}$

For questions 32 through 35, match each formula with its name: (12.5)

A. cyclopentene
B. methylpropene
C. cyclohexene
D. ethene
E. 3-methylcyclopentene

32. $H_2C=CH_2$

33. $CH_3-\overset{\overset{\displaystyle CH_3}{|}}{C}=CH_2$

34. 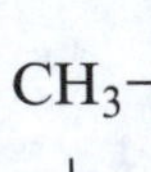 (cyclopentene ring)

35. 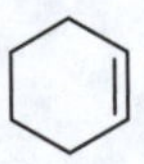 (cyclohexene ring)

36. What is the IUPAC name of $CH_3-CH_2-C\equiv CH$? (12.5)

    A. methylacetylene   **B.** propyne   **C.** propylene

    D. 4-butyne   **E.** 1-butyne

37. The cis isomer of 2-butene is (12.6)

    A. $H_2C=CH-CH_2-CH_3$ **B.** $CH_3-CH=CH-CH_3$

    C. $\begin{array}{c} CH_3 \quad\quad H \\ \diagdown\quad\diagup \\ C=C \\ \diagup\quad\diagdown \\ H\quad\quad CH_3 \end{array}$
    D. $\begin{array}{c} CH_3 \quad\quad CH_3 \\ \diagdown\quad\diagup \\ C=C \\ \diagup\quad\diagdown \\ H\quad\quad H \end{array}$

    E. $\begin{array}{c} CH_3\ \ CH_3 \\ |\quad\ \ | \\ CH=CH \end{array}$

38. The name of this compound is (12.6)

$$\begin{array}{c} Cl \quad\quad\quad H \\ \diagdown\quad\quad\diagup \\ C=C \\ \diagup\quad\quad\diagdown \\ H\quad\quad\quad Cl \end{array}$$

    A. dichloroethene   **B.** *cis*-1,2-dichloroethene

    C. *trans*-1,2-dichloroethene   **D.** *cis*-chloroethene

    E. *trans*-chloroethene

39. Hydrogenation of $CH_3-CH=CH_2$ gives (12.7)

    A. $3CO_2 + 6H_2$ **B.** $CH_3-CH_2-CH_3$

    C. $H_2C=CH-CH_3$ **D.** no reaction

    E. $CH_3-CH_2-CH_2-CH_3$

For problems 40 and 41, use the reaction $CH_3-CH=CH_2 + H_2O \xrightarrow{H^+}$ (12.7)

40. Choose the product of the reaction.

    A. $CH_3-CH_2-CH_2-OH$ **B.** no reaction   **C.** $CH_3-CH_2-CH_3$

    D. $CH_3-\overset{\displaystyle OH}{\underset{\displaystyle |}{CH}}-CH_2-OH$ **E.** $CH_3-\overset{\displaystyle OH}{\underset{\displaystyle |}{CH}}-CH_3$

41. The reaction is called

    A. hydrogenation   **B.** decomposition   **C.** reduction

    D. hydration   **E.** combustion

42. What is the structural formula for the monomer that forms the following polymer? (12.7)

$$\begin{array}{c} H \ \ Cl \ \ H \ \ Cl \ \ H \ \ Cl \\ |\quad |\quad |\quad |\quad |\quad | \\ -C-C-C-C-C-C- \\ |\quad |\quad |\quad |\quad |\quad | \\ H \ \ Cl \ \ H \ \ Cl \ \ H \ \ Cl \end{array}$$

    A. $H-\overset{\displaystyle Cl}{\underset{\displaystyle |}{C}}=CH_2$ **B.** $Cl-\overset{\displaystyle Cl}{\underset{\displaystyle |}{C}}=CCl_2$ **C.** $H_2C=\overset{\displaystyle Cl}{\underset{\displaystyle |}{C}}-Cl$

    D. $H_2C=CH_2$ **E.** $H\overset{\displaystyle Cl}{\underset{\displaystyle |}{C}}=CCl_2$

For questions 43 through 46, match the name of each of the following aromatic compounds with the correct structure: (12.8)

A.

B.

C.

D.

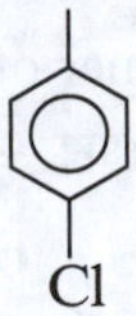

43. _____ chlorobenzene

44. _____ toluene

45. _____ 4-chlorotoluene

46. _____ 1,3-dimethylbenzene

Answers to the Practice Test

1. I	**2.** I	**3.** O	**4.** I	**5.** I
6. O	**7.** O	**8.** O	**9.** C	**10.** E
11. A	**12.** D	**13.** B	**14.** B	**15.** D
16. A	**17.** C	**18.** A	**19.** E	**20.** B
21. F	**22.** C	**23.** A	**24.** D	**25.** C
26. B	**27.** A	**28.** B	**29.** D	**30.** B
31. D	**32.** D	**33.** B	**34.** E	**35.** C
36. E	**37.** D	**38.** C	**39.** B	**40.** E
41. D	**42.** C	**43.** C	**44.** A	**45.** D
46. B				

Selected Answers and Solutions to Text Problems

12.1 Organic compounds always contain C and H and sometimes O, S, N, P, or a halogen atom. Inorganic compounds usually contain elements other than C and H.
  a. inorganic
  c. organic, expanded structural formula
  e. inorganic
  b. organic, condensed structural formula
  d. inorganic
  f. organic, molecular formula

12.3 a. Inorganic compounds are usually soluble in water.
  b. Organic compounds have lower boiling points than most inorganic compounds.
  c. Organic compounds contain carbon and hydrogen.
  d. Inorganic compounds contain ionic bonds.

12.5 a. Ethane boils at $-89\,°C$.
  b. Ethane burns vigorously in air.
  c. NaBr is a solid at $250\,°C$.
  d. NaBr dissolves in water.

12.7 a. Pentane has a chain of five carbon atoms.
  b. Ethane has a chain of two carbon atoms.
  c. Hexane has a chain of six carbon atoms.
  d. Cycloheptane has a ring of seven carbon atoms.

12.9 a. CH_4
  b. $CH_3{-}CH_3$
  c. $CH_3{-}CH_2{-}CH_2{-}CH_3$
  d. △

12.11 Two structures are isomers if they have the same molecular formula, but different arrangements of atoms.
  a. These condensed structural formulas represent the same molecule; the only difference is due to rotation of the structure. Each has a $CH_3{-}$ group attached to the middle carbon in a three-carbon chain.
  b. The molecular formula of both these condensed structural formulas is C_5H_{12}. However, they represent structural isomers because the C atoms are bonded in a different order; they have different arrangements. In the first, there is a $CH_3{-}$ group attached to carbon 2 of a four-carbon chain, and in the other, there is a five-carbon chain.
  c. The molecular formula of both these line-angle formulas is C_6H_{14}. However, they represent structural isomers because the C atoms are bonded in a different order; they have different arrangements. In the first, there is a $CH_3{-}$ group attached to carbon 3 of a five-carbon chain, and in the other, there is a $CH_3{-}$ group on carbon 2 and carbon 3 of a four-carbon chain.

12.13 a. dimethylpropane
  c. 4-*tert*-butyl-2,6-dimethylheptane
  e. 1-bromo-3-chlorocyclohexane
  b. 2,3-dimethylpentane
  d. methylcyclobutane

12.15 Draw the main chain with the number of carbon atoms in the ending of the name. For example, butane has a main chain of four carbon atoms, and hexane has a main chain of six carbon atoms. Attach substituents on the carbon atoms indicated. For example, in 3-methylpentane, a $CH_3{-}$ group is attached to carbon 3 of a five-carbon chain.

  a.
$$CH_3{-}CH_2{-}\overset{\overset{\textstyle CH_3}{|}}{\underset{\underset{\textstyle CH_3}{|}}{C}}{-}CH_2{-}CH_3$$

b.
$$CH_3-\overset{\overset{\displaystyle CH_3}{|}}{CH}-\overset{\overset{\displaystyle CH_3}{|}}{CH}-CH_2-\overset{\overset{\displaystyle CH_3}{|}}{CH}-CH_3$$

c.
$$CH_3-\overset{\overset{\displaystyle CH_3}{|}}{CH}-\overset{\overset{\displaystyle CH_2-CH_3}{|}}{CH}-CH_2-\overset{\overset{\displaystyle CH_3}{|}}{CH}-CH_2-CH_2-CH_3$$

d. $Br-CH_2-CH_2-Cl$

12.17 a. **b.**

c. **d.**

12.19 Alkanes with greater molar masses have higher boiling points. The boiling points of straight-chain alkanes are usually higher than their branched-chain isomers. Cycloalkanes have higher boiling points than their straight-chain counterparts.

  a. Heptane has a greater molar mass and so will have a higher boiling point than pentane.
  b. The cyclic alkane cyclopropane will have a higher boiling point than its straight-chain counterpart propane.
  c. The straight-chain alkane hexane will have a higher boiling point than its branched isomer 2-methylpentane.

12.21 a. $CH_3-CH_2-CH_2-CH_2-CH_2-CH_2-CH_3$
  b. Heptane is a liquid at room temperature since it has a boiling point of 98 °C.
  c. Heptane contains only nonpolar $C-H$ bonds, which makes it insoluble in water.
  d. Since the density of heptane (0.68 g/mL) is less than that of water, heptane will float on water.
  e. $C_7H_{16}(l) + 11O_2(g) \xrightarrow{\Delta} 7CO_2(g) + 8H_2O(g) + \text{energy}$

12.23 In combustion, a hydrocarbon reacts with oxygen to yield CO_2 and H_2O.

  a. $2C_2H_6(g) + 7O_2(g) \xrightarrow{\Delta} 4CO_2(g) + 6H_2O(g) + \text{energy}$
  b. $2C_3H_6(g) + 9O_2(g) \xrightarrow{\Delta} 6CO_2(g) + 6H_2O(g) + \text{energy}$
  c. $2C_8H_{18}(l) + 25O_2(g) \xrightarrow{\Delta} 16CO_2(g) + 18H_2O(g) + \text{energy}$

12.25 a. A condensed structural formula with a carbon–carbon double bond is an alkene.
  b. A condensed structural formula with a carbon–carbon triple bond is an alkyne.
  c. A line-angle formula with a carbon–carbon double bond is an alkene.
  d. A line-angle formula with a carbon–carbon double bond in a ring is a cycloalkene.

12.27 a. This alkene has a three-carbon chain with the double bond between carbon 1 and carbon 2, and a one-carbon methyl group attached to carbon 2. The IUPAC name is 2-methylpropene.
  b. This alkyne has a five-carbon chain with the triple bond between carbon 2 and carbon 3, and a bromine atom attached to carbon 4. The IUPAC name is 4-bromo-2-pentyne.
  c. This is a five-carbon cyclic structure with a double bond (between carbon 1 and 2 of the ring), and a two-carbon ethyl group attached to carbon 4 of the ring. The IUPAC name is 4-ethylcyclopentene.
  d. This alkene has a six-carbon chain with the double bond between carbon 2 and carbon 3, and a two-carbon ethyl group attached to carbon 4. The IUPAC name is 4-ethyl-2-hexene.

12.29 a. 1-Pentene is the five-carbon compound with a double bond between carbon 1 and carbon 2.
  $$H_2C=CH-CH_2-CH_2-CH_3$$

b. 2-Methyl-1-butene has a four-carbon chain with a double bond between carbon 1 and carbon 2 and a methyl group attached to carbon 2.

$$H_2C=\overset{\overset{\displaystyle CH_3}{|}}{C}-CH_2-CH_3$$

c. 3-Methylcyclohexene is a six-carbon cyclic compound with a double bond and a methyl group attached to carbon 3 of the ring.

d. 3-Chloro-1-butyne is a four-carbon compound with a triple bond between carbon 1 and carbon 2 and a chlorine atom attached to carbon 3.

$$HC\equiv C-\overset{\overset{\displaystyle Cl}{|}}{C}H-CH_3$$

12.31 a. *cis*-3-Heptene; this is a seven-carbon compound with a double bond between carbon 3 and carbon 4. Both alkyl groups are on the same side of the double bond; it is the cis isomer.
b. *trans*-3-Octene; this compound has eight carbons with a double bond between carbon 3 and carbon 4. The alkyl groups are on opposite sides of the double bond; it is the trans isomer.
c. Propene; this is a three-carbon compound with a double bond between carbon 1 and carbon 2. This compound does not have cis–trans isomers since there are 2 Hs attached to carbon 1 in the double bond.

12.33 a. *trans*-1-Bromo-2-chloroethene has a two-carbon chain with a double bond. The trans isomer has the attached groups on opposite sides of the double bond.

b. *cis*-2-Hexene has a six-carbon chain with a double bond between carbon 2 and carbon 3. The cis isomer has the alkyl groups on the same side of the double bond.

c. *cis*-4-Octene has an eight-carbon chain with a double bond between carbon 4 and carbon 5. The cis isomer has alkyl groups on the same side of the double bond.

12.35 a. Hydrogenation of an alkene gives the saturated compound, the alkane.

$$CH_3-CH_2-CH_2-CH_2-CH_3$$

b. In a hydration reaction, the H— and —OH from water add to the carbon atoms in the double bond to form an alcohol. The H— adds to the carbon with more hydrogens, and the —OH bonds to the carbon with fewer hydrogen atoms.

$$CH_3-\overset{\overset{\displaystyle CH_3}{|}}{\underset{\underset{\displaystyle OH}{|}}{C}}-CH_2-CH_3$$

c. Hydrogenation of an alkene gives the saturated compound, the alkane.

d. The H— and —OH from water add to the carbon atoms in the double bond to form an alcohol. The H— adds to the carbon with more hydrogens, and the —OH bonds to the carbon with fewer hydrogen atoms.

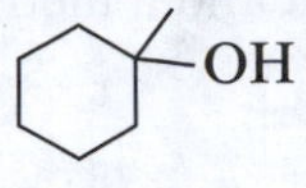

12.37 A polymer is a very large molecule composed of small units (monomers) that are repeated many times.

12.39

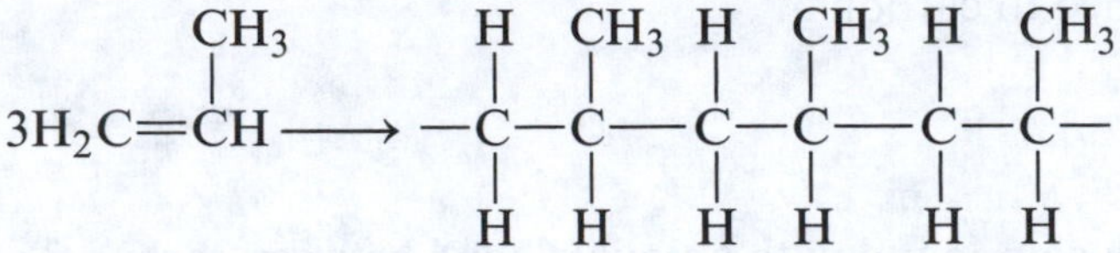

12.41

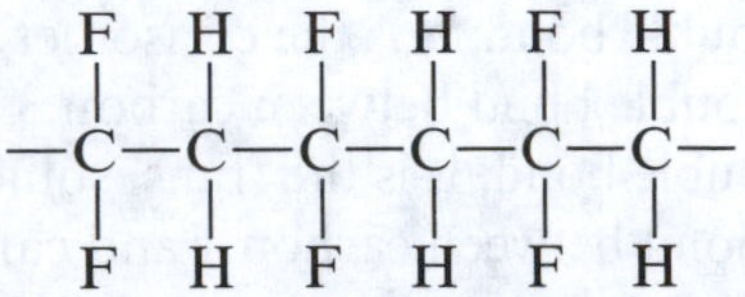

12.43 Aromatic compounds that contain a benzene ring with a single substituent are usually named as benzene derivatives. A benzene ring with a methyl substituent is named toluene. The methyl group is attached to carbon 1, and the ring is numbered to give the lower numbers to other substituents. If two groups are in the 1,2-position, this is *ortho-* (*o-*); 1,3- is *meta-* (*m-*); and 1,4- is *para-* (*p-*).

a. 2-chlorotoluene (*o*-chlorotoluene)   **b.** ethylbenzene
c. 1,3,5-trichlorobenzene   **d.** 1,3-dimethylbenzene (*m*-dimethylbenzene)
e. 3-bromo-5-chlorotoluene   **f.** isopropylbenzene

12.45 a.

b.

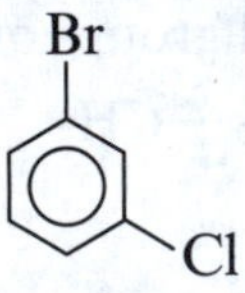

c.

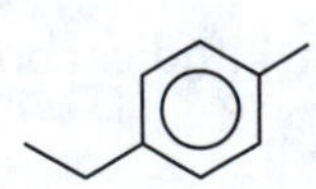

d.

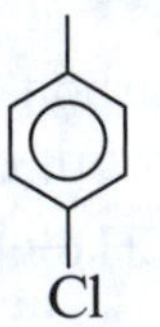

12.47

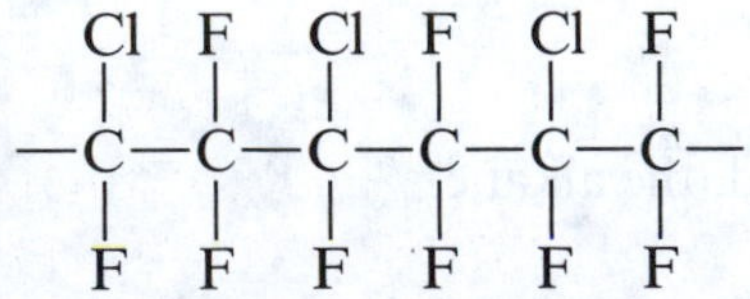

12.49 a. $2C_8H_{18}(l) + 25O_2(g) \xrightarrow{\Delta} 16CO_2(g) + 18H_2O(g) + \text{energy}$
   b. $C_5H_{12}(l) + 8O_2(g) \xrightarrow{\Delta} 5CO_2(g) + 6H_2O(g) + \text{energy}$
   c. $C_9H_{12}(l) + 12O_2(g) \xrightarrow{\Delta} 9CO_2(g) + 6H_2O(g) + \text{energy}$

12.51 a. Butane melts at $-138\ °C$.   **b.** Butane burns vigorously in air.
   c. Potassium chloride melts at 770 °C.   **d.** Potassium chloride contains ionic bonds.
   e. Butane is a gas at room temperature.

12.53 Two structures are isomers if they have the same molecular formula, but different arrangements of atoms.

    a. The molecular formula of both these line-angle formulas is C_6H_{12}. However, they represent structural isomers because the C atoms are bonded in a different order; they have different arrangements. In the first, there is a methyl group attached to a five-carbon ring, and in the other, there is a six-carbon ring.

    b. The molecular formula of the first line-angle structure is C_5H_{12}; the molecular formula of the second is C_6H_{14}. They are not structural isomers.

12.55 a.
$$CH_3-CH_2-CH_2-\underset{\underset{CH_3}{|}}{CH}-\underset{\underset{CH_3}{|}}{CH}-CH_3 \qquad \text{2,3-Dimethylhexane}$$

    b.
$$CH_3-\underset{\underset{CH_3}{|}}{CH}-\underset{\underset{CH_3}{|}}{CH}-\underset{\underset{CH_3}{|}}{CH}-CH_3 \qquad \text{2,3,4-Trimethylpentane}$$

12.57
$$-\overset{\overset{F}{|}}{\underset{\underset{F}{|}}{C}}-\overset{\overset{F}{|}}{\underset{\underset{F}{|}}{C}}-\overset{\overset{F}{|}}{\underset{\underset{F}{|}}{C}}-\overset{\overset{F}{|}}{\underset{\underset{F}{|}}{C}}-\overset{\overset{F}{|}}{\underset{\underset{F}{|}}{C}}-\overset{\overset{F}{|}}{\underset{\underset{F}{|}}{C}}-$$

12.59 a. Propane has three carbon atoms and two carbon–carbon single bonds.

    b. Cyclopropane has three carbon atoms and three carbon–carbon single bonds in a ring.

    c. Propene has three carbon atoms, one carbon–carbon single bond, and one carbon–carbon double bond.

    d. Propyne has three carbon atoms, one carbon–carbon single bond, and one carbon–carbon triple bond.

12.61 a. 2,2-dimethylbutane   **b.** chloroethane

    c. 2-bromo-4-ethylhexane   **d.** methylcyclopentane

12.63 a. [cyclopentene with Br on C1 and Br on C2]

    b. $CH_3-C\equiv C-CH_2-CH_2-CH_3$

    c.
$$\underset{H}{\overset{CH_3}{\diagdown}}C=C\underset{H}{\overset{CH_2-CH_2-CH_2-CH_3}{\diagup}}$$

12.65 a. This compound contains a five-carbon chain with a double bond between carbon 2 and carbon 3. The alkyl groups are on opposite sides of the double bond. The IUPAC name is *trans*-2-pentene.

    b. This alkene has a six-carbon chain with the double bond between carbon 1 and carbon 2. The substituents are a bromine atom attached to carbon 4 and a methyl group on carbon 5. The IUPAC name is 4-bromo-5-methyl-1-hexene.

    c. This is a five-carbon cyclic structure with a double bond and a methyl group attached to carbon 1 of the ring. The name is 1-methylcyclopentene.

12.67 a. These structures represent a pair of structural isomers. Both have the molecular formula C_5H_7Cl. In one isomer, the chlorine is attached to one of the carbons in the double bond; in the other isomer, the carbon to which the chlorine is attached is not part of the double bond.

    b. These structures are cis–trans isomers. In the cis isomer, the two methyl groups are on the same side of the double bond. In the trans isomer, the methyl groups are on opposite sides of the double bond.

12.69 a. *cis*-3-heptene
 b. *trans*-2-methyl-3-hexene
 c. *trans*-2-heptene

12.71 a.

$$\underset{H}{\overset{CH_3}{\diagup}} C = C \underset{H}{\overset{CH_2-CH_3}{\diagdown}}$$

cis-2-Pentene; both alkyl groups are on the same side of the double bond.

$$\underset{H}{\overset{CH_3}{\diagup}} C = C \underset{CH_2-CH_3}{\overset{H}{\diagdown}}$$

trans-2-Pentene; the alkyl groups are on opposite sides of the double bond.

 b.

$$\underset{H}{\overset{CH_3-CH_2}{\diagup}} C = C \underset{H}{\overset{CH_2-CH_3}{\diagdown}}$$

cis-3-Hexene; both alkyl groups are on the same side of the double bond.

$$\underset{H}{\overset{CH_3-CH_2}{\diagup}} C = C \underset{CH_2-CH_3}{\overset{H}{\diagdown}}$$

trans-3-Hexene; the alkyl groups are on opposite sides of the double bond.

12.73 a. $C_5H_{12}(l) + 8O_2(g) \xrightarrow{\Delta} 5CO_2(g) + 6H_2O(g) + \text{energy}$
 b. $C_4H_8(g) + 6O_2(g) \xrightarrow{\Delta} 4CO_2(g) + 4H_2O(g) + \text{energy}$
 c. $2C_7H_{14}(l) + 21O_2(g) \xrightarrow{\Delta} 14CO_2(g) + 14H_2O(g) + \text{energy}$

12.75 During hydrogenation, the multiple bonds are converted to single bonds and two H atoms are added for each bond converted.
 a. 3-methylpentane   **b.** methylcyclohexane   **c.** butane

12.77 a. Hydrogenation of a cycloalkene gives the saturated compound, the cycloalkane.

 b. The H— and —OH from water add to the carbon atoms in the double bond to form an alcohol. The H— adds to the carbon with more hydrogens (carbon 1), and the —OH bonds to the carbon with fewer hydrogen atoms (carbon 2).

OH

 c. The H— and —OH from water add to the carbon atoms in the double bond to form an alcohol. The H— adds to the carbon with more hydrogens (carbon 1), and the —OH bonds to the carbon with fewer hydrogen atoms (carbon 2).

$$CH_3-CH_2-\underset{\underset{CH_3}{|}}{\overset{\overset{OH}{|}}{C}}-CH_3$$

12.79 Styrene is $H_2C{=}CH$ and acrylonitrile is $H_2C{=}CH$. A section of the copolymer of styrene and acrylonitrile would be

12.81 Aromatic compounds that contain a benzene ring with a single substituent are usually named as benzene derivatives. A benzene ring with a methyl or amino substituent is named toluene or aniline, respectively. The methyl or amino group is attached to carbon 1, and the ring is numbered to give the lower numbers to other substituents.

    a. 4-methylaniline (*p*-methylaniline)

    b. 2-chlorotoluene (*o*-chlorotoluene)

    c. 1,2-difluorobenzene (*o*-difluorobenzene)

12.83 a. The prefix *p*- means that the two groups are in the 1 and 4 positions on the ring.

12.85 a. $C_5H_{12}(l) + 8O_2(g) \xrightarrow{\Delta} 5CO_2(g) + 6H_2O(g) +$ energy

    b. Molar mass of pentane (C_5H_{12}) = 5(12.01 g) + 12(1.008 g) = 72.15 g/mole (4 SFs)

    c. $1 \text{ gal} \times \dfrac{4 \text{ qt}}{1 \text{ gal}} \times \dfrac{946 \text{ mL}}{1 \text{ qt}} \times \dfrac{0.63 \text{ g}}{1 \text{ mL}} \times \dfrac{1 \text{ mole } C_5H_{12}}{72.15 \text{ g } C_5H_{12}} \times \dfrac{845 \text{ kcal}}{1 \text{ mole } C_5H_{12}}$

        = 2.8×10^4 kcal (2 SFs)

    d. $1 \text{ gal} \times \dfrac{4 \text{ qt}}{1 \text{ gal}} \times \dfrac{946 \text{ mL}}{1 \text{ qt}} \times \dfrac{0.63 \text{ g}}{1 \text{ mL}} \times \dfrac{1 \text{ mole } C_5H_{12}}{72.15 \text{ g } C_5H_{12}} \times \dfrac{5 \text{ moles } CO_2}{1 \text{ mole } C_5H_{12}} \times \dfrac{22.4 \text{ L } CO_2 \text{ (STP)}}{1 \text{ mole } CO_2}$

        = 3.7×10^3 L of CO_2 at STP (2 SFs)

12.87

$$\underset{\displaystyle CH_3-\underset{\displaystyle |}{\overset{\displaystyle |}{CH}}-\underset{\displaystyle |}{\overset{\displaystyle |}{CH}}-CH_3}{\overset{\displaystyle CH_3 \quad CH_3}{}} \qquad \underset{\displaystyle CH_3-\underset{\displaystyle |}{\overset{\displaystyle |}{C}}-CH_2-CH_3}{\overset{\displaystyle CH_3}{\underset{\displaystyle CH_3}{}}}$$

12.89 a. $CH_3-CH_2-CH_3$

    b. $C_3H_8(g) + 5O_2(g) \xrightarrow{\Delta} 3CO_2(g) + 4H_2O(g) +$ energy

    c. $12.0 \text{ L } C_3H_8 \times \dfrac{1 \text{ mole } C_3H_8}{22.4 \text{ L } C_3H_8 \text{ (STP)}} \times \dfrac{5 \text{ moles } O_2}{1 \text{ mole } C_3H_8} \times \dfrac{32.00 \text{ g } O_2}{1 \text{ mole } O_2} = 85.7$ g of O_2 (3 SFs)

    d. $12.0 \text{ L } C_3H_8 \times \dfrac{1 \text{ mole } C_3H_8}{22.4 \text{ L } C_3H_8 \text{ (STP)}} \times \dfrac{3 \text{ moles } CO_2}{1 \text{ mole } C_3H_8} \times \dfrac{44.01 \text{ g } CO_2}{1 \text{ mole } CO_2} = 70.7$ g of CO_2 (3 SFs)

12.91 Molar mass of bombykol ($C_{16}H_{30}O$) = 16(12.01 g) + 30(1.008 g) + 16.00 g = 238.4 g/mole (4 SFs)

$$50 \text{ ng} \times \dfrac{1 \text{ g}}{10^9 \text{ ng}} \times \dfrac{1 \text{ mole}}{238.4 \text{ g}} \times \dfrac{6.02 \times 10^{23} \text{ molecules}}{1 \text{ mole}} = 1 \times 10^{14} \text{ molecules of bombykol (1 SF)}$$

12.93 $H_2C{=}CH{-}CH_2{-}CH_2{-}CH_3$ 1-Pentene

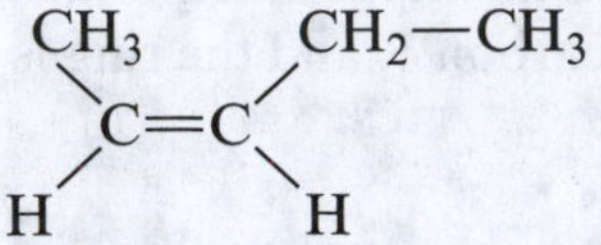

cis-2-Pentene

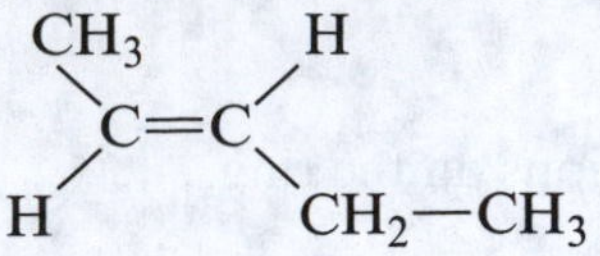

trans-2-Pentene

12.95 a.

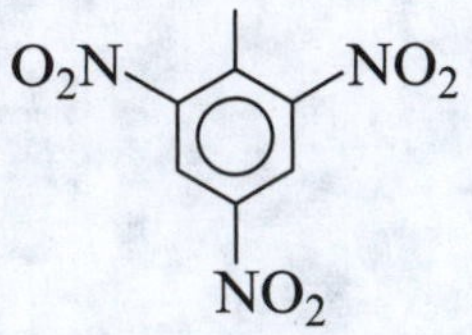

b.

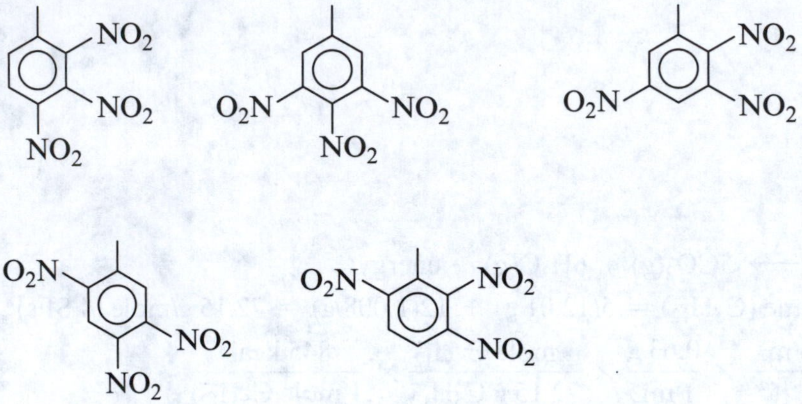

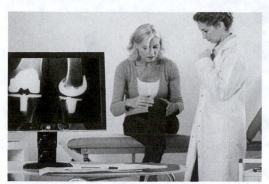

Credit: Picture Partners/Alamy

After a ski accident in which she injured her knee, Janet's doctor recommends that she have surgery to repair her torn ACL (anterior cruciate ligament). In preparation for the surgery, Liz, a nurse anesthetist, gives Janet 50 mg of propofol (Diprivan) to relax her. During surgery, propofol, a general anesthetic, is given by IV drip, which causes unconsciousness by blocking pain within the brain. After surgery, Janet begins physical therapy. Her therapist recommends that Janet also lose weight to improve her recovery.

Personnel involved in the administration of anesthesia include an anesthesiologist as well as a nurse anesthetist. An anesthetist may also specialize in the administration of anesthesia for a particular type of surgery such as obstetrical, orthopedic, cardiothoracic, or pediatric. What is the functional group in $CH_3—CH_2—O—CH_2—CH_3$, a compound previously used as a general anesthetic?

LOOKING AHEAD

 The Health icon indicates a question that is related to health and medicine.

13.1 Alcohols, Phenols, and Thiols

Learning Goal: Write the IUPAC and common names for alcohols, phenols, and thiols. Draw their condensed structural and line-angle formulas.

> **REVIEW**
> Naming and Drawing Alkanes (12.2)

- An alcohol contains the hydroxyl group (—OH) attached to a carbon chain.
- In the IUPAC system, alcohols are named by replacing the *e* of the alkane name with *ol*. The location of the —OH group is given by numbering the carbon chain.
- Simple alcohols are generally named by their common names with the alkyl name preceding the term *alcohol*. For example, $CH_3—OH$ is methyl alcohol, and $CH_3—CH_2—OH$ is ethyl alcohol.

> **CORE CHEMISTRY SKILL**
> Identifying Alcohols, Phenols, and Thiols

$$CH_3—OH \qquad CH_3—CH_2—OH \qquad CH_3—CH_2—CH_2—OH$$

Methanol   Ethanol   1-Propanol
(methyl alcohol)   (ethyl alcohol)   (propyl alcohol)

- When a hydroxyl group is attached to a benzene ring, the compound is a phenol.
- Thiols are similar to alcohols, except they have an —SH group in place of the —OH group.

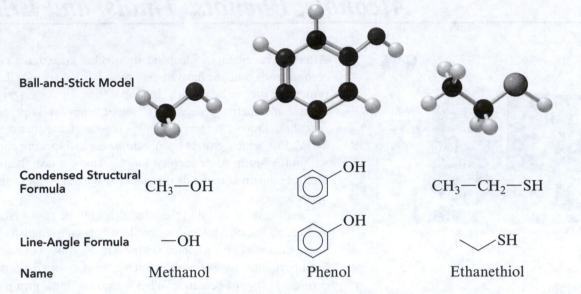

Ball-and-Stick Model			
Condensed Structural Formula	CH_3—OH	OH	CH_3—CH_2—SH
Line-Angle Formula	—OH	OH	SH
Name	Methanol	Phenol	Ethanethiol

Naming Alcohols and Phenols	
STEP 1	Name the longest carbon chain attached to the —OH group by replacing the *e* in the corresponding alkane name with *ol*. Name an aromatic alcohol as a *phenol*.
STEP 2	Number the chain starting at the end nearer to the —OH group.
STEP 3	Give the location and name for each substituent relative to the —OH group.

♦ **Learning Exercise 13.1A**

Write the correct IUPAC and common name (if any) for each of the following compounds:

1. CH_3—CH_2—OH

2. CH_3—CH_2—CH_2—OH

3. CH_3—$\overset{\overset{\displaystyle OH}{|}}{CH}$—$CH_2$—$CH_2$—$CH_3$

4.

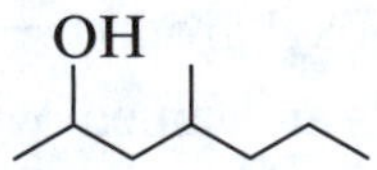

5.

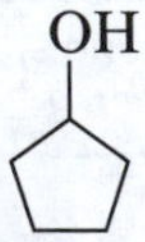

6.

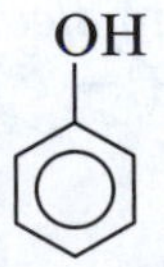

Answers
1. ethanol (ethyl alcohol)
2. 1-propanol (propyl alcohol)
3. 2-pentanol
4. 4-methyl-2-heptanol
5. cyclopentanol
6. phenol

◆ **Learning Exercise 13.1B**

Draw the condensed structural formulas for **1**, **2**, and **3** and the line-angle formulas for **4**, **5**, and **6**.

1. 2-butanol

2. 2-propanethiol

3. 2,4-dimethyl-1-pentanol

4. 2-bromoethanol

5. cyclohexanol

6. 2,3-dichlorophenol

Answers

1. $CH_3-\overset{\overset{\displaystyle OH}{|}}{CH}-CH_2-CH_3$

2. $CH_3-\overset{\overset{\displaystyle SH}{|}}{CH}-CH_3$

3. $CH_3-\overset{\overset{\displaystyle CH_3}{|}}{CH}-CH_2-\overset{\overset{\displaystyle CH_3}{|}}{CH}-CH_2-OH$

4.

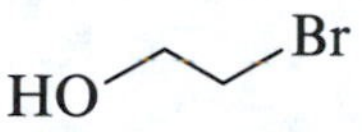

5.

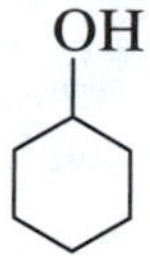

6.

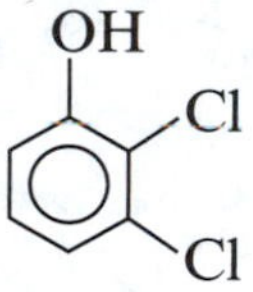

13.2 Ethers

Learning Goal: Write the IUPAC and common names for ethers; draw their condensed structural and line-angle formulas.

- In ethers, an oxygen atom is connected by two single bonds to alkyl or aromatic groups.
- In the IUPAC name, the smaller alkyl group and the oxygen are named as an *alkoxy group* attached to the longer alkane chain, which is numbered to give the location of the alkoxy group.
- In the common names of ethers, the alkyl groups are listed alphabetically, followed by the word *ether*.

SAMPLE PROBLEM   Writing IUPAC Names for Ethers

Write the IUPAC and common names for $CH_3-CH_2-CH_2-O-CH_3$.

Solution:

Analyze the Problem	Given	Need	Connect
	ether	IUPAC and common name	alkoxy group, carbon chain

STEP 1 **Write the alkane name of the longer carbon chain.**

$CH_3—CH_2—CH_2—O—CH_3$ propane

STEP 2 **Name the oxygen and smaller alkyl group as an alkoxy group.**

Methoxy

$CH_3—CH_2—CH_2—O—CH_3$ methoxypropane

STEP 3 **Number the longer carbon chain from the end nearer the alkoxy group and give its location.**

$CH_3—CH_2—CH_2—O—CH_3$ 1-methoxypropane
  3   2   1

The common name lists the alkyl groups alphabetically before the word ether.

Propyl   *Methyl*

$CH_3—CH_2—CH_2—O—CH_3$ methyl propyl ether

♦ **Learning Exercise 13.2A**

Write an IUPAC and common name for each of the following:

1. $CH_3—O—CH_3$

2.

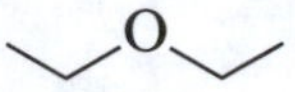

3. $CH_3—CH_2—CH_2—CH_2—O—CH_3$

4. $CH_3—O—CH_2—CH_3$

5.

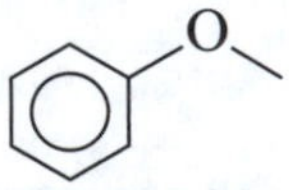

Answers
1. methoxymethane (dimethyl ether)
2. ethoxyethane (diethyl ether)
3. 1-methoxybutane (butyl methyl ether)
4. methoxyethane (ethyl methyl ether)
5. methoxybenzene (methyl phenyl ether)

♦ **Learning Exercise 13.2B**

Draw the condensed structural formulas for **1** and **2** and the line-angle formulas for **3** and **4**.

1. ethyl propyl ether

2. 2-methoxypropane

3. dipropyl ether

4. 3-ethoxypentane

Answers   1. $CH_3-CH_2-O-CH_2-CH_2-CH_3$ 2. $CH_3-\overset{\overset{\displaystyle O-CH_3}{|}}{CH}-CH_3$

3. 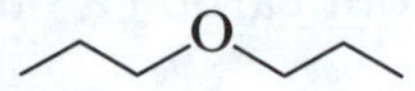 ~~~O~~~   4. 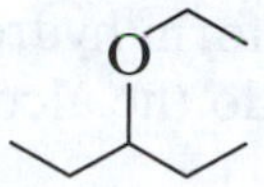 (structure)

♦ **Learning Exercise 13.2C**

Identify the functional groups of alcohol, phenol, thiol, and/or ether in each of the following compounds:

1. 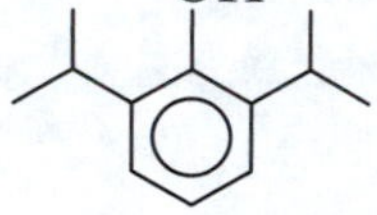  (structure with OH)

an anesthetic

_____

2. $CH_3-\overset{\overset{\displaystyle CH_3}{|}}{CH}-CH_2-CH_2-SH$

a component of skunk odor

_____

3. $CH_3-O-CH_2-CH_2-OH$

a solvent

_____

4. $CH_3-O-CH_2-CH_2-O-CH_3$

used in lithium batteries

_____

5. (structure)

the odor of fennel

_____

6. 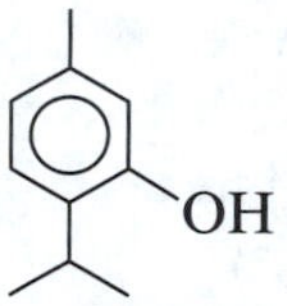 (structure with OH)

the odor of thyme

_____

Answers   1. phenol   2. thiol   3. alcohol, ether
4. ether   5. ether   6. phenol

13.3 Physical Properties of Alcohols, Phenols, and Ethers

Learning Goal: Describe the classification of alcohols; describe the boiling points and solubility of alcohols, phenols, and ethers.

• Alcohols are classified according to the number of alkyl groups attached to the carbon bonded to the —OH group.
• In a primary alcohol, there is one alkyl group attached to the carbon atom bonded to the —OH. In a secondary alcohol, there are two alkyl groups, and in a tertiary alcohol, there are three alkyl groups attached to the carbon atom with the —OH group.

CH_3-CH_2-OH $CH_3-\overset{\overset{\displaystyle CH_3}{|}}{CH}-OH$ $CH_3-\overset{\overset{\displaystyle CH_3}{|}}{\underset{\underset{\displaystyle CH_3}{|}}{C}}-OH$

Primary (1°)   Secondary (2°)   Tertiary (3°)

- The polar —OH group gives alcohols higher boiling points than alkanes and ethers of similar mass.
- Alcohols with one to three carbons are completely soluble in water because the —OH group forms hydrogen bonds with water molecules.
- Although ethers can form hydrogen bonds with water, they do not form as many hydrogen bonds with water as do the alcohols. Ethers containing one to four carbon atoms are slightly soluble in water.
- Phenol is slightly soluble in water and acts as a weak acid.

♦ **Learning Exercise 13.3A**

Classify each of the following alcohols as primary (1°), secondary (2°), or tertiary (3°):

1. _____ CH_3—CH_2—OH

2. _____

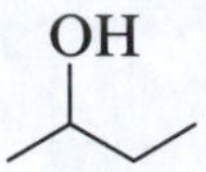

3. _____

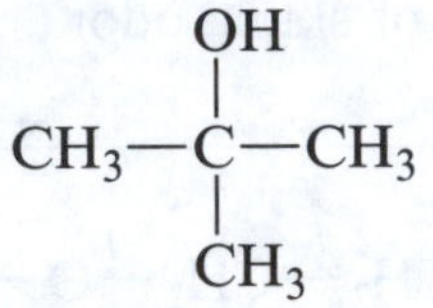

4. _____

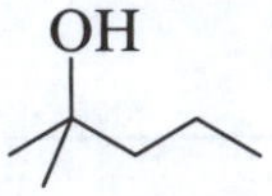

5. _____

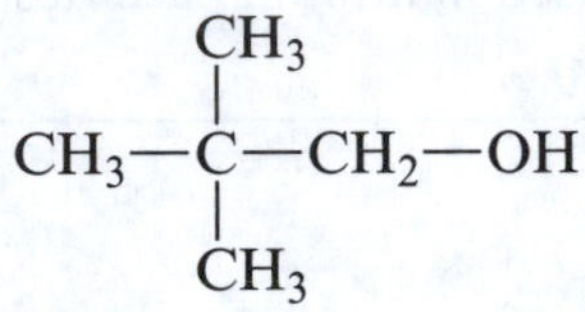

6. _____

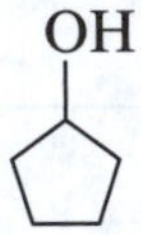

Answers   1. primary (1°)   2. secondary (2°)   3. tertiary (3°)
   4. tertiary (3°)   5. primary (1°)   6. secondary (2°)

♦ **Learning Exercise 13.3B**

Select the compound in each pair with the higher boiling point.

1. CH_3—CH_2—CH_3 or CH_3—CH_2—OH

2. CH_3—O—CH_2—CH_3 or CH_3—CH_2—CH_2—OH

3. CH_3—CH_2—CH_2—OH or CH_3—CH_2—CH_2—CH_3

Answers   1. CH_3—CH_2—OH   2. CH_3—CH_2—CH_2—OH   3. CH_3—CH_2—CH_2—OH

♦ **Learning Exercise 13.3C**

Select the compound in each pair that is more soluble in water.

1. CH_3—CH_3 or CH_3—CH_2—OH
2. CH_3—CH_2—CH_2—OH or CH_3—CH_2—CH_2—CH_2—CH_2—OH
3. 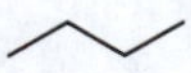 〜 or 〜OH
4. benzene or phenol
5. methoxymethane or methoxypentane

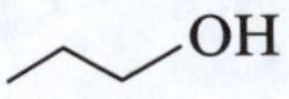

Answers   1. CH_3-CH_2-OH 2. $CH_3-CH_2-CH_2-OH$

3. ⌀⌀⌀OH   4. phenol

5. methoxymethane

13.4   Reactions of Alcohols and Thiols

Learning Goal: Write balanced chemical equations for the combustion, dehydration, and oxidation of alcohols and thiols.

- Alcohols undergo combustion with O_2 to form CO_2, H_2O, and energy.
- At high temperatures, an alcohol dehydrates in the presence of an acid to yield an alkene and water.

$$CH_3-CH_2-OH \xrightarrow{H^+, \text{ heat}} H_2C{=}CH_2 + H_2O$$

- Using an oxidizing agent [O], primary alcohols oxidize to aldehydes, which usually oxidize further to carboxylic acids. Secondary alcohols oxidize to ketones, but tertiary alcohols do not oxidize.

$$CH_3-CH_2-OH \xrightarrow{[O]} CH_3-\overset{\overset{\displaystyle O}{\|}}{C}-H + H_2O$$

1° Alcohol   Aldehyde

$$CH_3-\overset{\overset{\displaystyle OH}{|}}{CH}-CH_3 \xrightarrow{[O]} CH_3-\overset{\overset{\displaystyle O}{\|}}{C}-CH_3 + H_2O$$

2° Alcohol   Ketone

- Thiols undergo oxidation and lose hydrogen from the $-SH$ group to form disulfides, $-S-S-$.

Key Terms for Sections 13.1 to 13.4

Match each of the following key terms with the correct description:

  a. primary alcohol   **b.** thiol   **c.** dehydration

  d. phenol   **e.** tertiary alcohol   **f.** ether

1. _____ has one alkyl group bonded to the carbon with the $-OH$ group

2. _____ contains an $-SH$ group

3. _____ contains an oxygen atom attached to two carbon groups that are alkyl or aromatic

4. _____ has three alkyl groups bonded to the carbon with the $-OH$ group

5. _____ contains a benzene ring bonded to a hydroxyl group

6. _____ a reaction that removes water from an alcohol in the presence of an acid to form an alkene

Answers   **1.** a   **2.** b   **3.** f   **4.** e   **5.** d   **6.** c

---

Study Note

When a secondary alcohol is dehydrated, two products may be formed. The major product is the alkene formed by removing the hydrogen atom from the carbon atom that has fewer hydrogen atoms.

♦ **Learning Exercise 13.4A**

Write the balanced chemical equation for the complete combustion of the following:

1. ethanol

2. 1-hexanol

Answers   1. $C_2H_6O + 3O_2 \longrightarrow 2CO_2 + 3H_2O + energy$
            2. $C_6H_{14}O + 9O_2 \longrightarrow 6CO_2 + 7H_2O + energy$

♦ **Learning Exercise 13.4B**

CORE CHEMISTRY SKILL
Writing Equations for the Dehydration of Alcohols

Draw the condensed structural or line-angle formulas for the products expected from dehydration of each of the following alcohols:

1. $CH_3-CH_2-CH_2-CH_2-OH \xrightarrow{\text{H}^+,\text{ heat}}$

2. 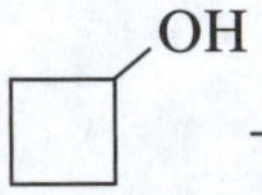 $\xrightarrow{\text{H}^+,\text{ heat}}$

3. 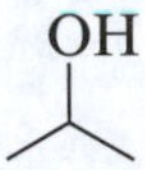 $\xrightarrow{\text{H}^+,\text{ heat}}$

4. $CH_3-CH_2-\overset{\overset{\displaystyle OH}{|}}{C}H-CH_2-CH_3 \xrightarrow{\text{H}^+,\text{ heat}}$

Answers   1. $CH_3-CH_2-CH=CH_2$ 2.

        3. 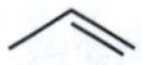   4. $CH_3-CH_2-CH=CH-CH_3$

♦ **Learning Exercise 13.4C**

Draw the condensed structural or line-angle formula for the major product expected from dehydration of each of the following:

1. 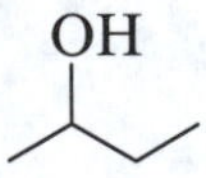 $\xrightarrow{\text{H}^+,\text{ heat}}$

2. $CH_3-CH_2-\overset{\overset{\displaystyle OH}{|}}{C}H-\overset{\overset{\displaystyle CH_3}{|}}{C}H-CH_3 \xrightarrow{\text{H}^+,\text{ heat}}$

3. 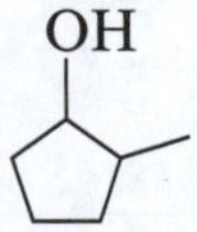 $\xrightarrow{\text{H}^+,\text{ heat}}$

Answers   1.    2. $CH_3-CH_2-CH=\overset{\overset{\displaystyle CH_3}{|}}{C}-CH_3$ 3.

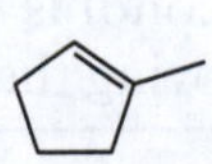

♦ **Learning Exercise 13.4D**

Draw the condensed structural or line-angle formula for the aldehyde
or ketone expected in the oxidation of each of the following:

1. $CH_3-CH_2-CH_2-CH_2-OH \xrightarrow{[O]}$

2. 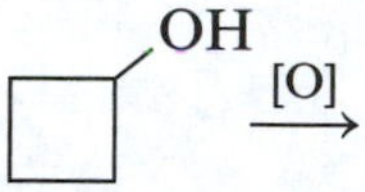 $\xrightarrow{[O]}$

3. 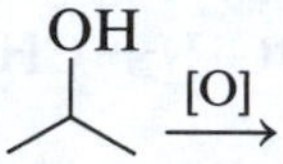 $\xrightarrow{[O]}$

4. $CH_3-CH_2-\overset{\overset{\displaystyle OH}{|}}{CH}-CH_2-CH_3 \xrightarrow{[O]}$

Answers

1. $CH_3-CH_2-CH_2-\overset{\overset{\displaystyle O}{\|}}{C}-H$

2.

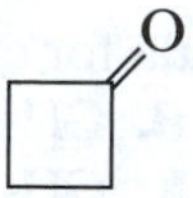

3.

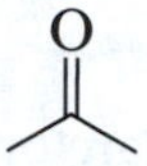

4. $CH_3-CH_2-\overset{\overset{\displaystyle O}{\|}}{C}-CH_2-CH_3$

Checklist for Chapter 13

You are ready to take the Practice Test for Chapter 13. Be sure you have accomplished the following
learning goals for this chapter. If not, review the Section listed at the end of the goal. Then apply your
new skills and understanding to the Practice Test.

After studying Chapter 13, I can successfully:

_____ Write the IUPAC or common name for an alcohol, phenol, or thiol and draw the condensed
structural or line-angle formula. (13.1)

_____ Write the IUPAC or common name for an ether and draw the condensed structural or line-angle
formula. (13.2)

_____ Classify an alcohol as primary, secondary, or tertiary. (13.3)

_____ Describe the solubility of alcohols, phenols, and ethers in water and compare their boiling points. (13.3)

_____ Write balanced equations for the combustion of alcohols. (13.4)

_____ Draw the products of alcohols that undergo dehydration and oxidation. (13.4)

Practice Test for Chapter 13

The chapter Sections to review are shown in parentheses at the end of each question.

For questions 1 through 4, identify each of the following condensed structural formulas as: (13.1)

    A. alcohol   **B.** ether   **C.** thiol

1. $CH_3-CH_2-\overset{\overset{\displaystyle OH}{|}}{CH}-CH_3$

2. $CH_3-\overset{\overset{\displaystyle SH}{|}}{\underset{\underset{\displaystyle CH_3}{|}}{C}}-CH_3$

3. CH$_3$—CH$_2$—CH with CH$_2$—OH above and CH$_3$ below

4. CH$_3$—CH$_2$—O—CH$_3$

For questions 5 through 9, match the names of the following compounds with their correct structure: (13.1, 13.2)

    A. 1-propanol   **B.** cyclobutanol   **C.** 2-propanol
    D. ethyl methyl ether   **E.** diethyl ether

5. 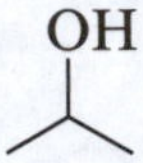 (structure: isopropyl with OH)

6. CH$_3$—CH$_2$—CH$_2$—OH

7. CH$_3$—O—CH$_2$—CH$_3$

8. 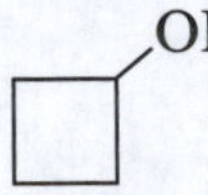 (cyclobutane with OH)

9. CH$_3$—CH$_2$—O—CH$_2$—CH$_3$

10. Phenol is (13.1)
    A. the alcohol of benzene   **B.** the aldehyde of benzene   **C.** the phenyl group of benzene
    D. the ketone of benzene   **E.** another name for cyclohexanol

11. The condensed structural formula for ethanethiol is (13.1)
    A. CH$_3$—SH   **B.** CH$_3$—CH$_2$—OH   **C.** CH$_3$—CH$_2$—SH
    D. CH$_3$—CH$_2$—S—CH$_3$ **E.** CH$_3$—S—OH

12. Why are short-chain alcohols water-soluble? (13.3)
    A. They are nonpolar.   **B.** They can hydrogen bond.   **C.** They are organic.
    D. They are bases.   **E.** They are acids.

For questions 13 through 17, classify each alcohol as (13.3)

    A. primary (1°)   **B.** secondary (2°)   **C.** tertiary (3°)

13. CH$_3$—CH$_2$—CH$_2$—OH

14. 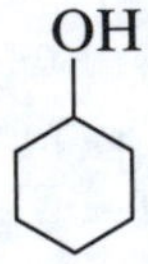 (cyclohexane with OH)

15. (cyclopentane with CH$_3$ and OH)

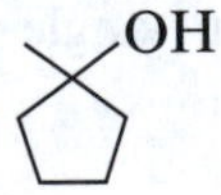

16. CH$_3$—C—CH$_2$—CH$_2$—CH$_3$ with OH above and CH$_3$ below

17. CH$_3$—CH—CH$_2$—CH$_2$—CH$_2$—CH$_3$ with OH above

18. When (structure with OH) undergoes oxidation, the product is (13.4)
    A. an alkane   **B.** an aldehyde   **C.** a ketone
    D. an ether   **E.** a phenol

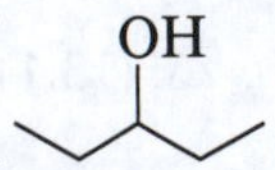

$$\overset{\displaystyle O}{\underset{\displaystyle \|}{}}$$

19. The compound $CH_3 - \overset{O}{\overset{\|}{C}} - CH_3$ is formed by the oxidation of (13.4)

    A. 2-propanol   **B.** propane   **C.** 1-propanol

    D. dimethyl ether   **E.** ethyl methyl ketone

20. The dehydration of cyclohexanol gives (13.4)

    A. cyclohexane   **B.** cyclohexene   **C.** cyclohexyne

    D. benzene   **E.** phenol

Complete questions 21 through 23 by indicating the product (A to E) formed in each of the following reactions: (13.4)

    A. primary alcohol   **B.** secondary alcohol   **C.** aldehyde   **D.** ketone   **E.** alkene

21. _____ oxidation of a primary alcohol

22. _____ oxidation of a secondary alcohol

23. _____ dehydration of 1-propanol

24. The major product from the dehydration of 2-methylcyclobutanol is: (13.4)

    A. cyclobutene   **B.** 1-methylcyclobutene   **C.** 2-methylcyclobutene

    D. 3-methylcyclobutene   **E.** 1-methylcyclobutane

Answers to the Practice Test

1. A	**2.** C	**3.** A	**4.** B	**5.** C
6. A	**7.** D	**8.** B	**9.** E	**10.** A
11. C	**12.** B	**13.** A	**14.** B	**15.** C
16. C	**17.** B	**18.** C	**19.** A	**20.** B
21. C	**22.** D	**23.** E	**24.** B	

Selected Answers and Solutions to Text Problems

13.1 a. This compound has a two-carbon chain. The final *e* from ethane is dropped, and *ol* is added to indicate an alcohol. The IUPAC name is ethanol. The common name is ethyl alcohol.

b. This compound has a four-carbon chain with a hydroxyl group attached to carbon 2. The IUPAC name is 2-butanol. The common name is *sec*-butyl alcohol.

c. This is the line-angle formula of a five-carbon chain with a thiol group attached to carbon 2. The IUPAC name is 2-pentanethiol.

d. This compound has a six-carbon ring with a hydroxyl group attached to carbon 1, and a methyl group attached to carbon 4 of the ring. The IUPAC name is 4-methylcyclohexanol.

e. This compound is a phenol because the —OH group is attached to a benzene ring. For a phenol, the carbon atom attached to the —OH group is understood to be carbon 1; no number is needed to give the location of the —OH group. This compound also has a fluorine atom attached to carbon 3 of the ring; the IUPAC name is 3-fluorophenol. The common name is *m*-fluorophenol.

13.3 a. 1-Propanol has a three-carbon chain with a hydroxyl group attached to carbon 1.

$$CH_3—CH_2—CH_2—OH$$

b. 3-Pentanethiol has a five-carbon chain with a thiol group attached to carbon 3.

$$CH_3—CH_2—\overset{\overset{\displaystyle SH}{|}}{CH}—CH_2—CH_3$$

c. 2-Methyl-2-butanol has a four-carbon chain with a methyl group and a hydroxyl group attached to carbon 2.

$$CH_3—\overset{\overset{\displaystyle OH}{|}}{\underset{\underset{\displaystyle CH_3}{|}}{C}}—CH_2—CH_3$$

d. *p*-Chlorophenol has an —OH group attached to carbon 1 of a benzene ring, and a chlorine attached to carbon 4 of the ring (*para*- position).

e. 2-Bromo-5-chlorophenol has an —OH group attached to carbon 1 of a benzene ring, a bromine atom attached to carbon 2, and a chlorine attached to carbon 5 of the ring.

13.5 a. The IUPAC name of the ether with a one-carbon alkyl group and a two-carbon alkyl group attached to an oxygen atom is methoxyethane. The common name is ethyl methyl ether.

b. The IUPAC name of the ether with a one-carbon alkyl group and a six-carbon aromatic group attached to an oxygen atom is methoxybenzene. The common name is methyl phenyl ether.

c. The IUPAC name of the ether with a two-carbon alkyl group and a four-carbon cycloalkyl group attached to an oxygen atom is ethoxycyclobutane. The common name is cyclobutyl ethyl ether.

d. The IUPAC name of the ether with a one-carbon alkyl group and a three-carbon alkyl group attached to an oxygen atom is 1-methoxypropane. The common name is methyl propyl ether.

13.7 a. Ethyl propyl ether has a two-carbon group and a three-carbon group attached to oxygen by single bonds.

$$CH_3-CH_2-O-CH_2-CH_2-CH_3$$

b. Cyclopropyl ethyl ether has a two-carbon group and a three-carbon cycloalkyl group attached to oxygen by single bonds.

c. Methoxycyclopentane has a one-carbon group and a five-carbon cycloalkyl group attached to oxygen by single bonds.

d. 1-Ethoxy-2-methylbutane has a four-carbon chain with a methyl group attached to carbon 2, and an ethoxy group attached to carbon 1.

$$\begin{array}{c} \qquad\qquad\quad CH_3 \\ \qquad\qquad\quad | \\ CH_3-CH_2-O-CH_2-CH-CH_2-CH_3 \end{array}$$

e. 2,3-Dimethoxypentane has a five-carbon chain with two methoxy groups attached, one to carbon 2 and the other to carbon 3.

$$\begin{array}{c} \quad O-CH_3 \\ \quad | \\ CH_3-CH-CH-CH_2-CH_3 \\ \qquad\quad | \\ \qquad\quad O-CH_3 \end{array}$$

13.9 The carbon bonded to the hydroxyl group ($-$OH) is attached to one alkyl group in a primary (1°) alcohol, except for methanol; to two alkyl groups in a secondary (2°) alcohol; and to three alkyl groups in a tertiary (3°) alcohol.

a. primary (1°) alcohol **b.** primary (1°) alcohol
c. tertiary (3°) alcohol **d.** secondary (2°) alcohol

13.11 a. Methanol molecules can form hydrogen bonds and will have a higher boiling point than the alkane ethane.

b. 1-Butanol molecules can form hydrogen bonds and will have a higher boiling point than diethyl ether, which cannot form hydrogen bonds with other ether molecules.

c. 1-Butanol molecules can form hydrogen bonds and will have a higher boiling point than the alkane pentane.

13.13 a. Soluble; ethanol with a short carbon chain is soluble because the hydroxyl group forms hydrogen bonds with water.

b. Slightly soluble; ethers with up to four carbon atoms are slightly soluble in water because they can form a few hydrogen bonds with water.

c. Insoluble; an alcohol with a carbon chain of five or more carbon atoms is not soluble in water.

13.15 a. Methanol has a polar $-$OH group that can form hydrogen bonds with water, but the alkane ethane does not.

b. 2-Propanol is more soluble in water than 1-butanol because 2-propanol has a shorter carbon chain.

c. 1-Propanol is more soluble because it can form more hydrogen bonds with water than the ether can.

13.17 a. CH_3-OH Molecular formula of methanol $= CH_4O$

$$2CH_4O + 3O_2 \xrightarrow{\Delta} 2CO_2 + 4H_2O + \text{energy}$$

b. $\underset{\underset{CH_3-\overset{\displaystyle OH}{\overset{|}{CH}}-CH_2-CH_3}{}}{}$ Molecular formula of 2-butanol = $C_4H_{10}O$

$C_4H_{10}O + 6O_2 \overset{\Delta}{\longrightarrow} 4CO_2 + 5H_2O + energy$

13.19 Dehydration is the removal of an H— and an —OH from adjacent carbon atoms of an alcohol to form a water molecule and the corresponding alkene.

a. $CH_3-\overset{\displaystyle CH_3}{\overset{|}{CH}}-CH=CH_2$

b.

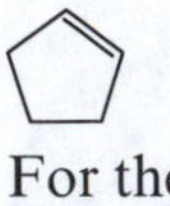

c. For the dehydration of this asymmetrical alcohol, the —OH is removed from carbon 1 of the ring and H— from carbon 2, which has the smaller number of H atoms.

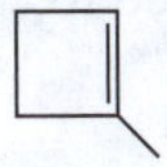

d. For the dehydration of this asymmetrical alcohol, the —OH is removed from carbon 2 and H— from carbon 3, which has the smaller number of H atoms.

$CH_3-\overset{\displaystyle CH_3}{\overset{|}{CH}}-CH_2-CH=CH-CH_3$

13.21 Alcohols can produce alkenes by the loss of water (dehydration).

a. CH_3-CH_2-OH b. 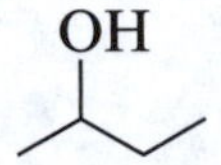 (structure with OH)   c. 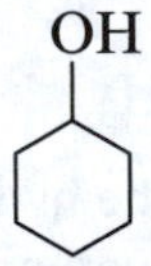 (cyclohexanol structure with OH)

13.23 A primary alcohol oxidizes to an aldehyde, and a secondary alcohol oxidizes to a ketone.

a. $CH_3-CH_2-CH_2-CH_2-\overset{\displaystyle O}{\overset{||}{C}}-H$

b. $CH_3-CH_2-\overset{\displaystyle O}{\overset{||}{C}}-CH_3$

c. 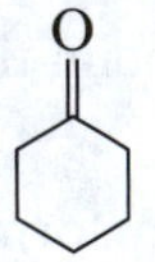 (cyclohexanone structure)

d. 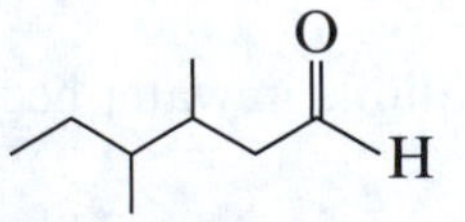 (branched aldehyde structure)

e. $CH_3-\overset{\displaystyle CH_3}{\overset{|}{CH}}-CH_2-\overset{\displaystyle O}{\overset{||}{C}}-H$

13.25 a. An aldehyde is the product of the oxidation of a primary alcohol.

CH_3-OH

b. A ketone is the product of the oxidation of a secondary alcohol.

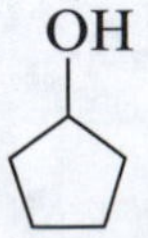

(cyclopentanol structure with OH)

c. A ketone is the product of the oxidation of a secondary alcohol.

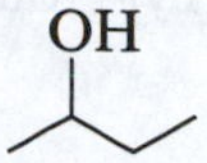

d. An aldehyde is the product of the oxidation of a primary alcohol.

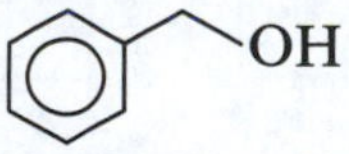

e. A ketone is the product of the oxidation of a secondary alcohol.

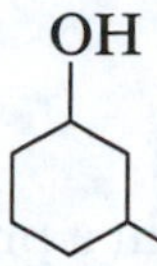

13.27 a. Capsaicin contains alkene, phenol, and ether functional groups.
b. 2-Propene-1-thiol contains alkene and thiol functional groups.
c. Resveratrol contains phenol and alkene functional groups.

13.29 a. alcohol   **b.** ether
c. thiol   **d.** alcohol

13.31 a. 2-chloro-4-methylcyclohexanol
b. methoxybenzene (methyl phenyl ether)
c. 2-propanethiol
d. 2,4-dimethyl-2-pentanol

13.33 a. ether   **b.** thiol
c. alcohol   **d.** phenol

13.35 a. 1-methoxypropane (methyl propyl ether)
b. 2-butanethiol
c. 4-bromo-2-pentanol
d. 3-methylphenol (*m*-methylphenol, *m*-cresol)

13.37 a.

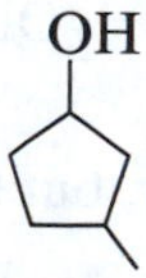

b.

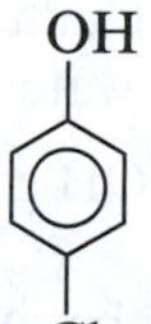

c. $CH_3-CH-CH-CH_2-CH_3$ with CH_3 and OH substituents

d.

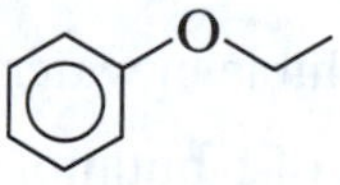

13.39 a. $CH_3-CH_2-CH-CH_2-CH_3$ with SH substituent

b. $CH_3-CH-CH_2-CH_2-CH_3$ with $O-CH_3$ substituent

c.

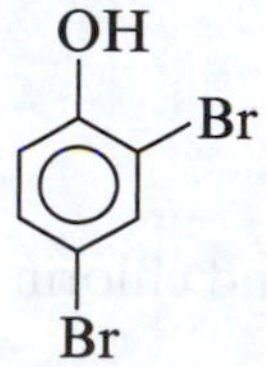

d. $CH_3-C-CH-CH_3$ with CH_3, CH_3, and OH substituents

13.41 a. 2-methyl-2-propanol
c. 3-bromophenol
e. methoxycyclopentane

b. 2,4-dimethyl-2-pentanol
d. 2-ethoxypentane

13.43 Draw the carbon chain first, and place the —OH group on the carbon atoms in the chain to give different structural formulas. Shorten the chain by one carbon, and attach a methyl group and —OH group to give different compounds.

$$CH_3-CH_2-CH_2-CH_2-OH \qquad CH_3-\overset{\overset{\displaystyle OH}{|}}{CH}-CH_2-CH_3$$

$$CH_3-\overset{\overset{\displaystyle CH_3}{|}}{CH}-CH_2-OH \qquad CH_3-\overset{\overset{\displaystyle OH}{|}}{\underset{\underset{\displaystyle CH_3}{|}}{C}}-CH_3$$

13.45 The carbon bonded to the hydroxyl group (—OH) is attached to one alkyl group in a primary (1°) alcohol, except for methanol; to two alkyl groups in a secondary (2°) alcohol; and to three alkyl groups in a tertiary (3°) alcohol.
  a. secondary (2°) alcohol
  b. primary (1°) alcohol
  c. primary (1°) alcohol
  d. secondary (2°) alcohol
  e. primary (1°) alcohol
  f. tertiary (3°) alcohol

13.47 a. 1-Propanol molecules can form hydrogen bonds and will have a higher boiling point than the alkane butane.
  b. 1-Propanol molecules can form hydrogen bonds and will have a higher boiling point than ethyl methyl ether, which cannot form hydrogen bonds with other ether molecules.
  c. 1-Butanol has a greater molar mass than ethanol and will have a higher boiling point.

13.49 a. Soluble; 2-propanol with a short carbon chain is soluble because the hydroxyl group forms hydrogen bonds with water.
  b. Insoluble; the long carbon chain in dipropyl ether diminishes the effect of hydrogen bonding of water to the —O— group.
  c. Insoluble; the long carbon chain in 1-hexanol diminishes the effect of the polar —OH group on hydrogen bonding with water.

13.51 a. 1-Propanol has a polar —OH group that can form hydrogen bonds with water, but the alkane butane does not.
  b. 1-Propanol is more soluble because it can form more hydrogen bonds with water than the ether can.
  c. Ethanol is more soluble in water than 1-hexanol because it has a shorter carbon chain.

13.53 a. Molecular formula of 1-butanol = $C_4H_{10}O$

$$C_4H_{10}O + 6O_2 \overset{\Delta}{\longrightarrow} 4CO_2 + 5H_2O + \text{energy}$$

  b. Molecular formula of 1-propanol = C_3H_8O

$$2C_3H_8O + 9O_2 \overset{\Delta}{\longrightarrow} 6CO_2 + 8H_2O + \text{energy}$$

  c. Molecular formula of cyclopentanol = $C_5H_{10}O$

$$C_5H_{10}O + 7O_2 \overset{\Delta}{\longrightarrow} 5CO_2 + 5H_2O + \text{energy}$$

  13.55 a. 2,5-Dichlorophenol is a benzene ring with a hydroxyl group on carbon 1, and chlorine atoms on carbon 2 and carbon 5.

b. 3-Methyl-1-butanethiol is a four-carbon chain with a methyl group on carbon 3, and a thiol group on carbon 1.

$$CH_3-\underset{\underset{\displaystyle CH_3}{|}}{CH}-CH_2-CH_2-SH$$

trans-2-Butene-1-thiol is similar in structure to 3-methyl-1-butanethiol except that there is a double bond between carbon 2 and carbon 3, and the carbon chains on the double bond are on opposite sides of the double bond.

c. Pentachlorophenol is a benzene ring with one hydroxyl group and five chlorine atoms attached to the ring.

13.57 a. Dehydration of an alcohol produces an alkene.

$$CH_3-CH=CH_2$$

b. Oxidation of a primary alcohol produces an aldehyde.

$$CH_3-CH_2-\overset{\overset{\displaystyle O}{||}}{C}-H$$

c. Dehydration of an alcohol produces an alkene.

d. Dehydration of an alcohol produces an alkene.

e. Oxidation of a secondary alcohol produces a ketone.

13.59 The name 4-hexyl-1,3-benzenediol tells us that there is a six-carbon alkyl group attached to carbon 4 of a benzene ring, and hydroxyl groups attached to carbon 1 and carbon 3 of the ring.

13.61 Since the compound is synthesized from a primary alcohol and oxidizes to give a carboxylic acid, it must be an aldehyde.

$$CH_3-\underset{\underset{\displaystyle CH_3}{|}}{CH}-\overset{\overset{\displaystyle O}{||}}{C}-H$$

13.63 A $CH_3-CH_2-CH_2-OH$ 1-Propanol

B $CH_3-CH=CH_2$ Propene

C $CH_3-CH_2-\overset{\displaystyle O}{\overset{\|}{C}}-H$ Propanal

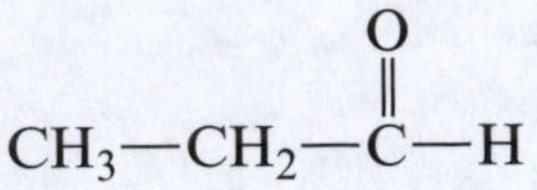

13.65 a. 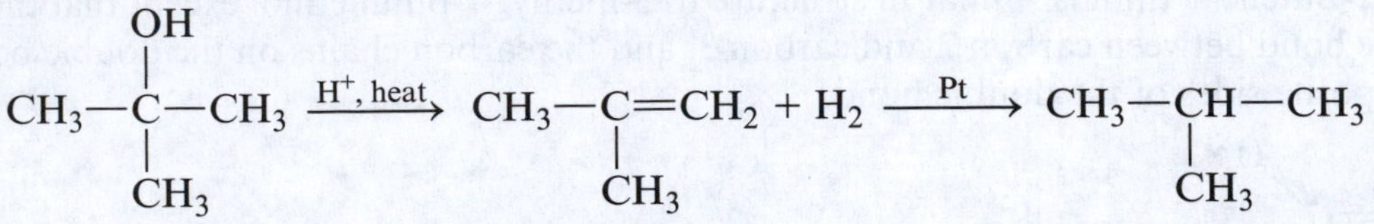 $CH_3-\underset{\underset{CH_3}{|}}{\overset{\overset{OH}{|}}{C}}-CH_3 \xrightarrow{H^+, heat} CH_3-\underset{\underset{CH_3}{|}}{C}=CH_2 + H_2 \xrightarrow{Pt} CH_3-\underset{\underset{CH_3}{|}}{CH}-CH_3$

b. 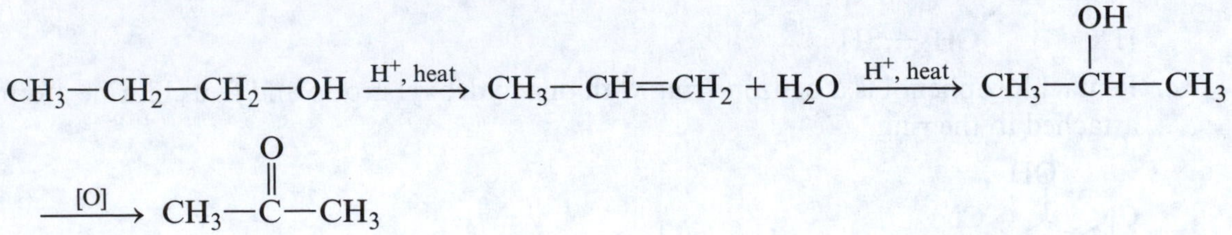 $CH_3-CH_2-CH_2-OH \xrightarrow{H^+, heat} CH_3-CH=CH_2 + H_2O \xrightarrow{H^+, heat} CH_3-\overset{\overset{OH}{|}}{CH}-CH_3$

$\xrightarrow{[O]} CH_3-\overset{\displaystyle O}{\overset{\|}{C}}-CH_3$

13.67 a. $CH_3-CH_2-CH_2-CH_2-CH_2-CH_2-CH_2-CH_2-OH$

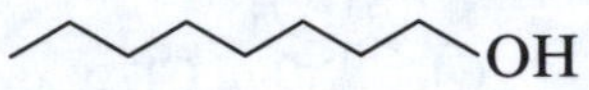

b. 1°

c.

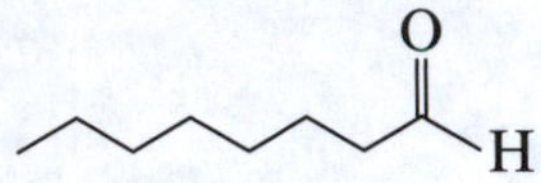

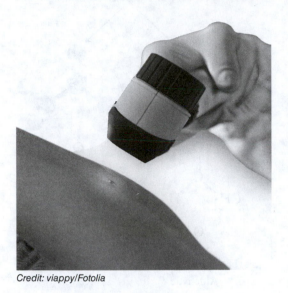

Credit: viappy/Fotolia

Diana recently noticed a change in color of a small mole on her arm, which her dermatologist determines is melanoma. A surgeon excises the mole including subcutaneous fat.

Skin cancer begins in the cells that make up the outer layer (epidermis) of the skin. Limiting exposure to UV (ultraviolet) radiation from the Sun or from tanning salons can help to reduce the risk of developing skin cancer. Sunscreen absorbs UV light radiation when the skin is exposed to sunlight and thus helps protect against sunburn. The sun protection factor (SPF) number gives the amount of time for protected skin to sunburn compared to the time for unprotected skin to sunburn. One of the principal ingredients in sunscreens is oxybenzone. Identify the functional groups in oxybenzone.

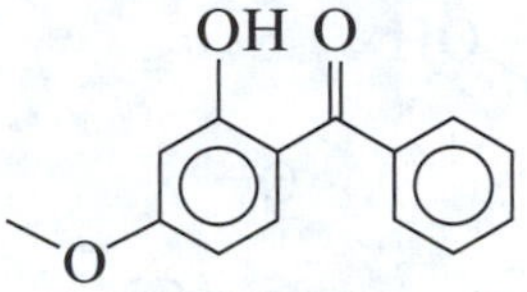

Oxybenzone

LOOKING AHEAD

14.1 Aldehydes and Ketones
14.2 Physical Properties of Aldehydes and Ketones

14.3 Oxidation and Reduction of Aldehydes and Ketones

14.4 Addition of Alcohols: Hemiacetals and Acetals

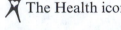 The Health icon indicates a question that is related to health and medicine.

14.1 Aldehydes and Ketones

Learning Goal: Identify compounds with carbonyl groups as aldehydes and ketones. Write the IUPAC and common names for aldehydes and ketones; draw their condensed structural and line-angle formulas.

REVIEW

Naming and Drawing Alkanes (12.2)

- In an aldehyde, the carbonyl group appears at the end of a carbon chain attached to at least one hydrogen atom.
- In a ketone, the carbonyl group occurs between carbon groups.

- In the IUPAC system, aldehydes and ketones are named by replacing the *e* in the longest chain containing the carbonyl group with *al* for aldehydes, and *one* for ketones. The location of the carbonyl group in a ketone is given if there are more than four carbon atoms in the chain.

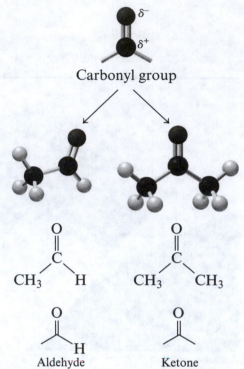

Carbonyl group

CH₃—C—H   CH₃—C—CH₃

Ethanal   Propanone
(acetaldehyde)   (dimethyl ketone)

♦ **Learning Exercise 14.1A**

Classify each of the following compounds:

    a. alcohol   **b.** aldehyde
    c. ketone   **d.** ether

Aldehyde   Ketone

The carbonyl group is found in aldehydes and ketones.

1. ____ CH₃—CH₂—CH₂—C—H

2. ____ 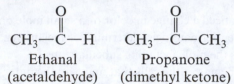 OH

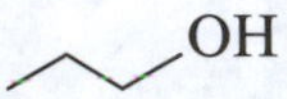

3. ____ CH₃—CH₂—C—CH₂—CH₃   **4.** ____ CH₃—CH₂—O—CH₃

5. ____ (cyclohexanone structure)   **6.** ____ CH₃—C—H

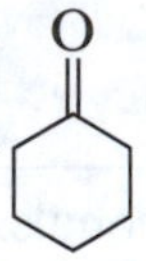

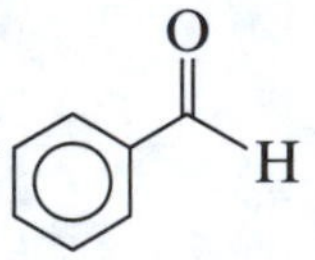

7. ____ (benzaldehyde structure) H   **8.** ____ (structure with OH)

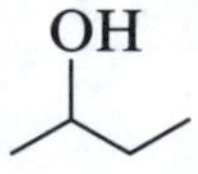

Answers	**1.** b	**2.** a	**3.** c	**4.** d
	5. c	**6.** b	**7.** b	**8.** a

Naming Aldehydes	
STEP 1	Name the longest carbon chain by replacing the *e* in the alkane name with *al*.
STEP 2	Name and number any substituents by counting the carbonyl group as carbon 1.

♦ **Learning Exercise 14.1B**

Write the correct IUPAC name and common name, if any, for the following aldehydes:

CORE CHEMISTRY SKILL

Naming Aldehydes and Ketones

1. CH₃—C—H   **2.** 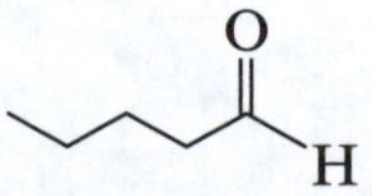 (pentanal structure) H

3. 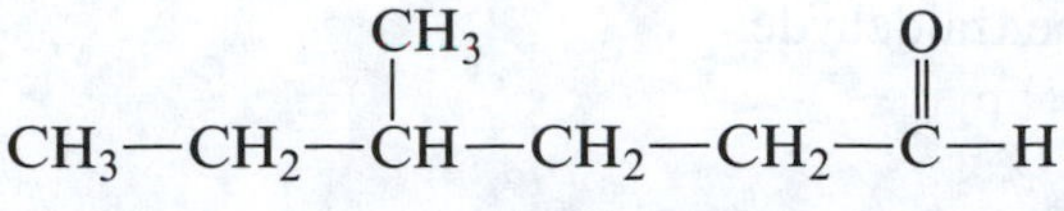 $CH_3-CH_2-\overset{\overset{\displaystyle CH_3}{|}}{CH}-CH_2-CH_2-\overset{\overset{\displaystyle O}{||}}{C}-H$

4. 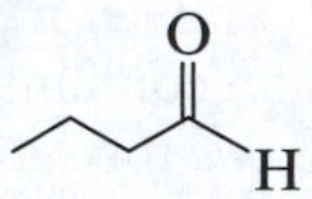 line-angle structure with aldehyde H

5. 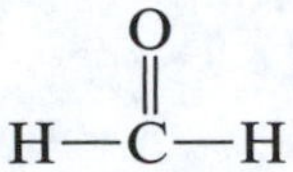 $H-\overset{\overset{\displaystyle O}{||}}{C}-H$

Answers   **1.** ethanal (acetaldehyde)   **2.** pentanal   **3.** 4-methylhexanal
                    4. butanal (butyraldehyde)   **5.** methanal (formaldehyde)

Naming Ketones	
STEP 1	Name the longest carbon chain by replacing the *e* in the alkane name with *one*.
STEP 2	Number the carbon chain starting from the end nearer the carbonyl group and indicate its location.
STEP 3	Name and number any substituents on the carbon chain.

♦ **Learning Exercise 14.1C**

Write the IUPAC name and common name, if any, for the following ketones:

1. 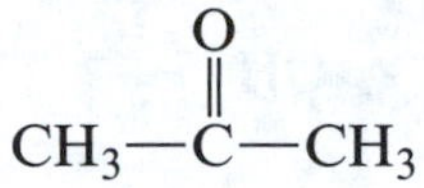 $CH_3-\overset{\overset{\displaystyle O}{||}}{C}-CH_3$ **2.** 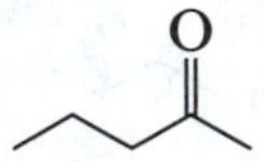 line-angle ketone structure

3. $CH_3-CH_2-\overset{\overset{\displaystyle O}{||}}{C}-CH_2-CH_3$ 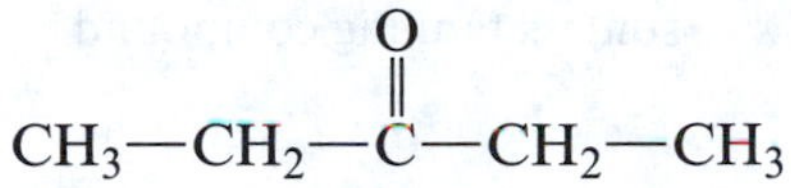   **4.** 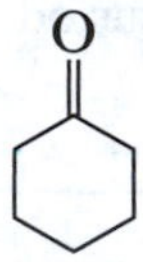 cyclohexanone structure

5. 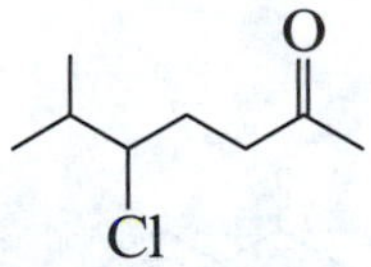 line-angle structure with Cl substituent and ketone

Answers   **1.** propanone (dimethyl ketone, acetone)   **2.** 2-pentanone (methyl propyl ketone)
                    3. 3-pentanone (diethyl ketone)   **4.** cyclohexanone
                    5. 5-chloro-6-methyl-2-heptanone

♦ **Learning Exercise 14.1D**

Draw the condensed structural formulas for **1** to **4** and the line-angle formulas for **5** and **6**.

  1. ethanal   **2.** 2-methylbutanal

  3. 2-chloropropanal   **4.** ethyl methyl ketone

5. 2-hexanone

6. benzaldehyde

Answers

1. $CH_3-\overset{\overset{\displaystyle O}{\|}}{C}-H$

2. $CH_3-CH_2-\overset{\overset{\displaystyle CH_3}{|}}{CH}-\overset{\overset{\displaystyle O}{\|}}{C}-H$

3. $CH_3-\overset{\overset{\displaystyle Cl}{|}}{CH}-\overset{\overset{\displaystyle O}{\|}}{C}-H$

4. $CH_3-CH_2-\overset{\overset{\displaystyle O}{\|}}{C}-CH_3$

5.

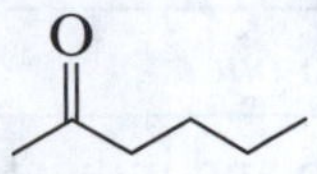

6.

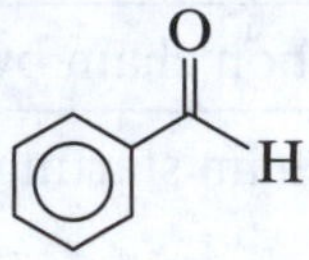

♦ **Learning Exercise 14.1E**

Identify the functional groups of alcohol, phenol, ether, aldehyde, and/or ketone in each of the following compounds:

1.

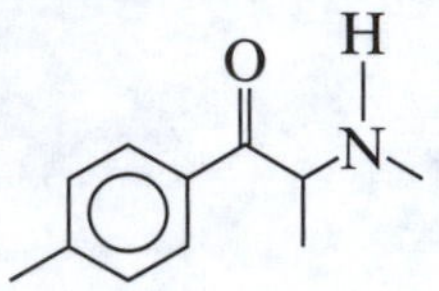

designer drug, illegal
stimulant

_____

2.

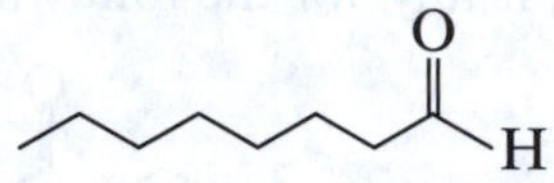

citrus odor

_____

3.

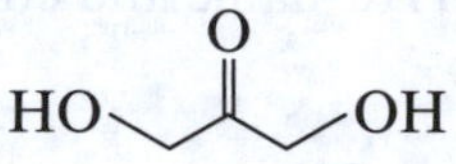

sunless tanning compound

_____

4.

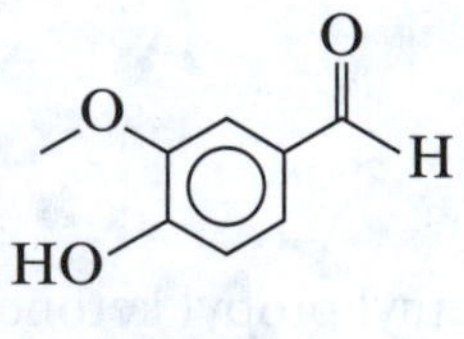

vanilla flavor

_____

5.

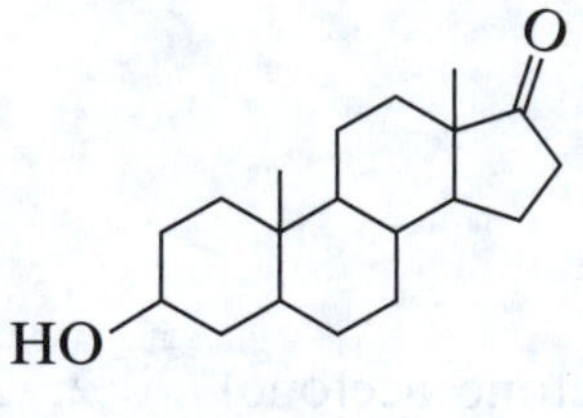

male sex hormone

_____

6.

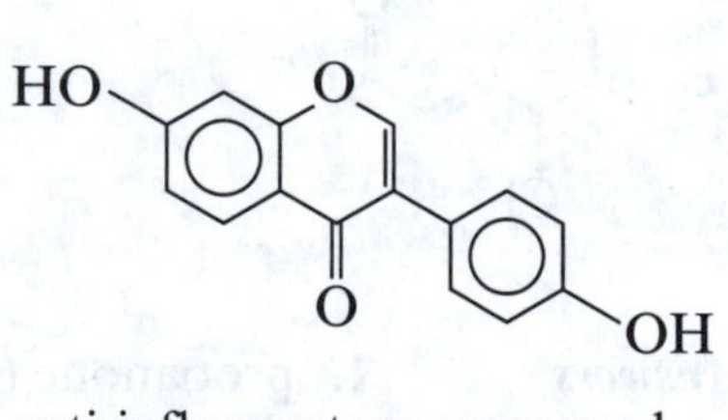

anti-inflammatory compound
from soybeans

_____

Answers **1.** ketone   **2.** aldehyde   **3.** alcohol, ketone
           4. phenol, ether, aldehyde   **5.** alcohol, ketone   **6.** phenol, ether, ketone

14.2 Physical Properties of Aldehydes and Ketones

Hydrogen bonds

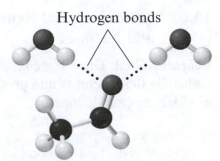

Learning Goal: Describe the boiling points and solubilities of aldehydes and ketones.

- The polarity of the carbonyl group provides aldehydes and ketones with dipole–dipole attractions that alkanes do not have. Thus, aldehydes and ketones have higher boiling points than alkanes. However, aldehydes and ketones cannot form hydrogen bonds with each other as do alcohols. Thus, alcohols have higher boiling points than aldehydes and ketones of similar molar mass.

- The polarity of the carbonyl group makes aldehydes and ketones with one to four carbon atoms soluble in water.

- For aldehydes and ketones, the boiling points increase as the number of carbon atoms in the chain increases. As the molecules become larger, there are more temporary dipoles (dispersion forces), which give higher boiling points.

The polar carbonyl group hydrogen bonds with water, which makes aldehydes and ketones with one to four carbons soluble in water.

♦ **Learning Exercise 14.2A**

Indicate the compound with the highest boiling point in each of the following groups of compounds:

1. CH_3—CH_2—$\overset{\overset{\displaystyle O}{\|}}{C}$—H, CH_3—CH_2—CH_2—OH, or CH_3—$\overset{\overset{\displaystyle O}{\|}}{C}$—$CH_3$

2. acetaldehyde or propionaldehyde   3. propanone or butanone

4. methylcyclohexane or cyclohexanone

Answers   1. CH_3—CH_2—CH_2—OH   2. propionaldehyde
                   3. butanone   4. cyclohexanone

♦ **Learning Exercise 14.2B**

Indicate whether each of the following compounds is soluble (S) or not soluble (NS) in water:

1. _____ 3-hexanone   2. _____ propanal   3. _____ acetaldehyde

4. _____ butanal   5. _____ cyclohexanone

Answers   1. NS   2. S   3. S   4. S   5. NS

14.3   Oxidation and Reduction of Aldehydes and Ketones

REVIEW

Writing Equations for the
Oxidation of Alcohols (13.4)

Learning Goal: Draw the condensed structural and line-angle formulas for the reactants and products in the oxidation or reduction of aldehydes and ketones.

- Using an oxidizing agent, primary alcohols are oxidized to aldehydes, which usually oxidize further to carboxylic acids. Secondary alcohols are oxidized to ketones, but tertiary alcohols do not oxidize.

$$CH_3-CH_2-OH \xrightarrow{[O]} CH_3-\overset{\overset{\displaystyle O}{\|}}{C}-H + H_2O$$

1° Alcohol   Aldehyde

$$CH_3-\overset{\overset{\displaystyle OH}{|}}{CH}-CH_3 \xrightarrow{[O]} CH_3-\overset{\overset{\displaystyle O}{\|}}{C}-CH_3 + H_2O$$

2° Alcohol   Ketone

- In the Tollens' test, an aldehyde is oxidized as Ag^+ is reduced to a silver mirror.
- In the Benedict's test, an aldehyde is oxidized and Cu^{2+} is reduced to Cu_2O.
- Aldehydes and ketones are reduced by hydrogen (H_2), using a catalyst such as nickel, platinum, or palladium, to produce primary or secondary alcohols.

$$CH_3-\overset{\overset{\displaystyle O}{\|}}{C}-H + H_2 \xrightarrow{Pt} CH_3-CH_2-OH$$

Aldehyde   Primary alcohol

$$CH_3-\overset{\overset{\displaystyle O}{\|}}{C}-CH_3 + H_2 \xrightarrow{Pt} CH_3-\overset{\overset{\displaystyle OH}{|}}{CH}-CH_3$$

Ketone   Secondary alcohol

♦ Learning Exercise 14.3A

Draw the condensed structural or line-angle formula for the aldehyde or ketone expected from the oxidation of each of the following:

1. $CH_3-CH_2-OH \xrightarrow{[O]}$

2. [cyclohexanol structure with OH] $\xrightarrow{[O]}$

3. [line-angle structure with OH] $\xrightarrow{[O]}$

4. $CH_3-\overset{\overset{\displaystyle OH}{|}}{CH}-CH_3 \xrightarrow{[O]}$

Answers

1. $CH_3-\overset{\overset{\displaystyle O}{\|}}{C}-H$

2. [cyclohexanone structure]

3. [line-angle ketone structure]

4. $CH_3-\overset{\overset{\displaystyle O}{\|}}{C}-CH_3$

♦ **Learning Exercise 14.3B**

Draw the condensed structural or line-angle formula for the alcohol that was oxidized to give each of the following compounds:

1. $CH_3-\overset{\overset{\displaystyle O}{\|}}{C}-CH_2-CH_3$

2. 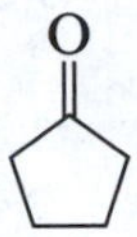 [cyclopentanone line-angle structure]

3. $CH_3-CH_2-\overset{\overset{\displaystyle O}{\|}}{C}-H$

Answers

1. $CH_3-\overset{\overset{\displaystyle OH}{|}}{CH}-CH_2-CH_3$

2. 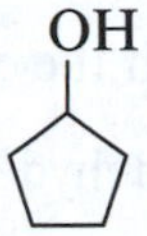 [cyclopentanol line-angle structure with OH]

3. $CH_3-CH_2-CH_2-OH$

♦ **Learning Exercise 14.3C**

Draw the condensed structural or line-angle formula for the reduction product from each of the following:

1. $CH_3-\overset{\overset{\displaystyle O}{\|}}{C}-CH_2-CH_3 + H_2 \xrightarrow{\text{Pt}}$

2. $CH_3-CH_2-\overset{\overset{\displaystyle O}{\|}}{C}-H + H_2 \xrightarrow{\text{Pt}}$

3. 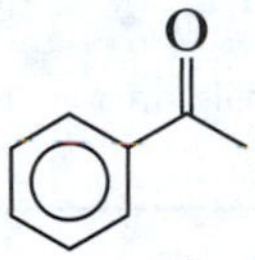 [benzaldehyde structure] $H + H_2 \xrightarrow{\text{Pt}}$

4. 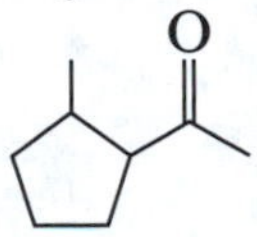 [methylcyclopentane carbaldehyde structure] $H + H_2 \xrightarrow{\text{Pt}}$

Answers

1. $CH_3-\overset{\overset{\displaystyle OH}{|}}{CH}-CH_2-CH_3$

2. $CH_3-CH_2-CH_2-OH$

3. 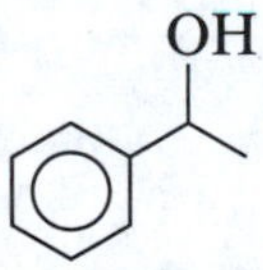 [1-phenylethanol line-angle structure with OH]

4. 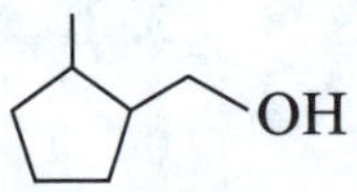 [cyclopentane with CH$_2$OH and methyl substituents]

14.4 Hemiacetals and Acetals

Learning Goal: Draw the condensed structural and line-angle formulas for the products of the addition of alcohols to the carbonyl group of aldehydes and ketones.

- A hemiacetal forms when one alcohol adds to the carbonyl group of an aldehyde or ketone in the presence of an acid catalyst.

- An acetal forms when two alcohols add to the carbonyl group of an aldehyde or ketone in the presence of an acid catalyst.

- A hemiacetal contains two functional groups, a hydroxyl group (—OH) and an alkoxy group (—OR), bonded to the same C atom.

- An acetal contains two alkoxy groups (—OR) bonded to the same C atom.

| Aldehyde or ketone | Alcohol | | Hemiacetal | Alcohol | | Acetal |

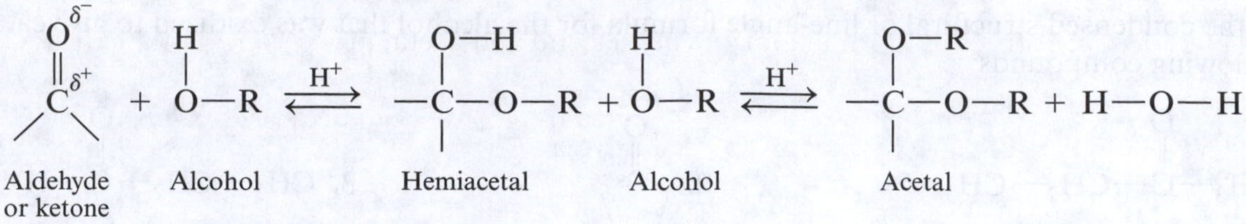

Key Terms for Sections 14.1 to 14.4

Match each of the following key terms with the correct description:

    a. acetal   **b.** hemiacetal   **c.** aldehyde   **d.** ketone

1. _____ an organic compound with a carbonyl group attached to two carbon groups

2. _____ an organic compound that contains two alkoxy groups bonded to the same C atom

3. _____ the product that forms when an alcohol adds to an aldehyde or a ketone

4. _____ an organic compound that contains a carbonyl group and a hydrogen atom at the end of the carbon chain

Answers   **1.** d   **2.** a   **3.** b   **4.** c

◆ **Learning Exercise 14.4A**

| CORE CHEMISTRY SKILL |
| Forming Hemiacetals and Acetals |

Match the descriptions shown below with the following types of compounds:

    A. hemiacetal   **B.** acetal

1. _____ the product from the addition of one alcohol to an aldehyde

2. _____ the product from the addition of one alcohol to a ketone

3. _____ a compound that contains two alkoxy groups

4. _____ a compound that contains of one alkoxy group and one alcohol group

Answers   **1.** A   **2.** A   **3.** B   **4.** A

◆ **Learning Exercise 14.4B**

Identify each of the following condensed structural formulas as a hemiacetal, acetal, or neither:

1. CH_3-O-CH_2-OH

2. 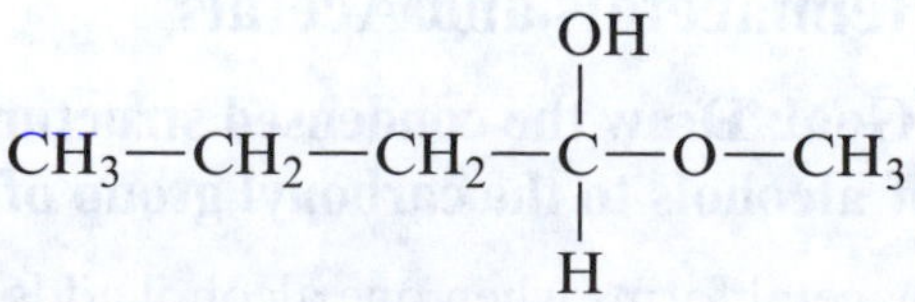 $CH_3-CH_2-CH_2-\overset{\displaystyle OH}{\underset{\displaystyle H}{C}}-O-CH_3$

3. $CH_3-\overset{\displaystyle O-CH_2-CH_3}{\underset{\displaystyle CH_3}{C}}-O-CH_2-CH_3$

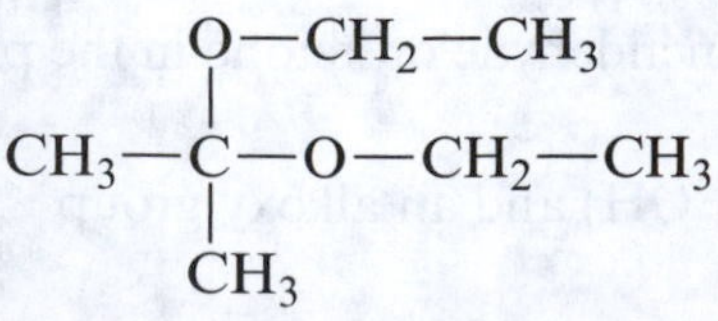

4. 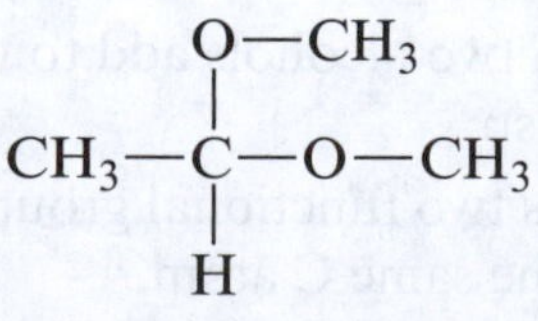 $CH_3-\overset{\displaystyle O-CH_3}{\underset{\displaystyle H}{C}}-O-CH_3$

Answers   **1.** hemiacetal   **2.** hemiacetal   **3.** acetal   **4.** acetal

♦ **Learning Exercise 14.4C**

Draw the condensed structural formula for the hemiacetal and acetal products when methanol adds to propanone.

Answers

$$CH_3-\underset{\underset{\displaystyle CH_3}{|}}{\overset{\overset{\displaystyle OH}{|}}{C}}-O-CH_3 \qquad CH_3-\underset{\underset{\displaystyle CH_3}{|}}{\overset{\overset{\displaystyle O-CH_3}{|}}{C}}-O-CH_3$$

Hemiacetal   Acetal

Checklist for Chapter 14

You are ready to take the Practice Test for Chapter 14. Be sure you have accomplished the following learning goals for this chapter. If not, review the Section listed at the end of the goal. Then apply your new skills and understanding to the Practice Test.

After studying Chapter 14, I can successfully:

_____ Identify condensed structural and line-angle formulas as aldehydes and ketones. (14.1)

_____ Write the IUPAC and common names for an aldehyde or ketone; draw the condensed structural and line-angle formula from the name. (14.1)

_____ Describe the boiling points of aldehydes and ketones. (14.2)

_____ Describe the solubility of aldehydes and ketones in water. (14.2)

_____ Draw the condensed structural and line-angle formulas for the reactants and products in the oxidation or reduction of aldehydes and ketones. (14.3)

_____ Draw the condensed structural and line-angle formulas for the hemiacetals and acetals that form when alcohols add to aldehydes or ketones. (14.4)

Practice Test for Chapter 14

The chapter Sections to review are shown in parentheses at the end of each question.

For questions 1 through 5, match each of the following compounds with the names given: (14.1)

   A. butanal   **B.** acetaldehyde   **C.** methanal   **D.** dimethyl ketone   **E.** propanal

1. _____ $H-\overset{\overset{\displaystyle O}{\|}}{C}-H$

2. _____ (line-angle structure with terminal $\overset{\overset{\displaystyle O}{\|}}{C}-H$)

3. _____ $CH_3-\overset{\overset{\displaystyle O}{\|}}{C}-CH_3$

4. _____ $CH_3-\overset{\overset{\displaystyle O}{\|}}{C}-H$

5. _____ $CH_3-CH_2-\overset{\overset{\displaystyle O}{\|}}{C}-H$

6. The line-angle formula for 4-bromo-3-methylcyclohexanone is (14.1)

A.

B.

C.

D.

E.

7. The name of the following compound is: (14.1)

$$CH_3-\overset{\overset{\displaystyle CH_3}{|}}{CH}-CH_2-\overset{\overset{\displaystyle CH_3}{|}}{CH}-CH_2-\overset{\overset{\displaystyle O}{||}}{C}-H$$

A. 3,5-dimethyl-1-hexanal   B. 2,4-dimethyl-6-hexanal   C. 3,5-dimethylhexanal
D. 1-aldo-3,5-dimethylhexane   E. 2,4-dimethylhexanal

8. The compound with the highest boiling point is (14.2)

A. $CH_3-CH_2-CH_2-CH_3$ B. $CH_3-CH_2-CH_2-OH$

C. $CH_3-\overset{\overset{\displaystyle O}{||}}{C}-CH_3$ D. $CH_3-CH_2-\overset{\overset{\displaystyle O}{||}}{C}-H$

E. $CH_3-CH_2-O-CH_3$

For questions 9 through 13, indicate the product (A to E) formed in each of the following reactions: (14.3)

A. primary alcohol   B. secondary alcohol   C. aldehyde
D. ketone   E. carboxylic acid

9. _____ oxidation of a primary alcohol   10. _____ oxidation of a secondary alcohol

11. _____ oxidation of an aldehyde   12. _____ reduction of a ketone

13. _____ reduction of an aldehyde

14. Benedict's reagent will oxidize (14.3)

A. $CH_3-\overset{\overset{\displaystyle O}{||}}{C}-CH_3$ B. $CH_3-CH_2-CH_2-CH_2-OH$

C.

D. $CH_3-\overset{\overset{\displaystyle OH}{|}}{CH}-\overset{\overset{\displaystyle O}{||}}{C}-H$

E. $CH_3-\overset{\overset{\displaystyle OH}{|}}{CH}-\overset{\overset{\displaystyle O}{||}}{C}-OH$

15. In the Tollens' test (14.3)

    A. an aldehyde is oxidized and Ag^+ is reduced

    B. an aldehyde is reduced and Ag^+ is oxidized

    C. a ketone is oxidized and Ag^+ is reduced

    D. a ketone is reduced and Ag^+ is oxidized

    E. all of these

For questions 16 through 19, match the name with one of the following condensed structural formulas: (14.4)

16. alcohol   **17.** ether   **18.** hemiacetal   **19.** acetal

             OH

A. 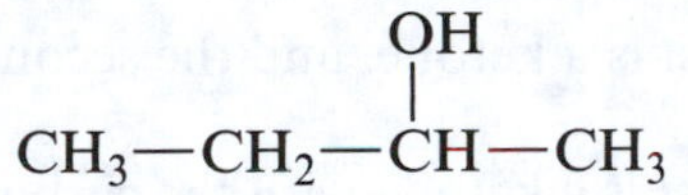 $CH_3-CH_2-CH-CH_3$

             OH

B. 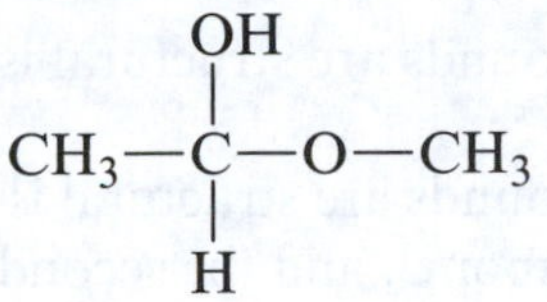 $CH_3-C-O-CH_3$

              H

          O—CH_3

C. $H-C-O-CH_3$

        H

             CH_3

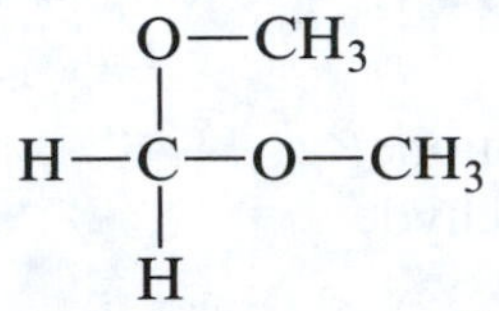

D. 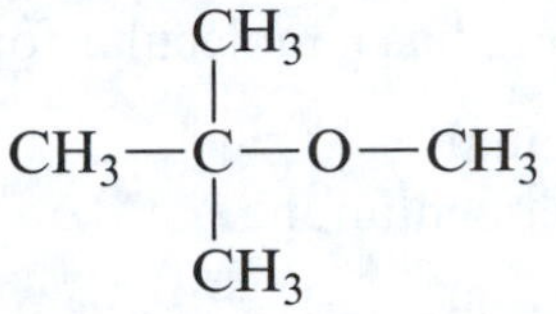 $CH_3-C-O-CH_3$

              CH_3

20. The reaction of an alcohol with an aldehyde is called (14.4)

    A. elimination   **B.** addition   **C.** substitution   **D.** hydrolysis   **E.** oxidation

Answers to the Practice Test

1. C	**2.** A	**3.** D	**4.** B	**5.** E
6. B	**7.** C	**8.** B	**9.** C, E	**10.** D
11. E	**12.** B	**13.** A	**14.** D	**15.** A
16. A	**17.** D	**18.** B	**19.** C	**20.** B

Selected Answers and Solutions to Text Problems

14.1 **a.** A carbonyl group (C=O) attached to two carbon atoms within the carbon chain makes this compound a ketone.
b. A carbonyl group (C=O) attached to a hydrogen atom at the end of the carbon chain makes this compound an aldehyde.
c. A carbonyl group (C=O) attached to two carbon atoms within the carbon chain makes this compound a ketone.
d. A carbonyl group (C=O) attached to a hydrogen atom at the end of the carbon chain makes this compound an aldehyde.

14.3 **a.** These compounds are structural isomers of C_3H_6O: The first is a ketone, and the second is an aldehyde.
b. These compounds are structural isomers of $C_5H_{10}O$: The first is a ketone with the carbonyl group on carbon 3, and the second is a ketone with the carbonyl group on carbon 2.
c. These compounds are not structural isomers: The first has a molecular formula of C_4H_8O, and the second has a molecular formula of $C_5H_{10}O$.

14.5 **a.** 3-bromobutanal **b.** 4-chloro-2-pentanone
c. 2-methylcyclopentanone **d.** 2-bromobenzaldehyde

14.7 **a.** acetaldehyde
b. methyl propyl ketone
c. formaldehyde

14.9 **a.** $CH_3-\overset{\overset{\displaystyle O}{\|}}{C}-H$ 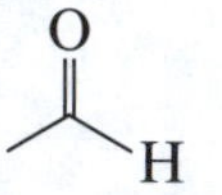

b. $CH_3-\overset{\overset{\displaystyle CH_3}{|}}{CH}-\overset{\overset{\displaystyle O}{\|}}{C}-CH_2-CH_3$ 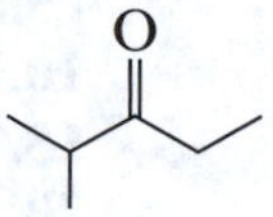

c. $CH_3-\overset{\overset{\displaystyle O}{\|}}{C}-CH_2-CH_2-CH_2-CH_3$ 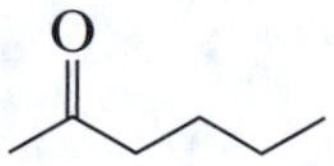

d. $CH_3-CH_2-CH_2-\overset{\overset{\displaystyle CH_3}{|}}{CH}-CH_2-\overset{\overset{\displaystyle O}{\|}}{C}-H$ 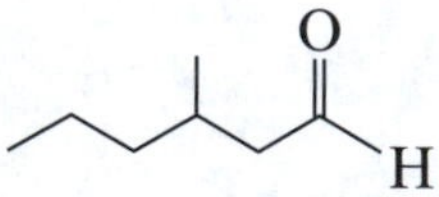

14.11 The name 4-methoxybenzaldehyde tells us that there is a methyl ether at carbon 4 of the ring in benzaldehyde.

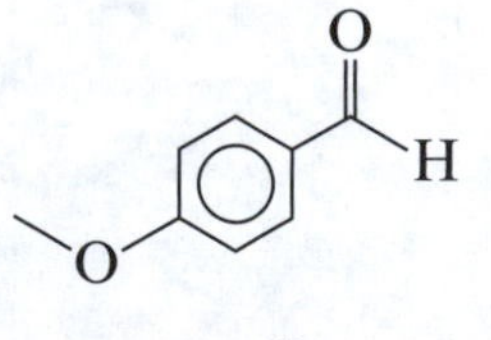

14.13 **a.** $CH_3-\overset{\overset{\displaystyle O}{\|}}{C}-H$ will have the higher boiling point since aldehydes have a polar carbonyl group and form dipole–dipole attractions that alkanes like propane do not.
b. Pentanal has a longer carbon chain, more electrons, and more dispersion forces than propanal, which give pentanal a higher boiling point.
c. 1-Butanol has a polar hydroxyl group and can form hydrogen bonds, which are stronger than the dipole–dipole attractions in butanal; 1-butanol will have the higher boiling point.

14.15 a. $CH_3-\overset{\overset{O}{\|}}{C}-\overset{\overset{O}{\|}}{C}-CH_2-CH_3$ is more soluble in water because it has two polar carbonyl groups and can form more hydrogen bonds with water.

b. Propanal is more soluble in water because it has a shorter carbon chain than pentanal, in which the longer hydrocarbon chain diminishes the effect of the polar carbonyl group on solubility.

c. Acetone is more soluble in water because it has a shorter carbon chain than 2-pentanone, in which the longer hydrocarbon chain diminishes the effect of the polar carbonyl group on solubility.

14.17 No. A hydrocarbon chain of eight carbon atoms diminishes the effect of the polar carbonyl group on solubility.

14.19 a. An aldehyde can be oxidized to a carboxylic acid.

$H-\overset{\overset{O}{\|}}{C}-OH$

b. None; a ketone cannot be further oxidized.

c. An aldehyde can be oxidized to a carboxylic acid.

d. None; a ketone cannot be further oxidized.

14.21 a. An aldehyde will react with Tollens' reagent.

b. A ketone will not react with Tollens' reagent.

c. An aldehyde will react with Tollens' reagent.

14.23 In reduction, an aldehyde will give a primary alcohol, and a ketone will give a secondary alcohol.

a. Butyraldehyde is the four-carbon aldehyde; it will be reduced to a four-carbon primary alcohol.

$CH_3-CH_2-CH_2-CH_2-OH$

b. Acetone is a three-carbon ketone; it will be reduced to a three-carbon secondary alcohol.

$CH_3-\overset{\overset{OH}{|}}{CH}-CH_3$

c. 3-Bromohexanal is a six-carbon aldehyde with a bromine atom on carbon 3; it reduces to the corresponding six-carbon primary alcohol with a bromine atom on carbon 3.

$CH_3-CH_2-CH_2-\overset{\overset{Br}{|}}{CH}-CH_2-CH_2-OH$

d. 2-Methyl-3-pentanone is a five-carbon ketone with a methyl group attached to carbon 2. It will be reduced to a five-carbon secondary alcohol with a methyl group attached to carbon 2.

$CH_3-\overset{\overset{CH_3}{|}}{CH}-\overset{\overset{OH}{|}}{CH}-CH_2-CH_3$

14.25 a. neither **b.** hemiacetal
  c. acetal **d.** hemiacetal
  e. acetal

14.27 A hemiacetal forms when an alcohol is added to the carbonyl of an aldehyde or ketone.

a.
$$\text{CH}_3-\underset{\underset{\text{H}}{|}}{\overset{\overset{\text{OH}}{|}}{\text{C}}}-\text{O}-\text{CH}_3$$

b.
$$\text{CH}_3-\underset{\underset{\text{CH}_3}{|}}{\overset{\overset{\text{OH}}{|}}{\text{C}}}-\text{O}-\text{CH}_3$$

c.
$$\text{CH}_3-\text{CH}_2-\text{CH}_2-\underset{\underset{\text{H}}{|}}{\overset{\overset{\text{OH}}{|}}{\text{C}}}-\text{O}-\text{CH}_3$$

14.29 An acetal forms when a second molecule of alcohol reacts with a hemiacetal.

a.
$$\text{CH}_3-\underset{\underset{\text{H}}{|}}{\overset{\overset{\text{O}-\text{CH}_3}{|}}{\text{C}}}-\text{O}-\text{CH}_3$$

b.
$$\text{CH}_3-\underset{\underset{\text{CH}_3}{|}}{\overset{\overset{\text{O}-\text{CH}_3}{|}}{\text{C}}}-\text{O}-\text{CH}_3$$

c.
$$\text{CH}_3-\text{CH}_2-\text{CH}_2-\underset{\underset{\text{H}}{|}}{\overset{\overset{\text{O}-\text{CH}_3}{|}}{\text{C}}}-\text{O}-\text{CH}_3$$

14.31 a. Oxybenzone contains ether, phenol, ketone, and aromatic functional groups.
 b. Molecular formula of oxybenzone = $C_{14}H_{12}O_3$
  Molar mass of oxybenzone ($C_{14}H_{12}O_3$)
  $= 14(12.01 \text{ g}) + 12(1.008 \text{ g}) + 3(16.00 \text{ g}) = 228.2 \text{ g/mole (4 SFs)}$
 c. $178 \text{ mL sunscreen} \times \dfrac{6.0 \text{ g oxybenzone}}{100 \text{ mL sunscreen}} = 11 \text{ g of oxybenzone (2 SFs)}$

14.33 Cinnamaldehyde contains aromatic, alkene, and aldehyde functional groups.

14.35 a.
$$\text{CH}_3-\text{CH}_2-\text{CH}_2 \quad \underset{\text{H}}{\overset{\text{H}}{\underset{\diagdown}{\overset{\diagup}{\text{C}=\text{C}}}}}\quad \overset{\overset{\text{O}}{\|}}{\text{C}}-\text{H}$$

 b.
$$\text{CH}_3-\underset{\underset{\text{CH}_3}{|}}{\text{C}}=\text{CH}-\text{CH}_2-\text{CH}_2-\underset{\underset{\text{CH}_3}{|}}{\text{CH}}-\overset{\overset{\text{O}}{\|}}{\text{C}}-\text{H}$$

14.37 The $C=O$ double bond has a dipole because the oxygen atom is highly electronegative compared to the carbon atom. In the $C=C$ double bond, both atoms have the same electronegativity, and there is no dipole.

14.39 a. An aldehyde will react with Tollens' reagent and produce a silver mirror.
 b. An aldehyde will react with Tollens' reagent and produce a silver mirror.
 c. An ether will not react with Tollens' reagent.

14.41 a. 2-bromo-4-chlorocyclopentanone
 c. 3-chloropropanal
 e. 5-chloro-3-hexanone
 b. 2,4-dibromobenzaldehyde
 d. 2-chloro-3-pentanone

14.43 a.   **b.** $CH_3-CH_2-CH_2-CH_2-\overset{\overset{\textstyle O}{\|}}{C}-H$

c. $CH_3-CH_2-\overset{\overset{\textstyle O}{\|}}{C}-CH_3$ **d.** $CH_3-CH_2-\overset{\overset{\textstyle CH_3}{|}}{CH}-CH_2-CH_2-\overset{\overset{\textstyle O}{\|}}{C}-H$

14.45 Compounds **a** and **b** are soluble in water because they have a polar group with an oxygen atom that hydrogen bonds with water and fewer than five carbon atoms.

14.47 a. $CH_3-CH_2-CH_2-OH$ **b.** $CH_3-CH_2-\overset{\overset{\textstyle O}{\|}}{C}-H$ **c.** CH_3-CH_2-OH

14.49 Aldehydes oxidize to form carboxylic acids; ketones do not oxidize further.

a. $CH_3-CH_2-\overset{\overset{\textstyle O}{\|}}{C}-OH$ **b.** $CH_3-CH_2-CH_2-\overset{\overset{\textstyle O}{\|}}{C}-OH$

c. no reaction

14.51 In reduction, an aldehyde will give a primary alcohol, and a ketone will give a secondary alcohol.

a. $CH_3-\overset{\overset{\textstyle OH}{|}}{CH}-CH_3$ **b.**

c. $CH_3-\overset{\overset{\textstyle CH_3}{|}}{CH}-CH_2-\overset{\overset{\textstyle OH}{|}}{CH}-CH_3$

14.53 a. propanal   **b.** 2-pentanone
c. 2-butanol   **d.** cyclohexanone

14.55 a. acetal; propanal and methanol   **b.** hemiacetal; butanone and ethanol
c. acetal; cyclohexanone and ethanol

14.57 a. True; both **A** and **B** have the molecular formula $C_5H_{10}O$, but a different arrangement of atoms.
b. True; both **D** and **F** contain a carbonyl group attached to one hydrogen atom.
c. True; both **B** and **C** are 2-pentanone.
d. True; both **C** and **D** have the molecular formula $C_5H_{10}O$, but a different arrangement of atoms.

14.59 Since the compound is synthesized from a secondary alcohol and cannot be further oxidized, it must be a ketone.

$CH_3-\overset{\overset{\textstyle O}{\|}}{C}-CH_2-CH_3$ Butanone

14.61 $CH_3-CH_2-CH_2-\overset{\overset{\textstyle O}{\|}}{C}-H$ Butanal

$CH_3-\overset{\overset{\textstyle CH_3}{|}}{CH}-\overset{\overset{\textstyle O}{\|}}{C}-H$ 2-Methylpropanal

$CH_3-\overset{\overset{\textstyle O}{\|}}{C}-CH_2-CH_3$ Butanone

Selected Answers to Combining Ideas from Chapters 12 to 14

CI.27 a.

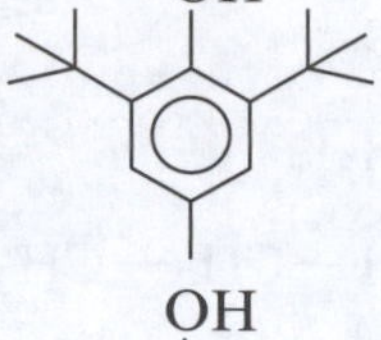

OH

b.

OH

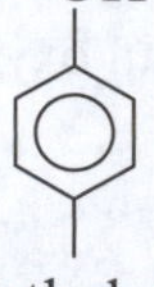

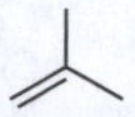

4-Methylphenol   2-Methylpropene

c. Molecular formula $= C_{15}H_{24}O$

Molar mass of BHT $(C_{15}H_{24}O)$

$= 15(12.01\ g) + 24(1.008\ g) + 16.00\ g = 220.4\ g/mole\ (4\ SFs)$

d. 50. ppm BHT $=$ 50. mg BHT/kg cereal

$$15\ \cancel{oz} \times \frac{1\ \cancel{lb}}{16\ \cancel{oz}} \times \frac{1\ kg\ cereal}{2.20\ \cancel{lb}} \times \frac{50.\ mg\ BHT}{1\ kg\ cereal} = 21\ mg\ of\ BHT\ (2\ SFs)$$

CI.29 a.

$$CH_3{-}\overset{\overset{\displaystyle O}{\|}}{C}{-}CH_3$$

b. Molecular formula $= C_3H_6O$

Molar mass of propanone (C_3H_6O)

$= 3(12.01\ g) + 6(1.008\ g) + 16.00\ g = 58.08\ g/mole\ (4\ SFs)$

c.

$$CH_3{-}\overset{\overset{\displaystyle OH}{|}}{CH}{-}CH_3$$

CI.31 a. $2C_8H_{18}(l) + 25O_2(g) \xrightarrow{\Delta} 16CO_2(g) + 18H_2O(g) + 11\ 020\ kJ$

b. The combustion of octane is an exothermic reaction.

c. Molar mass of octane $(C_8H_{18}) = 8(12.01\ g) + 18(1.008\ g) = 114.22\ g/mole\ (5\ SFs)$

$$1\ tank \times \frac{11.9\ gal}{1\ tank} \times \frac{4\ qt}{1\ gal} \times \frac{946\ mL}{1\ qt} \times \frac{0.803\ g\ C_8H_{18}}{1\ mL\ C_8H_{18}} \times \frac{1\ mole\ C_8H_{18}}{114.22\ g\ C_8H_{18}}$$

$= 317\ moles\ of\ C_8H_{18}\ (3\ SFs)$

d. $24\ 500\ mi \times \dfrac{1\ gal}{45\ mi} \times \dfrac{4\ qt}{1\ gal} \times \dfrac{946\ mL}{1\ qt} \times \dfrac{0.803\ g\ C_8H_{18}}{1\ mL\ C_8H_{18}} \times \dfrac{1\ mole\ C_8H_{18}}{114.22\ g\ C_8H_{18}}$

$$\times \frac{16\ moles\ CO_2}{2\ moles\ C_8H_{18}} \times \frac{44.01\ g\ CO_2}{1\ mole\ CO_2} \times \frac{1\ kg\ CO_2}{1000\ g\ CO_2} = 5.10 \times 10^3\ kg\ of\ CO_2\ (3\ SFs)$$

CI.33 a. $CH_3{-}CH_2{-}CH_2{-}\overset{\overset{\displaystyle O}{\|}}{C}{-}H$ **b.**

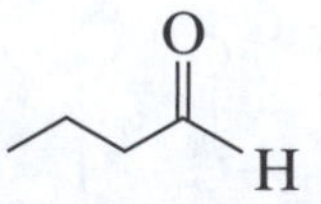

c. The IUPAC name of butyraldehyde is butanal.   **d.** $CH_3{-}CH_2{-}CH_2{-}CH_2{-}OH$

Carbohydrates

Kate has completed her diabetes educational class taught by Paula, a diabetes nurse. Kate now follows a healthy meal plan, which includes 45 to 60 g of carbohydrates in the form of fruits and vegetables. She is walking for 30 minutes twice a day and has lost 10 lb. Kate measures her blood glucose level before and after a meal using a blood glucose meter. She records the results in a log, which helps determine if she is maintaining a consistent blood sugar level or if her blood glucose is high or low. Her blood sugar level should be 70 to 110 mg/dL before a meal, and less than 180 mg/dL two hours after a meal. Draw the Fischer projection for D-glucose.

Credit: Rolf Bruderer/Alamy

LOOKING AHEAD

15.1 Carbohydrates

15.2 Chiral Molecules

15.3 Fischer Projections
     of Monosaccharides

15.4 Haworth Structures
     of Monosaccharides

15.5 Chemical Properties
     of Monosaccharides

15.6 Disaccharides

15.7 Polysaccharides

 The Health icon indicates a question that is related to health and medicine.

15.1 Carbohydrates

Learning Goal: Classify a monosaccharide as an aldose or a ketose, and indicate the number of carbon atoms.

- Photosynthesis is the process of using water, carbon dioxide, and the energy from the Sun to form monosaccharides and oxygen.

- Carbohydrates are classified as monosaccharides (simple sugars), disaccharides (two monosaccharide units), and polysaccharides (many monosaccharide units).

- Monosaccharides are polyhydroxy aldehydes (aldoses) or ketones (ketoses).

- Monosaccharides are classified as *aldo* for an aldehyde or *keto* for a ketone and by the number of carbon atoms as *trioses*, *tetroses*, *pentoses*, or *hexoses*.

♦ **Learning Exercise 15.1A**

Complete and balance the equations for the photosynthesis of

1. Glucose: _____ + _____ + energy $\longrightarrow$ $C_6H_{12}O_6$ + _____

2. Ribose: _____ + _____ + energy $\longrightarrow$ $C_5H_{10}O_5$ + _____

Answers   **1.** $6CO_2 + 6H_2O$ + energy $\longrightarrow$ $C_6H_{12}O_6 + 6O_2$
           2. $5CO_2 + 5H_2O$ + energy $\longrightarrow$ $C_5H_{10}O_5 + 5O_2$

♦ **Learning Exercise 15.1B** 🏃

Indicate the number of monosaccharide units (one, two, or many) in each of the following carbohydrates:

1. sucrose, a disaccharide   _____   **2.** cellulose, a polysaccharide _____

3. glucose, a monosaccharide _____   **4.** amylose, a polysaccharide _____

5. maltose, a disaccharide   _____

Answers   **1.** two   **2.** many   **3.** one   **4.** many   **5.** two

♦ **Learning Exercise 15.1C**

Identify each of the following monosaccharides as an aldotriose, a ketotriose, an aldotetrose, a ketotetrose, an aldopentose, a ketopentose, an aldohexose, or a ketohexose:

1. 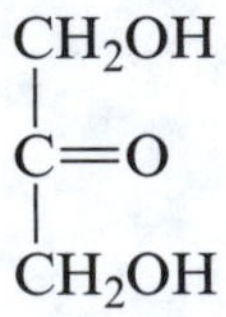
CH_2OH
|
$C=O$
|
CH_2OH

Dihydroxy-
acetone

2. 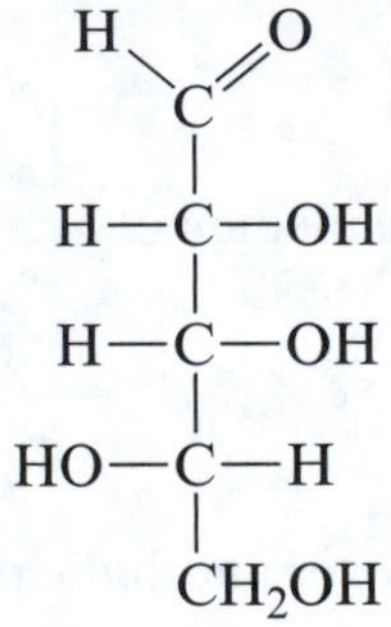
H   O
  \\C
H—C—OH
|
H—C—OH
|
HO—C—H
|
CH_2OH

Lyxose

3. 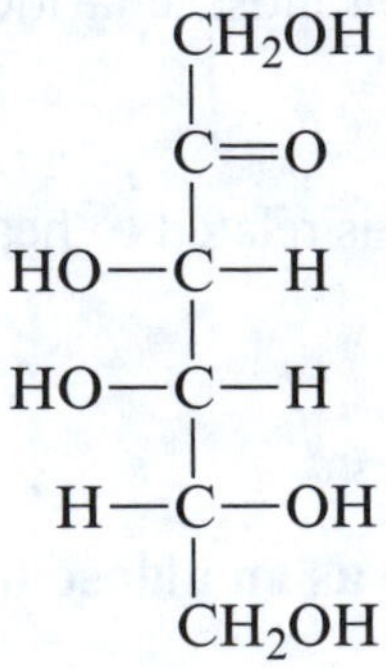
CH_2OH
|
$C=O$
|
HO—C—H
|
HO—C—H
|
H—C—OH
|
CH_2OH

Tagatose

4. 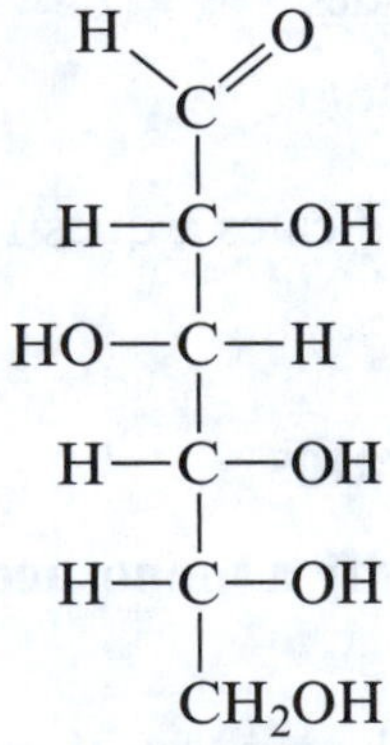
H   O
  \\C
H—C—OH
|
HO—C—H
|
H—C—OH
|
H—C—OH
|
CH_2OH

Glucose

5. 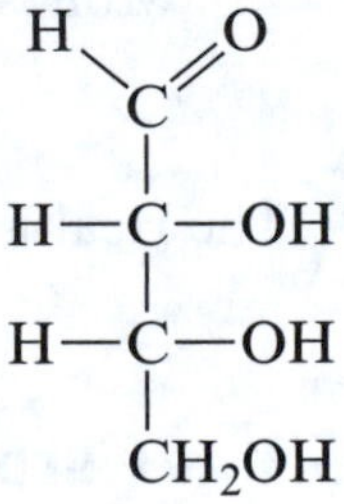
H   O
  \\C
H—C—OH
|
H—C—OH
|
CH_2OH

Erythrose

1. _____   **2.** _____   **3.** _____

4. _____   **5.** _____

Answers   **1.** ketotriose   **2.** aldopentose   **3.** ketohexose
           4. aldohexose   **5.** aldotetrose

15.2 Chiral Molecules

Learning Goal: Identify chiral and achiral carbon atoms in an organic molecule. Use Fischer projections to identify the D and L enantiomers of monosaccharides.

- Chiral molecules have mirror images that cannot be superimposed.
- When the mirror image of an object can be superimposed on the original, it is achiral.
- In a chiral molecule, at least one carbon atom is attached to four different atoms or groups.
- The mirror images of a chiral molecule represent two different molecules called *enantiomers.*
- In a Fischer projection, the prefixes D and L indicate the position of the —OH group attached to the chiral carbon.
- In D-glyceraldehyde, the —OH group is on the right of the chiral carbon; it is on the left in L-glyceraldehyde.

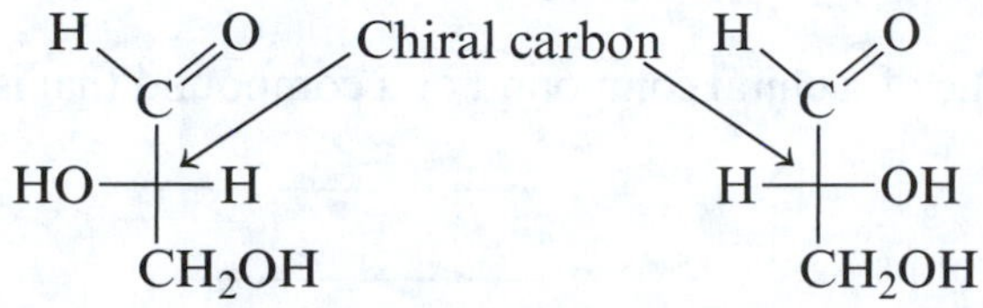

L-Glyceraldehyde D-Glyceraldehyde

- For compounds with two or more chiral carbons, the designation as a D and L isomer is determined by the position of the —OH group attached to the chiral carbon *farthest from the carbonyl group*.

L Isomer 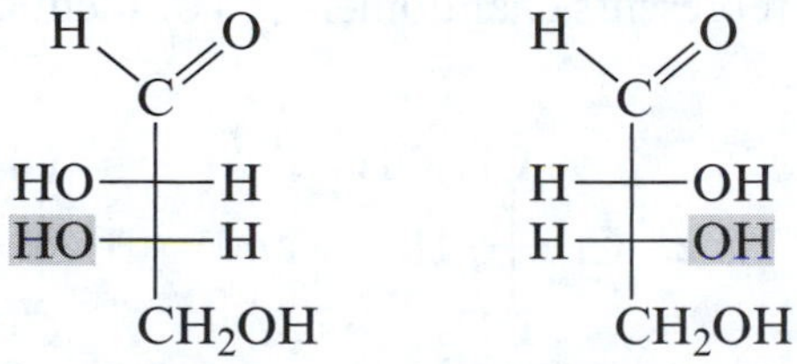 D Isomer

L-Erythrose D-Erythrose

♦ Learning Exercise 15.2A

CORE CHEMISTRY SKILL Identifying Chiral Molecules

Indicate whether each of the following objects is chiral or achiral:

1. a blank piece of paper __________
2. a glove __________
3. a plain baseball cap __________
4. a volleyball net __________
5. a left foot __________

Answers **1.** achiral **2.** chiral **3.** achiral **4.** achiral **5.** chiral

◆ **Learning Exercise 15.2B**

State whether each of the following molecules is chiral or achiral:

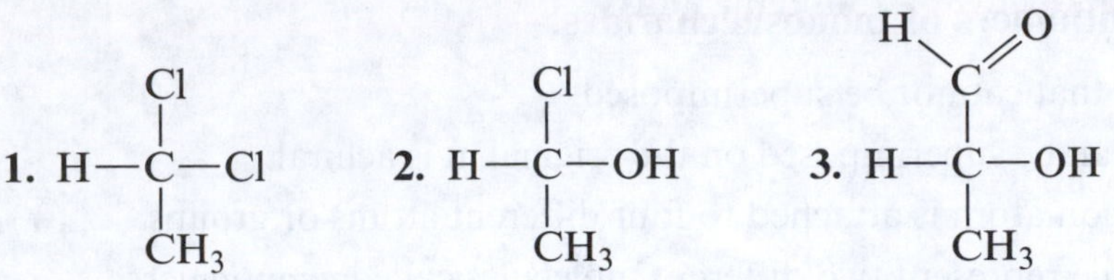

1. H—C—Cl with Cl above and CH₃ below, H on left

2. H—C—OH with Cl above and CH₃ below, H on left

3. H—C—OH with CHO (H—C=O) above and CH₃ below, H on left

_____ _____ _____

Answers   **1.** achiral   **2.** chiral   **3.** chiral

◆ **Learning Exercise 15.2C**

Identify each of the following features as characteristic of a chiral compound or a compound that is achiral:

1. central atom attached to two identical groups   _____

2. contains a carbon attached to four different groups   _____

3. has identical mirror images   _____

Answers   **1.** achiral   **2.** chiral   **3.** achiral

◆ **Learning Exercise 15.2D**

Indicate whether each pair of Fischer projections represents enantiomers (E) or identical structures (I).

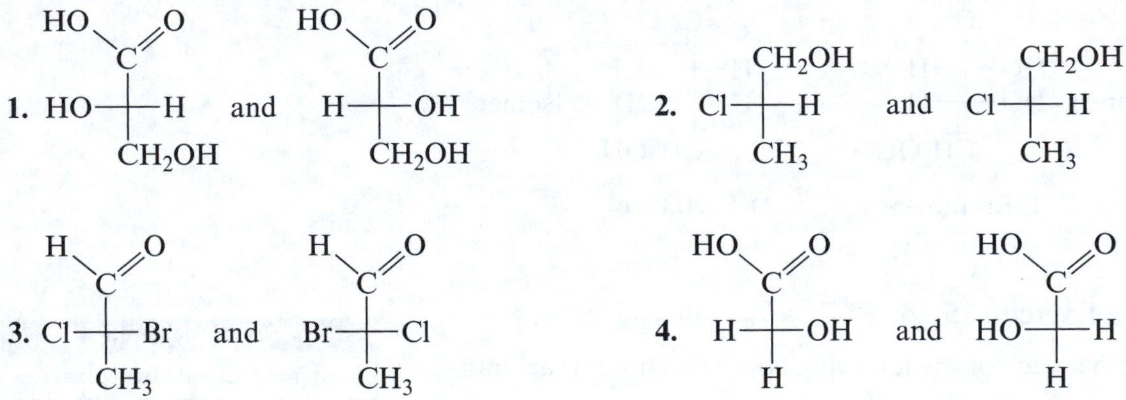

1. HO—C=O top; HO—H; CH₂OH   and   HO—C=O top; H—OH; CH₂OH

2. CH₂OH top; Cl—H; CH₃   and   CH₂OH top; Cl—H; CH₃

3. H—C=O top; Cl—Br; CH₃   and   H—C=O top; Br—Cl; CH₃

4. HO—C=O top; H—OH; H   and   HO—C=O top; HO—H; H

Answers   **1.** E   **2.** I   **3.** E   **4.** I

15.3   Fischer Projections of Monosaccharides

Learning Goal: Identify or draw the D and L configurations of the Fischer projections for common monosaccharides.

- In a Fischer projection, the carbon chain is drawn vertically, with the most oxidized carbon (usually an aldehyde or ketone) at the top.
- In the Fischer projection for a monosaccharide, the chiral —OH group *farthest* from the carbonyl group (C=O) is drawn on the left side in the L isomer and on the right side in the D isomer.

- The carbon atom in the —CH_2OH group at the bottom of the Fischer projection is not chiral because it does not have four different groups bonded to it.
- Important monosaccharides are the aldopentose ribose, the aldohexoses glucose and galactose, and the ketohexose fructose.

◆ Learning Exercise 15.3A

Identify each of the following Fischer projections of monosaccharides as the D or L isomer:

CORE CHEMISTRY SKILL

Identifying D and L Fischer Projections for Carbohydrates

1.
```
      CH2OH
       |
       C=O
  HO───┼───H
   H───┼───OH
      CH2OH
```

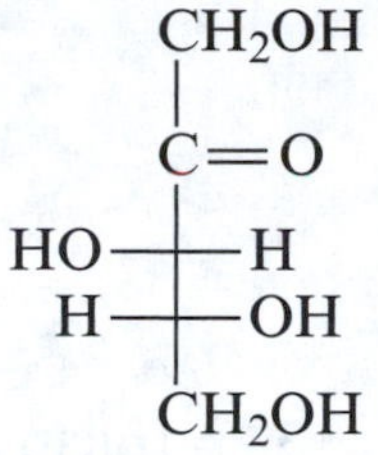

2.
```
    H     O
     \   /
      C
   H──┼──OH
   H──┼──OH
  HO──┼──H
  HO──┼──H
      CH2OH
```

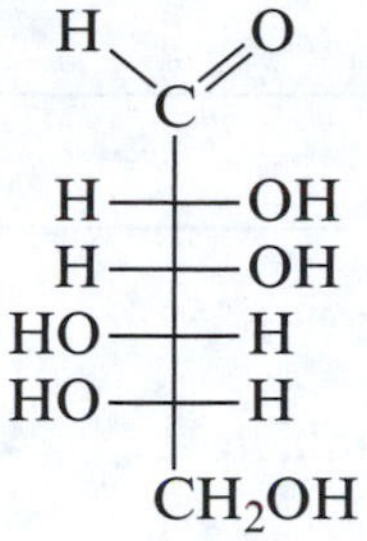

3.
```
    H     O
     \   /
      C
  HO──┼──H
   H──┼──OH
      CH2OH
```

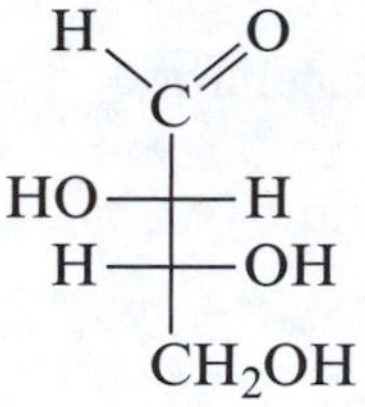

4.
```
      CH2OH
       |
       C=O
  HO───┼───H
  HO───┼───H
      CH2OH
```

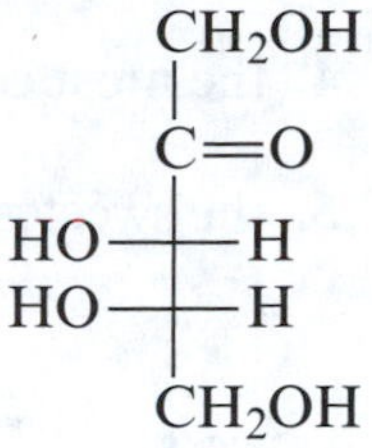

_____ Xylulose   _____ Mannose   _____ Threose   _____ Ribulose

Answers   **1.** D-Xylulose   **2.** L-Mannose   **3.** D-Threose   **4.** L-Ribulose

◆ Learning Exercise 15.3B

Draw the mirror image of each of the monosaccharides in Learning Exercise 15.3A and give the D or L name.

1.   **2.**   **3.**   **4.**

Answers

1. 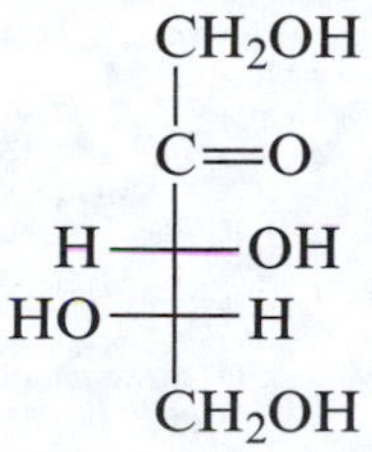
```
      CH2OH
       |
       C=O
   H───┼───OH
  HO───┼───H
      CH2OH
```
L-Xylulose

2. 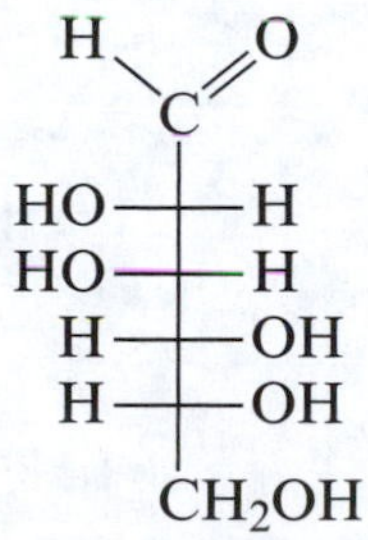
```
    H     O
     \   /
      C
  HO──┼──H
  HO──┼──H
   H──┼──OH
   H──┼──OH
      CH2OH
```
D-Mannose

3. 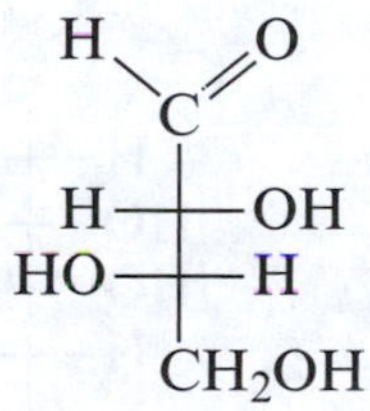
```
    H     O
     \   /
      C
   H──┼──OH
  HO──┼──H
      CH2OH
```
L-Threose

4. 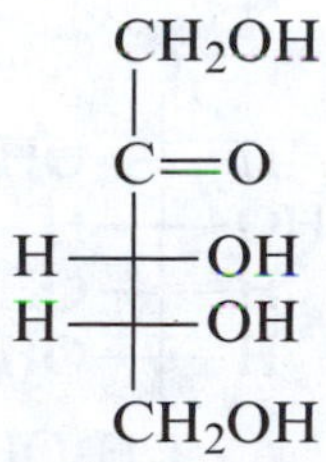
```
      CH2OH
       |
       C=O
   H───┼───OH
   H───┼───OH
      CH2OH
```
D-Ribulose

♦ **Learning Exercise 15.3C** ⅄

Identify the monosaccharide (D-glucose, D-fructose, or D-galactose) that fits each of the following descriptions:

1. a building block in cellulose   _____

2. also known as fruit sugar   _____

3. accumulates in the disease known as *galactosemia* _____

4. the most common monosaccharide   _____

5. the sweetest monosaccharide   _____

Answers   **1.** D-glucose   **2.** D-fructose   **3.** D-galactose   **4.** D-glucose   **5.** D-fructose

♦ **Learning Exercise 15.3D**

Draw the Fischer projection for each of the following monosaccharides:

D-Glucose   D-Galactose   D-Fructose

Answers

D-Glucose   D-Galactose   D-Fructose

15.4 Haworth Structures of Monosaccharides

Learning Goal: Draw and identify the Haworth structures for monosaccharides.

- The Haworth structure is a representation of the cyclic, stable form of monosaccharides, which are rings of five or six atoms.
- The Haworth structure forms by a reaction between the —OH group on carbon 5 of hexoses and the carbonyl group on carbon 1 or 2 of the same molecule.
- The formation of a new —OH group on carbon 1 (or 2 in fructose) gives α and β isomers of the cyclic monosaccharide. Because the molecule opens and closes continuously in solution, both α and β isomers are present.

	Drawing Haworth Structures
Step 1	Turn the Fischer projection clockwise by 90°.
Step 2	Fold the horizontal chain into a hexagon, rotate the groups on carbon 5, and bond the O on carbon 5 to carbon 1.
Step 3	Draw the new —OH group on carbon 1 below the ring to give the α isomer or above the ring to give the β isomer.

♦ **Learning Exercise 15.4**

Draw the Haworth structure for the α isomer of each of the following:

CORE CHEMISTRY SKILL
Drawing Haworth Structures

1. D-Glucose   **2.** D-Galactose   **3.** D-Fructose

Answers

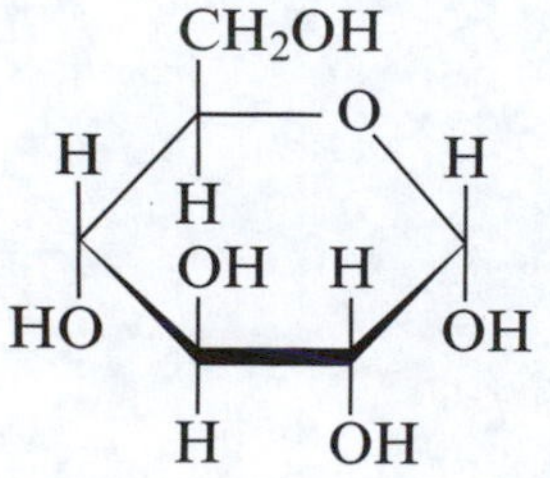

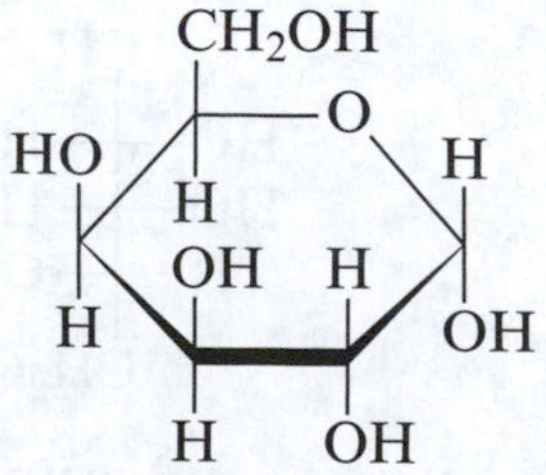

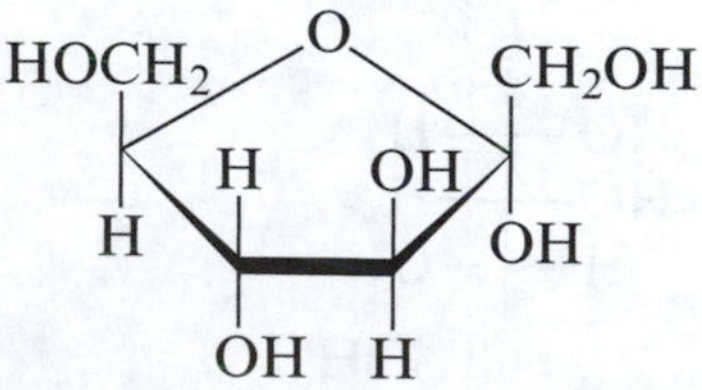

15.5   Chemical Properties of Monosaccharides

REVIEW

Writing Equations for the Oxidation of Alcohols (13.4)

Learning Goal: Identify the products of oxidation or reduction of monosaccharides; determine if a carbohydrate is a reducing sugar.

- Monosaccharides are called *reducing sugars* because an aldehyde group is oxidized to a carboxylic acid by Cu^{2+} in Benedict's solution, which is reduced.
- Monosaccharides are also reduced to give sugar alcohols.

♦ **Learning Exercise 15.5A**

What changes occur when a reducing sugar reacts with Benedict's reagent?

Answer   The carbonyl group of the reducing sugar is oxidized to a carboxylic acid group; the Cu^{2+} ion in Benedict's reagent is reduced to Cu^+, which forms a brick-red solid Cu_2O.

♦ **Learning Exercise 15.5B**

Draw and name the products from **(a)** the oxidation and **(b)** the reduction of D-lyxose.

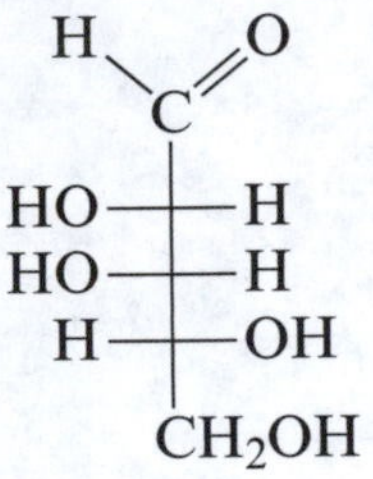

a. Oxidation product: **b.** Reduction product:

$$\begin{array}{c} H \diagdown \ \diagup O \\ C \\ HO-\!\!\!-H \\ HO-\!\!\!-H \\ H-\!\!\!-OH \\ CH_2OH \end{array}$$

D-Lyxose

Answers

a. Oxidation product:

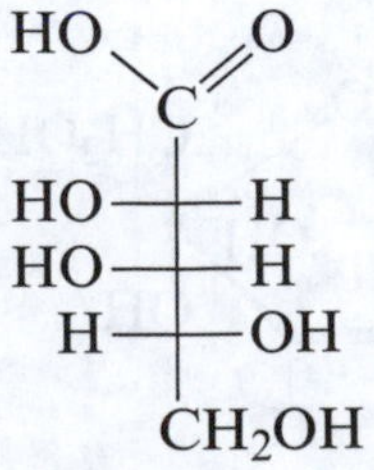

$$\begin{array}{c} HO \diagdown \ \diagup O \\ C \\ HO-\!\!\!-H \\ HO-\!\!\!-H \\ H-\!\!\!-OH \\ CH_2OH \end{array}$$

D-Lyxonic acid

b. Reduction product:

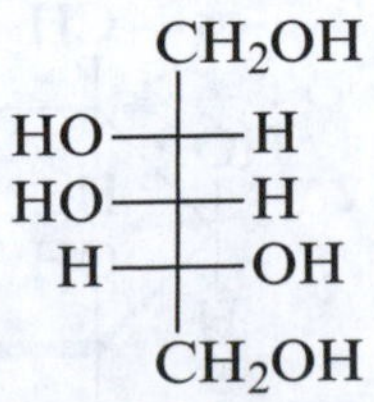

$$\begin{array}{c} CH_2OH \\ HO-\!\!\!-H \\ HO-\!\!\!-H \\ H-\!\!\!-OH \\ CH_2OH \end{array}$$

D-Lyxitol

15.6 Disaccharides

Learning Goal: Describe the monosaccharide units and linkages in disaccharides.

- Disaccharides have two monosaccharide units joined together by a glycosidic bond.

    Monosaccharide (1) + monosaccharide (2) $\longrightarrow$ disaccharide + H_2O

- In the most common disaccharides, maltose, lactose, and sucrose, there is at least one glucose unit.
- In the disaccharide maltose, two glucose units are linked by an $\alpha(1\rightarrow4)$ bond. The $\alpha(1\rightarrow4)$ indicates that the —OH group on carbon 1 of α-D-glucose is bonded to carbon 4 of the other glucose molecule.
- In lactose there is a $\beta(1\rightarrow4)$-glycosidic bond because the —OH group on carbon 1 of β-D-galactose forms a glycosidic bond with the —OH group on carbon 4 of a D-glucose molecule.
- Sucrose has an α-D-glucose and a β-D-fructose molecule joined by an $\alpha,\beta(1\rightarrow2)$-glycosidic bond.

    Glucose + glucose $\longrightarrow$ maltose + H_2O

    Glucose + galactose $\longrightarrow$ lactose + H_2O

    Glucose + fructose $\longrightarrow$ sucrose + H_2O

Lactose is a disaccharide found in milk and milk products.

Credit: Pearson Education, Inc.

♦ **Learning Exercise 15.6**

For the following disaccharides, state (**a**) the monosaccharide units, (**b**) the type of glycosidic bond, and (**c**) the name of the disaccharide:

1.

2.

3.

4.

	a. Monosaccharide(s)	b. Type of Glycosidic Bond	c. Name of Disaccharide
1.			
2.			
3.			
4.			

Answers   **1. a.** two glucose units   **b.** $\alpha(1 \rightarrow 4)$-glycosidic bond   **c.** β-maltose

2. a. galactose + glucose   **b.** $\beta(1 \rightarrow 4)$-glycosidic bond   **c.** α-lactose

3. a. fructose + glucose   **b.** $\alpha,\beta(1 \rightarrow 2)$-glycosidic bond   **c.** sucrose

4. a. two glucose units   **b.** $\alpha(1 \rightarrow 4)$-glycosidic bond   **c.** α-maltose

15.7   Polysaccharides

Learning Goal: Describe the structural features of amylose, amylopectin, glycogen, and cellulose.

The polysaccharide cellulose is the structural material in plants such as cotton.

Credit: Danny E Hooks/Shutterstock

- Polysaccharides are polymers of monosaccharide units.
- Starches consist of amylose and amylopectin. Amylose is an unbranched chain of glucose connected by $\alpha(1 \rightarrow 4)$-bonds.
- Amylopectin is a branched-chain polysaccharide; the glucose molecules are connected by $\alpha(1 \rightarrow 4)$-glycosidic bonds. However, at about every 25 glucose units, there is a branch of glucose molecules attached by an $\alpha(1 \rightarrow 6)$-glycosidic bond between carbon 1 of the branch and carbon 6 in the main chain.
- Glycogen, the storage form of glucose in animals, is similar to amylopectin, but has more branching.
- Cellulose is also a polymer of glucose, but in cellulose the glycosidic bonds are $\beta(1 \rightarrow 4)$-bonds rather than α bonds as in the starches. Humans can digest starches, but not cellulose, to obtain energy. However, cellulose is important as a source of fiber in our diets.

Key Terms for Sections 15.1 to 15.7

Match each of the following key terms with the correct description:

   a. carbohydrate   **b.** glucose   **c.** chiral carbon   **d.** Haworth structure
   e. disaccharide   **f.** cellulose   **g.** Fischer projection

1. _____ a simple or complex sugar composed of carbon, hydrogen, and oxygen

2. _____ a carbon that is bonded to four different groups

3. _____ a cyclic structure that represents the closed-chain form of a monosaccharide

4. _____ an unbranched polysaccharide that cannot be digested by humans

5. _____ an aldohexose that is the most prevalent monosaccharide in the diet

6. _____ a carbohydrate that contains two monosaccharides linked by a glycosidic bond

7. _____ a system for drawing chiral molecules that uses horizontal lines for bonds coming forward and vertical lines for bonds going back, with the chiral atom at the center

Answers   **1.** a   **2.** c   **3.** d   **4.** f   **5.** b   **6.** e   **7.** g

♦ **Learning Exercise 15.7** 🏃

List the monosaccharides and describe the glycosidic bonds in each of the following carbohydrates:

	Monosaccharides	**Type(s) of glycosidic bonds**
a. amylose	_____	_____
b. amylopectin	_____	_____
c. glycogen	_____	_____
d. cellulose	_____	_____

Answers
    a. glucose; $\alpha(1\rightarrow4)$-glycosidic bonds
    b. glucose; $\alpha(1\rightarrow4)$- and $\alpha(1\rightarrow6)$-glycosidic bonds
    c. glucose; $\alpha(1\rightarrow4)$- and $\alpha(1\rightarrow6)$-glycosidic bonds
    d. glucose; $\beta(1\rightarrow4)$-glycosidic bonds

Checklist for Chapter 15

You are ready to take the Practice Test for Chapter 15. Be sure you have accomplished the following learning goals for this chapter. If not, review the Section listed at the end of the goal. Then apply your new skills and understanding to the Practice Test.

After studying Chapter 15, I can successfully:

_____ Classify carbohydrates as monosaccharides, disaccharides, and polysaccharides. (15.1)

_____ Classify a monosaccharide as an aldose or ketose, and indicate the number of carbon atoms. (15.1)

_____ Identify a molecule as chiral or achiral; draw the D- and L-Fischer projections. (15.2)

_____ Draw and identify D- and L-Fischer projections for carbohydrate molecules. (15.3)

_____ Draw the open-chain structures for D-glucose, D-galactose, and D-fructose. (15.3)

_____ Draw or identify the Haworth structures of monosaccharides. (15.4)

_____ Describe some chemical properties of carbohydrates. (15.5)

_____ Describe the monosaccharide units and linkages in disaccharides. (15.6)

_____ Describe the structural features of amylose, amylopectin, glycogen, and cellulose. (15.7)

Practice Test for Chapter 15

The chapter Sections to review are shown in parentheses at the end of each question.

1. The requirements for photosynthesis are (15.1)
  A. sun   **B.** sun and water
  C. water and carbon dioxide   **D.** sun, water, and carbon dioxide
  E. carbon dioxide and sun

2. What are the products of photosynthesis? (15.1)
  A. carbohydrates   **B.** carbohydrates and oxygen
  C. carbon dioxide and oxygen   **D.** carbohydrates and carbon dioxide
  E. water and oxygen

3. The name "carbohydrate" came from the fact that (15.1)
   A. carbohydrates are hydrates of water
   B. carbohydrates contain hydrogen and oxygen in a 2:1 ratio
   C. carbohydrates contain a great quantity of water
   D. all plants produce carbohydrates
   E. carbon and hydrogen atoms are abundant in carbohydrates

4. What functional groups are in the open-chain forms of monosaccharides? (15.1)
   A. hydroxyl groups
   B. aldehyde groups
   C. ketone groups
   D. hydroxyl and aldehyde or ketone groups
   E. hydroxyl and ether groups

5. What is the classification of the following monosaccharide? (15.1)

$$CH_2OH$$
$$|$$
$$C=O$$
$$|$$
$$CH_2OH$$

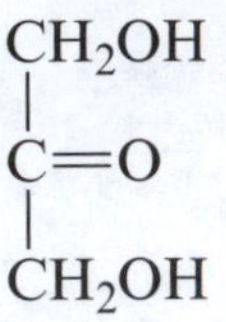

   A. aldotriose   B. ketotriose   C. aldotetrose   D. ketotetrose   E. ketopentose

For questions 6 through 9, identify each of the following pairs of Fischer projections as enantiomers (E) or identical compounds (I): (15.2)

6. 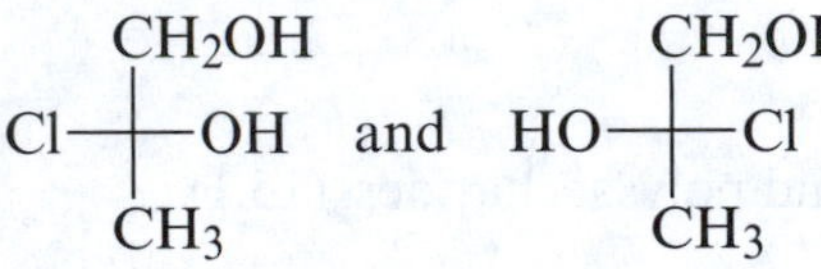 Cl—OH and HO—Cl (with CH₂OH top, CH₃ bottom)

7. 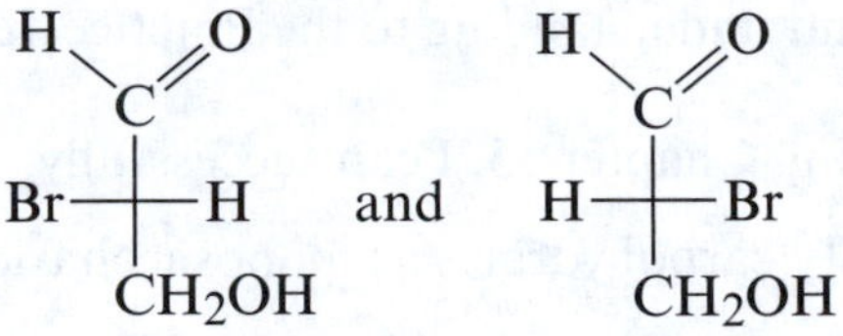 Br—H and H—Br (with CHO top, CH₂OH bottom)

8. 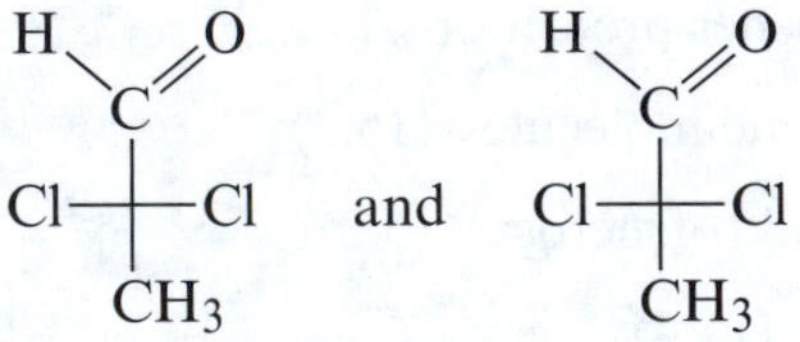 Cl—Cl and Cl—Cl (with CHO top, CH₃ bottom)

9. 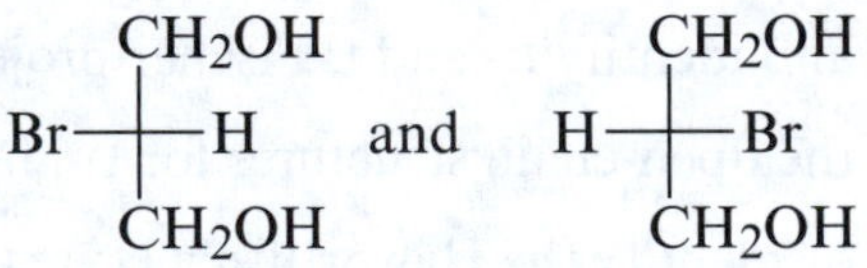 Br—H and H—Br (with CH₂OH top and bottom)

10. Identify the following as the D or L isomer: (15.3)

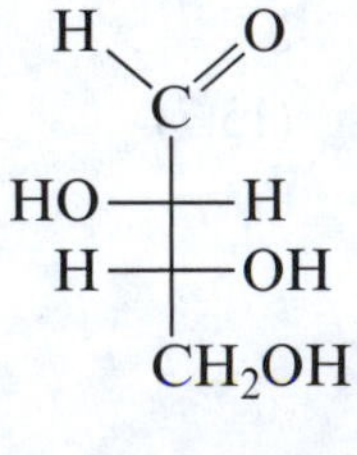

CHO / HO—H / H—OH / CH₂OH

For questions 11 through 15, refer to the following monosaccharide: (15.1, 15.4, 15.5, 15.7)

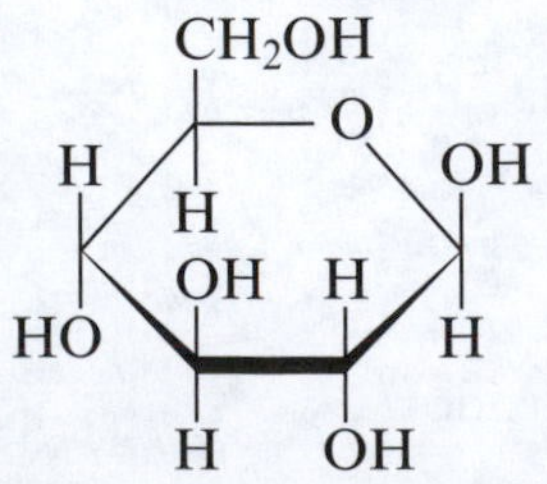

11. It is the Haworth structure of a(n)
    A. aldotriose   **B.** ketopentose   **C.** aldopentose   **D.** aldohexose   **E.** aldoheptose

12. It is the Haworth structure for
    A. fructose   **B.** glucose   **C.** ribose   **D.** glyceraldehyde   **E.** galactose

13. It is one of the products of the complete hydrolysis of
    A. maltose   **B.** sucrose   **C.** lactose   **D.** glycogen   **E.** all of these

14. A Benedict's test with this sugar would
    A. be positive   **B.** be negative   **C.** produce a blue precipitate
    D. give no color change   **E.** produce a silver mirror

15. It is the monosaccharide unit used to build the following polymer:
    A. amylose   **B.** amylopectin   **C.** cellulose   **D.** glycogen   **E.** all of these

For questions 16 through 20, identify each carbohydrate described as one of the following: (15.6, 15.7)
    A. maltose   **B.** sucrose   **C.** cellulose   **D.** amylopectin   **E.** glycogen

16. _____ a disaccharide that is not a reducing sugar

17. _____ a disaccharide that occurs as a breakdown product of amylose

18. _____ a carbohydrate that is produced as a storage form of energy in plants

19. _____ the storage form of energy in humans

20. _____ a carbohydrate that is used for structural purposes by plants

For questions 21 through 25, identify each carbohydrate described as one of the following: (15.6, 15.7)
    A. amylose   **B.** cellulose   **C.** glycogen   **D.** lactose   **E.** sucrose

21. _____ a polysaccharide composed of many glucose units linked only by $\alpha(1\rightarrow4)$-glycosidic bonds

22. _____ a sugar containing both glucose and galactose

23. _____ a polysaccharide composed of glucose units joined by both $\alpha(1\rightarrow4)$- and $\alpha(1\rightarrow6)$-glycosidic bonds

24. _____ a disaccharide that is a reducing sugar

25. _____ a carbohydrate composed of glucose units joined by $\beta(1\rightarrow4)$-glycosidic bonds

For questions 26 through 30, identify each carbohydrate described as one of the following: (15.6)
    A. glucose   **B.** lactose   **C.** sucrose   **D.** maltose

26. _____ a sugar composed of glucose and fructose

27. _____ also called table sugar

28. _____ found in milk and milk products

29. _____ gives glucitol upon reduction

30. _____ gives galactose upon hydrolysis

Answers to the Practice Test

1. D	**2.** B	**3.** B	**4.** D	**5.** B
6. E	**7.** E	**8.** I	**9.** I	**10.** D
11. D	**12.** B	**13.** E	**14.** A	**15.** E
16. B	**17.** A	**18.** D	**19.** E	**20.** C
21. A	**22.** D	**23.** C	**24.** D	**25.** B
26. C	**27.** C	**28.** B	**29.** A	**30.** B

Selected Answers and Solutions to Text Problems

15.1 Photosynthesis requires CO_2, H_2O, and the energy from the Sun. Respiration requires O_2 from the air and glucose from our foods.

15.3 Monosaccharides can be a chain of three to eight carbon atoms—one in a carbonyl group as an aldehyde or ketone, and the rest attached to hydroxyl groups. A monosaccharide cannot be split or hydrolyzed into smaller carbohydrates. A disaccharide consists of two monosaccharide units joined by a glycosidic bond. A disaccharide can be hydrolyzed into two monosaccharide units.

15.5 Hydroxyl groups are found in all monosaccharides, along with a carbonyl on the first or second carbon that gives an aldehyde or ketone functional group.

15.7 A ketopentose contains hydroxyl and ketone functional groups and has five carbon atoms.

15.9 **a.** This six-carbon monosaccharide has a carbonyl group on carbon 2; it is a ketohexose.
 b. This five-carbon monosaccharide has a carbonyl group on carbon 1; it is an aldopentose.

15.11 a. Achiral; there are no carbon atoms attached to four different groups.

 b. Chiral;

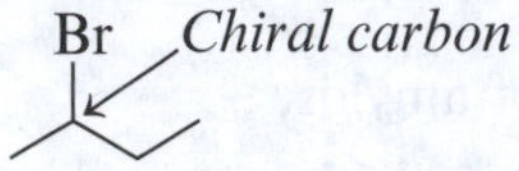

 c. Chiral;

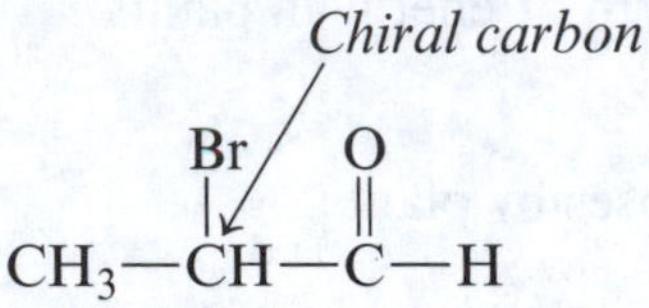

 d. Achiral; there are no carbon atoms attached to four different groups.

15.13 a.

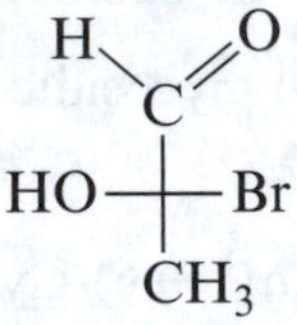

 b.

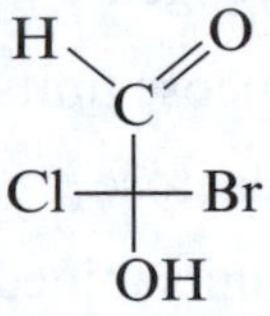

 c.

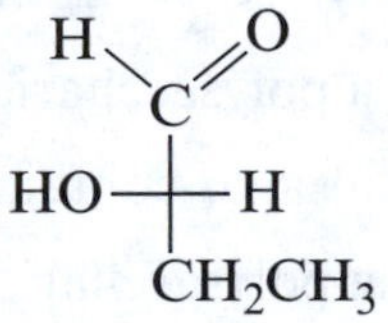

15.15 Enantiomers are nonsuperimposable mirror images.
 a. identical structures (no chiral carbons)
 b. enantiomers
 c. enantiomers
 d. enantiomers

15.17 a. When the —OH group is drawn to the right of the chiral carbon, it is the D isomer.
 b. When the —OH group is drawn to the right of the chiral carbon, it is the D isomer.
 c. When the —OH group is drawn to the left of the chiral carbon farthest from the top of the Fischer projection, it is the L isomer.

15.19 a. 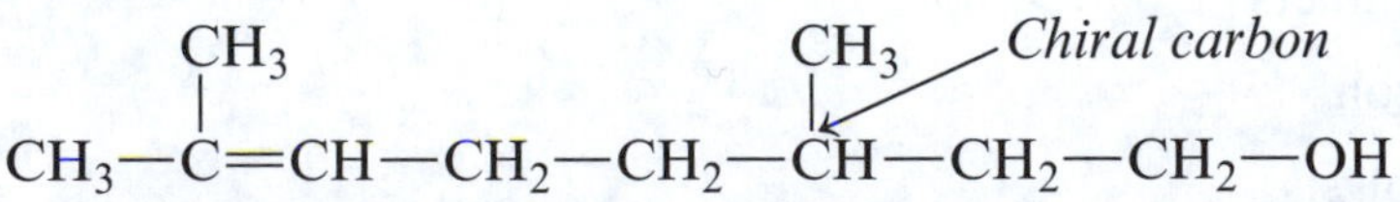 $CH_3-C=CH-CH_2-CH_2-CH-CH_2-CH_2-OH$

 b. 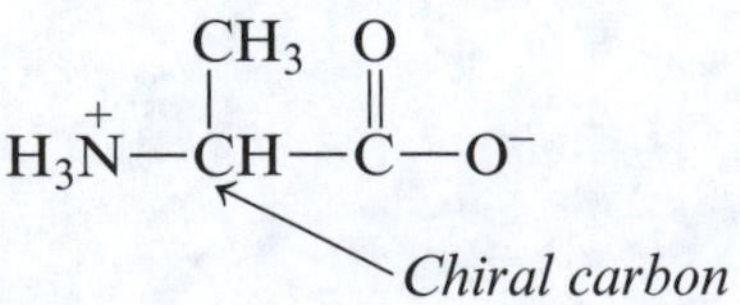 $H_3\overset{+}{N}-CH-C-O^-$

15.21 a. This structure is the D enantiomer since the hydroxyl group on the chiral carbon farthest from the carbonyl group is on the right.

b. This structure is the D enantiomer since the hydroxyl group on the chiral carbon farthest from the carbonyl group is on the right.

c. This structure is the L enantiomer since the hydroxyl group on the chiral carbon farthest from the carbonyl group is on the left.

d. This structure is the D enantiomer since the hydroxyl group on the chiral carbon farthest from the carbonyl group is on the right.

15.23 a.

$$
\begin{array}{c}
H \diagdown C \diagup^{O} \\
H \!-\!\!|\!-\! OH \\
HO \!-\!\!|\!-\! H \\
CH_2OH
\end{array}
$$

b.

$$
\begin{array}{c}
CH_2OH \\
C\!=\!O \\
H \!-\!\!|\!-\! OH \\
HO \!-\!\!|\!-\! H \\
CH_2OH
\end{array}
$$

c.

$$
\begin{array}{c}
H \diagdown C \diagup^{O} \\
HO \!-\!\!|\!-\! H \\
HO \!-\!\!|\!-\! H \\
H \!-\!\!|\!-\! OH \\
H \!-\!\!|\!-\! OH \\
CH_2OH
\end{array}
$$

d.

$$
\begin{array}{c}
H \diagdown C \diagup^{O} \\
HO \!-\!\!|\!-\! H \\
HO \!-\!\!|\!-\! H \\
HO \!-\!\!|\!-\! H \\
HO \!-\!\!|\!-\! H \\
CH_2OH
\end{array}
$$

15.25 L-Glucose is the mirror image of D-glucose.

$$
\begin{array}{c}
H \diagdown C \diagup^{O} \\
H \!-\!\!|\!-\! OH \\
HO \!-\!\!|\!-\! H \\
H \!-\!\!|\!-\! OH \\
H \!-\!\!|\!-\! OH \\
CH_2OH
\end{array}
$$

D-Glucose

$$
\begin{array}{c}
H \diagdown C \diagup^{O} \\
HO \!-\!\!|\!-\! H \\
H \!-\!\!|\!-\! OH \\
HO \!-\!\!|\!-\! H \\
HO \!-\!\!|\!-\! H \\
CH_2OH
\end{array}
$$

L-Glucose

15.27 In D-galactose, the —OH group on carbon 4 extends to the left; in D-glucose, this —OH group extends to the right.

15.29 a. Glucose is also called blood sugar.
b. Galactose is not metabolized in the condition called galactosemia.

15.31 In the cyclic structure of glucose, there are five carbon atoms and an oxygen atom.

15.33 In the α isomer, the —OH group on carbon 1 is drawn down; in the β isomer, the —OH group on carbon 1 is drawn up.

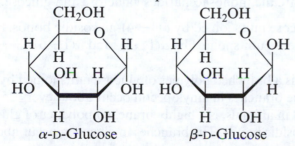

α-D-Glucose   β-D-Glucose

15.35 a. This is the α isomer because the —OH group on carbon 2 is down.
b. This is the α isomer because the —OH group on carbon 1 is down.

15.37 Oxidation product (sugar acid):   Reduction product (sugar alcohol):

Oxidation product:

```
    HO    O
      \  //
       C
   H ──┼── OH
  HO ──┼── H
   H ──┼── OH
      CH₂OH
```

D-Xylonic acid

Reduction product:

```
      CH₂OH
   H ──┼── OH
  HO ──┼── H
   H ──┼── OH
      CH₂OH
```

D-Xylitol

15.39 Oxidation product (sugar acid):   Reduction product (sugar alcohol):

Oxidation product:

```
    HO    O
      \  //
       C
  HO ──┼── H
   H ──┼── OH
   H ──┼── OH
      CH₂OH
```

D-Arabinonic acid

Reduction product:

```
      CH₂OH
  HO ──┼── H
   H ──┼── OH
   H ──┼── OH
      CH₂OH
```

D-Arabitol

15.41 a. When this disaccharide is hydrolyzed, galactose and glucose are produced. The glycosidic bond is a $\beta(1\rightarrow4)$ bond since the ether bond is drawn up from carbon 1 of the galactose unit, which is on the left in the drawing, to carbon 4 of the glucose on the right. β-Lactose is the name of this disaccharide since the —OH group on carbon 1 of the glucose unit is drawn up.

b. When this disaccharide is hydrolyzed, two molecules of glucose are produced. The glycosidic bond is an $\alpha(1\rightarrow4)$ bond since the ether bond is drawn down from carbon 1 of the glucose unit on the left to carbon 4 of the glucose on the right. α-Maltose is the name of this disaccharide since the —OH group on the rightmost glucose unit is drawn down.

15.43 a. β-Lactose is a reducing sugar; the ring on the right can open up to form an aldehyde that can undergo oxidation.

b. α-Maltose is a reducing sugar; the ring on the right can open up to form an aldehyde that can undergo oxidation.

15.45 a. Another name for table sugar is sucrose.

b. Lactose is the disaccharide found in milk and milk products.

c. Maltose is also called malt sugar.

d. When lactose is hydrolyzed, the products are the monosaccharides galactose and glucose.

15.47 a. Amylose is an unbranched polymer of glucose units joined by $\alpha(1\rightarrow4)$-glycosidic bonds. Amylopectin is a branched polymer of glucose units joined by $\alpha(1\rightarrow4)$- and $\alpha(1\rightarrow6)$-glycosidic bonds.

b. Amylopectin, which is produced in plants, is a branched polymer of glucose units joined by $\alpha(1\rightarrow4)$- and $\alpha(1\rightarrow6)$-glycosidic bonds. The branches in amylopectin occur about every 25 glucose units. Glycogen, which is produced in animals, is a highly branched polymer of glucose units joined by $\alpha(1\rightarrow4)$- and $\alpha(1\rightarrow6)$-glycosidic bonds. The branches in glycogen occur about every 10 to 15 glucose units.

15.49 a. Cellulose is not digestible by humans.

b. Amylose and amylopectin are the storage forms of carbohydrates in plants.

c. Amylose is the polysaccharide that contains only $\alpha(1\rightarrow4)$-glycosidic bonds.

d. Glycogen is the most highly branched polysaccharide.

15.51 $3.9 \text{ L blood} \times \dfrac{10 \text{ dL blood}}{1 \text{ L blood}} \times \dfrac{178 \text{ mg glucose}}{1 \text{ dL blood}} \times \dfrac{1 \text{ g glucose}}{1000 \text{ mg glucose}} = 6.9 \text{ g of glucose (2 SFs)}$

15.53 a. Total carbohydrate $= 23 \text{ g} + 24 \text{ g} + 26 \text{ g} + 0 \text{ g} = 73 \text{ g}$ (2 SFs)

Therefore, Kate has exceeded her limit of 45 to 60 g of carbohydrate.

b. $73 \text{ g carbohydrate} \times \dfrac{4 \text{ kcal}}{1 \text{ g carbohydrate}} = 290 \text{ kcal}$ (rounded off to the tens place)

15.55 a. Isomaltose is a disaccharide.
b. Isomaltose consists of two α-D-glucose units.
c. The glycosidic link in isomaltose is an $\alpha(1\rightarrow6)$-glycosidic bond.
d. The structure shown is α-isomaltose.
e. α-Isomaltose is a reducing sugar; the ring on the right can open up to form an aldehyde that can undergo oxidation.

15.57 a. Melezitose is a trisaccharide.
b. Melezitose contains two units of the aldohexose α-D-glucose and one unit of the ketohexose β-D-fructose.
c. Melezitose, like sucrose, is not a reducing sugar; the rings on the right cannot open up to form an aldehyde that can undergo oxidation.

15.59 a. Sucrose is the disaccharide found in sugarcane and table sugar.
b. Cellulose is the structural polysaccharide found in cotton and other plants.

15.61 A chiral carbon is bonded to four different groups.

a. 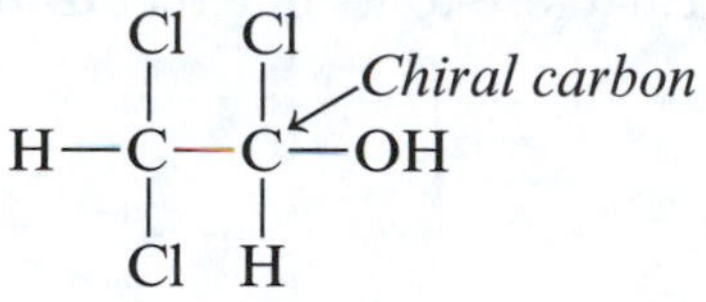 (structure with chiral carbon) **b.** none

c. none **d.** 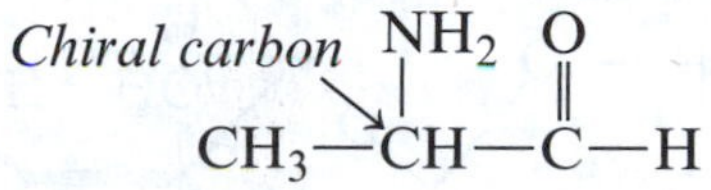 (structure with chiral carbon)

e. 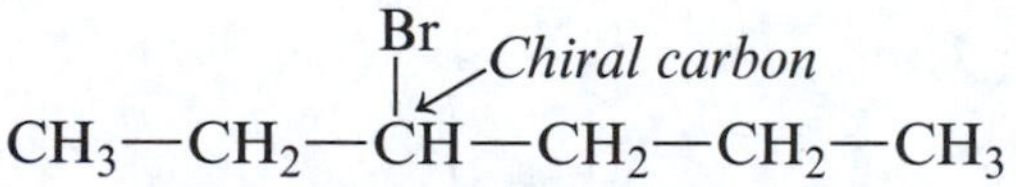 (structure with chiral carbon)

15.63 Enantiomers are nonsuperimposable mirror images.
a. identical compounds (no chiral carbons) **b.** enantiomers
c. enantiomers **d.** identical compounds (no chiral carbons)

15.65 D-Fructose is a ketohexose with the carbonyl group on carbon 2; D-galactose is an aldohexose where the carbonyl group is on carbon 1. In the Fischer projection of D-galactose, the —OH group on carbon 4 is drawn on the left; in fructose, the —OH group on carbon 4 is drawn on the right.

15.67 D-Galactose is the mirror image of L-galactose. In the Fischer projection of D-galactose, the —OH groups on carbon 2 and carbon 5 are drawn on the right side, but they are drawn on the left for carbon 3 and carbon 4. In L-galactose, the —OH groups are reversed: carbon 2 and carbon 5 have —OH groups drawn on the left, and carbon 3 and carbon 4 have —OH groups drawn on the right.

15.69 a.

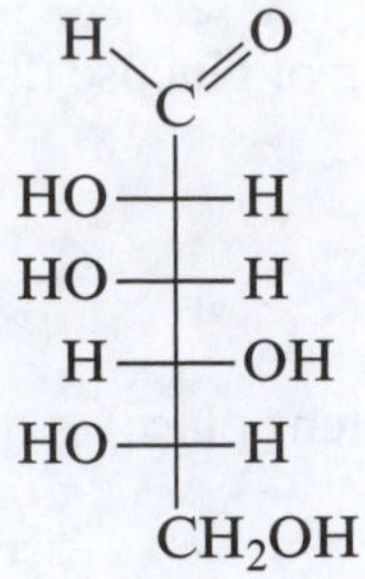

L-Gulose

b.

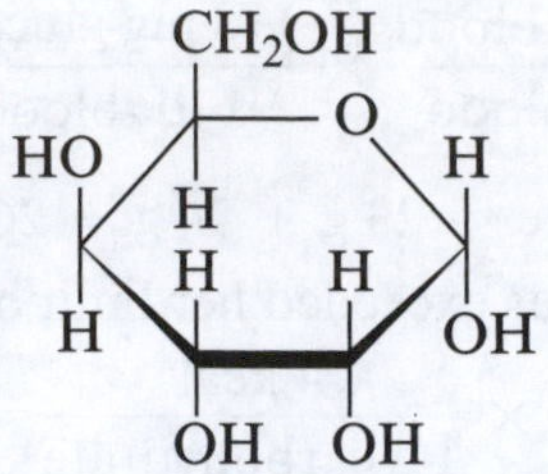

α-D-Gulose β-D-Gulose

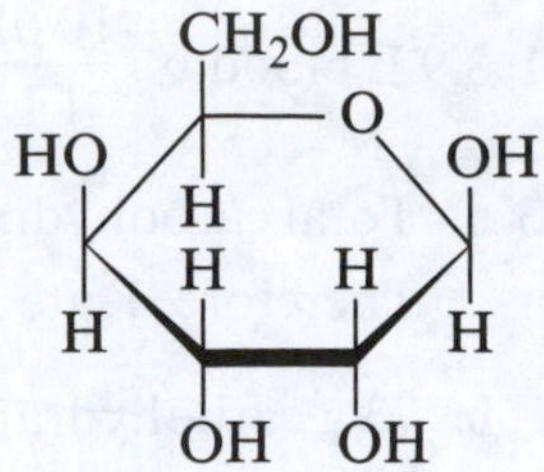

15.71 Since D-sorbitol can be oxidized to D-glucose, it must contain the same number of carbons with the same groups attached as glucose. The difference is that sorbitol has only hydroxyl groups, while glucose has an aldehyde group. In sorbitol, the aldehyde group is changed to a hydroxyl group.

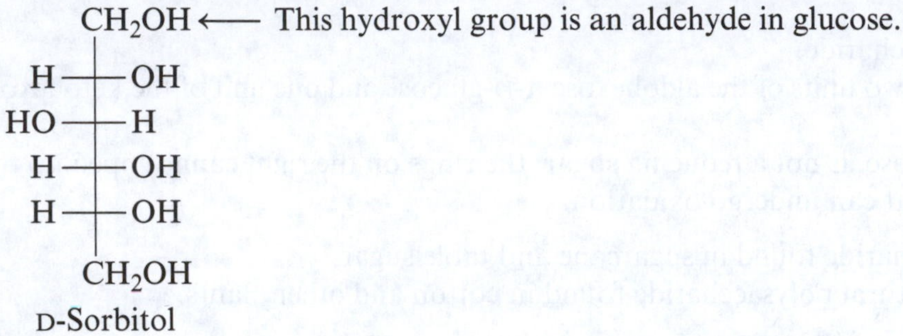

D-Sorbitol

15.73 When α-galactose forms an open-chain structure in water, it can close to form either α- or β-galactose.

15.75

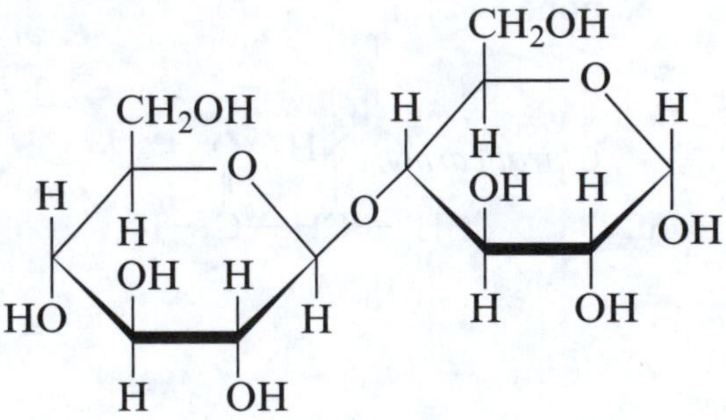

15.77 a.

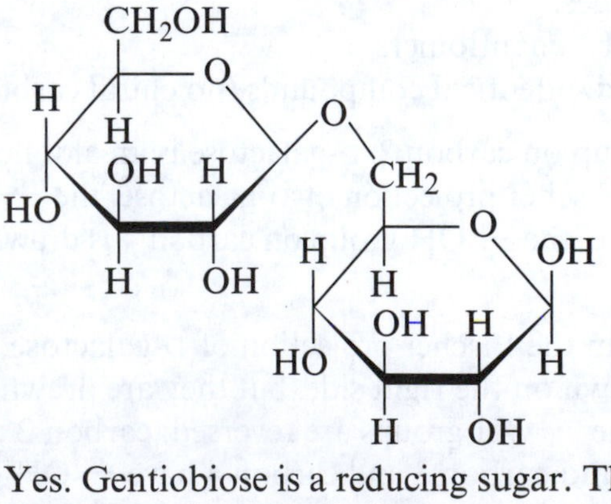

b. Yes. Gentiobiose is a reducing sugar. The ring on the right can open up to form an aldehyde that can undergo oxidation.

16

Carboxylic Acids and Esters

Robert has recuperated well from his heart surgery. After his operation, Maureen, a surgical technician, applies Indermil, a liquid adhesive, to close Robert's incision sites. The adhesive, which contains an ester, is applied to the edges of his skin. Once applied, it hardens quickly to close incisions and keeps the edges together. Within 8 to 10 days after Robert's surgery, the clear, dry adhesive is peeled from the incision sites, leaving his skin with minimal scarring.

Use line-angle structural formulas to write the reaction for the formation of Indermil by the esterification of 2-cyanopropenoic acid and 1-butanol.

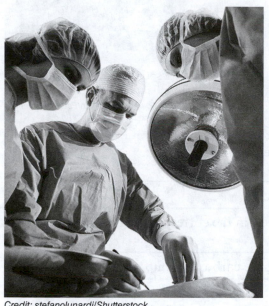

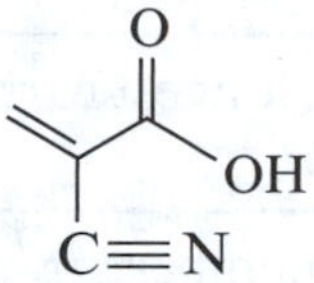

Credit: stefanolunardi/Shutterstock

2-Cyanopropenoic acid

LOOKING AHEAD

16.1 Carboxylic Acids
16.2 Properties of Carboxylic Acids

16.3 Esters

16.4 Properties of Esters

 The Health icon indicates a question that is related to health and medicine.

16.1 Carboxylic Acids

Learning Goal: Write the IUPAC and common names for carboxylic acids; draw their condensed structural and line-angle formulas.

REVIEW
Naming and Drawing Alkanes (12.2)

- The IUPAC names of carboxylic acids replace the *e* of the corresponding alkane with *oic acid*. Simple acids usually are named by the common system using the prefixes *form* (1C), *acet* (2C), *propion* (3C), or *butyr* (4C), followed by *ic acid*.

Methanoic acid
(formic acid)

Ethanoic acid
(acetic acid)

Butanoic acid
(butyric acid)

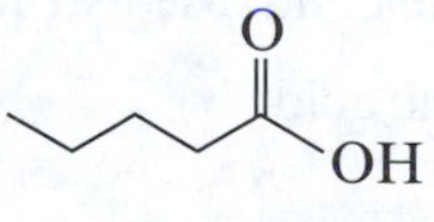

Pentanoic acid

- The name of the carboxylic acid of benzene is benzoic acid. When substituents are bonded to benzene, the ring is numbered from the carboxylic acid on carbon 1 in the direction that gives the substituents the smallest possible numbers. For common names, the prefixes *ortho*, *meta*, and *para* are used to show the position of substituents.

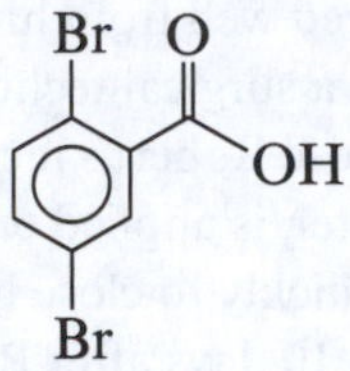

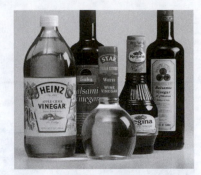

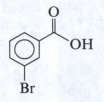

Benzoic acid | 2,5-Dibromobenzoic acid | 3-Bromobenzoic acid (*m*-bromobenzoic acid)

The sour taste of vinegar is due to ethanoic acid (acetic acid).
Credit: Pearson Education/Pearson Science

Naming Carboxylic Acids	
STEP 1	Identify the longest carbon chain and replace the *e* in the corresponding alkane name with *oic acid.*
STEP 2	Name and number any substituents by counting the carboxyl group as carbon 1.

♦ Learning Exercise 16.1A

CORE CHEMISTRY SKILL
Naming Carboxylic Acids

Write the IUPAC name (and common name, if any) for each of the following carboxylic acids:

1. $CH_3-\overset{\displaystyle O}{\overset{\|}{C}}-OH$

2. $CH_3-\overset{\displaystyle Br}{\overset{|}{CH}}-\overset{\displaystyle O}{\overset{\|}{C}}-OH$

3.

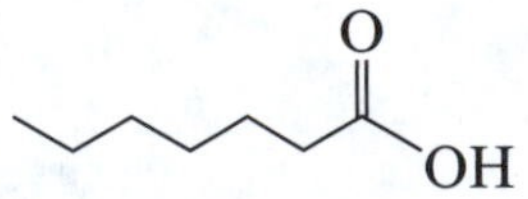

4.

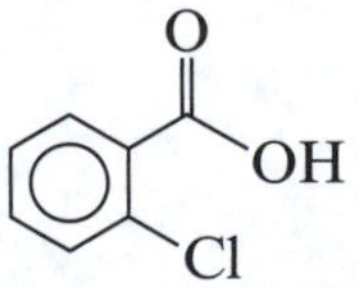

Answers
1. ethanoic acid (acetic acid)
2. 2-bromopropanoic acid (α-bromopropionic acid)
3. heptanoic acid
4. 2-chlorobenzoic acid (*o*-chlorobenzoic acid)

♦ Learning Exercise 16.1B

Draw the condensed structural formulas for **1**, **2**, and **3** and the line-angle formulas for **4**, **5**, and **6**.

1. acetic acid
2. β-chloropropanoic acid
3. methanoic acid

4. 2-bromobutanoic acid
5. 3-iodobenzoic acid
6. 3-methylpentanoic acid

Answers

1. $CH_3-\overset{\displaystyle O}{\overset{\|}{C}}-OH$

2. $Cl-CH_2-CH_2-\overset{\displaystyle O}{\overset{\|}{C}}-OH$

3. $H-\overset{\displaystyle O}{\overset{\|}{C}}-OH$

4.

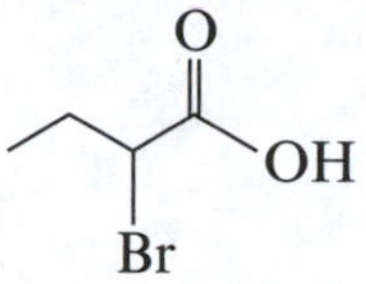

5.

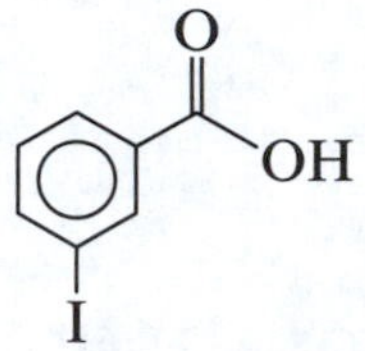

6.

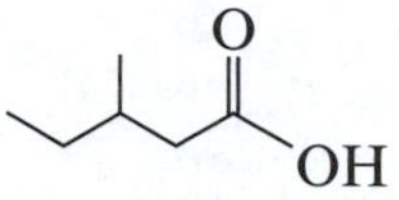

16.2 Properties of Carboxylic Acids

Learning Goal: Describe the boiling points, solubility, dissociation, and neutralization of carboxylic acids.

REVIEW

Writing Equations for Reactions of Acids and Bases (11.7)

- Carboxylic acids have higher boiling points than other polar compounds such as alcohols.
- Because they have two polar groups, carboxylic acids form dimers, which contain two sets of hydrogen bonds.
- Carboxylic acids with one to five carbon atoms are soluble in water.
- As weak acids, carboxylic acids dissociate slightly in water to form acidic solutions of H_3O^+ and carboxylate ions.
- When bases neutralize carboxylic acids, the products are carboxylate salts and water.

◆ **Learning Exercise 16.2A**

Identify the compound in each pair that has the higher boiling point:

1. acetic acid or butyric acid

2. propanoic acid or 2-propanol

3. propanoic acid or propanone

4. acetic acid or acetaldehyde

Answers   1. butyric acid   2. propanoic acid   3. propanoic acid   4. acetic acid

◆ **Learning Exercise 16.2B**

Indicate whether each of the following carboxylic acids is soluble (S) or not soluble (NS) in water:

1. _____ hexanoic acid

2. _____ acetic acid

3. _____ propanoic acid

4. _____ benzoic acid

5. _____ formic acid

6. _____ octanoic acid

Answers   1. NS   2. S   3. S   4. NS   5. S   6. NS

♦ **Learning Exercise 16.2C**

Draw the condensed structural formulas for the products from the dissociation of each of the following carboxylic acids in water:

1.
$$CH_3-\overset{\displaystyle O}{\overset{\|}{C}}-OH + H_2O \rightleftharpoons$$

2. 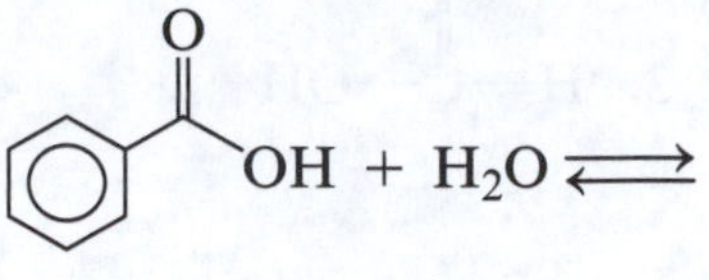
$OH + H_2O \rightleftharpoons$

Answers

1. $CH_3-\overset{\displaystyle O}{\overset{\|}{C}}-O^- + H_3O^+$

2. 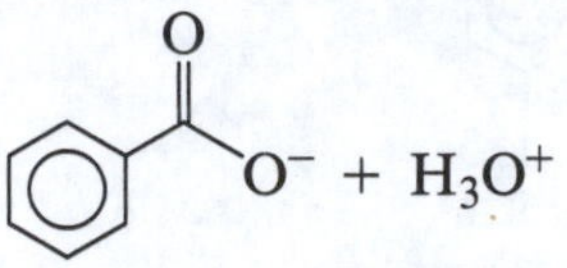 $O^- + H_3O^+$

♦ **Learning Exercise 16.2D**

Draw the condensed structural formulas for the products and write the name of the carboxylate salt produced in each of the following reactions:

1.
$$CH_3-CH_2-\overset{\displaystyle O}{\overset{\|}{C}}-OH + NaOH \longrightarrow$$

2.
$$H-\overset{\displaystyle O}{\overset{\|}{C}}-OH + KOH \longrightarrow$$

Answers

1. $CH_3-CH_2-\overset{\displaystyle O}{\overset{\|}{C}}-O^- Na^+ + H_2O$

   Sodium propanoate
   (sodium propionate)

2. $H-\overset{\displaystyle O}{\overset{\|}{C}}-O^- K^+ + H_2O$

   Potassium methanoate
   (potassium formate)

16.3   Esters

Learning Goal: Write the IUPAC and common names for esters; draw condensed structural and line-angle formulas.

- In the presence of a strong acid, a carboxylic acid reacts with an alcohol to produce an ester and water.

$$CH_3-\overset{\displaystyle O}{\overset{\|}{C}}-OH + HO-CH_2-CH_3 \underset{}{\overset{H^+, \text{ heat}}{\rightleftharpoons}} CH_3-\overset{\displaystyle O}{\overset{\|}{C}}-O-CH_2-CH_3 + H_2O$$

- The names of esters consist of two words, one from the alcohol and the other from the carboxylic acid, with the *ic acid* ending replaced by *ate*.

$$CH_3-\overset{\displaystyle O}{\overset{\|}{C}}-O-CH_3$$
Methyl ethanoate
(methyl acetate)

$$CH_3-CH_2-\overset{\displaystyle O}{\overset{\|}{C}}-O-CH_2-CH_3$$
Ethyl propanoate
(ethyl propionate)

♦ **Learning Exercise 16.3A**

Draw the condensed structural or line-angle formulas for the products of each of the following esterification reactions:

1. $CH_3-\overset{\displaystyle O}{\overset{\|}{C}}-OH + HO-CH_3 \underset{\longleftarrow}{\overset{H^+,\ heat}{\longrightarrow}}$

2. $H-\overset{\displaystyle O}{\overset{\|}{C}}-OH + HO-CH_2-CH_3 \underset{\longleftarrow}{\overset{H^+,\ heat}{\longrightarrow}}$

3. 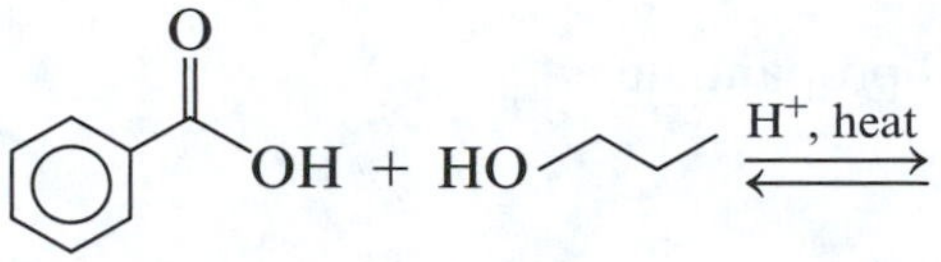 (benzene ring)$-\overset{\displaystyle O}{\overset{\|}{C}}-OH + HO$⁀⁀ $\underset{\longleftarrow}{\overset{H^+,\ heat}{\longrightarrow}}$

4. propanoic acid and ethanol $\underset{\longleftarrow}{\overset{H^+,\ heat}{\longrightarrow}}$

Answers

1. $CH_3-\overset{\displaystyle O}{\overset{\|}{C}}-O-CH_3 + H_2O$

2. $H-\overset{\displaystyle O}{\overset{\|}{C}}-O-CH_2-CH_3 + H_2O$

3. 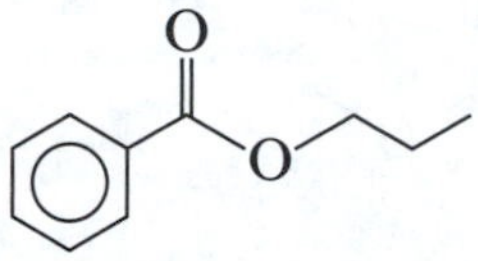 (benzene ring)$-\overset{\displaystyle O}{\overset{\|}{C}}-O$⁀⁀ $+ H_2O$

4. $CH_3-CH_2-\overset{\displaystyle O}{\overset{\|}{C}}-O-CH_2-CH_3 + H_2O$

Naming Esters	
STEP 1	Write the name for the carbon chain from the alcohol as an *alkyl* group.
STEP 2	Change the *ic acid* of the acid name to *ate*.

The odor of grapes is due to an ester.

Credit: Lynn Watson/ Shutterstock

♦ **Learning Exercise 16.3B**

Write the IUPAC and common name, if any, for each of the following:

1. 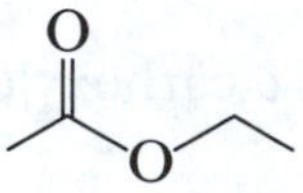 (line-angle ester structure)

2. $CH_3-CH_2-CH_2-\overset{\displaystyle O}{\overset{\|}{C}}-O-CH_3$

3. $CH_3-CH_2-\overset{\displaystyle O}{\overset{\|}{C}}-O-CH_2-CH_2-CH_3$

4. 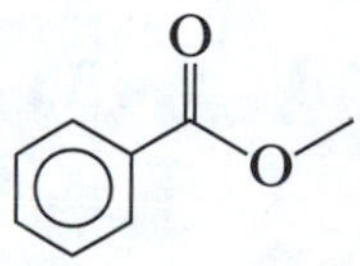 (benzene ring with ester structure)

Answers
1. ethyl ethanoate (ethyl acetate)
3. propyl propanoate (propyl propionate)
2. methyl butanoate (methyl butyrate)
4. methyl benzoate

♦ **Learning Exercise 16.3C**

Draw the condensed structural formulas for **1** and **2** and the line-angle formulas for **3** and **4**.

1. propyl acetate

2. ethyl butyrate

3. ethyl benzoate

4. methyl propanoate

Answers

1. $CH_3-\overset{\overset{\displaystyle O}{\|}}{C}-O-CH_2-CH_2-CH_3$ 2. $CH_3-CH_2-CH_2-\overset{\overset{\displaystyle O}{\|}}{C}-O-CH_2-CH_3$

3.

4.

16.4 Properties of Esters

Learning Goal: Describe the boiling points and solubility of esters; draw the condensed structural and line-angle formulas for the products from acid and base hydrolysis of esters.

- Esters have higher boiling points than alkanes but lower boiling points than alcohols and carboxylic acids of similar mass.

- In hydrolysis, esters are split apart by a reaction with water. When the catalyst is an acid, the products are a carboxylic acid and an alcohol.

$$CH_3-\overset{\overset{\displaystyle O}{\|}}{C}-O-CH_3 + H_2O \overset{H^+,\ heat}{\rightleftharpoons} CH_3-\overset{\overset{\displaystyle O}{\|}}{C}-OH + HO-CH_3$$

Methyl ethanoate   Ethanoic acid   Methanol
(methyl acetate)   (acetic acid)   (methyl alcohol)

- Saponification is the hydrolysis of an ester in the presence of a base, which produces a carboxylate salt and an alcohol.

$$CH_3-\overset{\overset{\displaystyle O}{\|}}{C}-O-CH_3 + NaOH \overset{Heat}{\longrightarrow} CH_3-\overset{\overset{\displaystyle O}{\|}}{C}-O^- Na^+ + HO-CH_3$$

Methyl ethanoate   Sodium ethanoate   Methanol
(methyl acetate)   (sodium acetate)   (methyl alcohol)

Key Terms for Sections 16.1 to 16.4

Match each of the following key terms with the correct description:

     a. carboxylic acid   **b.** saponification   **c.** esterification   **d.** hydrolysis   **e.** ester

1. _____ an organic compound containing the carboxyl group ($-COOH$)

2. _____ a reaction of a carboxylic acid and an alcohol in the presence of an acid catalyst

3. _____ a type of organic compound that produces pleasant aromas in flowers and fruits

4. _____ the hydrolysis of an ester with a strong base, producing a carboxylate salt and an alcohol

5. _____ the splitting of a molecule such as an ester by the addition of water in the presence of an acid

Answers   **1.** a   **2.** c   **3.** e   **4.** b   **5.** d

♦ Learning Exercise 16.4A

Identify the compound with the higher boiling point in each of the following pairs of compounds:

1. CH_3-CH_2-OH or $\underset{\displaystyle \overset{\displaystyle O}{\|}}{H-C}-O-CH_3$

2. $CH_3-\underset{\displaystyle \overset{\displaystyle O}{\|}}{C}-O-CH_3$ or $CH_3-\underset{\displaystyle \overset{\displaystyle OH}{|}}{CH}-CH_2-CH_3$

3. 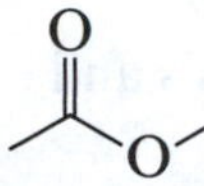 (line-angle structure) or (line-angle structure)

Answers   **1.** CH_3-CH_2-OH **2.** $CH_3-\underset{\displaystyle \overset{\displaystyle OH}{|}}{CH}-CH_2-CH_3$ **3.** 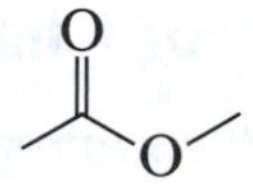 (line-angle structure)

♦ Learning Exercise 16.4B

CORE CHEMISTRY SKILL
Hydrolyzing Esters

Draw the condensed or line-angle structural formulas for the products of hydrolysis or saponification for each of the following esters:

1. $CH_3-CH_2-CH_2-\underset{\displaystyle \overset{\displaystyle O}{\|}}{C}-O-CH_3 + H_2O \xrightarrow{\text{H}^+,\ \text{heat}}$

2. $CH_3-\underset{\displaystyle \overset{\displaystyle O}{\|}}{C}-O-CH_3 + NaOH \xrightarrow{\text{Heat}}$

3. 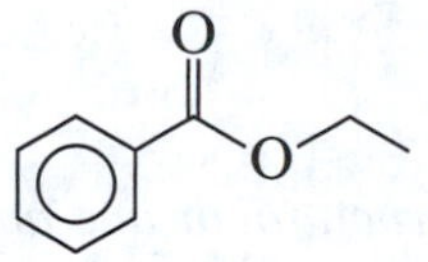 (line-angle structure) $+ KOH \xrightarrow{\text{Heat}}$

4. 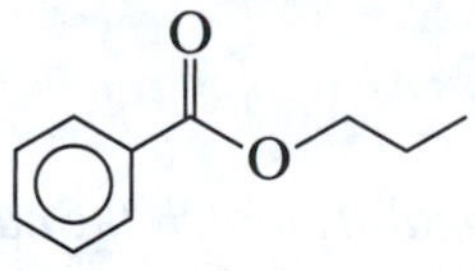 (line-angle structure) $+ H_2O \xrightarrow{\text{H}^+,\ \text{heat}}$

Answers

1. $CH_3-CH_2-CH_2-\underset{\displaystyle \overset{\displaystyle O}{\|}}{C}-OH + HO-CH_3$

2. $CH_3-\underset{\displaystyle \overset{\displaystyle O}{\|}}{C}-O^-\ Na^+ + HO-CH_3$

3. 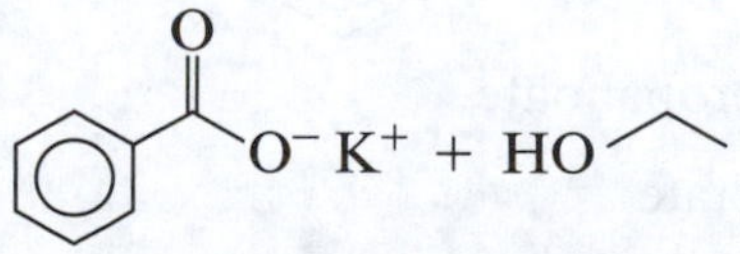 (line-angle structure) $O^-\ K^+ + HO$ (line-angle structure)

4. 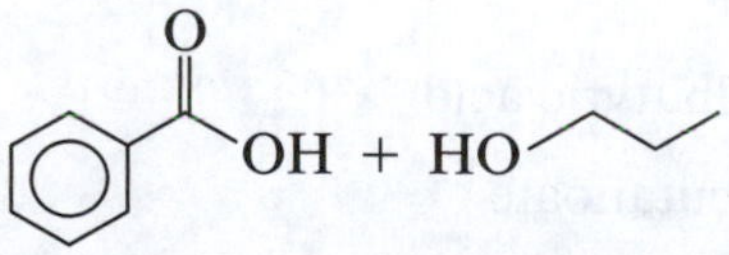 (line-angle structure) $OH + HO$ (line-angle structure)

Checklist for Chapter 16

You are ready to take the Practice Test for Chapter 16. Be sure you have accomplished the following learning goals for this chapter. If not, review the Section listed at the end of the goal. Then apply your new skills and understanding to the Practice Test.

After studying Chapter 16, I can successfully:

_____ Write the IUPAC and common names, and draw condensed and line-angle structural formulas for carboxylic acids. (16.1)

_____ Describe the boiling points, solubility, dissociation, and neutralization of carboxylic acids. (16.2)

_____ Draw the condensed structural and line-angle formulas for the products of neutralization of carboxylic acids. (16.2)

_____ Write equations for the preparation of esters. (16.3)

_____ Write the IUPAC or common names and draw the condensed structural and line-angle formulas for esters. (16.3)

_____ Describe the boiling points and solubility of esters. (16.4)

_____ Draw the condensed structural and line-angle formulas for the products from hydrolysis and saponification of esters. (16.4)

Practice Test for Chapter 16

The chapter Sections to review are shown in parentheses at the end of each question.

For questions 1 through 5, match each condensed structural or line-angle formula to its functional group: (16.1, 16.3)

    A. alcohol   **B.** aldehyde   **C.** carboxylic acid   **D.** ester   **E.** ketone

1. _____ $CH_3-CH-CH_2-OH$ (with CH_3 branch)

2. _____ $CH_3-CH_2-\overset{O}{\overset{\|}{C}}-OH$

3. _____ $CH_3-CH_2-\overset{O}{\overset{\|}{C}}-H$

4. _____ $CH_3-\overset{O}{\overset{\|}{C}}-O-CH_3$

5. _____

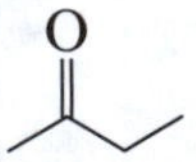

For questions 6 through 11, match the names of the compounds with their condensed structural or line-angle formulas: (16.1, 16.4)

A. $CH_3-\overset{O}{\overset{\|}{C}}-O-CH_2-CH_3$ **B.** $CH_3-CH_2-CH_2-\overset{O}{\overset{\|}{C}}-O^- Na^+$ **C.** $CH_3-\overset{O}{\overset{\|}{C}}-O^- Na^+$

D. 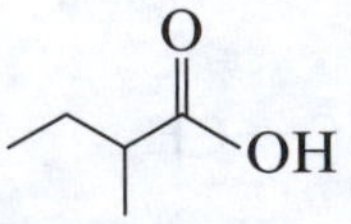   **E.** $CH_3-CH_2-\overset{O}{\overset{\|}{C}}-O-CH_3$ **F.**

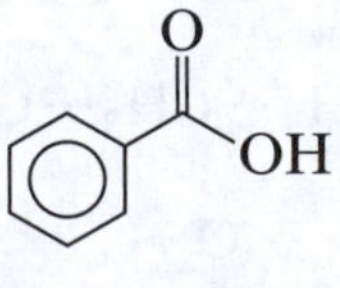

6. _____ α-methylbutyric acid   **7.** _____ methyl propanoate

8. _____ sodium butanoate   **9.** _____ ethyl acetate

10. _____ sodium acetate   **11.** _____ benzoic acid

12. The name of the following compound is: 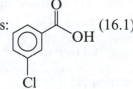 (16.1)

    A. *p*-chlorobenzoic acid   **B.** chlorobenzoic acid   **C.** *m*-chlorobenzoic acid
    D. 4-chlorobenzoic acid   **E.** benzoic acid chloride

13. The products of the following reaction are: (16.2)

$$H-\overset{\overset{\displaystyle O}{\|}}{C}-OH + H_2O \rightleftarrows$$

    A. $H-\overset{\overset{\displaystyle O}{\|}}{C}-O^- + H_3O^+$
    B. $H-\overset{\overset{\displaystyle O}{\|}}{C}-\overset{+}{O}H_2 + OH^-$
    C. $H-\overset{\overset{\displaystyle O}{\|}}{C}-O-CH_3$

    D. $CH_3-\overset{\overset{\displaystyle O}{\|}}{C}-OH$
    E. $H-\overset{\overset{\displaystyle O}{\|}}{C}-O^-Na^+ + H_2O$

14. The name of the following compound is: (16.2)

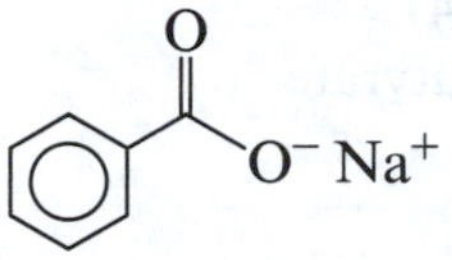

    A. benzene sodium   **B.** sodium benzoate   **C.** *o*-benzoic acid
    D. sodium benzene carboxylate   **E.** benzoic acid

15. What is the product when a carboxylic acid reacts with sodium hydroxide? (16.2)
    A. carboxylate salt   **B.** alcohol   **C.** ester
    D. aldehyde   **E.** no reaction

16. Carboxylic acids are water-soluble due to their (16.2)
    A. nonpolar nature   **B.** ionic bonds   **C.** ability to lower pH
    D. ability to hydrogen bond   **E.** high melting points

For questions 17 through 20, refer to the following reactions: (16.2, 16.4)

    A. $CH_3-\overset{\overset{\displaystyle O}{\|}}{C}-OH + HO-CH_3 \underset{\xrightarrow{H^+,\ heat}}{\rightleftarrows} CH_3-\overset{\overset{\displaystyle O}{\|}}{C}-O-CH_3 + H_2O$

    B. $CH_3-\overset{\overset{\displaystyle O}{\|}}{C}-OH + NaOH \longrightarrow CH_3-\overset{\overset{\displaystyle O}{\|}}{C}-O^-Na^+ + H_2O$

    C. $CH_3-\overset{\overset{\displaystyle O}{\|}}{C}-O-CH_3 + H_2O \underset{\xrightarrow{H^+,\ heat}}{\rightleftarrows} CH_3-\overset{\overset{\displaystyle O}{\|}}{C}-OH + HO-CH_3$

    D. $CH_3-\overset{\overset{\displaystyle O}{\|}}{C}-O-CH_3 + NaOH \xrightarrow{Heat} CH_3-\overset{\overset{\displaystyle O}{\|}}{C}-O^-Na^+ + HO-CH_3$

17. _____ is an acid hydrolysis   18. _____ is a neutralization

19. _____ is a saponification   20. _____ is an esterification

21. What is the name of the organic product in the following reaction? (16.3)

$$CH_3-\overset{\overset{\text{O}}{\|}}{C}-OH + HO-CH_3 \underset{\xrightarrow{\hspace{0.8cm}}}{\overset{H^+, \text{ heat}}{\rightleftharpoons}} CH_3-\overset{\overset{\text{O}}{\|}}{C}-O-CH_3 + H_2O$$

    A. methyl acetate   **B.** acetic acid   **C.** methyl alcohol
    D. acetaldehyde   **E.** ethyl methanoate

22. Identify the carboxylic acid and alcohol needed to produce the compound shown below: (16.3)

$$CH_3-CH_2-CH_2-\overset{\overset{\text{O}}{\|}}{C}-O-CH_2-CH_3$$

    A. propanoic acid and ethanol   **B.** acetic acid and 1-pentanol
    C. acetic acid and 1-butanol   **D.** butanoic acid and ethanol
    E. hexanoic acid and methanol

23. The name of $CH_3-CH_2-\overset{\overset{\text{O}}{\|}}{C}-O-CH_2-CH_3$ is: (16.3)

    A. ethyl acetate   **B.** ethyl ethanoate   **C.** ethyl propanoate
    D. propyl ethanoate   **E.** ethyl butyrate

24. The ester produced from the reaction of 1-butanol and propanoic acid is: (16.4)
    A. butyl propanoate   **B.** butyl propanone   **C.** propyl butyrate
    D. propyl butanone   **E.** heptanoate

25. In a hydrolysis reaction, (16.4)
    A. an acid reacts with an alcohol   **B.** an ester reacts with NaOH
    C. an ester reacts with H_2O **D.** an acid neutralizes a base
    E. water is added to an alkene

26. Esters (16.3, 16.4)
    A. have pleasant odors
    B. can undergo hydrolysis
    C. are formed from alcohols and carboxylic acids
    D. have a lower boiling point than the corresponding acid
    E. all of the above

27. The compound with the highest boiling point is: (16.2, 16.4)
    A. formic acid   **B.** acetic acid   **C.** propanol
    D. propanoic acid   **E.** ethyl acetate

28. The reaction of methyl acetate with NaOH produces (16.4)
    A. ethanol and formic acid   **B.** ethanol and sodium formate   **C.** ethanol and sodium ethanoate
    D. methanol and acetic acid   **E.** methanol and sodium acetate

Answers to the Practice Test

1. A	**2.** C	**3.** B	**4.** D	**5.** E
6. D	**7.** E	**8.** B	**9.** A	**10.** C
11. F	**12.** C	**13.** A	**14.** B	**15.** A
16. D	**17.** C	**18.** B	**19.** D	**20.** A
21. A	**22.** D	**23.** C	**24.** A	**25.** C
26. E	**27.** D	**28.** E		

Selected Answers and Solutions to Text Problems

16.1 Methanoic acid (formic acid) is the carboxylic acid that is responsible for the pain associated with ant stings.

16.3 **a.** $CH_3-CH_2-CH_2-CH_2-CH_2-\overset{\overset{\displaystyle O}{\|}}{C}-OH$ Hexanoic acid

b. $CH_3-CH_2-\overset{\overset{\displaystyle CH_3-CH_2}{|}}{CH}-\overset{\overset{\displaystyle O}{\|}}{C}-OH$ 2-Ethylbutanoic acid

16.5 **a.** Ethanoic acid (acetic acid) is the carboxylic acid with two carbons.
b. Butanoic acid (butyric acid) is the carboxylic acid with four carbons.
c. 3-Methylhexanoic acid is a six-carbon carboxylic acid with a methyl group on carbon 3 of the chain.
d. 3,4-Dibromobenzoic acid is an aromatic carboxylic acid with bromine atoms on carbon 3 and carbon 4 of the ring.

16.7 **a.** 2-Chloroethanoic acid is a carboxylic acid that has a two-carbon chain with a chlorine atom on carbon 2.

$Cl-CH_2-\overset{\overset{\displaystyle O}{\|}}{C}-OH$

b. 3-Hydroxypropanoic acid is a carboxylic acid that has a three-carbon chain with a hydroxyl group on carbon 3.

$HO-CH_2-CH_2-\overset{\overset{\displaystyle O}{\|}}{C}-OH$

c. α-Methylbutyric acid is a carboxylic acid that has a four-carbon chain with a methyl group on the carbon adjacent to the carboxyl group.

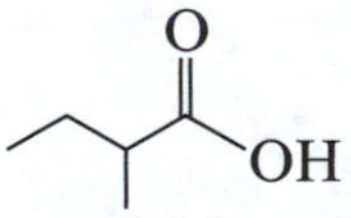

d. 3,5-Dibromoheptanoic acid is a carboxylic acid that has a seven-carbon chain with bromine atoms attached to carbon 3 and carbon 5.

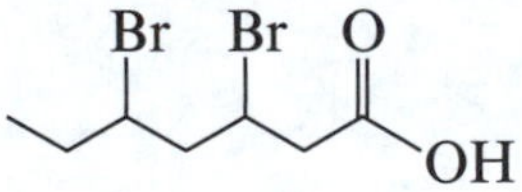

16.9 Aldehydes and primary alcohols oxidize to produce the corresponding carboxylic acid.

a. $H-\overset{\overset{\displaystyle O}{\|}}{C}-OH$

b. $CH_3-\overset{\overset{\displaystyle O}{\|}}{C}-OH$

c.

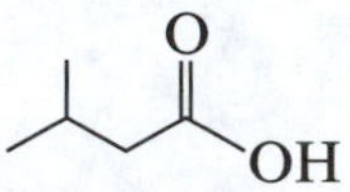

d.

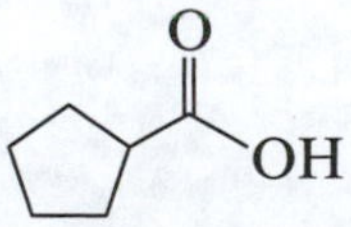

16.11 a. Butanoic acid has a greater molar mass and would have a higher boiling point than ethanoic acid.

    b. Propanoic acid can form dimers, effectively doubling the molar mass, which gives propanoic acid a higher boiling point than 1-propanol.

    c. Butanoic acid can form dimers, effectively doubling the molar mass, which gives butanoic acid a higher boiling point than butanone.

16.13 a. Propanoic acid is the most soluble of the group because it has the fewest number of carbon atoms in its hydrocarbon chain. Solubility of carboxylic acids decreases as the number of carbon atoms in the hydrocarbon chain increases.

    b. Propanoic acid is more soluble than 1-hexanol because it has fewer carbon atoms in its hydrocarbon chain. Propanoic acid is also more soluble than 1-hexanol because the carboxyl group forms more hydrogen bonds with water than does the hydroxyl group of an alcohol. An alkane is not soluble in water.

16.15 a. $CH_3-CH_2-CH_2-\overset{\overset{\displaystyle O}{\|}}{C}-OH + H_2O \rightleftharpoons CH_3-CH_2-CH_2-\overset{\overset{\displaystyle O}{\|}}{C}-O^- + H_3O^+$

    b. $CH_3-\overset{\overset{\displaystyle CH_3}{|}}{CH}-\overset{\overset{\displaystyle O}{\|}}{C}-OH + H_2O \rightleftharpoons CH_3-\overset{\overset{\displaystyle CH_3}{|}}{CH}-\overset{\overset{\displaystyle O}{\|}}{C}-O^- + H_3O^+$

16.17 a. $CH_3-CH_2-CH_2-CH_2-\overset{\overset{\displaystyle O}{\|}}{C}-OH + NaOH \longrightarrow$

        $CH_3-CH_2-CH_2-CH_2-\overset{\overset{\displaystyle O}{\|}}{C}-O^- Na^+ + H_2O$

    b. $CH_3-\overset{\overset{\displaystyle Cl}{|}}{CH}-\overset{\overset{\displaystyle O}{\|}}{C}-OH + NaOH \longrightarrow CH_3-\overset{\overset{\displaystyle Cl}{|}}{CH}-\overset{\overset{\displaystyle O}{\|}}{C}-O^- Na^+ + H_2O$

    c. $OH + NaOH \longrightarrow$ $O^- Na^+ + H_2O$

16.19 A carboxylic acid salt is named using the name of the metal cation followed by the name of the carboxylate anion, obtained by replacing the *ic acid* ending of the acid name with *ate*.

    a. sodium pentanoate

    b. sodium 2-chloropropanoate (sodium α-chloropropionate)

    c. sodium benzoate

16.21 A carboxylic acid reacts with an alcohol to form an ester and water. In an ester, the —H of the carboxylic acid is replaced by an alkyl group.

    a. $CH_3-\overset{\overset{\displaystyle O}{\|}}{C}-O-CH_2-CH_3$

    b. $CH_3-CH_2-CH_2-\overset{\overset{\displaystyle O}{\|}}{C}-O-CH_2-CH_3$

    c. $\overset{\overset{\displaystyle O}{\|}}{C}-O-CH_2-CH_3$

16.23 A carboxylic acid and an alcohol react to give an ester with the elimination of water.

a. (structure)   b. (structure)

16.25 a. The carboxylic acid part of the ester is from methanoic acid (formic acid), and the alcohol part is from methanol (methyl alcohol).

b. The carboxylic acid part of the ester is from propanoic acid (propionic acid), and the alcohol part is from ethanol (ethyl alcohol).

c. The carboxylic acid part of the ester is from butanoic acid (butyric acid), and the alcohol part is from methanol (methyl alcohol).

d. The carboxylic acid part of the ester is from 3-methylbutanoic acid (β-methylbutyric acid), and the alcohol part is from ethanol (ethyl alcohol).

16.27 a. The alcohol part of the ester is from methanol (methyl alcohol), and the carboxylic acid part is from methanoic acid (formic acid). The ester is named methyl methanoate (methyl formate).

b. The alcohol part of the ester is from ethanol (ethyl alcohol), and the carboxylic acid part is from propanoic acid (propionic acid). The ester is named ethyl propanoate (ethyl propionate).

c. The alcohol part of the ester is from methanol (methyl alcohol), and the carboxylic acid part is from butanoic acid (butyric acid). The ester is named methyl butanoate (methyl butyrate).

d. The alcohol part of the ester is from 2-methyl-1-propanol, and the carboxylic acid part is from pentanoic acid. The ester is named 2-methylpropyl pentanoate.

16.29 a. The alcohol part of the ester comes from the three-carbon 1-propanol, and the carboxylate part comes from the four-carbon butanoic acid.

$$CH_3-CH_2-CH_2-\overset{\overset{\displaystyle O}{\|}}{C}-O-CH_2-CH_2-CH_3$$

b. The alcohol part of the ester comes from the four-carbon 1-butanol, and the carboxylate part comes from the one-carbon formic acid.

$$H-\overset{\overset{\displaystyle O}{\|}}{C}-O-CH_2-CH_2-CH_2-CH_3$$

c. The alcohol part of the ester comes from the two-carbon ethanol, and the carboxylate part comes from the five-carbon pentanoic acid.

(structure)

d. The alcohol part of the ester comes from the one-carbon methanol, and the carboxylate part comes from the three-carbon propanoic acid.

(structure)

16.31 a. The flavor and odor of bananas is due to pentyl ethanoate (pentyl acetate).

b. The flavor and odor of oranges is due to octyl ethanoate (octyl acetate).

c. The flavor and odor of apricots is due to pentyl butanoate (pentyl butyrate).

16.33 a. $CH_3-CH_2-\overset{\overset{\displaystyle O}{\|}}{C}-OH$

b. $CH_3-CH_2-CH_2-CH_2-OH$

c. (structure)

16.35 a. Yes, esters with two to five carbon atoms are soluble in water (ethyl acetate has four carbon atoms).

b. No, esters with six or more carbon atoms are not soluble in water (pentyl butanoate has nine carbon atoms).

16.37 Acid hydrolysis of an ester gives the carboxylic acid and the alcohol that were combined to form the ester; base hydrolysis of an ester gives the salt of the carboxylic acid and the alcohol that were combined to form the ester.

a. $CH_3-CH_2-\overset{\overset{\displaystyle O}{\|}}{C}-O^-\ Na^+ + HO-CH_3$

b. 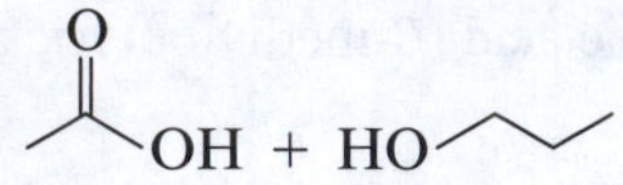 OH + HO⌒

c. $CH_3-CH_2-CH_2-\overset{\overset{\displaystyle O}{\|}}{C}-OH + HO-CH_2-CH_3$

d. 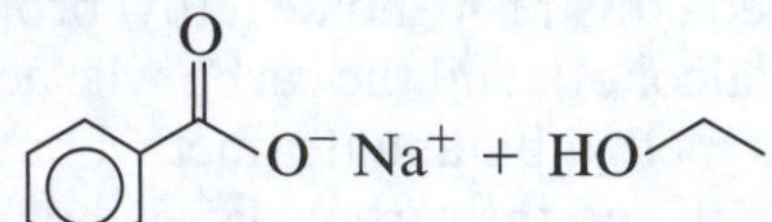 O⁻ Na⁺ + HO⌒

e. 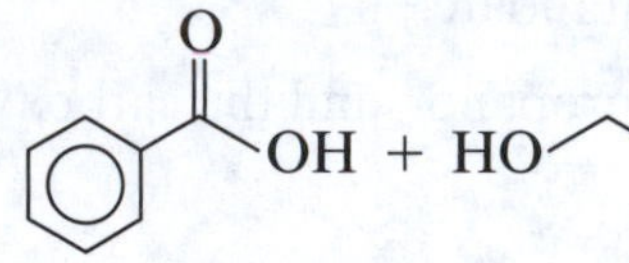 OH + HO⌒

16.39 a. $H_2C{=}\overset{\overset{\displaystyle CN}{|}}{C}-\overset{\overset{\displaystyle O}{\|}}{C}-OH + HO-CH_2-CH_2-CH_2-CH_3 \underset{\longleftarrow}{\overset{H^+,\ heat}{\longrightarrow}}$

$H_2C{=}\overset{\overset{\displaystyle CN}{|}}{C}-\overset{\overset{\displaystyle O}{\|}}{C}-O-CH_2-CH_2-CH_2-CH_3 + H_2O$

b. butyl 2-cyano-2-propenoate

16.41 $CH_3-CH_2-CH_2-\overset{\overset{\displaystyle O}{\|}}{C}-OH$ Butanoic acid

$CH_3-\overset{\overset{\displaystyle CH_3}{|}}{CH}-\overset{\overset{\displaystyle O}{\|}}{C}-OH$ 2-Methylpropanoic acid

16.43 $CH_3-\overset{\overset{\displaystyle O}{\|}}{C}-O-CH_3$ Methyl ethanoate

$H-\overset{\overset{\displaystyle O}{\|}}{C}-O-CH_2-CH_3$ Ethyl methanoate

16.45 a. $CH_3-CH_2-CH_2-\overset{\overset{\displaystyle O}{\|}}{C}-O-CH_3$

b. butanoic acid and methanol

c.
$$CH_3-CH_2-CH_2-\overset{\displaystyle O}{\overset{\|}{C}}-O-CH_3 + H_2O \underset{\longleftarrow}{\overset{H^+,\ heat}{\longrightarrow}}$$

$$CH_3-CH_2-CH_2-\overset{\displaystyle O}{\overset{\|}{C}}-OH + HO-CH_3$$

d.
$$CH_3-CH_2-CH_2-\overset{\displaystyle O}{\overset{\|}{C}}-O-CH_3 + NaOH \xrightarrow{Heat}$$

$$CH_3-CH_2-CH_2-\overset{\displaystyle O}{\overset{\|}{C}}-O^-\ Na^+ + HO-CH_3$$

16.47 a. 3-methylbutanoic acid (β-methylbutyric acid)
  b. ethyl benzoate
  c. ethyl propanoate (ethyl propionate)
  d. 2-chlorobenzoic acid (*o*-chlorobenzoic acid)
  e. pentanoic acid
  f. 2-propyl ethanoate (isopropyl acetate)

16.49

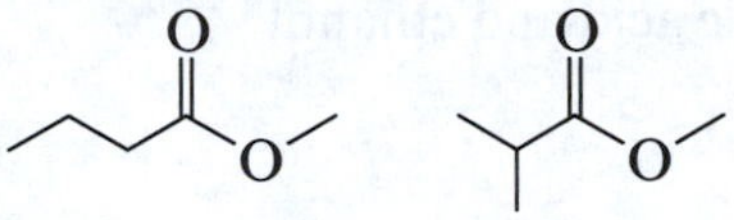

16.51 a.
$$CH_3-CH_2-CH_2-CH_2-CH_2-\overset{\displaystyle O}{\overset{\|}{C}}-O-CH_3$$

  b.

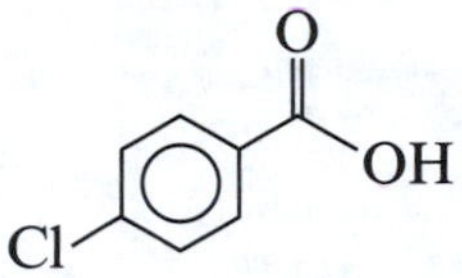

  c.
$$Cl-CH_2-CH_2-\overset{\displaystyle O}{\overset{\|}{C}}-OH$$

  d.
$$CH_3-CH_2-CH_2-\overset{\displaystyle O}{\overset{\|}{C}}-O-CH_2-CH_3$$

  e.
$$CH_3-CH_2-\overset{\displaystyle CH_3}{\overset{|}{CH}}-CH_2-\overset{\displaystyle O}{\overset{\|}{C}}-OH$$

  f.

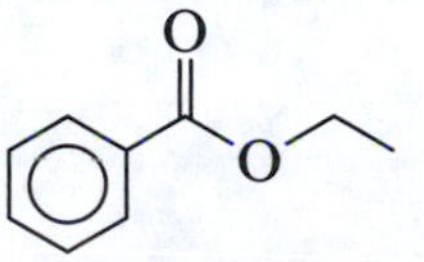

16.53 a. Ethanoic acid has a higher boiling point than 1-propanol because, although both molecules form hydrogen bonds, two molecules of ethanoic acid hydrogen bond to form a dimer, which effectively doubles the molar mass and requires a higher temperature to reach the boiling point.
  b. Ethanoic acid has the higher boiling point because it forms hydrogen bonds, whereas butane does not.

16.55 Of the three compounds, methyl formate would have the lowest boiling point since it has only dipole–dipole attractions. Both acetic acid and 1-propanol can form hydrogen bonds, but because acetic acid can form dimers and double the effective molar mass, it has the highest boiling point. Methyl formate, 32 °C; 1-propanol, 97 °C; acetic acid, 118 °C.

16.57 a, c, and **d** are soluble in water.

16.59 a. $CH_3-CH_2-\overset{\displaystyle O}{\overset{\|}{C}}-O^- + H_3O^+$

b. $CH_3-CH_2-\overset{\displaystyle O}{\overset{\|}{C}}-O^- K^+ + H_2O$

c. (structure of propanoate phenyl ester) $+ H_2O$

d. (structure of ethyl benzoate) $+ H_2O$

16.61 a. 3-methylbutanoic acid and methanol   **b.** benzoic acid and ethanol
c. hexanoic acid and ethanol

16.63 a. $CH_3-CH_2-\overset{\displaystyle O}{\overset{\|}{C}}-OH + HO-\overset{\displaystyle CH_3}{\overset{|}{C}H}-CH_3$

b. $CH_3-\overset{\displaystyle CH_3}{\overset{|}{C}H}-\overset{\displaystyle O}{\overset{\|}{C}}-O^- Na^+ + HO-CH_2-CH_2-CH_3$

16.65 a. $H_2C{=}CH_2 + H_2O \xrightarrow{H^+} CH_3-CH_2-OH \xrightarrow{[O]} CH_3-\overset{\displaystyle O}{\overset{\|}{C}}-OH$

b. $CH_3-CH_2-CH_2-CH_2-OH \xrightarrow{[O]} CH_3-CH_2-CH_2-\overset{\displaystyle O}{\overset{\|}{C}}-OH$

16.67 (benzene ring)$-\overset{\displaystyle O}{\overset{\|}{C}}-O-CH_3 + KOH \xrightarrow{\text{Heat}}$ (benzene ring)$-\overset{\displaystyle O}{\overset{\|}{C}}-O^- K^+ + HO-CH_3$

In KOH solution, the ester undergoes saponification to form the carboxylate salt, potassium benzoate, and methanol, which are soluble in water. When HCl is added, the salt is converted to benzoic acid, which is insoluble.

16.69 a. $CH_3-\overset{\displaystyle O}{\overset{\|}{C}}-O-CH_2-CH_2-CH_3$

b. $CH_3-\overset{\displaystyle O}{\overset{\|}{C}}-OH + HO-CH_2-CH_2-CH_3 \underset{\xrightarrow{\hspace{1cm}}}{\overset{H^+,\ heat}{\rightleftharpoons}}$

$CH_3-\overset{\displaystyle O}{\overset{\|}{C}}-O-CH_2-CH_2-CH_3 + H_2O$

c.
$$CH_3-\overset{\overset{\displaystyle O}{\|}}{C}-O-CH_2-CH_2-CH_3 \ + \ H_2O \ \underset{\longleftarrow}{\overset{H^+, \ heat}{\rightleftharpoons}}$$

$$CH_3-\overset{\overset{\displaystyle O}{\|}}{C}-OH \ + \ HO-CH_2-CH_2-CH_3$$

d.
$$CH_3-\overset{\overset{\displaystyle O}{\|}}{C}-O-CH_2-CH_2-CH_3 \ + \ NaOH \ \overset{Heat}{\longrightarrow}$$

$$CH_3-\overset{\overset{\displaystyle O}{\|}}{C}-O^- \ Na^+ \ + \ HO-CH_2-CH_2-CH_3$$

e. Molar mass of propyl acetate ($C_5H_{10}O_2$)
= 5(12.01 g) + 10(1.008 g) + 2(16.00 g) = 102.13 g/mole (5 SFs)

$$1.58 \ g \ \cancel{C_5H_{10}O_2} \times \frac{1 \ mole \ \cancel{C_5H_{10}O_2}}{102.13 \ g \ \cancel{C_5H_{10}O_2}} \times \frac{1 \ mole \ \cancel{NaOH}}{1 \ mole \ \cancel{C_5H_{10}O_2}} \times \frac{1 \ L \ \cancel{solution}}{0.208 \ \cancel{mole \ NaOH}} \times \frac{1000 \ mL \ solution}{1 \ L \ \cancel{solution}}$$

= 74.4 mL of a 0.208 M NaOH solution (3 SFs)

17
Lipids

Rebecca changes her diet by decreasing the fat content and increasing the amount of vegetables for more dietary fiber. She drinks six glasses of water every day, has a regular exercise program, and walks every day. Recently, Rebecca had a gallbladder attack. In the body, the gallbladder stores bile, which is used in the intestine for the digestion of fats. If the bile is not moving out of the gallbladder, it can form gallstones, which are mostly cholesterol. If a gallstone blocks the bile duct, gallbladder disease can occur. Rebecca's doctor prescribes a medication that dissolves gallstones and recommends a liver flush to remove small gallstones.

Draw the line-angle formula for cholesterol.

Credit: michaeljung/Fotolia

LOOKING AHEAD

17.1 Lipids
17.2 Fatty Acids
17.3 Waxes and Triacylglycerols

17.4 Chemical Properties of Triacylglycerols
17.5 Phospholipids

17.6 Steroids: Cholesterol, Bile Salts, and Steroid Hormones
17.7 Cell Membranes

The Health icon indicates a question that is related to health and medicine.

17.1 Lipids

Learning Goal: Describe the classes of lipids.

- Lipids are biomolecules that are not soluble in water but are soluble in organic solvents.
- Classes of lipids include waxes, triacylglycerols, glycerophospholipids, sphingolipids, and steroids.

♦ **Learning Exercise 17.1**

Match the correct class of lipids with the compositions below:

    a. wax   **b.** triacylglycerol   **c.** glycerophospholipid   **d.** steroid   **e.** sphingomyelin

1. _____ a fused structure of four cycloalkanes

2. _____ a long-chain alcohol and a long-chain saturated fatty acid

3. _____ glycerol and three fatty acids

4. _____ sphingosine, fatty acid, phosphate group, and an amino alcohol

5. _____ glycerol, two fatty acids, phosphate group, and an amino alcohol

Answers   **1.** d   **2.** a   **3.** b   **4.** e   **5.** c

17.2 Fatty Acids

Learning Goal: Draw the condensed structural and line-angle formulas for a fatty acid, and identify it as saturated or unsaturated.

- Fatty acids are unbranched carboxylic acids that typically contain an even number (12 to 20) of carbon atoms.
- Fatty acids may be saturated, monounsaturated with one carbon–carbon double bond, or polyunsaturated with two or more double bonds. The double bonds in naturally occurring unsaturated fatty acids are almost always cis.

♦ **Learning Exercise 17.2A**

Draw the condensed structural and line-angle formulas for each of the following fatty acids:

CORE CHEMISTRY SKILL
Identifying Fatty Acids

1. linoleic acid (18:2)

2. stearic acid (18:0)

3. palmitoleic acid (16:1)

Answers

1. $CH_3-(CH_2)_4-CH=CH-CH_2-CH=CH-(CH_2)_7-\overset{\overset{\displaystyle O}{\|}}{C}-OH$

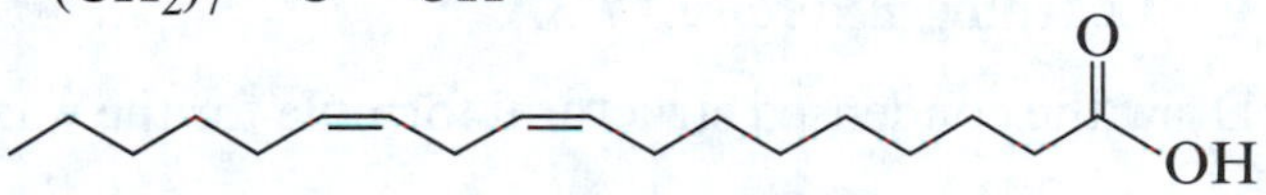

2. $CH_3-(CH_2)_{16}-\overset{\overset{\displaystyle O}{\|}}{C}-OH$

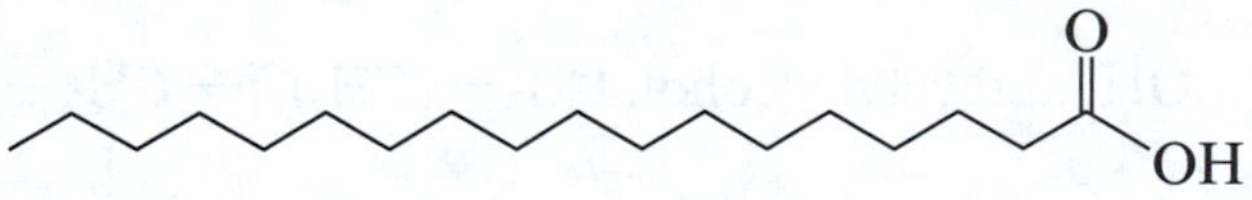

3. $CH_3-(CH_2)_5-CH=CH-(CH_2)_7-\overset{\overset{\displaystyle O}{\|}}{C}-OH$

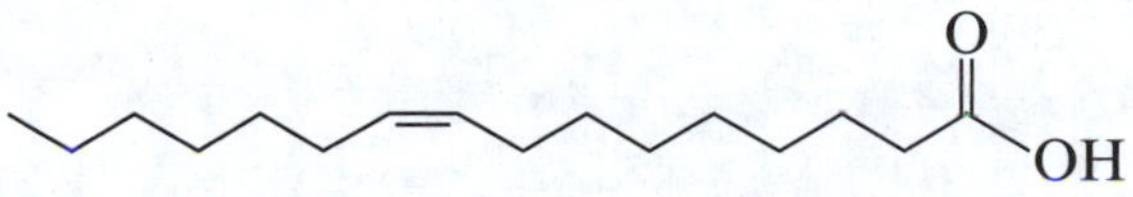

♦ **Learning Exercise 17.2B**

For the fatty acids in Learning Exercise 17.2A, identify which

a. _____ is the most saturated   **b.** _____ is the most unsaturated
c. _____ has the lowest melting point   **d.** _____ has the highest melting point
e. _____ is/are found in vegetables   **f.** _____ is/are from animal sources

Answers   **a.** 2   **b.** 1   **c.** 1   **d.** 2   **e.** 1, 3   **f.** 2

♦ **Learning Exercise 17.2C**

For the following questions, refer to the line-angle formula for the fatty acid:

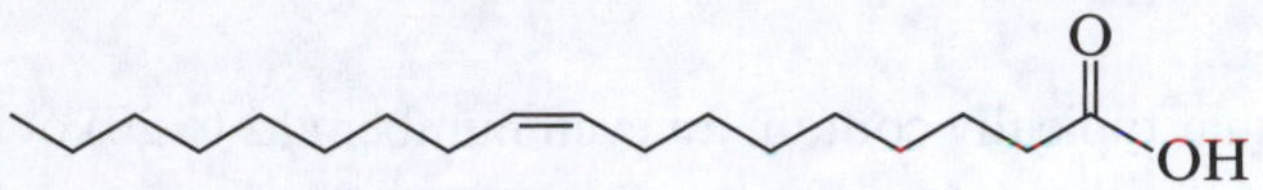

1. What is the name of this fatty acid? _____

2. Is it a saturated or an unsaturated compound? Why? _____

3. Is the double bond cis or trans? _____

4. Is it likely to be a solid or a liquid at room temperature? _____

5. Why is it insoluble in water? _____

Answers   1. oleic acid   2. unsaturated; double bond   3. cis
             4. liquid   5. It has a long hydrocarbon chain.

17.3   Waxes and Triacylglycerols

Learning Goal: Draw the condensed structural and line-angle formulas for a wax or triacylglycerol produced by the reaction of a fatty acid and an alcohol or glycerol.

* A wax is an ester of a long-chain saturated fatty acid and a long-chain alcohol.
* The triacylglycerols in fats and oils are esters of glycerol with three fatty acids.
* Fats from animal sources contain more saturated fatty acids and have higher melting points than most vegetable oils.

♦ **Learning Exercise 17.3A**

Draw the condensed structural formula for the wax formed by the reaction of palmitic acid,

$$CH_3-(CH_2)_{14}-\overset{\overset{\displaystyle O}{\|}}{C}-OH,$$ and cetyl alcohol, $CH_3-(CH_2)_{14}-CH_2-OH$.

Answer $$CH_3-(CH_2)_{14}-\overset{\overset{\displaystyle O}{\|}}{C}-O-CH_2-(CH_2)_{14}-CH_3$$

	Drawing the Structure for a Triacylglycerol
STEP 1	Draw the condensed structural formulas for glycerol and the fatty acids.
STEP 2	Form ester bonds between the hydroxyl groups on glycerol and the carboxyl groups on each fatty acid.

♦ **Learning Exercise 17.3B**

Draw the condensed structural formula for the triacylglycerol formed from glycerol and three molecules of each of the following fatty acids and give the name:

1. palmitic acid, $CH_3-(CH_2)_{14}-\overset{\displaystyle O}{\overset{\|}{C}}-OH$

2. myristic acid, $CH_3-(CH_2)_{12}-\overset{\displaystyle O}{\overset{\|}{C}}-OH$

Answers

1. 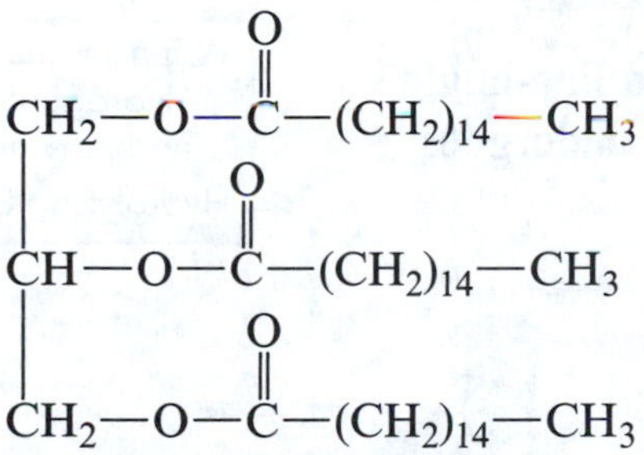
$$CH_2-O-\overset{\overset{\displaystyle O}{\|}}{C}-(CH_2)_{14}-CH_3$$
$$CH-O-\overset{\overset{\displaystyle O}{\|}}{C}-(CH_2)_{14}-CH_3$$
$$CH_2-O-\overset{\overset{\displaystyle O}{\|}}{C}-(CH_2)_{14}-CH_3$$
Glyceryl tripalmitate
(tripalmitin)

2. 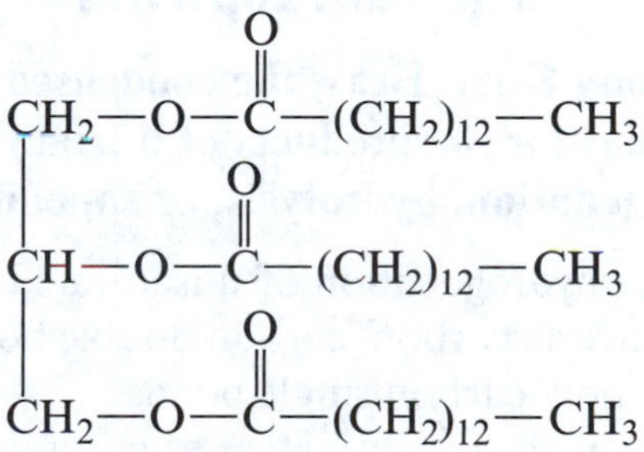
$$CH_2-O-\overset{\overset{\displaystyle O}{\|}}{C}-(CH_2)_{12}-CH_3$$
$$CH-O-\overset{\overset{\displaystyle O}{\|}}{C}-(CH_2)_{12}-CH_3$$
$$CH_2-O-\overset{\overset{\displaystyle O}{\|}}{C}-(CH_2)_{12}-CH_3$$
Glyceryl trimyristate
(trimyristin)

♦ **Learning Exercise 17.3C**

Draw the line-angle formula for each of the following triacylglycerols:

1. glyceryl tristearate (tristearin)

2. glyceryl trioleate (triolein)

Answers

1.

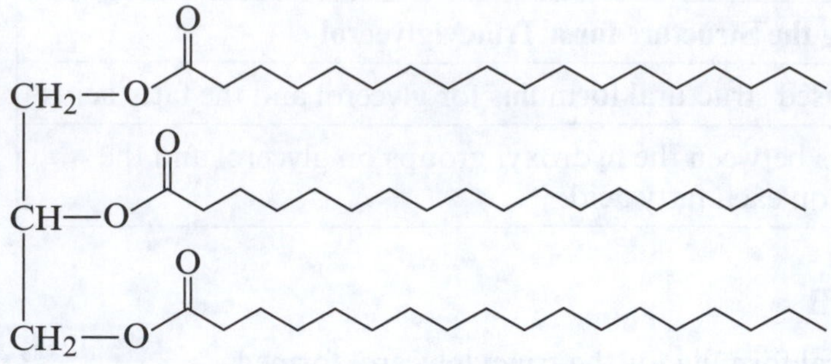

Glyceryl tristearate (tristearin)

2.

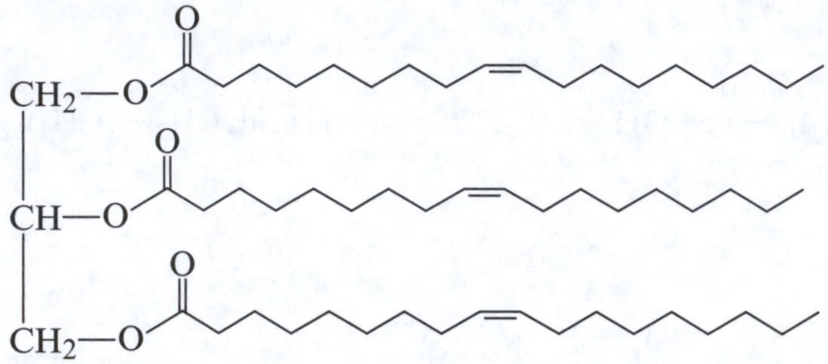

Glyceryl trioleate (triolein)

17.4 Chemical Properties of Triacylglycerols

Learning Goal: Draw the condensed structural and line-angle formulas for the products of a triacylglycerol that undergoes hydrogenation, hydrolysis, or saponification.

REVIEW

Writing Equations for
 Hydrogenation, Hydration,
 and Polymerization (12.7)
Hydrolyzing Esters (16.4)

- The hydrogenation of unsaturated fatty acids converts carbon–carbon double bonds to carbon–carbon single bonds.

- The hydrolysis of the ester bonds in fats or oils, in the presence of acids or enzymes, produces glycerol and fatty acids.

- In saponification, a triacylglycerol heated with a strong base produces glycerol and the salts of the fatty acids (soaps).

H_2

Shortening (soft)

Vegetable oils (liquids)   Tub (soft) margarine   Stick margarine (solid)

Many soft margarines, stick margarines, and solid shortenings are produced by the partial hydrogenation of vegetable oils.

Credit: Pearson Education/Pearson Science

♦ **Learning Exercise 17.4**

Use condensed structural formulas to write the balanced chemical equation for the following reactions of glyceryl trioleate (triolein):

1. hydrogenation with a nickel catalyst

2. acid hydrolysis

3. saponification with NaOH

Answers

1.

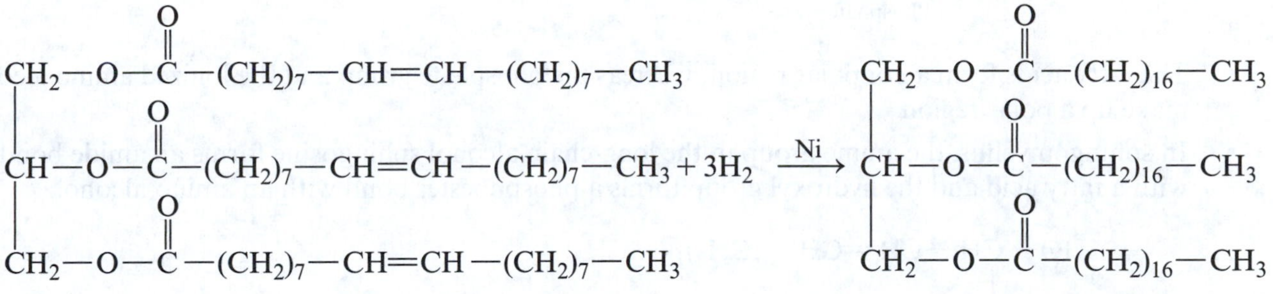

$$\text{CH}_2\text{—O—}\overset{\displaystyle O}{\overset{\|}{\text{C}}}\text{—(CH}_2)_7\text{—CH}\text{=}\text{CH—(CH}_2)_7\text{—CH}_3$$

$$\text{CH—O—}\overset{\displaystyle O}{\overset{\|}{\text{C}}}\text{—(CH}_2)_7\text{—CH}\text{=}\text{CH—(CH}_2)_7\text{—CH}_3 + 3\text{H}_2 \quad \xrightarrow{\text{Ni}}$$

$$\text{CH}_2\text{—O—}\overset{\displaystyle O}{\overset{\|}{\text{C}}}\text{—(CH}_2)_7\text{—CH}\text{=}\text{CH—(CH}_2)_7\text{—CH}_3$$

$$\text{CH}_2\text{—O—}\overset{\displaystyle O}{\overset{\|}{\text{C}}}\text{—(CH}_2)_{16}\text{—CH}_3$$

$$\text{CH—O—}\overset{\displaystyle O}{\overset{\|}{\text{C}}}\text{—(CH}_2)_{16}\text{—CH}_3$$

$$\text{CH}_2\text{—O—}\overset{\displaystyle O}{\overset{\|}{\text{C}}}\text{—(CH}_2)_{16}\text{—CH}_3$$

2.

$$\text{CH}_2\text{—O—}\overset{\displaystyle O}{\overset{\|}{\text{C}}}\text{—(CH}_2)_7\text{—CH}\text{=}\text{CH—(CH}_2)_7\text{—CH}_3$$

$$\text{CH—O—}\overset{\displaystyle O}{\overset{\|}{\text{C}}}\text{—(CH}_2)_7\text{—CH}\text{=}\text{CH—(CH}_2)_7\text{—CH}_3 + 3\text{H}_2\text{O} \quad \xrightarrow{\text{H}^+,\ \text{heat}}$$

$$\text{CH}_2\text{—O—}\overset{\displaystyle O}{\overset{\|}{\text{C}}}\text{—(CH}_2)_7\text{—CH}\text{=}\text{CH—(CH}_2)_7\text{—CH}_3$$

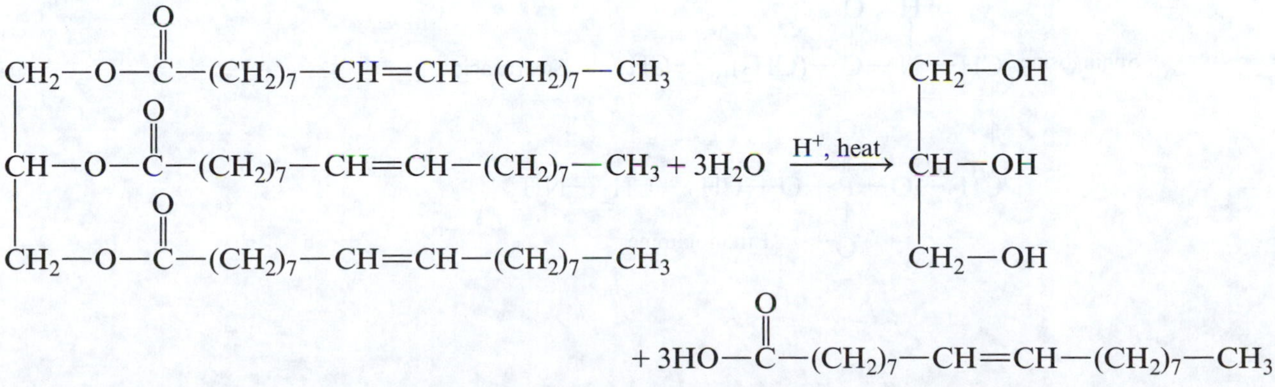

$$\text{CH}_2\text{—OH}$$
$$\text{CH—OH}$$
$$\text{CH}_2\text{—OH}$$

$$+ 3\text{HO—}\overset{\displaystyle O}{\overset{\|}{\text{C}}}\text{—(CH}_2)_7\text{—CH}\text{=}\text{CH—(CH}_2)_7\text{—CH}_3$$

3.

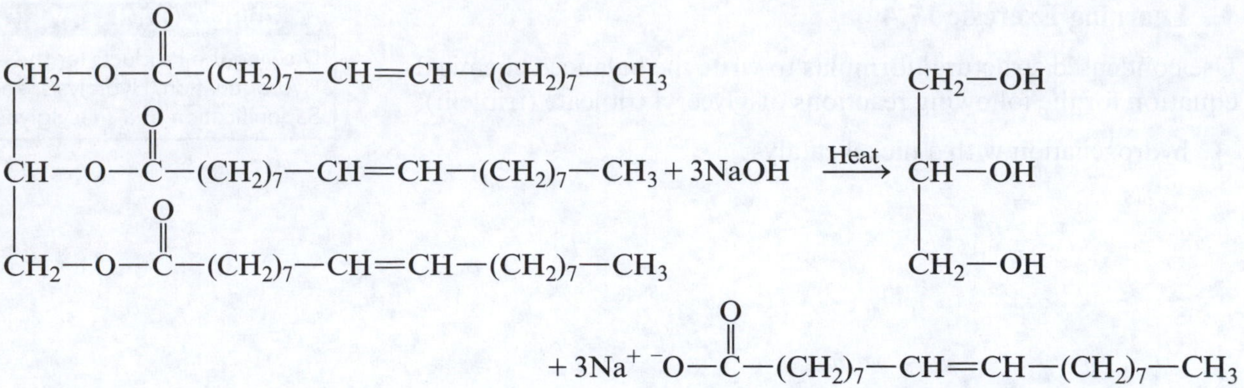

$$CH_2-O-\overset{\overset{O}{\|}}{C}-(CH_2)_7-CH=CH-(CH_2)_7-CH_3$$

$$CH-O-\overset{\overset{O}{\|}}{C}-(CH_2)_7-CH=CH-(CH_2)_7-CH_3 + 3NaOH \xrightarrow{\text{Heat}}$$

$$CH_2-O-\overset{\overset{O}{\|}}{C}-(CH_2)_7-CH=CH-(CH_2)_7-CH_3$$

$$CH_2-OH$$
$$CH-OH$$
$$CH_2-OH$$

$$+ 3Na^+ \; {}^-O-\overset{\overset{O}{\|}}{C}-(CH_2)_7-CH=CH-(CH_2)_7-CH_3$$

17.5   Phospholipids

Learning Goal: Draw the structure of a phospholipid that contains glycerol or sphingosine.

- The phospholipids are a family of lipids similar in structure to triacylglycerols.
- Glycerophospholipids are esters of glycerol with two fatty acids and a phosphate group attached to an amino alcohol.

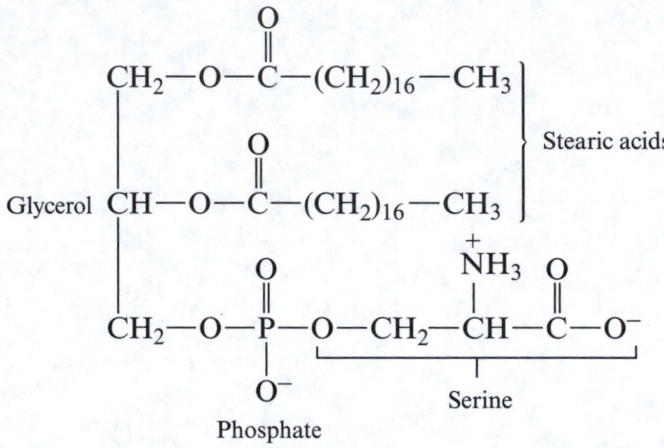

Glycerol
$$CH_2-O-\overset{\overset{O}{\|}}{C}-(CH_2)_{16}-CH_3$$
$$CH-O-\overset{\overset{O}{\|}}{C}-(CH_2)_{16}-CH_3$$ Stearic acids
$$CH_2-O-\overset{\overset{O}{\|}}{\underset{\underset{O^-}{|}}{P}}-O-CH_2-\overset{\overset{+}{\underset{}{NH_3}}}{\underset{}{CH}}-\overset{\overset{O}{\|}}{C}-O^-$$
Phosphate   Serine

- The fatty acids form a nonpolar region, whereas the phosphate group and the ionized amino alcohol make up a polar region.
- In sphingomyelins, the amine group in the long-chain alcohol sphingosine forms an amide bond with a fatty acid and the hydroxyl group forms a phosphoester bond with an amino alcohol.

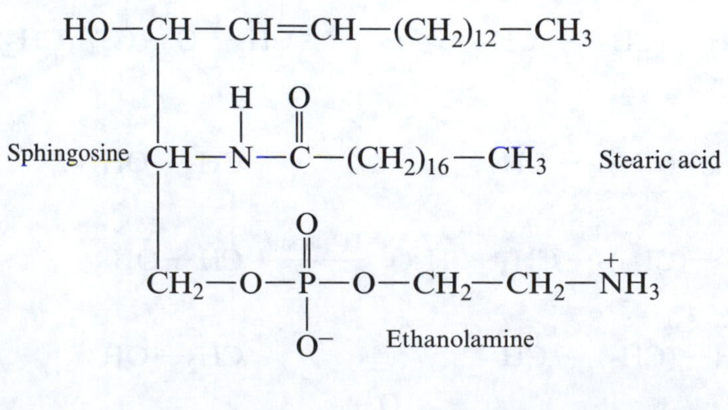

$$HO-CH-CH=CH-(CH_2)_{12}-CH_3$$

Sphingosine
$$CH-\overset{\overset{H}{|}}{N}-\overset{\overset{O}{\|}}{C}-(CH_2)_{16}-CH_3 \quad \text{Stearic acid}$$

$$CH_2-O-\overset{\overset{O}{\|}}{\underset{\underset{O^-}{|}}{P}}-O-CH_2-CH_2-\overset{+}{NH_3}$$
Ethanolamine

♦ **Learning Exercise 17.5A**

Draw the condensed structural formula for a glycerophospholipid that contains two palmitic acids, phosphate, and serine (ionized).

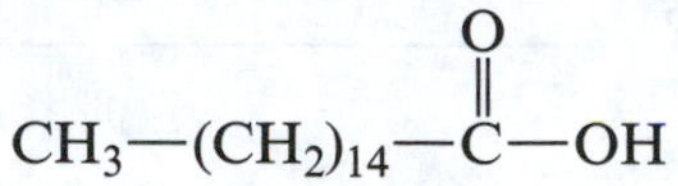

$$
\begin{array}{c}
\quad\quad\quad O \\
\quad\quad\quad \| \\
CH_3-(CH_2)_{14}-C-OH \\
\text{Palmitic acid}
\end{array}
\qquad\qquad
\begin{array}{c}
\quad\quad\quad \overset{+}{N}H_3 \ \ O \\
\quad\quad\quad | \quad\quad \| \\
HO-CH_2-CH-C-O^- \\
\text{Serine}
\end{array}
$$

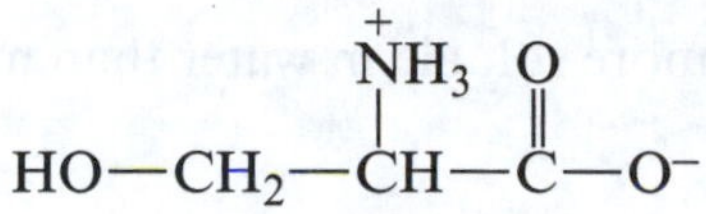

Answer

$$
\begin{array}{l}
\quad\quad\quad\quad\quad\quad O \\
\quad\quad\quad\quad\quad\quad \| \\
CH_2-O-C-(CH_2)_{14}-CH_3 \\
| \quad\quad\quad\quad O \\
| \quad\quad\quad\quad \| \\
CH-O-C-(CH_2)_{14}-CH_3 \\
| \quad\quad\quad\quad O \quad\quad\quad \overset{+}{N}H_3 \ O \\
| \quad\quad\quad\quad \| \quad\quad\quad | \quad\quad \| \\
CH_2-O-P-O-CH_2-CH-C-O^- \\
\quad\quad\quad\quad\quad | \\
\quad\quad\quad\quad\quad O^-
\end{array}
$$

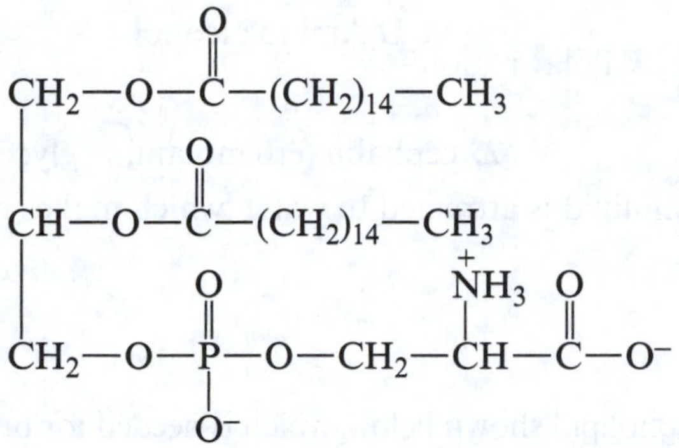

♦ **Learning Exercise 17.5B**

Use the following glycerophospholipid to answer **A** to **F** and **1** to **3**:

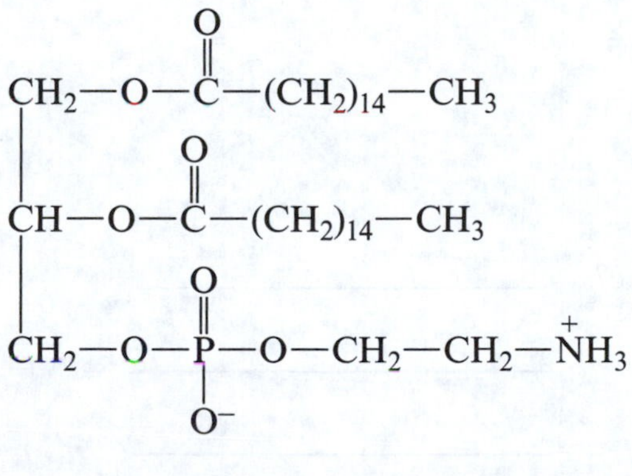

$$
\begin{array}{l}
\quad\quad\quad\quad\quad\quad O \\
\quad\quad\quad\quad\quad\quad \| \\
CH_2-O-C-(CH_2)_{14}-CH_3 \\
| \quad\quad\quad\quad O \\
| \quad\quad\quad\quad \| \\
CH-O-C-(CH_2)_{14}-CH_3 \\
| \quad\quad\quad\quad O \\
| \quad\quad\quad\quad \| \\
CH_2-O-P-O-CH_2-CH_2-\overset{+}{N}H_3 \\
\quad\quad\quad\quad\quad | \\
\quad\quad\quad\quad\quad O^-
\end{array}
$$

On the above condensed structural formula, indicate the:

A. two fatty acids   **B.** glycerol

C. phosphate group   **D.** amino alcohol

E. nonpolar region   **F.** polar region

1. What is the name of the amino alcohol? _____

2. What type of glycerophospholipid is it? _____

3. Why is a glycerophospholipid more soluble in water than most lipids? _____

Answers **B** Glycerol

$$CH_2-O-\overset{\overset{\displaystyle O}{\|}}{C}-(CH_2)_{14}-CH_3$$

$$CH-O-\overset{\overset{\displaystyle O}{\|}}{C}-(CH_2)_{14}-CH_3$$

A Fatty acids

E Nonpolar region

$$CH_2-O-\overset{\overset{\displaystyle O}{\|}}{\underset{\underset{\displaystyle O^-}{\|}}{P}}-O-CH_2-CH_2-\overset{+}{N}H_3$$

C Phosphate group

F Polar region

D Amino alcohol

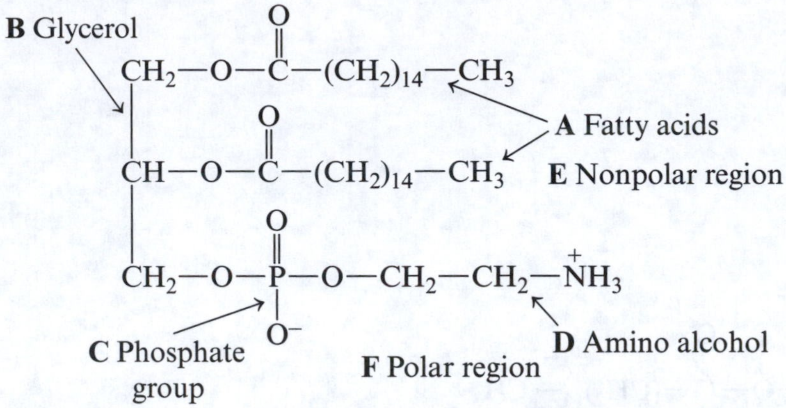

1. ethanolamine   2. cephalin (ethanolamine glycerophospholipid)
3. The polar portion of the glycerophospholipid is attracted to water, which makes it more soluble in water than other lipids.

♦ **Learning Exercise 17.5C**

Answer the following questions for the phospholipid shown below, which is needed for brain and nerve tissues:

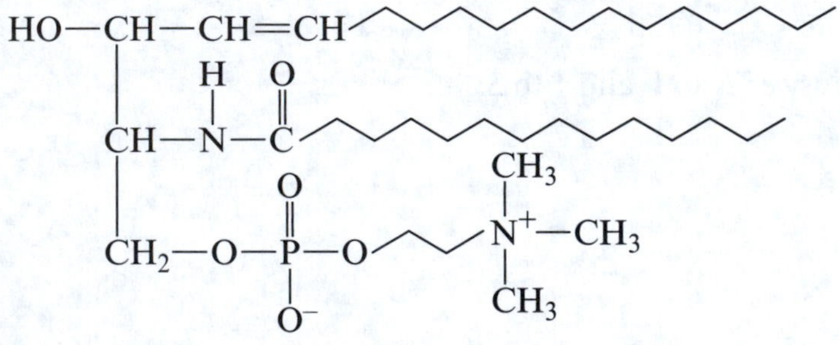

$$HO-CH-CH=CH\sim\sim\sim\sim\sim\sim$$

$$\underset{\underset{\displaystyle CH_2-O-\overset{\overset{\displaystyle O}{\|}}{\underset{\underset{\displaystyle O^-}{\|}}{P}}-O\sim\overset{CH_3}{\underset{CH_3}{\overset{|}{N}{}^+}-CH_3}}{\overset{\overset{\displaystyle H\ \ O}{|\ \ \|}}{CH-N-C\sim\sim\sim\sim\sim\sim}}}$$

1. Is the phospholipid formed from glycerol or sphingosine? _____

2. What is the fatty acid? _____

3. What type of bond connects the fatty acid? _____

4. What is the name of the amino alcohol? _____

Answers   1. sphingosine   2. myristic acid   3. an amide bond   4. choline

17.6 Steroids: Cholesterol, Bile Salts, and Steroid Hormones

Learning Goal: Draw the structure of the steroid nucleus; compare the structures and functions of steroid compounds.

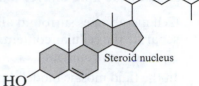

Steroid nucleus

Cholesterol

- Steroids are lipids containing the steroid nucleus, which is a fused structure of three cyclohexane rings and one cyclopentane ring.
- Steroids include cholesterol, bile salts, and steroid hormones.
- Bile salts, synthesized from cholesterol, mix with water-insoluble fats and break them apart during digestion.
- There are a variety of lipoproteins, which differ in density, lipid composition, and function. They include chylomicrons, very-low-density lipoproteins (VLDLs), low-density lipoproteins (LDLs), and high-density lipoproteins (HDLs).
- LDL carries cholesterol to the tissues where it can be used for the synthesis of cell membranes and steroid hormones.
- HDL picks up cholesterol from the tissues and carries it to the liver, where it can be converted to bile salts, which are eliminated from the body.
- The steroid hormones are closely related in structure to cholesterol and depend on cholesterol for their synthesis.
- The sex hormones such as estrogen and testosterone are responsible for sexual characteristics and reproduction.
- The adrenal corticosteroids include aldosterone, which regulates water balance, and cortisone, which regulates glucose levels in the blood.

> **CORE CHEMISTRY SKILL**
> Identifying the Steroid Nucleus

♦ Learning Exercise 17.6A

1. Draw the line-angle structure for the steroid nucleus.   2. Draw the line-angle structure for cholesterol.

Answers   1.   2.

HO

♦ Learning Exercise 17.6B

Match each of the following compounds with the correct phrase below:

    a. estrogen   **b.** testosterone   **c.** cortisone   **d.** aldosterone   **e.** bile salts

1. _____ increases the blood glucose level

2. _____ increases the reabsorption of Na^+ by the kidneys

3. _____ stimulates the development of secondary sex characteristics in females

4. _____ stimulates the retention of water by the kidneys

5. _____ stimulates the development of secondary sex characteristics in males

6. _____ secreted by the gallbladder into the small intestine to emulsify fats in the diet

Answers   **1.** c   **2.** d   **3.** a   **4.** d   **5.** b   **6.** e

17.7 Cell Membranes

Learning Goal: Describe the composition and function of the lipid bilayer in cell membranes.

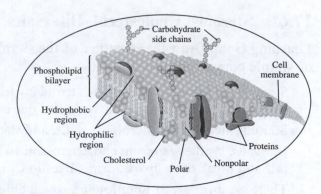

The fluid mosaic model of a cell membrane includes a bilayer of phospholipids.

- Cell membranes surround all of our cells and separate the cellular contents from the external aqueous environment.

- In the fluid mosaic model (lipid bilayer) of a cell membrane, two layers of phospholipids are arranged with their hydrophilic heads at the outer and inner surfaces of the membrane, and their hydrophobic tails in the center.

- Proteins and cholesterol are embedded in the lipid bilayer, and carbohydrates are attached to its surface.

- Nutrients and waste products move through the cell membrane using diffusion (passive transport), facilitated transport, or active transport.

Key Terms for Sections 17.1 to 17.7

    a. lipid   **b.** fatty acid   **c.** triacylglycerol   **d.** lipid bilayer
    e. saponification   **f.** glycerophospholipid   **g.** steroid

Match each of the following key terms with the correct description:

1. _____ a compound consisting of glycerol bonded to two fatty acids and a phosphate group that is attached to an amino alcohol

2. _____ a type of compound that is not soluble in water, but is soluble in nonpolar solvents

3. _____ the hydrolysis of a triacylglycerol with a strong base, producing salts called soaps and glycerol

4. _____ a compound consisting of glycerol bonded to three fatty acids

5. _____ a compound composed of a multicyclic ring system

6. _____ a model for cell membranes in which phospholipids are arranged in two layers

7. _____ a long-chain carboxylic acid found in triacylglycerols

Answers   **1.** f   **2.** a   **3.** e   **4.** c   **5.** g   **6.** d   **7.** b

◆ **Learning Exercise 17.7A**

1. What is the function of the lipid bilayer in cell membranes?

2. What type of lipid makes up the lipid bilayer?

3. What is the arrangement of the lipids in a lipid bilayer?

Answers

1. The lipid bilayer separates the contents of a cell from the surrounding aqueous environment.
2. The lipid bilayer is primarily composed of glycerophospholipids.
3. The nonpolar hydrocarbon tails are in the center of the bilayer, whereas the polar sections are aligned along the outside of the bilayer.

♦ **Learning Exercise 17.7B**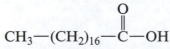

Identify the type of transport described by each of the following:

1. K^+ ions move from low to high concentration in a cell. _____

2. Water moves through a cell membrane. _____

3. A molecule moves through a protein channel. _____

Answers   **1.** active transport   **2.** diffusion (passive transport)   **3.** facilitated transport

Checklist for Chapter 17

You are ready to take the Practice Test for Chapter 17. Be sure you have accomplished the following learning goals for this chapter. If not, review the Section listed at the end of the goal. Then apply your new skills and understanding to the Practice Test.

After studying Chapter 17, I can successfully:

_____ Describe the classes of lipids. (17.1)

_____ Identify a fatty acid as saturated or unsaturated. (17.2)

_____ Draw the condensed structural and line-angle formulas for a wax or triacylglycerol produced by the reaction of a fatty acid and an alcohol or glycerol. (17.3)

_____ Draw the condensed structural and line-angle formulas for the product from the reaction of a triacylglycerol with hydrogen, an acid, or a base. (17.4)

_____ Describe the components of glycerophospholipids and sphingomyelins. (17.5)

_____ Draw the structures of a steroid and cholesterol. (17.6)

_____ Describe the composition and function of the lipid bilayer in cell membranes. (17.7)

Practice Test for Chapter 17

The chapter Sections to review are shown in parentheses at the end of each question.

1. A biomolecule that is soluble in organic solvents but not in water is a (17.1)
   A. carbohydrate   **B.** lipid   **C.** protein   **D.** nucleic acid   **E.** soap

2. A fatty acid that is unsaturated is usually (17.2)
   A. from animal sources and liquid at room temperature
   B. from animal sources and solid at room temperature
   C. from vegetable sources and liquid at room temperature
   D. from vegetable sources and solid at room temperature
   E. from both vegetable and animal sources and solid at room temperature

3. The following condensed structural formula is a(an): (17.1)

$$CH_3-(CH_2)_{16}-\overset{\displaystyle O}{\overset{\displaystyle \|}{C}}-OH$$

   A. unsaturated fatty acid   **B.** saturated fatty acid   **C.** wax
   D. triacylglycerol   **E.** sphingomyelin

4. The following compound is a (17.2)

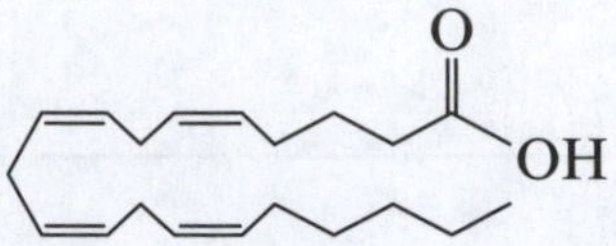

A. cholesterol   **B.** sphingosine   **C.** fatty acid
D. sphingomyelin   **E.** steroid

For questions 5 through 8, consider the following compound: (17.3, 17.4)

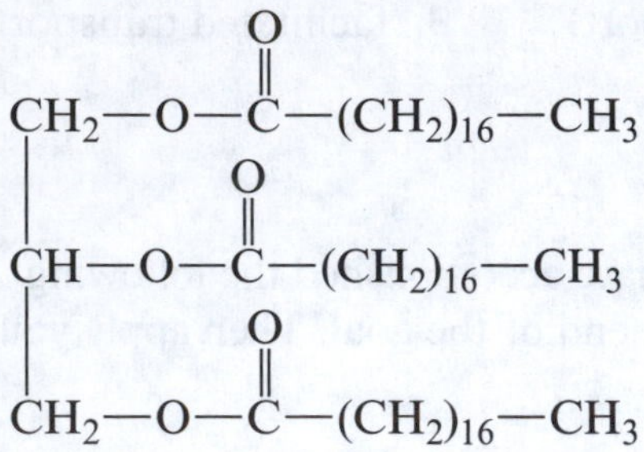

5. This compound belongs to the family called
  A. waxes   **B.** triacylglycerols   **C.** glycerophospholipids
  D. sphingomyelins   **E.** steroids

6. The compound was formed by
  A. esterification   **B.** acid hydrolysis   **C.** saponification
  D. emulsification   **E.** oxidation

7. The compound would be expected to be
  A. saturated and a solid at room temperature
  B. saturated and a liquid at room temperature
  C. unsaturated and a solid at room temperature
  D. unsaturated and a liquid at room temperature
  E. supersaturated and a liquid at room temperature

8. If this compound reacts with a strong base such as NaOH, the products would be
  A. glycerol and fatty acid
  B. glycerol and water
  C. glycerol and the sodium salt of the fatty acid
  D. an ester and the sodium salt of the fatty acid
  E. an ester and fatty acid

For questions 9 and 10, consider the following reaction: (17.4)

$$\text{Triacylglycerol} + 3\text{NaOH} \xrightarrow{\text{Heat}} \text{glycerol} + 3 \text{ sodium salts of fatty acids}$$

9. The reaction of a triacylglycerol with a strong base such as NaOH is called
  A. esterification   **B.** lipogenesis   **C.** hydrolysis   **D.** saponification   **E.** β oxidation

10. What is another name for the sodium salts of the fatty acids?
  A. margarines   **B.** fat substitutes   **C.** soaps   **D.** perfumes   **E.** vitamins

For questions 11 through 16, consider the following phospholipid:

A

$$CH_2-O-\overset{\overset{\displaystyle O}{\|}}{C}-(CH_2)_{14}-CH_3$$

$$CH-O-\overset{\overset{\displaystyle O}{\|}}{C}-(CH_2)_{14}-CH_3$$
B

$$CH_2-O-\overset{\overset{\displaystyle O}{\|}}{\underset{\underset{\displaystyle O^-}{|}}{P}}-O-CH_2-CH_2-\overset{+}{N}H_3$$
D   C

Match the labels, **A** to **D**, with the following: (17.5)

11. _____ glycerol   **12.** _____ the phosphate group

13. _____ the amino alcohol   **14.** _____ the polar region

15. _____ the nonpolar region

16. This phospholipid belongs to the family called (17.5)
    A. cholines   **B.** cephalins   **C.** sphingomyelins
    D. glycolipids   **E.** lecithins

17. Which of these are found in glycerophospholipids? (17.5)
    A. fatty acids   **B.** glycerol   **C.** amino alcohol
    D. phosphate   **E.** all of these

For questions 18 through 21, classify each lipid as a: (17.1, 17.2, 17.3, 17.5, 17.6)

    A. wax   **B.** triacylglycerol   **C.** glycerophospholipid
    D. steroid   **E.** fatty acid

18. _____ cholesterol

19. _____ $CH_3-(CH_2)_{14}-\overset{\overset{\displaystyle O}{\|}}{C}-OH$

20. _____ $CH_3-(CH_2)_{14}-\overset{\overset{\displaystyle O}{\|}}{C}-O-(CH_2)_{29}-CH_3$

21. _____ an ester of glycerol with three palmitic acid molecules

For questions 22 through 26, select answers from the following: (17.6)

    A. testosterone   **B.** estrogen   **C.** prednisone
    D. cortisone   **E.** aldosterone

22. _____ stimulates development of the female sexual characteristics

23. _____ regulates electrolytes and controls water balance by the kidneys

24. _____ stimulates development of the male sexual characteristics

25. _____ increases the blood glucose level

26. _____ used medically to reduce inflammation and treat asthma

27. The type of lipoprotein that transports cholesterol to the liver for elimination is called a (17.6)
   - **A.** chylomicron
   - **B.** high-density lipoprotein
   - **C.** low-density lipoprotein
   - **D.** very-low-density lipoprotein
   - **E.** all of these

28. The lipid bilayer of a cell consists of (17.7)

   - **A.** cholesterol
   - **B.** phospholipids
   - **C.** proteins
   - **D.** carbohydrates
   - **E.** all of these

29. The type of transport that allows chloride ions to move through the integral proteins in the cell membrane is (17.7)

   - **A.** passive transport
   - **B.** active transport
   - **C.** facilitated transport
   - **D.** all of these
   - **E.** none of these

30. The movement of small molecules through a cell membrane from a higher concentration to a lower concentration is (17.7)

   - **A.** passive transport
   - **B.** active transport
   - **C.** facilitated transport
   - **D.** all of these
   - **E.** none of these

Answers to the Practice Test

1. B	2. C	3. B	4. C	5. B
6. A	7. A	8. C	9. D	10. C
11. A	12. D	13. C	14. C, D	15. B
16. B	17. E	18. D	19. E	20. A
21. B	22. B	23. E	24. A	25. D
26. C	27. B	28. E	29. C	30. A

Selected Answers and Solutions to Text Problems

17.1 Because lipids are not soluble in water, a polar solvent, they are nonpolar molecules.

17.3 Lipids can store energy, protect and insulate internal organs, and act as chemical messengers. They are also important components of cell membranes.

17.5 All fatty acids contain a long chain of carbon atoms with a carboxylic acid group. Saturated fatty acids contain only carbon–carbon single bonds; unsaturated fatty acids contain one or more carbon–carbon double bonds.

17.7 **a.**

b.

17.9 **a.** Lauric acid (12:0) has only carbon–carbon single bonds; it is saturated.
b. Linolenic acid (18:3) has three carbon–carbon double bonds; it is polyunsaturated.
c. Palmitoleic acid (16:1) has one carbon–carbon double bond; it is monounsaturated.
d. Stearic acid (18:0) has only carbon–carbon single bonds; it is saturated.

17.11 In a cis fatty acid, the hydrogen atoms are on the same side of the double bond, which produces a kink in the carbon chain. In a trans fatty acid, the hydrogen atoms are on opposite sides of the double bond, which gives a carbon chain without any kink.

17.13 In an omega-3 fatty acid, there is a double bond beginning at carbon 3, counting from the methyl group. In an omega-6 fatty acid, there is a double bond beginning at carbon 6, counting from the methyl group.

17.15 Arachidonic acid and PGE_1 are both carboxylic acids with 20 carbon atoms. The differences are that arachidonic acid contains four cis double bonds and no other functional groups, whereas PGE_1 has one trans double bond, one ketone group, and two hydroxyl groups. In addition, a part of the PGE_1 chain forms a cyclopentane ring.

17.17 Prostaglandins raise or lower blood pressure, stimulate contraction and relaxation of smooth muscle, and may cause inflammation and pain.

17.19 Palmitic acid is the 16-carbon saturated fatty acid (16:0), and myricyl alcohol is a 30-carbon long-chain alcohol.

$$CH_3-(CH_2)_{14}-\overset{\overset{\textstyle O}{\|}}{C}-O-(CH_2)_{29}-CH_3$$

17.21 Triacylglycerols are composed of fatty acids and glycerol. In this case, the fatty acid is stearic acid, an 18-carbon saturated fatty acid (18:0).

$$
\begin{array}{l}
CH_2-O-\overset{\overset{\textstyle O}{\|}}{C}-(CH_2)_{16}-CH_3 \\[2pt]
| \quad\quad\quad\; \overset{\textstyle O}{\|} \\[-2pt]
CH-O-\overset{}{C}-(CH_2)_{16}-CH_3 \\[2pt]
| \quad\quad\quad\; \overset{\textstyle O}{\|} \\[-2pt]
CH_2-O-\overset{}{C}-(CH_2)_{16}-CH_3
\end{array}
$$

17.23 Glyceryl tricaprylate (tricaprylin) has three caprylic acids (an 8-carbon saturated fatty acid, 8:0) forming ester bonds with glycerol.

17.25 Safflower oil has a lower melting point because it contains mostly polyunsaturated fatty acids,
whereas olive oil contains a large amount of monounsaturated oleic acid. A polyunsaturated fatty
acid has two or more kinks in its carbon chain, which means it does not have as many dispersion
forces compared to the hydrocarbon chains in olive oil.

17.27 Sunflower oil has about 24% monounsaturated fats, whereas safflower oil has about 18%.
Sunflower oil has about 66% polyunsaturated fats, whereas safflower oil has about 73%.

17.29 a. The reaction of palm oil with KOH is saponification, and the products are glycerol and the
potassium salts of the fatty acids, which are soaps.
   b. The reaction of glyceryl trilinoleate from safflower oil with water and HCl is hydrolysis,
which splits the ester bonds to produce glycerol and three molecules of linoleic acid.

17.31 Hydrogenation of an unsaturated triacylglycerol adds H_2 to each of the double bonds, producing a
saturated triacylglycerol containing only carbon–carbon single bonds.

17.33 Acid hydrolysis of a fat gives glycerol and the fatty acids.

17.35 Basic hydrolysis (saponification) of a fat gives glycerol and the salts of the fatty acids.

$$
\begin{array}{l}
CH_2-O-\overset{\displaystyle O}{\overset{\|}{C}}-(CH_2)_{12}-CH_3 \\
CH-O-\overset{\displaystyle O}{\overset{\|}{C}}-(CH_2)_{12}-CH_3 + 3NaOH \xrightarrow{\text{Heat}} \\
CH_2-O-\overset{\displaystyle O}{\overset{\|}{C}}-(CH_2)_{12}-CH_3
\end{array}
\quad
\begin{array}{l}
CH_2-OH \\
CH-OH + 3Na^+ \; ^-O-\overset{\displaystyle O}{\overset{\|}{C}}-(CH_2)_{12}-CH_3 \\
CH_2-OH
\end{array}
$$

17.37

$$
\begin{array}{l}
CH_2-O-\overset{\displaystyle O}{\overset{\|}{C}}-(CH_2)_{16}-CH_3 \\
CH-O-\overset{\displaystyle O}{\overset{\|}{C}}-(CH_2)_{16}-CH_3 \\
CH_2-O-\overset{\displaystyle O}{\overset{\|}{C}}-(CH_2)_{16}-CH_3
\end{array}
$$

17.39 A triacylglycerol consists of glycerol and three fatty acids. A glycerophospholipid also contains glycerol, but has only two fatty acids, with the hydroxyl group on the third carbon attached by a phosphoester bond to phosphoric acid, which forms another phosphoester bond to an amino alcohol.

17.41

$$
\begin{array}{l}
CH_2-O-\overset{\displaystyle O}{\overset{\|}{C}}-(CH_2)_{14}-CH_3 \\
CH-O-\overset{\displaystyle O}{\overset{\|}{C}}-(CH_2)_{14}-CH_3 \\
CH_2-O-\overset{\displaystyle O}{\overset{\|}{P}}-O-CH_2-CH_2-\overset{+}{N}H_3 \\
O^-
\end{array}
$$

17.43 This glycerophospholipid is a cephalin. It contains glycerol, oleic acid, stearic acid, phosphate, and ethanolamine.

17.45 a. This phospholipid is formed from sphingosine.   **b.** The fatty acid is palmitic acid (16:0).
c. An amide bond connects the fatty acid to sphingosine.   **d.** The amino alcohol is ethanolamine.

17.47

17.49 Bile salts act to emulsify fat globules, allowing the fat to be more easily digested by lipases.

17.51 Chylomicrons have a lower density than VLDL (very-low-density lipoprotein). They pick up triacylglycerols from the intestine, whereas VLDL transports triacylglycerols synthesized in the liver.

17.53 "Bad" cholesterol is the cholesterol carried by LDL that can form deposits in the arteries called plaque, which narrows the arteries.

17.55 Both estradiol and testosterone contain the steroid nucleus and a hydroxyl group. Testosterone has a ketone group, a double bond, and two methyl groups. Estradiol has an aromatic ring, a hydroxyl group in place of the ketone, and a methyl group.

17.57 Cortisol (b), estradiol (c), and testosterone (d) are steroid hormones.

17.59 The function of the lipid bilayer in the cell membrane is to keep the cell contents separated from the outside environment and to allow the cell to regulate the movement of substances into and out of the cell.

17.61 Because the molecules of cholesterol are large and rigid, they reduce the flexibility of the lipid bilayer and add strength to the cell membrane.

17.63 The peripheral proteins in the membrane emerge on the inner or outer surface only, whereas the integral proteins extend through the membrane to both surfaces.

17.65 a. A molecule moving through a protein channel is an example of facilitated diffusion.
  b. A molecule, like O_2, moving from higher concentration to lower concentration is an example of diffusion.

17.67 The functional groups in Pravachol are ester, alcohol, alkene, and carboxylic acid.

17.69 a. $\dfrac{80.\ \text{mg Pravachol}}{1\ \text{day}} \times \dfrac{1\ \text{g Pravachol}}{1000\ \text{mg Pravachol}} \times \dfrac{7\ \text{days}}{1\ \text{week}} = 0.56\ \text{g of Pravachol/week (2 SFs)}$

  b. $\dfrac{18\ \text{mg cholesterol}}{5.0\ \text{mL blood}} \times \dfrac{100\ \text{mL blood}}{1\ \text{dL blood}} = 360\ \text{mg/dL (2 SFs)}$

17.71

$$
\begin{array}{l}
\quad\quad\quad\ \ \overset{\displaystyle O}{\overset{\displaystyle \|}{}} \\
CH_2\!-\!O\!-\!C\!-\!(CH_2)_{14}\!-\!CH_3 \\
|\quad\quad\ \ \overset{\displaystyle O}{\overset{\displaystyle \|}{}} \\
CH\!-\!O\!-\!C\!-\!(CH_2)_{14}\!-\!CH_3 \\
|\quad\quad\ \ \overset{\displaystyle O}{\overset{\displaystyle \|}{}} \\
CH_2\!-\!O\!-\!C\!-\!(CH_2)_{14}\!-\!CH_3
\end{array}
$$

17.73 a. monounsaturated fatty acid
  c. polyunsaturated omega-3 fatty acid
  b. polyunsaturated omega-6 fatty acid

17.75 a. Beeswax and carnauba are waxes. Vegetable oil and glyceryl tricaprate (tricaprin) are triacylglycerols.

  b.
$$
\begin{array}{l}
\quad\quad\quad\ \ \overset{\displaystyle O}{\overset{\displaystyle \|}{}} \\
CH_2\!-\!O\!-\!C\!-\!(CH_2)_{8}\!-\!CH_3 \\
|\quad\quad\ \ \overset{\displaystyle O}{\overset{\displaystyle \|}{}} \\
CH\!-\!O\!-\!C\!-\!(CH_2)_{8}\!-\!CH_3 \\
|\quad\quad\ \ \overset{\displaystyle O}{\overset{\displaystyle \|}{}} \\
CH_2\!-\!O\!-\!C\!-\!(CH_2)_{8}\!-\!CH_3
\end{array}
$$

17.77 a. $46\ \text{g fat} \times \dfrac{9\ \text{kcal}}{1\ \text{g fat}} = 410\ \text{kcal from fat}$ $\dfrac{410\ \text{kcal}}{830\ \text{kcal}} \times 100\% = 49\%\ \text{fat (2 SFs)}$

  b. $29\ \text{g fat} \times \dfrac{9\ \text{kcal}}{1\ \text{g fat}} = 260\ \text{kcal from fat}$ $\dfrac{260\ \text{kcal}}{520\ \text{kcal}} \times 100\% = 50.\%\ \text{fat (2 SFs)}$

  c. $18\ \text{g fat} \times \dfrac{9\ \text{kcal}}{1\ \text{g fat}} = 160\ \text{kcal from fat}$ $\dfrac{160\ \text{kcal}}{560\ \text{kcal}} \times 100\% = 29\%\ \text{fat (2 SFs)}$

17.79 a. Beeswax is a wax.
  b. Cholesterol is a steroid.
  c. Lecithin is a glycerophospholipid.

d. Glyceryl tripalmitate (tripalmitin) is a triacylglycerol.

e. Sodium stearate is a soap.

f. Safflower oil is a triacylglycerol.

17.81 a. Estrogen contains the steroid nucleus (5).

b. Cephalin contains glycerol (1), fatty acids (2), phosphate (3), and an amino alcohol (4).

c. Waxes contain fatty acid (2).

d. Triacylglycerols contain glycerol (1) and fatty acids (2).

17.83 Cholesterol (a) and carbohydrates (c) are found in cell membranes.

17.85

$$
\begin{array}{l}
CH_2-O-\overset{\overset{\displaystyle O}{\|}}{C}-(CH_2)_{16}-CH_3 \\[2pt]
CH-O-\overset{\overset{\displaystyle O}{\|}}{C}-(CH_2)_{16}-CH_3 \\[2pt]
CH_2-O-\overset{\overset{\displaystyle O}{\|}}{\underset{\underset{\displaystyle O^-}{|}}{P}}-O-CH_2-CH_2-\overset{+}{N}H_3
\end{array}
$$

17.87 a. HDL (4) is known as "good" cholesterol.

b. LDL (3) transports most of the cholesterol to the cells.

c. Chylomicrons (1) carry triacylglycerols from the intestine to the fat cells.

d. HDL (4) transports cholesterol to the liver.

17.89 a. Adding NaOH would saponify lipids such as glyceryl tristearate (tristearin), forming glycerol and salts of the fatty acids that are soluble in water and would wash down the drain.

b.

$$
\begin{array}{l}
CH_2-O-\overset{\overset{\displaystyle O}{\|}}{C}-(CH_2)_{16}-CH_3 \\[2pt]
CH-O-\overset{\overset{\displaystyle O}{\|}}{C}-(CH_2)_{16}-CH_3 \quad + \quad 3NaOH \quad \xrightarrow{\text{Heat}} \\[2pt]
CH_2-O-\overset{\overset{\displaystyle O}{\|}}{C}-(CH_2)_{16}-CH_3
\end{array}
$$

$$
\begin{array}{l}
CH_2-OH \\[2pt]
CH-OH \quad + \quad 3Na^+ \; {}^-O-\overset{\overset{\displaystyle O}{\|}}{C}-(CH_2)_{16}-CH_3 \\[2pt]
CH_2-OH
\end{array}
$$

c. Tristearin = $C_{57}H_{110}O_6$

Molar mass of tristearin ($C_{57}H_{110}O_6$)

= 57(12.01 g) + 110(1.008 g) + 6(16.00 g) = 891.5 g/mole (4 SFs)

$$10.0 \text{ g tristearin} \times \frac{1 \text{ mole tristearin}}{891.5 \text{ g tristearin}} \times \frac{3 \text{ moles NaOH}}{1 \text{ mole tristearin}} \times \frac{1000 \text{ mL solution}}{0.500 \text{ mole NaOH}}$$

= 67.3 mL of NaOH solution (3 SFs)

17.91 a.

$$\begin{array}{l}
CH_2\text{—}O\text{—}\overset{\displaystyle \overset{O}{\|}}{C}\text{—}(CH_2)_{16}\text{—}CH_3 \\[1em]
CH\text{—}O\text{—}\overset{\displaystyle \overset{O}{\|}}{C}\text{—}(CH_2)_{16}\text{—}CH_3 \\[1em]
CH_2\text{—}O\text{—}\overset{\displaystyle \overset{O}{\|}}{C}\text{—}(CH_2)_{16}\text{—}CH_3
\end{array}$$

b. Since triolein contains three carbon–carbon double bonds, each mole of triolein requires three moles of H_2 for complete hydrogenation.

$$1.00 \text{ mole triolein} \times \frac{3 \text{ moles } H_2}{1 \text{ mole triolein}} = 3.00 \text{ moles of } H_2 \text{ (3 SFs)}$$

c. $1.00 \text{ mole triolein} \times \dfrac{3.00 \text{ moles } H_2}{1 \text{ mole triolein}} \times \dfrac{2.016 \text{ g } H_2}{1 \text{ mole } H_2} = 6.05 \text{ g of } H_2 \text{ are needed (3 SFs).}$

d. $1.00 \text{ mole triolein} \times \dfrac{3.00 \text{ moles } H_2}{1 \text{ mole triolein}} \times \dfrac{22.4 \text{ L } H_2 \text{ (STP)}}{1 \text{ mole } H_2} = 67.2 \text{ L of } H_2 \text{ (at STP) are needed}$

$$(3 \text{ SFs}).$$

Lance, an environmental health practitioner, has returned to a farm where he found small amounts of fenbendazole, a sheep dewormer, in the soil. When he is working on-site, he wears protective clothing while he obtains samples of soil and water. His recent tests indicate that the presence of fenbendazole is now within the acceptable range in both the soil and the water in the lake near the farm. Lance specializes in measuring levels of contaminants in soil and water caused by manufacturing processes or farming. In his work, Lance also acts as an advisor, consultant, and safety officer to ensure that soil and water on the farm are safe and to maintain protection of the environment.

Circle the amine and amide functional groups in the line-angle formula of fenbendazole.

Fenbendazole

Credit: Bart Coenders/E+/Getty Images

LOOKING AHEAD

The Health icon indicates a question that is related to health and medicine.

18.1 Amines

Learning Goal: Write the IUPAC and common names for amines; draw the condensed structural and line-angle formulas when given their names. Classify amines as primary (1°), secondary (2°), or tertiary (3°).

REVIEW
Naming and Drawing Alkanes (12.2)

- Amines are derivatives of ammonia (NH_3), in which alkyl or aromatic groups replace one or more hydrogen atoms.

- In the IUPAC system, the *e* in the corresponding alkane name of the longest chain bonded to the N atom is replaced by *amine*. Alkyl groups or aromatic groups attached to the N atom are named with the prefix *N-*.

- In the common names of simple amines, the alkyl groups are listed alphabetically followed by *amine*.

- Amines are classified as primary, secondary, or tertiary when the nitrogen atom is bonded to one, two, or three alkyl or aromatic groups.

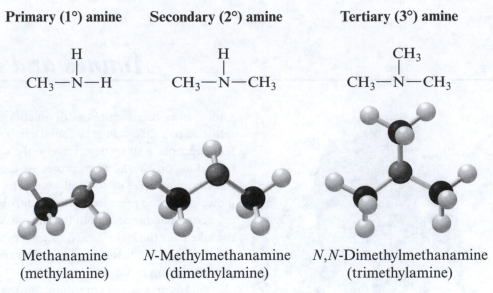

Primary (1°) amine

Secondary (2°) amine

Tertiary (3°) amine

Methanamine
(methylamine)

N-Methylmethanamine
(dimethylamine)

N,N-Dimethylmethanamine
(trimethylamine)

- The amine of benzene is named aniline.

Aniline

- Many amines, which are prevalent in synthetic and naturally occurring compounds, have physiological activity.

IUPAC Naming of Amines	
STEP 1	Name the longest carbon chain bonded to the N atom by replacing the *e* of its alkane name with *amine*.
STEP 2	Number the carbon chain to show the position of the amine group and other substituents.
STEP 3	Any alkyl group attached to the nitrogen atom is indicated by the prefix *N*- and the alkyl name, which is placed in front of the amine name.

♦ **Learning Exercise 18.1A**

Write the IUPAC name and common name, if any, for each of the following:

1. CH₃—N(H)—CH₂—CH₃

2.

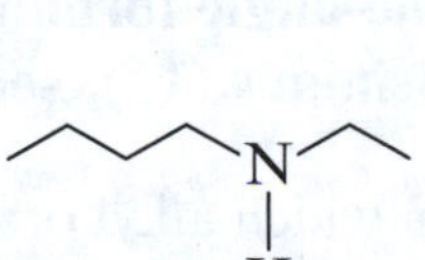

3. 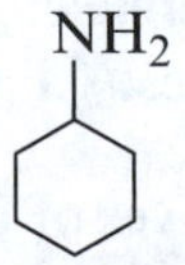 (cyclohexyl with NH₂)

4.

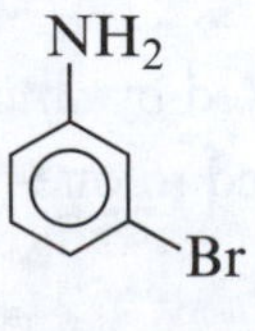

5. CH₃—CH₂—N(CH₃)—CH₃

Answers

1. *N*-methylethanamine (ethylmethylamine)   2. *N*-ethyl-1-butanamine (butylethylamine)
3. cyclohexanamine (cyclohexylamine)   4. 3-bromoaniline (*m*-bromoaniline)
5. *N*,*N*-dimethylethanamine (ethyldimethylamine)

♦ **Learning Exercise 18.1B**

Classify each of the amines in Learning Exercise 18.1A as primary (1°), secondary (2°), or tertiary (3°):

1. _____   2. _____   3. _____

4. _____   5. _____

Answers

1. secondary (2°)   2. secondary (2°)   3. primary (1°)
4. primary (1°)   5. tertiary (3°)

♦ **Learning Exercise 18.1C**

Draw the condensed structural formulas for **1** and **2** and the line-angle formulas for **3** and **4**.

1. 2-propanamine   2. *N*-ethyl-*N*-methyl-1-butanamine

3. 4-bromoaniline   4. *N*,*N*-dimethyl-3-pentanamine

Answers

$$\begin{array}{c} NH_2 \\ | \\ 1.\ CH_3{-}CH{-}CH_3 \end{array}$$

$$\begin{array}{c} CH_3 \\ | \\ 2.\ CH_3{-}CH_2{-}N{-}CH_2{-}CH_2{-}CH_2{-}CH_3 \end{array}$$

3.

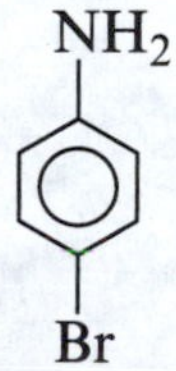

4.

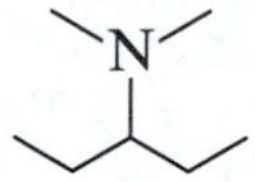

18.2 Properties of Amines

Learning Goal: Describe the boiling points and solubility of amines; write balanced chemical equations for the reaction of amines with water and with acids.

REVIEW

Writing Equations for Reactions of Acids and Bases (11.7)

• Primary amines can form more hydrogen bonds, and thus have higher boiling points than secondary amines of the same mass. Tertiary amines cannot hydrogen bond with each other because they have no N—H bonds.

- Amines have higher boiling points than hydrocarbons but lower boiling points than alcohols of similar mass because the N atom is not as electronegative as the O atom in alcohols.
- Hydrogen bonding allows amines with one to six carbon atoms to be soluble in water.
- In water, amines act as weak bases by accepting H^+ from water to produce ammonium and hydroxide ions.

$$CH_3-NH_2 + H_2O \rightleftharpoons CH_3-\overset{+}{N}H_3 + OH^-$$

Methylamine   Methylammonium   Hydroxide
                         ion   ion

- Strong acids neutralize amines to yield ammonium salts.

$$CH_3-NH_2 + HCl \longrightarrow CH_3-\overset{+}{N}H_3Cl^-$$

Methylamine   Methylammonium chloride

- In a quaternary ammonium salt, a nitrogen atom is bonded to four carbon groups.

♦ **Learning Exercise 18.2A**

Indicate the compound in each pair that has the higher boiling point.

1. CH_3-NH_2 or CH_3-OH _____

2. $CH_3-CH_2-CH_3$ or $CH_3-CH_2-NH_2$ _____

3. $CH_3-\underset{\underset{H}{|}}{N}-CH_3$ or $CH_3-CH_2-NH_2$ _____

Answers   1. CH_3-OH 2. $CH_3-CH_2-NH_2$ 3. $CH_3-CH_2-NH_2$

♦ **Learning Exercise 18.2B**

Draw the condensed structural or line-angle formula for the product of each of the following reactions:

1. $CH_3-CH_2-NH_2 + H_2O \rightleftharpoons$

2. $CH_3-CH_2-CH_2-NH_2 + HCl \longrightarrow$

3. 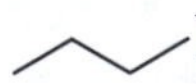 $\wedge\!\!\wedge NH_2 + H_2O \rightleftharpoons$

4. 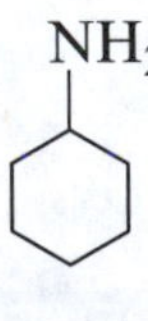 (cyclohexane with NH_2) $+ HBr \longrightarrow$

5. 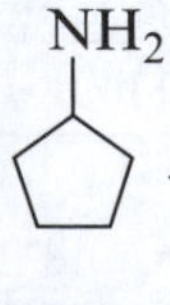 (cyclopentane with NH_2) $+ H_2O \rightleftharpoons$

The amines in fish react with the acid in lemon to neutralize the "fishy" odor.

Credit: Pearson Education/Pearson Science

6. $CH_3-CH_2-NH_3^+ Cl^- + NaOH \longrightarrow$

Answers

1. $CH_3-CH_2-\overset{+}{N}H_3 + OH^-$

2. $CH_3-CH_2-CH_2-\overset{+}{N}H_3\ Cl^-$

3. 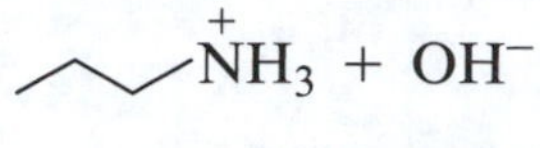 $\diagup\!\!\!\diagdown\overset{+}{N}H_3 + OH^-$

4. 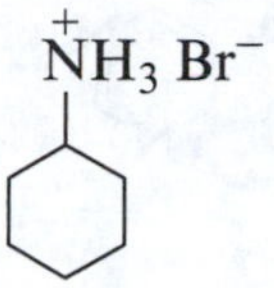 $\overset{\displaystyle \overset{+}{N}H_3\ Br^-}{\bigcirc}$ (cyclohexane ring with $\overset{+}{N}H_3\ Br^-$)

5. 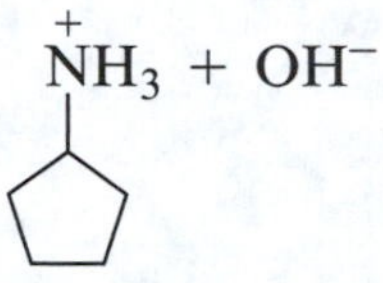 $\overset{\displaystyle \overset{+}{N}H_3 + OH^-}{\bigcirc}$ (cyclopentane ring with $\overset{+}{N}H_3 + OH^-$)

6. $CH_3-CH_2-NH_2 + NaCl + H_2O$

♦ **Learning Exercise 18.2C**

Indicate if each of the following is soluble in water. Explain.

1. $CH_3-CH_2-CH_2-\underset{\underset{\displaystyle CH_2-CH_3}{|}}{\overset{\overset{\displaystyle CH_2-CH_2-CH_3}{|}}{N}}-CH_2-CH_3$

2. $CH_3-CH_2-\underset{\underset{\displaystyle NH_2}{|}}{CH}-CH_2-CH_3$

3.  $\diagup\!\!\!\diagdown\!\!\!\diagup\!\!\!\diagdown_{NH_2}$

Answers

1. No, an amine with eight carbon atoms is mostly nonpolar and not soluble in water.
2. Yes, amines with five carbon atoms form hydrogen bonds, which makes them soluble in water.
3. Yes, amines with four carbon atoms form hydrogen bonds, which makes them soluble in water.

18.3 Heterocyclic Amines

Learning Goal: Identify heterocyclic amines; distinguish between the types of heterocyclic amines.

- A heterocyclic amine is a cyclic compound containing one or more nitrogen atoms in the ring.
- Most heterocyclic amines contain five or six atoms in the ring.
- An alkaloid is a physiologically active amine obtained from plants.

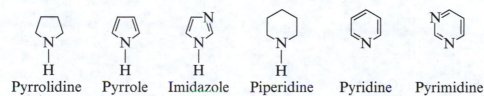

Pyrrolidine   Pyrrole   Imidazole   Piperidine   Pyridine   Pyrimidine

♦ **Learning Exercise 18.3**

Match each of the following heterocyclic structures with its correct name:

a.

b.

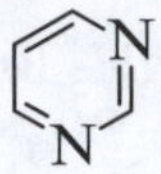

c.

d.

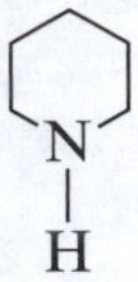

e.

f.

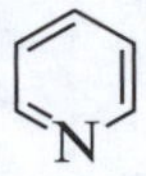

1. _____ pyrrolidine

2. _____ imidazole

3. _____ pyridine

4. _____ pyrrole

5. _____ pyrimidine

6. _____ piperidine

Answers   1. c   2. e   3. f   4. a   5. b   6. d

18.4 Neurotransmitters

Learning Goal: Describe the role of amines as neurotransmitters.

- Neurotransmitters are chemical compounds that transit impulses from a nerve cell (neuron) to a target cell in another nerve cell, a muscle cell, or a gland cell.
- A neuron consists of a cell body and filaments called axons and dendrites that form synapses.
- When an impulse arrives at the neuron, a neurotransmitter that is stored at the ends of the axon terminal is released into the synapse and diffuses to its receptor sites on the dendrites of adjacent nerve cells.
- Some examples of neurotransmitters are: acetylcholine, dopamine, epinephrine, norepinephrine, serotonin, histamine, glutamate, and gamma(γ)-aminobutyric acid (GABA).

♦ **Learning Exercise 18.4A**

Draw the condensed structural formula for each of the following neurotransmitters:
1. histamine

2. epinephrine

3. gamma(γ)-aminobutyric acid (GABA)

Answers

1. 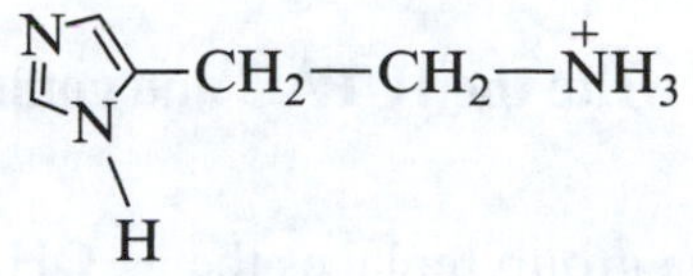 (imidazole ring)—CH_2—CH_2—$\overset{+}{N}H_3$

2. HO—(benzene ring with HO)—CH—CH_2—$\overset{+}{N}H_2$—CH_3 with OH on the CH

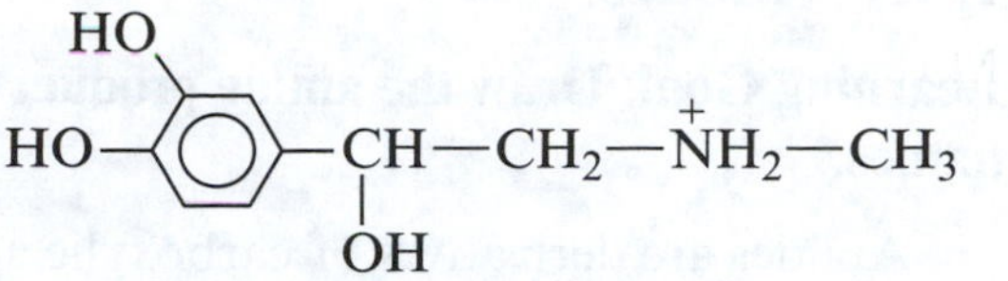

3. $H_3\overset{+}{N}$—CH_2—CH_2—CH_2—$\overset{O}{\overset{\|}{C}}$—$O^-$

♦ **Learning Exercise 18.4B**

Match the physiological process affected by each of the following neurotransmitters (**a** to **e**) with the correct physiological function or effect (**1** to **5**):

    a. dopamine   **b.** serotonin   **c.** GABA

    d. epinephrine   **e.** acetylcholine

1. _____ excitatory function that stimulates muscle contraction and plays a role in memory and sleep

2. _____ inhibitory function that blocks the firing by a neuron

3. _____ inhibitory function that lowers anxiety

4. _____ low levels can lead to depression, anxiety, and eating disorders

5. _____ an excitatory function that increases heart rate and blood pressure

Answers   **1.** a   **2.** e   **3.** c   **4.** b   **5.** d

♦ **Learning Exercise 18.4C**

Match the disorder (**1** to **4**) that may be caused by abnormal levels of each of the following neurotransmitters (**a** to **d**):

    a. dopamine   **b.** norepinephrine

    c. histamine   **d.** acetylcholine

1. _____ low levels can cause Parkinson's disease

2. _____ low levels can cause Alzheimer's disease

3. _____ low levels can lead to attention deficit disorder

4. _____ high levels can cause allergic reactions

Answers   **1.** a   **2.** d   **3.** b   **4.** c

18.5 Amides

Learning Goal: Draw the amide product from amidation, and write the IUPAC and common names.

- Amides are derivatives of carboxylic acids in which an amine group replaces the —OH group in the carboxylic acid.
- Amides are named by replacing the *ic acid* or *oic acid* ending of the corresponding carboxylic acid with *amide*. When an alkyl group is attached to the N atom, it is listed as *N-alkyl*.

$$CH_3{-}CH_2{-}CH_2{-}\overset{\overset{\displaystyle O}{\|}}{C}{-}NH_2$$

Butanamide
(butyramide)

$$CH_3{-}\overset{\overset{\displaystyle O}{\|}}{C}{-}\overset{\overset{\displaystyle H}{|}}{N}{-}CH_3$$

N-Methylethanamide
(*N*-methylacetamide)

- When a carboxylic acid reacts with ammonia or an amine, an amide is produced.

$$CH_3{-}\overset{\overset{\displaystyle O}{\|}}{C}{-}OH + NH_3 \xrightarrow{\text{Heat}} CH_3{-}\overset{\overset{\displaystyle O}{\|}}{C}{-}NH_2 + H_2O$$

$$CH_3{-}\overset{\overset{\displaystyle O}{\|}}{C}{-}OH + H_2N{-}CH_3 \xrightarrow{\text{Heat}} CH_3{-}\overset{\overset{\displaystyle O}{\|}}{C}{-}\overset{\overset{\displaystyle H}{|}}{N}{-}CH_3 + H_2O$$

- Amides with one to five carbons are soluble in water.

♦ **Learning Exercise 18.5A**

Draw the condensed structural or line-angle formula for the amide formed in each of the following reactions:

CORE CHEMISTRY SKILL
Forming Amides

1. $$CH_3{-}CH_2{-}\overset{\overset{\displaystyle O}{\|}}{C}{-}OH + NH_3 \xrightarrow{\text{Heat}}$$

2. 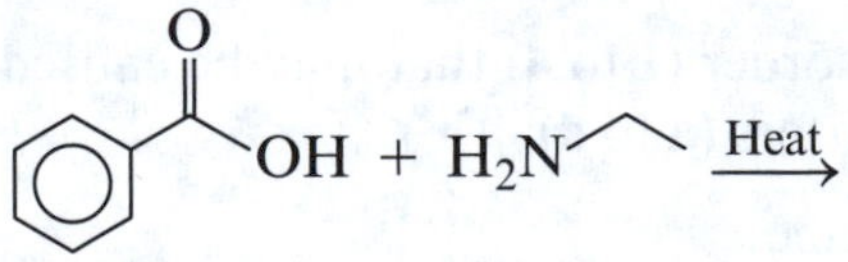 (benzoic acid) ⬡—C(=O)—OH + H_2N⌐ $\xrightarrow{\text{Heat}}$

3. $$CH_3{-}\overset{\overset{\displaystyle O}{\|}}{C}{-}OH + \overset{\overset{\displaystyle CH_3}{|}}{NH}{-}CH_3 \xrightarrow{\text{Heat}}$$

Answers

1. $$CH_3{-}CH_2{-}\overset{\overset{\displaystyle O}{\|}}{C}{-}NH_2$$

2. 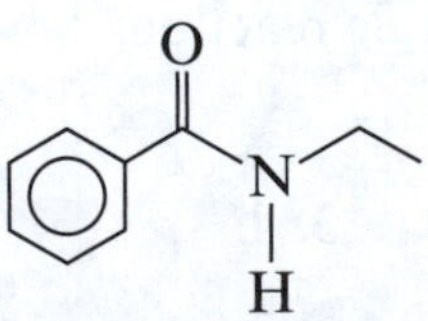 ⬡—C(=O)—N(H)⌐

3. $$CH_3{-}\overset{\overset{\displaystyle O}{\|}}{C}{-}\overset{\overset{\displaystyle CH_3}{|}}{N}{-}CH_3$$

Naming Amides	
STEP 1	Replace the *oic acid* in the carboxylic acid name with *amide*.
STEP 2	Name each substituent on the N atom using the prefix *N-* and the alkyl name.

◆ **Learning Exercise 18.5B**

Name each of the following amides:

1. $CH_3-CH_2-\overset{\displaystyle O}{\overset{\|}{C}}-NH_2$

2. 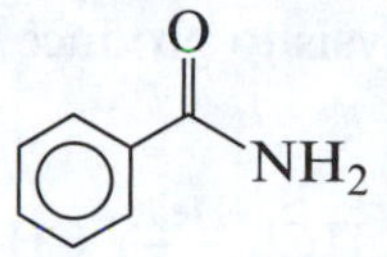 (benzene ring)$-\overset{\displaystyle O}{\overset{\|}{C}}-NH_2$

3. 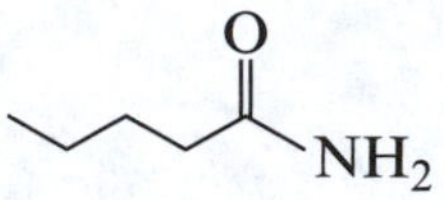 (line-angle structure)$-\overset{\displaystyle O}{\overset{\|}{C}}-NH_2$

4. $CH_3-\overset{\displaystyle O}{\overset{\|}{C}}-\overset{\displaystyle H}{\overset{\|}{N}}-CH_2-CH_3$

5. 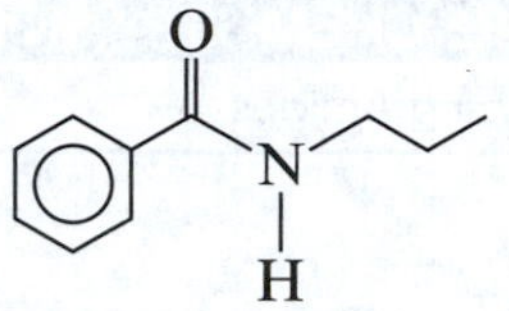 (benzene ring)$-\overset{\displaystyle O}{\overset{\|}{C}}-\overset{}{\underset{\displaystyle H}{N}}$(propyl)

Answers   1. propanamide (propionamide)   2. benzamide
                    3. pentanamide   4. *N*-ethylethanamide (*N*-ethylacetamide)
                    5. *N*-propylbenzamide

◆ **Learning Exercise 18.5C**

Draw the condensed structural formulas for **1** and **2** and the line-angle formulas for **3** and **4**.

1. methanamide   2. butanamide

3. 3-chloropentanamide   4. *N*-ethylbenzamide

Answers   1. $H-\overset{\displaystyle O}{\overset{\|}{C}}-NH_2$ 2. $CH_3-CH_2-CH_2-\overset{\displaystyle O}{\overset{\|}{C}}-NH_2$

                    3. 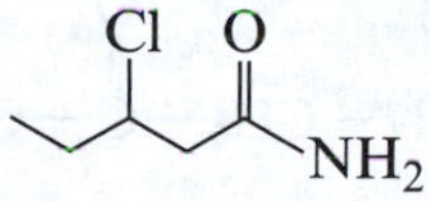 (line-angle structure with Cl)$\overset{Cl}{}\overset{\displaystyle O}{\overset{\|}{C}}-NH_2$ 4. 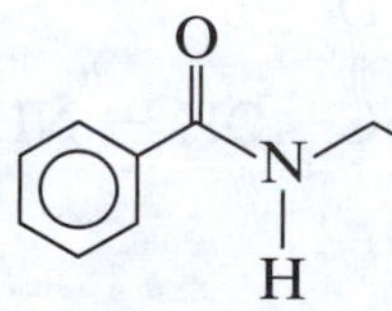 (benzene ring)$-\overset{\displaystyle O}{\overset{\|}{C}}-\overset{}{\underset{\displaystyle H}{N}}$(ethyl)

18.6 Hydrolysis of Amides

Learning Goal: Write balanced chemical equations for the hydrolysis of amides.

- Amides undergo acid hydrolysis to produce the carboxylic acid and an ammonium salt.

$$CH_3-\overset{O}{\overset{\|}{C}}-NH_2 + H_2O + HCl \xrightarrow{\text{Heat}} CH_3-\overset{O}{\overset{\|}{C}}-OH + NH_4^+\,Cl^-$$

- Amides undergo base hydrolysis to produce the carboxylate salt and an amine.

$$CH_3-\overset{O}{\overset{\|}{C}}-NH_2 + NaOH \xrightarrow{\text{Heat}} CH_3-\overset{O}{\overset{\|}{C}}-O^-\,Na^+ + NH_3$$

Key Terms for Sections 18.1 to 18.6

CORE CHEMISTRY SKILL

Hydrolyzing Amides

Match each of the following key terms with the correct description:

a. heterocyclic amine **b.** amidation **c.** amine
d. amide **e.** alkaloid **f.** neurotransmitter

1. ____ a compound that contains nitrogen, is active physiologically, and is produced by plants
2. ____ a cyclic organic compound that contains one or more nitrogen atoms
3. ____ the reaction of a carboxylic acid and an amine
4. ____ the hydrolysis of this compound produces a carboxylic acid and an amine
5. ____ an organic compound that contains an amino group
6. ____ a compound that transmits an impulse from a neuron to a target cell

Answers **1.** e **2.** a **3.** b **4.** d **5.** c **6.** f

♦ Learning Exercise 18.6

Draw the condensed structural formulas for the products formed by the hydrolysis of each of the following with HCl and with NaOH:

1. $CH_3-CH_2-\overset{O}{\overset{\|}{C}}-NH_2$

2. $CH_3-\overset{O}{\overset{\|}{C}}-\overset{H}{\overset{|}{N}}-CH_2-CH_3$

Answers

1. (HCl) $CH_3-CH_2-\overset{O}{\overset{\|}{C}}-OH + NH_4^+\,Cl^-$ (NaOH) $CH_3-CH_2-\overset{O}{\overset{\|}{C}}-O^-\,Na^+ + NH_3$

2. (HCl) $CH_3-\overset{O}{\overset{\|}{C}}-OH + CH_3-CH_2-\overset{+}{N}H_3\,Cl^-$ (NaOH) $CH_3-\overset{O}{\overset{\|}{C}}-O^-\,Na^+ + H_2N-CH_2-CH_3$

Checklist for Chapter 18

You are ready to take the Practice Test for Chapter 18. Be sure you have accomplished the following learning goals for this chapter. If not, review the Section listed at the end of the goal. Then apply your new skills and understanding to the Practice Test.

After studying Chapter 18, I can successfully:

_____ Write the IUPAC and common names of amines. (18.1)

_____ Draw the condensed structural and line-angle formulas of amines. (18.1)

_____ Classify amines as primary, secondary, or tertiary. (18.1)

_____ Compare the boiling points and solubility of alkanes, alcohols, and amines of similar mass. (18.2)

_____ Write equations for the ionization and neutralization of amines. (18.2)

_____ Identify heterocyclic amines. (18.3)

_____ Describe the role of amines as neurotransmitters. (18.4)

_____ Draw the condensed structural formulas of neurotransmitters. (18.4)

_____ Write the IUPAC and common names of amides, and draw their condensed structural formulas. (18.5)

_____ Write balanced chemical equations for the acidic and basic hydrolysis of amides. (18.6)

Practice Test for Chapter 18

The chapter Sections to review are shown in parentheses at the end of each question.

Classify each of the amines in questions 1 through 6 as a: (18.1)

    A. primary amine   **B.** secondary amine   **C.** tertiary amine

1. _____ $CH_3-\overset{\overset{\displaystyle CH_3}{|}}{CH}-NH_2$

2. _____ (line-angle structure with N and H)

3. _____ $CH_3-CH_2-\overset{\overset{\displaystyle NH_2}{|}}{CH}-CH_2-CH_3$

4. _____ $CH_3-\overset{\overset{\displaystyle H}{|}}{N}-CH_2-CH_3$

5. _____ $CH_3-\overset{\overset{\displaystyle CH_3}{|}}{CH}-CH_2-\overset{\overset{\displaystyle H}{|}}{N}-\overset{\overset{\displaystyle CH_3}{|}}{CH}-CH_3$

6. _____ $CH_3-\overset{\overset{\displaystyle CH_3}{|}}{\underset{\underset{\displaystyle CH_3}{|}}{C}}-CH_2-NH_2$

Match the amines and amides in questions 7 through 11 with the following names: (18.1, 18.5)

    A. ethyldimethylamine   **B.** butanamide   **C.** *N*-methylacetamide
    D. benzamide   **E.** *N*-ethylbutyramide

7. _____ $CH_3-CH_2-\overset{\overset{\displaystyle CH_3}{|}}{N}-CH_3$

8. _____ $CH_3-CH_2-CH_2-\overset{\overset{\displaystyle O}{||}}{C}-NH_2$

9. _____ (benzene ring)$-\overset{\overset{\displaystyle O}{||}}{C}-NH_2$

10. _____ $CH_3-CH_2-CH_2-\overset{\overset{\displaystyle O}{||}}{C}-\overset{\overset{\displaystyle H}{|}}{N}-CH_2-CH_3$

11. _____ $CH_3-\overset{\overset{\displaystyle O}{\|}}{C}-\overset{\overset{\displaystyle H}{|}}{N}-CH_3$

In questions 12 through 15, identify the compound with the higher boiling point: (18.2)

12. **A.** $CH_3-CH_2-NH_2$ or **B.** CH_3-CH_2-OH

13. **A.** $CH_3-NH-CH_3$ or **B.** $CH_3-CH_2-NH_2$

14. **A.** $CH_3-CH_2-CH_2-CH_3$ or **B.** $CH_3-CH_2-CH_2-NH_2$

15. **A.** $CH_3-CH_2-CH_2-OH$ or **B.** $CH_3-CH_2-CH_2-NH_2$

For questions 16 through 19, identify the product as **A** *to* **D**: (18.2, 18.5, 18.6)

A. $CH_3-CH_2-\overset{\overset{\displaystyle O}{\|}}{C}-OH + \overset{+}{N}H_4\,Cl^-$ **B.** $CH_3-CH_2-\overset{+}{N}H_3\,Cl^-$

C. $CH_3-CH_2-CH_2-\overset{+}{N}H_3 + OH^-$ **D.** $CH_3-\overset{\overset{\displaystyle O}{\|}}{C}-NH_2$

16. _____ dissociation of propylamine in water

17. _____ hydrolysis of propanamide with hydrochloric acid

18. _____ reaction of ethylamine and hydrochloric acid

19. _____ amidation of acetic acid

20. Amines used in drugs are converted to their amine salt because the salt is (18.2)
    A. a solid at room temperature   **B.** soluble in water   **C.** odorless
    D. soluble in body fluids   **E.** all of these

21. Piperidine is a heterocyclic amine that (18.3)
    A. has a ring of six atoms
    B. has a ring of five atoms
    C. contains two nitrogen atoms in a ring
    D. has one nitrogen atom in an aromatic ring system
    E. has two nitrogen atoms in an aromatic ring system

22. Alkaloids are (18.3)
    A. physiologically active nitrogen-containing compounds
    B. produced by plants
    C. used in anesthetics, in antidepressants, and as stimulants
    D. often habit-forming
    E. all of these

For questions 23 to 26, match the name of the alkaloid (**A** *to* **D**) *with the correct description:* (18.3)

    A. caffeine   **B.** nicotine   **C.** morphine   **D.** quinine

23. _____ a painkiller from the opium poppy plant

24. _____ obtained from the bark of the cinchona tree and used in the treatment of malaria

25. _____ a stimulant obtained from the leaves of tobacco plants

26. _____ a stimulant obtained from coffee beans and tea leaves

*For questions 27 to 30, match the name of the neurotransmitter (**A** to **D**) with its structure: (18.4)*

    A. dopamine   **B.** glutamate   **C.** acetylcholine   **D.** serotonin

27. _____ $CH_3-\overset{\overset{O}{\|}}{C}-O-CH_2-CH_2-\overset{\overset{CH_3}{|}}{\underset{\underset{CH_3}{|}}{\overset{+}{N}}}-CH_3$

28. _____ [structure: serotonin]

29 _____ $HO-\overset{HO}{\underset{}{\bigcirc}}-CH_2-CH_2-\overset{+}{N}H_3$

30. _____ $\overset{\overset{O}{\|}\;\;\overset{O^-}{}}{\underset{\overset{+}{H_3}N-CH-CH_2-CH_2-\overset{\overset{O}{\|}}{C}-O^-}{C}}$

31. The IUPAC name of this amide is: (18.5) $CH_3-\overset{\overset{O}{\|}}{C}-\overset{\overset{H}{|}}{N}-CH_3$
       A. methyl acetate   **B.** methylethanamide   **C.** *N*-methylethanamide
       D. *N*-methylacetamide   **E.** methylacetamide

32. What is the IUPAC name of the product from the reaction of propanoic acid and ammonia? (18.5)
       A. propionamide   **B.** *N*-methylpropanamide   **C.** *N*-methylpropionamide
       D. *N*-methylacetamide   **E.** propanamide

33. When *N*-ethylacetamide is hydrolyzed with HCl, the product(s) are: (18.6)
       A. acetic acid and ethylamine   **B.** acetic acid and ethylammonium chloride
       C. ethylamine   **D.** sodium acetate and ethylamine
       E. ethane and acetamide

34. When *N*-ethylacetamide is hydrolyzed with NaOH, the product(s) are: (18.6)
       A. acetic acid and ethylamine   **B.** acetic acid and ethylammonium chloride
       C. sodium acetate   **D.** sodium acetate and ethylamine
       E. ethane and acetamide

Answers to the Practice Test

1. A	2. C	3. A	4. B	5. B
6. A	7. A	8. B	9. D	10. E
11. C	12. B	13. B	14. B	15. A
16. C	17. A	18. B	19. D	20. E
21. A	22. E	23. C	24. D	25. B
26. A	27. C	28. D	29. A	30. B
31. C	32. E	33. B	34. D	

Selected Answers and Solutions to Text Problems

18.1 The common name of an amine consists of naming the alkyl groups attached to the nitrogen atom in alphabetical order. In the IUPAC name, the *e* in the name of the corresponding alkane is replaced with *amine*. Any alkyl group attached to the nitrogen atom is indicated by the prefix *N-* and the alkyl name, which is placed in front of the amine name.

    a. A two-carbon alkyl group attached to —NH$_2$ is ethylamine. In the IUPAC name, the *e* in ethane is replaced with *amine* to give ethanamine.

    b. A one-carbon and a three-carbon alkyl group attached to nitrogen form methylpropylamine. The IUPAC name is *N*-methyl-1-propanamine.

    c. A one-carbon, a two-carbon, and a three-carbon alkyl group attached to nitrogen form ethylmethylpropylamine. The IUPAC name is *N*-ethyl-*N*-methyl-1-propanamine.

    d. A three-carbon isopropyl group attached to nitrogen forms isopropylamine. The IUPAC name is 2-propanamine.

18.3 **a.**

    b.

    c.

18.5 In a primary (1°) amine, there is one alkyl group (and two hydrogens) attached to a nitrogen atom.

18.7 **a.** This is a primary (1°) amine; there is only one alkyl group (and two hydrogens) attached to the nitrogen atom.

    b. This is a secondary (2°) amine; there are two alkyl groups (and one hydrogen) attached to the nitrogen atom.

    c. This is a primary (1°) amine; there is only one alkyl group (and two hydrogens) attached to the nitrogen atom.

    d. This is a tertiary (3°) amine; there are two alkyl groups and one aromatic group (and no hydrogens) attached to the nitrogen atom.

    e. This is a tertiary (3°) amine; there are three alkyl groups (and no hydrogens) attached to the nitrogen atom.

18.9 Amines have higher boiling points than hydrocarbons but lower boiling points than alcohols of similar mass. Boiling point increases with molecular mass.

    a. CH$_3$—CH$_2$—OH has the higher boiling point because the —OH group forms stronger hydrogen bonds than the —NH$_2$ group.

    b. CH$_3$—CH$_2$—CH$_2$—NH$_2$ has the higher boiling point because it has a greater molar mass.

    c. CH$_3$—CH$_2$—CH$_2$—NH$_2$ has the higher boiling point because it is a primary amine that forms hydrogen bonds. A tertiary amine cannot form hydrogen bonds with other tertiary amines.

18.11 As a primary amine, propylamine can form two hydrogen bonds with other amine molecules, which gives it the highest boiling point. Ethylmethylamine, a secondary amine, can form one hydrogen bond with other amine molecules, and butane cannot form hydrogen bonds. Thus, butane has the lowest boiling point of the three compounds.

18.13 Amines with fewer than seven carbon atoms are soluble in water. In larger amines, the nonpolar hydrocarbon groups diminish the effect of hydrogen bonding from the amine group.

    a. Yes; amines with fewer than seven carbon atoms hydrogen bond with water molecules and are soluble in water.

    b. Yes; amines with fewer than seven carbon atoms hydrogen bond with water molecules and are soluble in water.

    c. No; an amine with nine carbon atoms has large hydrocarbon sections that make it insoluble in water.

    d. Yes; amines with fewer than seven carbon atoms hydrogen bond with water molecules and are soluble in water.

18.15 Amines, which are weak bases, accept H^+ from water to give a hydroxide ion and the related ammonium ion.

    a. $CH_3{-}NH_2 + H_2O \rightleftharpoons CH_3{-}\overset{+}{N}H_3 + OH^-$

    b. $CH_3{-}\underset{\underset{H}{|}}{N}{-}CH_3 + H_2O \rightleftharpoons CH_3{-}\overset{+}{N}H_2{-}CH_3 + OH^-$

    c. $\text{C}_6\text{H}_5\text{NH}_2 + H_2O \rightleftharpoons \text{C}_6\text{H}_5\overset{+}{N}H_3 + OH^-$

18.17 Amines, which are weak bases, combine with the H^+ from HCl to yield the ammonium chloride salt.

    a. $CH_3{-}\overset{+}{N}H_3\ Cl^-$

    b. $CH_3{-}\overset{+}{N}H_2{-}CH_3\ Cl^-$

    c. $\text{C}_6\text{H}_5\overset{+}{N}H_3\ Cl^-$

18.19 a.

    b. The ammonium salt (Novocain) is more soluble in water and body fluids than the amine procaine.

18.21 Heterocyclic amines contain one or more nitrogen atoms in a ring.
    a. Piperidine is a six-atom ring with one nitrogen atom and no double bonds.
    b. Pyrimidine is a six-atom ring with two nitrogen atoms and three double bonds.
    c. Pyrrole is a five-atom ring with one nitrogen atom and two double bonds.

18.23 The six-atom ring with one nitrogen atom and no double bonds in Ritalin is piperidine.

18.25 The five-atom ring with one nitrogen atom and two double bonds in serotonin is pyrrole.

18.27 A neurotransmitter is a chemical compound that transmits an impulse from a nerve cell to a target cell.

18.29 When a nerve impulse reaches the axon terminal, it stimulates the release of neurotransmitters into the synapse.

18.31 A neurotransmitter must be removed from its receptor so that new signals can come from the nerve cells.

18.33 Acetylcholine is a neurotransmitter that communicates between the nervous system and muscle cells.

18.35 Dopamine is a neurotransmitter that controls muscle movement, regulates the sleep–wake cycle, and helps to improve cognition, attention, memory, and learning.

18.37 Serotonin is a neurotransmitter that helps to decrease anxiety and improve mood, learning, and memory; it also reduces appetite, and induces sleep.

18.39 Histamine is a neurotransmitter that causes allergic reactions, which may include inflammation, watery eyes, itchy skin, and hay fever.

18.41 Excess glutamate in the synapse can lead to destruction of brain cells.

18.43 GABA is a neurotransmitter that regulates muscle tone, sleep, and anxiety.

18.45 Carboxylic acids react with amines to form amides with the elimination of water.

a. $CH_3-\overset{\overset{\displaystyle O}{\|}}{C}-NH_2$

b. $CH_3-\overset{\overset{\displaystyle O}{\|}}{C}-\overset{\overset{\displaystyle H}{|}}{N}-CH_2-CH_3$

c.

18.47 a. *N*-Methylethanamide (*N*-methylacetamide); the "*N*-methyl" means that there is a one-carbon alkyl group attached to the nitrogen. The "ethanamide" component tells us that the carbonyl portion has two carbon atoms.
  b. Butanamide (butyramide) has a four-carbon carbonyl portion attached to an amine group.
  c. Methanamide (formamide) has only the carbonyl carbon attached to the amine group.
  d. *N*-Methylbenzamide; the "*N*-methyl" means that there is a one-carbon alkyl group attached to the nitrogen. The "benzamide" component tells us that this is the amide of benzoic acid.

18.49 a. This is an amide of propionic acid, which has three carbon atoms.

$CH_3-CH_2-\overset{\overset{\displaystyle O}{\|}}{C}-NH_2$

  b. This is an amide of pentanoic acid, which has five carbon atoms.

$CH_3-CH_2-CH_2-CH_2-\overset{\overset{\displaystyle O}{\|}}{C}-NH_2$

  c. The nitrogen atom in *N*-ethylbenzamide is attached to a two-carbon ethyl group. The "benzamide" component tells us that this is the amide of benzoic acid.

  d. This is an amide of butyric acid, which has four carbon atoms. The "*N*-ethyl" means that there is a two-carbon alkyl group attached to the nitrogen.

18.51 a. Ethanamide has the higher melting point because it forms more hydrogen bonds as a primary amide than can *N*-methylethanamide, which is a secondary amide.
  b. Propionamide has the higher melting point because it forms hydrogen bonds, but butane does not.
  c. *N*-Methylpropanamide, a secondary amide, has the higher melting point because it can form hydrogen bonds, whereas *N,N*-dimethylpropanamide, a tertiary amide, cannot form hydrogen bonds.

18.53 Acid hydrolysis of an amide gives the carboxylic acid and the ammonium salt.

a.

$$CH_3\text{—C(=O)—OH} + NH_4^+\ Cl^-$$

b. $CH_3\text{—}CH_2\text{—}\overset{O}{\overset{\|}{C}}\text{—OH} + NH_4^+\ Cl^-$

c. $CH_3\text{—}CH_2\text{—}CH_2\text{—}\overset{O}{\overset{\|}{C}}\text{—OH} + CH_3\text{—}\overset{+}{N}H_3\ Cl^-$

d. benzoic acid $\text{—OH} + NH_4^+\ Cl^-$

18.55 amine, heterocyclic amine

18.57 $70.\ \cancel{\text{kg body mass}} \times \dfrac{65\ \cancel{\text{mg dicyclanil}}}{1\ \cancel{\text{kg body mass}}} \times \dfrac{1\ \text{mL spray}}{50.\ \cancel{\text{mg dicyclanil}}} = 91\ \text{mL of dicyclanil spray (2 SFs)}$

18.59 amine, carboxylic acid, amide, aromatic, ester

18.61 phenol, alcohol, amine

18.63 Excitatory neurotransmitters, like glutamate, open ion channels and stimulate the receptors to send more signals.

18.65 Dopamine, norepinephrine, and epinephrine all have catechol (3,4-dihydroxyphenyl) and amine components.

18.67 a. tetraethylammonium bromide; quaternary (4°) ammonium salt
    b. 1-pentanamine (pentylamine); primary (1°) amine
    c. *N*-ethyl-1-propanamine (ethylpropylamine); secondary (2°) amine

18.69 a. An amine group is attached to carbon 3 of a five-carbon alkane chain.

$$CH_3\text{—}CH_2\text{—}\underset{\underset{NH_2}{|}}{CH}\text{—}CH_2\text{—}CH_3$$

    b. Three two-carbon ethyl groups are attached to a nitrogen atom.

$$CH_3\text{—}CH_2\text{—}\underset{\underset{CH_2\text{—}CH_3}{|}}{N}\text{—}CH_2\text{—}CH_3$$

    c. This is an ammonium salt with two methyl groups attached to the nitrogen atom.

$$CH_3\text{—}\underset{\underset{}{|+}}{\overset{\overset{CH_3}{|}}{N}}H_2\ Cl^-$$

    d. A six-carbon cycloalkyl group (and two hydrogens) are attached to a nitrogen atom.

18.71 a. An alcohol with an —OH group such as 1-butanol forms stronger hydrogen bonds than an amine and has a higher boiling point than an amine.
    b. Ethylamine, a primary amine, forms more hydrogen bonds and has a higher boiling point than dimethylamine, which forms fewer hydrogen bonds as a secondary amine.

18.73 a. Ethylamine is a small amine that is soluble because it forms hydrogen bonds with water. Dibutylamine has two large nonpolar alkyl groups that decrease its solubility in water.

b. Trimethylamine is a small tertiary amine that is soluble because it hydrogen bonds with water. *N*-Ethylcyclohexylamine has a large nonpolar cycloalkyl group that decreases its solubility in water.

18.75 a. *N*-Ethylethanamide; the "*N*-ethyl" means that there is a two-carbon alkyl group attached to the nitrogen. The "ethanamide" component tells us that the carbonyl portion has two carbon atoms.

b. Propanamide has a three-carbon carbonyl portion attached to an amine group.

c. 3-Methylbutanamide has a four-carbon chain with a methyl substituent on carbon 3 as its carbonyl portion attached to an amine group.

18.77 a. An ammonium salt and a strong base produce the amine, a salt, and water.

$$CH_3{-}CH_2{-}\overset{\overset{\displaystyle H}{|}}{N}{-}CH_3 \ + \ NaCl \ + \ H_2O$$

b. An amine in water accepts H^+ from water, which produces an ammonium ion and OH^-.

$$CH_3{-}CH_2{-}\overset{+}{N}H_2{-}CH_3 \ + \ OH^-$$

18.79 aromatic, amine, carboxylate salt

18.81 a. The alkaloid from the bark of the cinchona tree used in treating malaria is quinine.

b. The alkaloid found in tobacco is nicotine.

c. The alkaloid found in coffee and tea is caffeine.

d. The alkaloids that are painkillers found in the opium poppy plant are morphine and codeine.

18.83 Tryptophan contains a carboxylic acid group that is not present in serotonin. Serotonin has a hydroxyl group (—OH) on the aromatic ring that is not present in tryptophan.

18.85 SSRI stands for selective serotonin reuptake inhibitor.

18.87 $CH_3{-}CH_2{-}CH_2{-}NH_2$ $CH_3{-}CH_2{-}\overset{\overset{\displaystyle H}{|}}{N}{-}CH_3$ $CH_3{-}\overset{\overset{\displaystyle CH_3}{|}}{N}{-}CH_3$

Propylamine (1°)   Ethylmethylamine (2°)   Trimethylamine (3°)

$CH_3{-}\overset{\overset{\displaystyle CH_3}{|}}{C}H{-}NH_2$ Isopropylamine (1°)

18.89 a. aromatic, amine, amide, carboxylic acid, cycloalkene

b. aromatic, ether, alcohol, amine

c. aromatic, carboxylic acid

d. phenol, amine, carboxylic acid

e. aromatic, heterocyclic amine, ester, amide

f. amine, heterocyclic amine, aromatic

18.91 Dopamine is needed in the brain, where it is important in controlling muscle movement. Since dopamine cannot cross the blood–brain barrier, persons with low levels of dopamine are given L-dopa, which can cross the blood–brain barrier, where it is converted to dopamine.

Selected Answers to Combining Ideas from Chapters 15 to 18

CI.35 a.

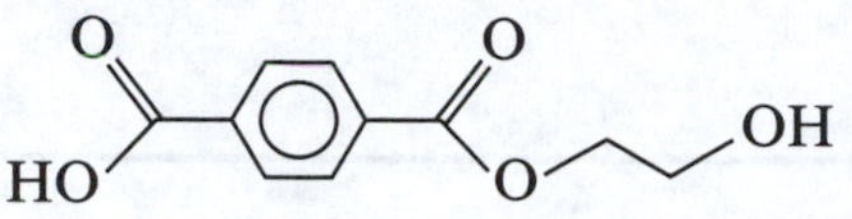

b.

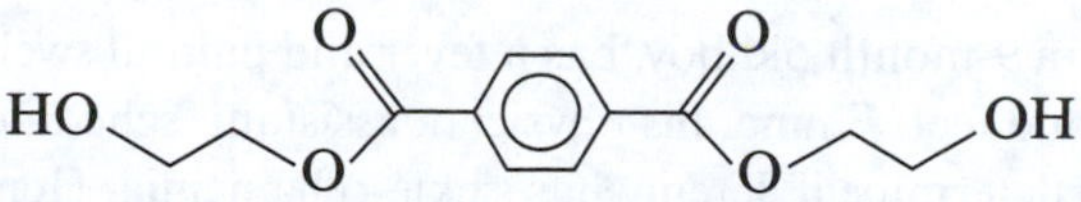

c. 1.80×10^9 lb PETE $\times \dfrac{1 \text{ kg PETE}}{2.20 \text{ lb PETE}} = 8.18 \times 10^8$ kg of PETE (3 SFs)

d. $\dfrac{8.18 \times 10^8 \text{ kg PETE recycled}}{2.71 \times 10^9 \text{ kg PETE sold}} \times 100\% = 30.2\%$ recycled (3 SFs)

e. 8.18×10^8 kg PETE $\times \dfrac{1000 \text{ g PETE}}{1 \text{ kg PETE}} \times \dfrac{1 \text{ mL PETE}}{1.38 \text{ g PETE}} \times \dfrac{1 \text{ L PETE}}{1000 \text{ mL PETE}}$

$= 5.93 \times 10^8$ L of PETE (3 SFs)

f. 5.93×10^8 L PETE $\times \dfrac{1 \text{ landfill}}{2.7 \times 10^7 \text{ L PETE}} = 22$ landfills (2 SFs)

CI.37 a. Metformin contains amine functional groups.
b. Because of its amine groups, metformin will act like a base in water.
c. The molecular formula of metformin is $C_4H_{11}N_5$.
d. Molar mass of metformin ($C_4H_{11}N_5$)
$= 4(12.01 \text{ g}) + 11(1.008 \text{ g}) + 5(14.01 \text{ g}) = 129.18$ g/mole (5 SFs)
e. 1 week $\times \dfrac{7 \text{ days}}{1 \text{ week}} \times \dfrac{2 \text{ doses}}{1 \text{ day}} \times \dfrac{1 \text{ tablet}}{1 \text{ dose}} \times \dfrac{500. \text{ mg metformin}}{1 \text{ tablet}} \times \dfrac{1 \text{ g}}{1000 \text{ mg}}$

$\times \dfrac{1 \text{ mole metformin}}{129.18 \text{ g metformin}}$

$= 0.0542$ mole of metformin (3 SFs)

CI.39 a. **A**, **B**, and **C** are all glucose units.
b. An $\alpha(1 \rightarrow 6)$-glycosidic bond connects monosaccharides **A** and **B**.
c. An $\alpha(1 \rightarrow 4)$-glycosidic bond connects monosaccharides **B** and **C**.
d. The structure drawn is β-panose.
e. Panose is a reducing sugar because it has a free hydroxyl group on carbon 1 of structure **C**, which allows glucose **C** to form an aldehyde.

19

Amino Acids and Proteins

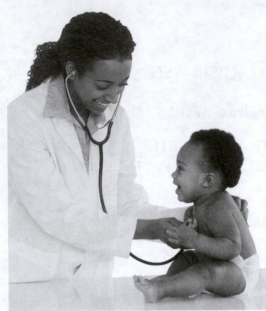

Jeremy, a 9-month old boy, has a fever and painful swelling of his hands and feet. Emma, his physician assistant, schedules blood tests to determine if Jeremy has sickle-cell anemia. Hemoglobin is a protein that carries oxygen in the blood. When a mutation occurs, resulting in a change in the amino acid sequence of hemoglobin, the shape of the molecule may be altered, affecting its ability to carry oxygen properly.

Jeremy's blood tests indicate that he has sickle-cell anemia. Emma prescribes hydroxyurea, which is a medication that stimulates the production of fetal hemoglobin. Fetal hemoglobin, which is not produced after birth, is a variation of the hemoglobin molecule that binds more tightly to oxygen molecules than normal adult hemoglobin. How is the function of hemoglobin changed in sickle-cell anemia?

Hydroxyurea

Credit: Tetra Images/Alamy Stock Photo

LOOKING AHEAD

19.1 Proteins and Amino Acids
19.2 Proteins: Primary Structure
19.3 Proteins: Secondary Structure

19.4 Proteins: Tertiary and Quaternary Structures

19.5 Protein Hydrolysis and Denaturation

 The Health icon indicates a question that is related to health and medicine.

19.1 Proteins and Amino Acids

Learning Goal: Classify proteins by their functions. Give the name and abbreviations for an amino acid and draw its structure at physiological pH.

- Some proteins are enzymes or hormones, while others are important in structure, transport, protection, storage, and contraction of muscles.

- A group of 20 amino acids provides the building blocks of proteins.

- In an amino acid, a central (alpha) carbon is attached to an ammonium group ($-NH_3^+$), a carboxylate group ($-COO^-$), a hydrogen atom ($-H$), and a side chain or R group, which is unique for each amino acid.

- Each specific R group determines whether an amino acid is nonpolar, polar neutral, acidic, or basic. Nonpolar amino acids contain hydrogen, alkyl, or aromatic side chains, whereas polar amino acids contain electronegative atoms such as oxygen or sulfur. Acidic side chains contain a carboxylate group ($-COO^-$), and basic side chains contain an ammonium group ($-NH_3^+$).

Ionized form of alanine

♦ **Learning Exercise 19.1A**

Match one of the following functions with each protein below:

a. structural b. contractile c. storage d. transport
e. hormone f. enzyme g. protection

1. ____ hemoglobin, carries oxygen in blood 2. ____ trypsin, hydrolyzes proteins

3. ____ casein, a protein in milk 4. ____ vasopressin, regulates blood pressure

5. ____ myosin, found in muscle tissue 6. ____ immunoglobulin, an antibody

7. ____ keratin, a major protein of hair 8. ____ lipoprotein, carries lipids in blood

Answers **1.** d **2.** f **3.** c **4.** e **5.** b **6.** g **7.** a **8.** d

♦ **Learning Exercise 19.1B**

Using the appropriate R group, complete the condensed structural formula of each of the following amino acids. Indicate whether the amino acid would be nonpolar, polar neutral, acidic, or basic, and give its three- and one-letter abbreviations.

> **CORE CHEMISTRY SKILL**
> Drawing the Structure for an Amino Acid at Physiological pH

1. 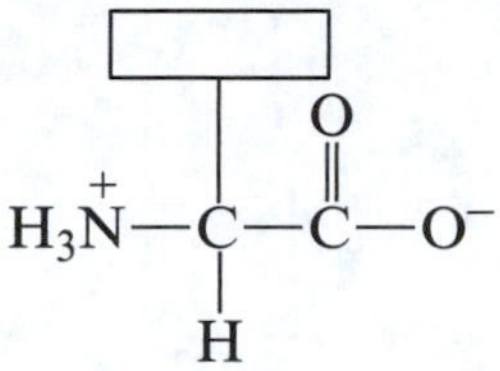 $H_3\overset{+}{N}-C-C-O^-$
   Glycine _____

2. 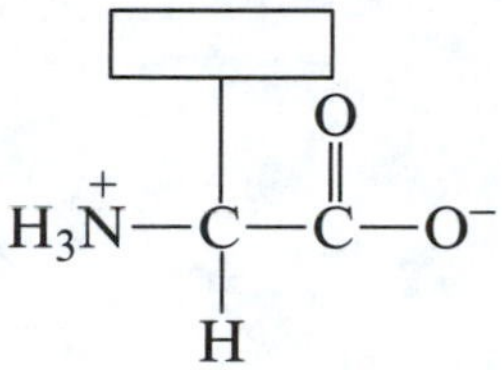 $H_3\overset{+}{N}-C-C-O^-$
   Alanine _____

3. 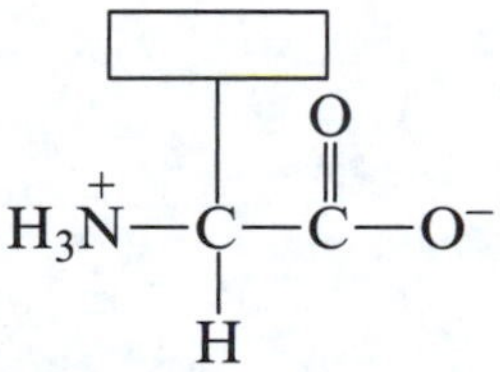 $H_3\overset{+}{N}-C-C-O^-$
   Serine _____

4. 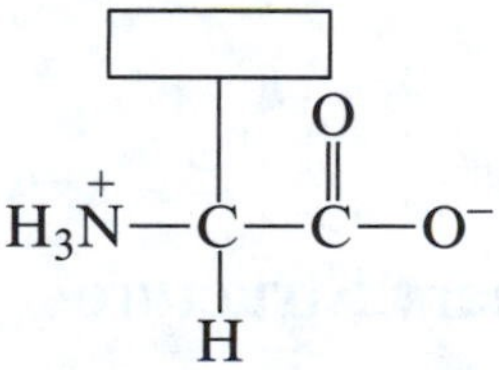 $H_3\overset{+}{N}-C-C-O^-$
   Aspartate _____

Answers

1. 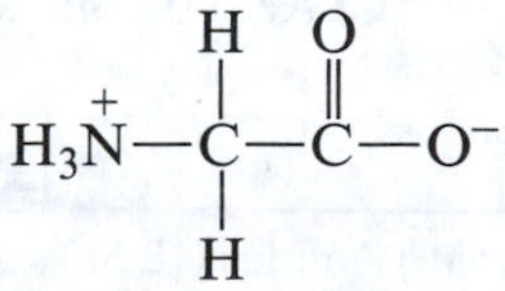 $H_3\overset{+}{N}-C-C-O^-$
   Glycine, nonpolar, Gly, G

2. 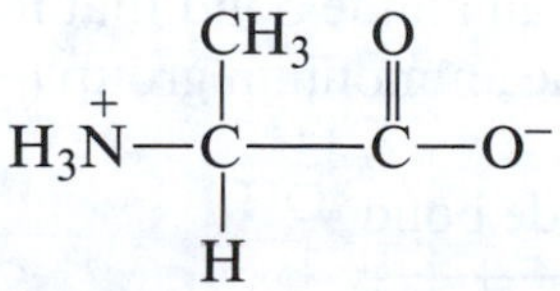 $H_3\overset{+}{N}-C-C-O^-$
   Alanine, nonpolar, Ala, A

3. 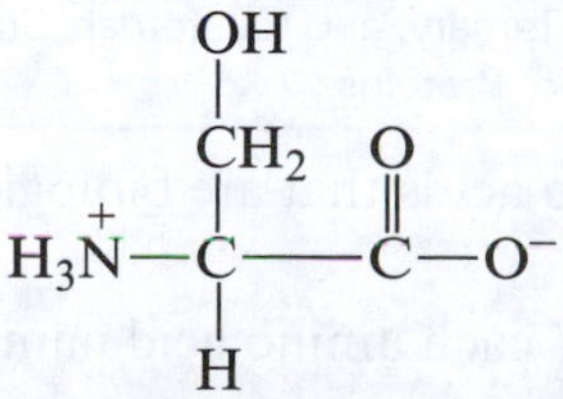 $H_3\overset{+}{N}-C-C-O^-$
   Serine, polar neutral, Ser, S

4. 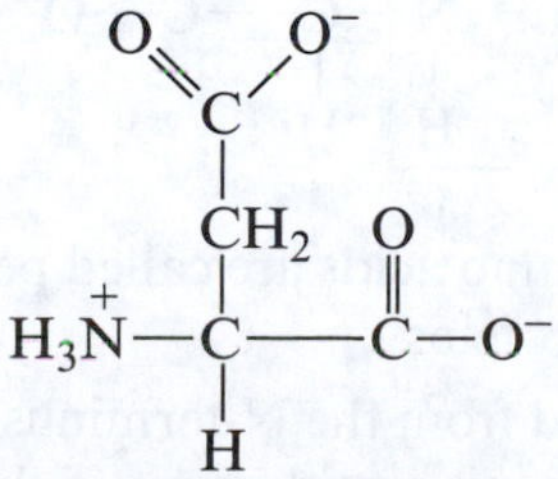 $H_3\overset{+}{N}-C-C-O^-$
   Aspartate, acidic, Asp, D

♦ **Learning Exercise 19.1C**

Draw the structure for each of the following amino acids at physiological pH:

a. Val   **b.** C

c. threonine   **d.** F

Answers

a.

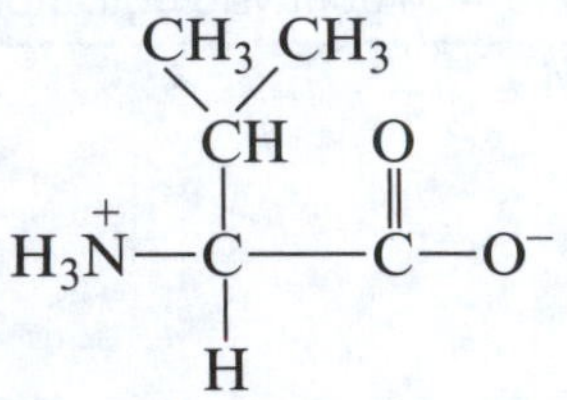

$$
\begin{array}{c}
CH_3 \; CH_3 \\
CH \quad\; O \\
H_3\overset{+}{N}-C-C-O^- \\
H
\end{array}
$$

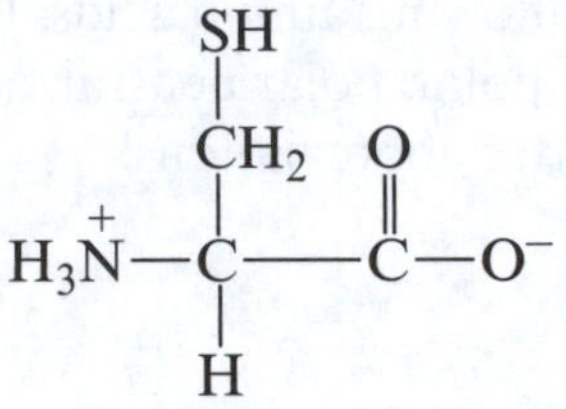

b.

$$
\begin{array}{c}
SH \\
CH_2 \quad\; O \\
H_3\overset{+}{N}-C-C-O^- \\
H
\end{array}
$$

c.

$$
\begin{array}{c}
CH_3 \; OH \\
CH \quad\; O \\
H_3\overset{+}{N}-C-C-O^- \\
H
\end{array}
$$

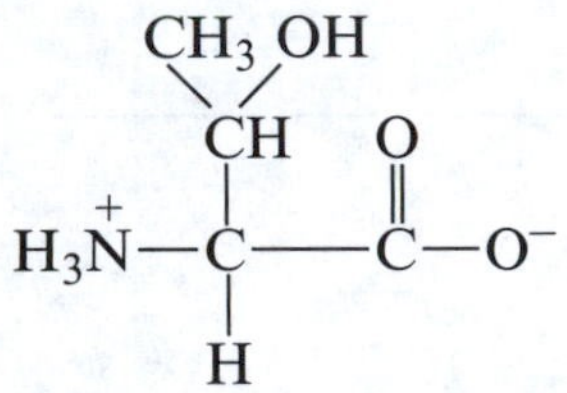

d.

$$
\begin{array}{c}
C_6H_5 \\
CH_2 \quad\; O \\
H_3\overset{+}{N}-C-C-O^- \\
H
\end{array}
$$

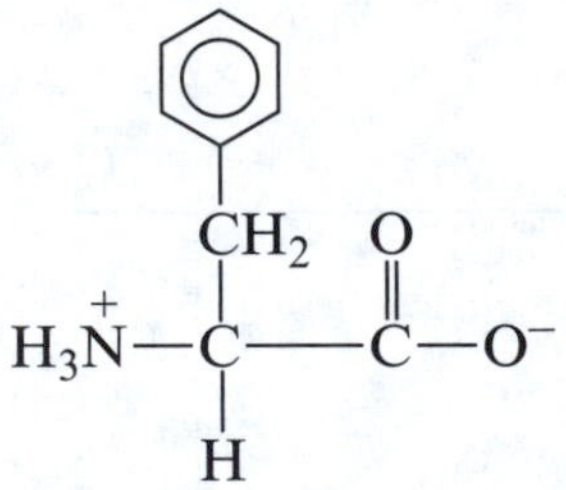

19.2 Proteins: Primary Structure

Learning Goal: Draw the condensed structural formula for a peptide and give its name. Describe the primary structure for a protein.

- The primary structure of a protein is the sequence of amino acids connected by peptide bonds.
- A peptide bond is an amide bond that forms when the carboxylate group ($-COO^-$) of one amino acid reacts with the ammonium group ($-NH_3^+$) of the next amino acid.

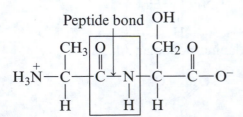

Peptide bond

REVIEW
Forming Amides (18.5)

CORE CHEMISTRY SKILL
Identifying the Primary, Secondary, Tertiary, and Quaternary Structures of Proteins

- Short chains of amino acids are called peptides. Long chains of amino acids that are biologically active are called proteins.
- Peptides are named from the N-terminus by replacing the *ine* or *ate* of each amino acid name with *yl* followed by the amino acid name of the C-terminus.

- Essential amino acids must be obtained from the proteins in the diet.
- Complete proteins, which are found in most animal products, contain all of the essential amino acids.

	Drawing a Peptide
STEP 1	Draw the structure for each amino acid in the peptide, starting with the N-terminus.
STEP 2	Remove the O atom from the carboxylate group of the N-terminal amino acid and two H atoms from the ammonium group in the adjacent amino acid.
STEP 3	Use peptide bonds to connect the amino acids.

♦ **Learning Exercise 19.2A**

Draw the condensed structural formulas at physiological pH, and write the name for each of the following:

1. Leu–Phe   **2.** YV

3. Gly–Ser–Cys

Answers

1. 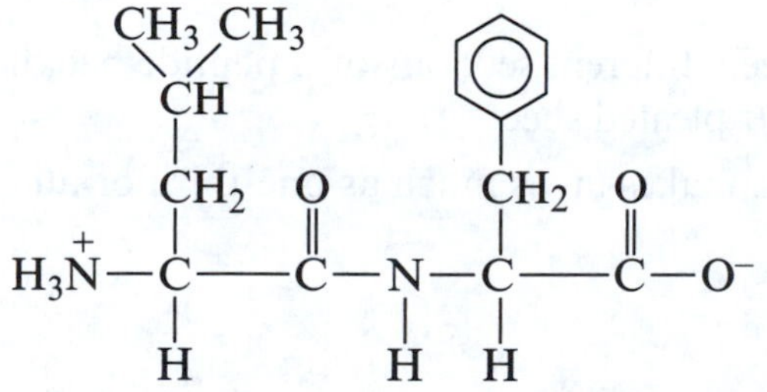 Leucylphenylalanine

2. 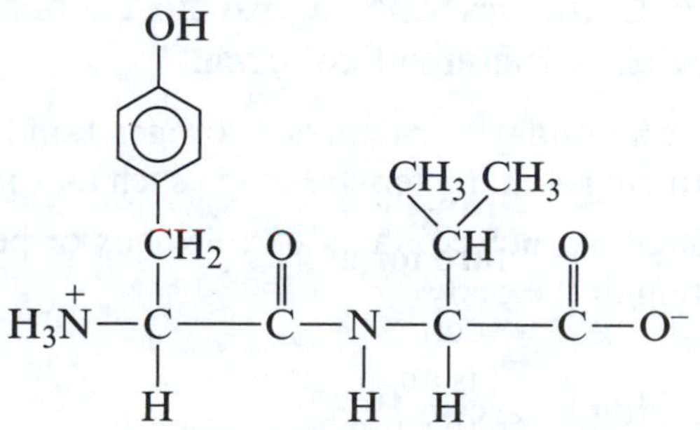 Tyrosylvaline

3. 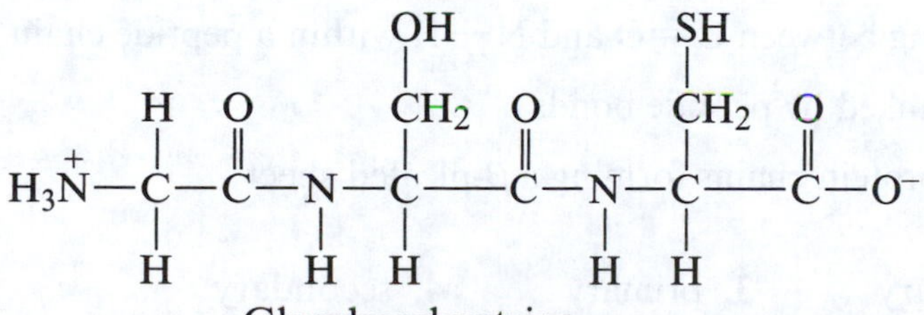 Glycylserylcysteine

♦ **Learning Exercise 19.2B** 🏃

Many seeds, vegetables, and legumes are low in one or more of the essential amino acids tryptophan, isoleucine, and lysine.

	Tryptophan	Isoleucine	Lysine
Sesame seeds	OK	low	low
Sunflower seeds	OK	OK	low
Garbanzo beans	low	OK	OK
Rice	OK	OK	low
Cornmeal	low	OK	low

Indicate whether the following protein combinations are complementary or not:

1. _____ sesame seeds and sunflower seeds

2. _____ sunflower seeds and garbanzo beans

3. _____ sesame seeds and rice

4. _____ sesame seeds and garbanzo beans

5. _____ garbanzo beans and rice

6. _____ cornmeal and garbanzo beans

Answers
1. not complementary; both are low in lysine   2. complementary
3. not complementary; both are low in lysine   4. complementary
5. complementary   6. not complementary; both are low in tryptophan

19.3   Proteins: Secondary Structure

Learning Goal: Describe the two most common types of secondary structures for a protein. Describe the structure of collagen.

- In the secondary structures, hydrogen bonds form between different sections of a peptide backbone producing a characteristic shape such as an α helix and β-pleated sheet.
- Collagen, which contains a triple helix of peptide chains, makes up as much as one-third of all protein in the body.

♦ **Learning Exercise 19.3**

Identify the following descriptions of protein structure as primary or secondary:

1. _____ hydrogen bonding forming an α helix

2. _____ hydrogen bonding occurring between C=O and N—H within a peptide chain

3. _____ the order of amino acids linked by peptide bonds

4. _____ hydrogen bonds between protein chains forming a β-pleated sheet

Answers   1. secondary   2. secondary   3. primary   4. secondary

19.4   Proteins: Tertiary and Quaternary Structures

Learning Goal: Describe the tertiary and quaternary structures of a protein.

- In a tertiary structure, hydrophobic R groups are found on the inside and hydrophilic R groups are found on the surface. The tertiary structure is stabilized by interactions between R groups.
- In a quaternary structure, two or more subunits combine for biological activity. They are held together by the same interactions found in tertiary structures.

♦ **Learning Exercise 19.4**

Identify the following descriptions of protein structure as tertiary, quaternary, or both:

1. _____ a disulfide bond joining distant parts of a protein chain

2. _____ the combination of two or more protein subunits

3. _____ hydrophilic side groups seeking contact with water

4. _____ a salt bridge forming between two oppositely charged side chains

5. _____ hydrophobic side groups forming a nonpolar center

Answers   1. tertiary   2. quaternary
               3. both tertiary and quaternary   4. both tertiary and quaternary
               5. both tertiary and quaternary

19.5   Protein Hydrolysis and Denaturation

Learning Goal: Describe the hydrolysis and denaturation of proteins.

- Hydrolysis of a protein occurs when peptide bonds in the primary structure are broken, producing amino acids.
- Denaturation of a protein occurs when denaturing agents destroy the secondary, tertiary, and quaternary structures (but not the primary structure) of the protein which causes a loss of biological activity.
- Denaturing agents include heat, acids, bases, organic solvents, agitation, and heavy metal ions.

Key Terms for Sections 19.1 to 19.5

Match each of the following key terms with the correct description:

   a. protein   **b.** tertiary structure   **c.** peptide
   d. alpha helix   **e.** denaturation

1. _____ the folding of a protein into a compact structure

2. _____ the loss of secondary, tertiary, and/or quaternary structure of a protein

3. _____ the combination of two or more amino acids

4. _____ a polypeptide that has biological activity

5. _____ a secondary level of protein structure

Answers   1. b   2. e   3. c   4. a   5. d

♦ **Learning Exercise 19.5** 🏃

Indicate the denaturing agent in each of the following examples:

   a. heat   **b.** pH change   **c.** organic solvent
   d. heavy metal ions   **e.** agitation

  1. _____ placing surgical instruments in a 120 °C autoclave

  2. _____ whipping cream to make a dessert

  3. _____ applying tannic acid to a burn

  4. _____ placing $AgNO_3$ drops in the eyes of newborns

  5. _____ using alcohol to disinfect a wound

  6. _____ using a *Lactobacillus* culture to produce acid that converts milk to yogurt

Answers   **1.** a   **2.** e   **3.** b   **4.** d   **5.** c   **6.** b

Checklist for Chapter 19

You are ready to take the Practice Test for Chapter 19. Be sure you have accomplished the following learning goals for this chapter. If not, review the Section listed at the end of the goal. Then apply your new skills and understanding to the Practice Test.

After studying Chapter 19, I can successfully:

_____ Classify proteins by their functions in the body. (19.1)

_____ Draw the ionized structural formula for an amino acid at physiological pH. (19.1)

_____ Describe a peptide bond; draw the structural formulas for a peptide. (19.2)

_____ Describe the primary structure of a protein. (19.2)

_____ Describe the secondary structures of a protein. Describe the structure of collagen. (19.3)

_____ Describe the tertiary and quaternary structures of a protein. (19.4)

_____ Identify the hydrolysis products of peptides. (19.5)

_____ Describe the two most common types of ways that denaturation affects the structure for a protein. (19.5)

Practice Test for Chapter 19

The chapter Sections to review are shown in parentheses at the end of each question.

  1. Which amino acid is nonpolar? (19.1)
    A. serine   **B.** aspartate   **C.** valine   **D.** cysteine   **E.** lysine

  2. Which amino acid has a basic side chain? (19.1)
    A. serine   **B.** aspartate   **C.** valine   **D.** cysteine   **E.** lysine

  3. All amino acids (19.1)
    A. have the same side chains
    B. have ionized structures
    C. have the same charge at physiological pH
    D. show hydrophobic tendencies
    E. are essential amino acids

For questions 4 through 7, match the function to the correct protein: (19.1)

    A. enzyme   **B.** structural   **C.** transport   **D.** storage

4. myoglobin in muscle

5. α-keratin in skin

6. peptidase for protein hydrolysis

7. ferritin in the liver

8. Essential amino acids (19.2)
    A. are the amino acids that must be supplied by the diet
    B. are not synthesized by the body
    C. are missing in incomplete proteins
    D. are present in proteins from animal sources
    E. all of the above

9. The sequence Tyr–Ala–Gly (19.2)
    A. is a tripeptide
    B. has two peptide bonds
    C. has tyrosine at the N-terminus
    D. has glycine at the C-terminus
    E. all of these

10. Which amino acid will form disulfide bonds in a tertiary structure? (19.3)
    A. serine   **B.** aspartate   **C.** valine   **D.** cysteine   **E.** lysine

11. What type of bond is used to form the α helix structure of a protein? (19.3)
    A. peptide bond   **B.** hydrogen bond   **C.** salt bridge
    D. disulfide bond   **E.** hydrophobic interaction

12. The bonding in a tertiary structure expected between lysine and aspartate is a (19.4)
    A. salt bridge   **B.** hydrogen bond   **C.** disulfide bond
    D. hydrophobic interaction   **E.** hydrophilic interaction

13. What type of interaction places portions of the protein chain in the center of a tertiary structure? (19.4)
    A. peptide bond   **B.** salt bridge   **C.** disulfide bond
    D. hydrophobic interaction   **E.** hydrophilic interaction

For questions 14 through 18, identify the protein structural level that each of the following phrases describes: (19.2, 19.3, 19.4)

    A. primary   **B.** secondary   **C.** tertiary   **D.** quaternary   **E.** pentenary

14. peptide bonds   **15.** a β-pleated sheet   **16.** two or more protein subunits

17. an α helix   **18.** disulfide bonds

19. Denaturation of a protein (19.5)
    A. occurs at a pH of 7   **B.** causes a change in protein structure
    C. hydrolyzes a protein   **D.** oxidizes the protein
    E. adds amino acids to a protein

20. Which of the following will not cause denaturation? (19.5)
    A. $0\,°C$ **B.** $AgNO_3$ **C.** $80\,°C$ **D.** ethanol   **E.** pH 1

Answers to the Practice Test

1. C	**2.** E	**3.** B	**4.** C	**5.** B
6. A	**7.** D	**8.** E	**9.** E	**10.** D
11. B	**12.** A	**13.** D	**14.** A	**15.** B
16. D	**17.** B	**18.** C, D	**19.** B	**20.** A

Selected Answers and Solutions to Text Problems

19.1 a. Hemoglobin, which carries oxygen in the blood, is a transport protein.
   b. Collagen, which is a major component of tendons and cartilage, is a structural protein.
   c. Keratin, which is found in hair, is a structural protein.
   d. Amylases, which catalyze the hydrolysis of starch, are enzymes.

19.3 All α-amino acids contain a carboxylate group and an ammonium group bonded to the alpha carbon.

19.5 a.   b.

c.   d.

19.7 a. Glycine has an H as its R group which makes it a nonpolar amino acid. It is hydrophobic.
   b. Threonine has an R group that contains the polar —OH group which makes threonine a polar neutral amino acid. It is hydrophilic.
   c. Glutamate has an R group containing a polar carboxylate group. Glutamate is a polar acidic amino acid. It is hydrophilic.
   d. Phenylalanine has an R group containing an aromatic ring; this makes phenylalanine a non-polar amino acid. It is hydrophobic.

19.9 Amino acids have both three-letter and one-letter abbreviations.
   a. Ala is the three-letter abbreviation for the amino acid alanine.
   b. Q is the one-letter abbreviation for the amino acid glutamine.
   c. K is the one-letter abbreviation for the amino acid lysine.
   d. Cys is the three-letter abbreviation for the amino acid cysteine.

19.11 In a peptide, the amino acids are joined by peptide bonds (amide bonds). The first amino acid has a free —NH_3^+ group, and the last one has a free —COO^- group.

a.   b.

Ala–Cys, AC   Ser–Phe, SF

c.

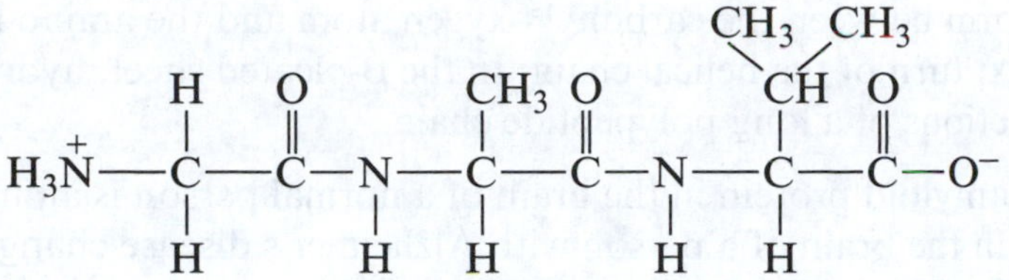

$$\text{H}_3\overset{+}{\text{N}}-\underset{\underset{\text{H}}{|}}{\overset{\overset{\text{H}}{|}}{\text{C}}}-\overset{\text{O}}{\overset{\|}{\text{C}}}-\underset{\underset{\text{H}}{|}}{\text{N}}-\underset{\underset{\text{H}}{|}}{\overset{\overset{\text{CH}_3}{|}}{\text{C}}}-\overset{\text{O}}{\overset{\|}{\text{C}}}-\underset{\underset{\text{H}}{|}}{\text{N}}-\underset{\underset{\text{H}}{|}}{\overset{\overset{\text{CH}}{|}}{\text{C}}}-\overset{\text{O}}{\overset{\|}{\text{C}}}-\text{O}^-$$

Gly–Ala–Val, GAV

d.

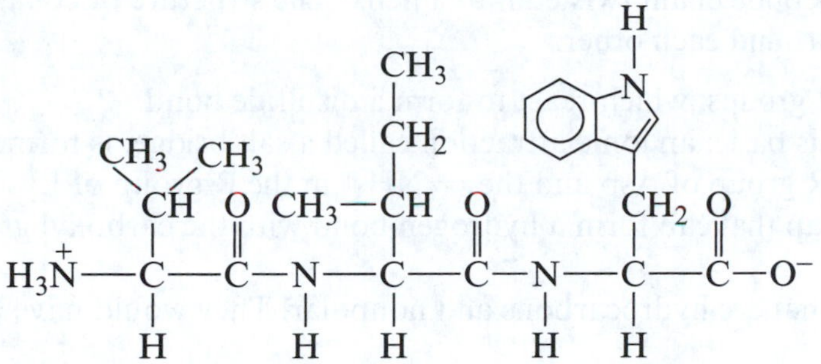

Val–Ile–Trp, VIW

19.13 a.

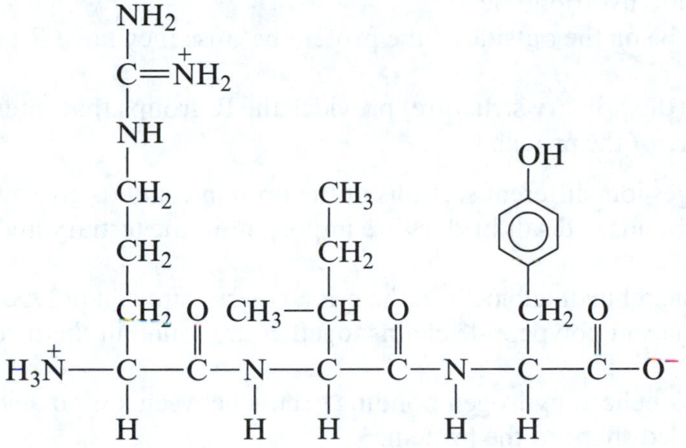

RIY

b.

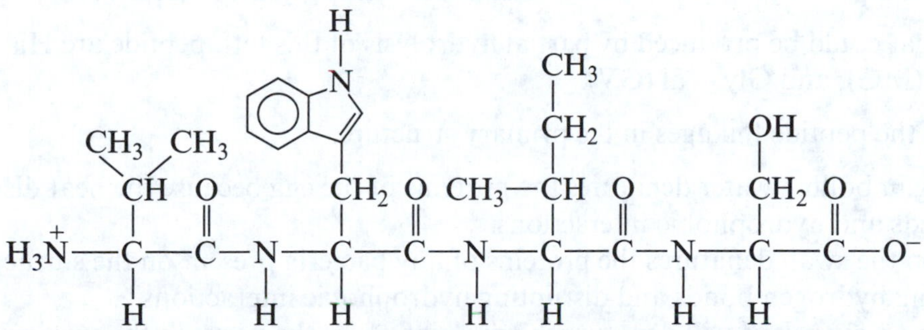

VWIS

19.15 a. Lysine and tryptophan are lacking in corn but supplied by peas; methionine is lacking in peas but supplied by corn.

   b. Lysine is lacking in rice but supplied by soy; methionine is lacking in soy but supplied by rice.

19.17 When atoms in the backbone of a protein or peptide form hydrogen bonds within a single polypeptide chain or between polypeptide chains, secondary structure results. The two most common secondary structures are the alpha helix and the beta-pleated sheet.

19.19 In the α helix, hydrogen bonds form between the carbonyl oxygen atom and the amino hydrogen atom of a peptide bond in the next turn of the helical chain. In the β-pleated sheet, hydrogen bonds occur between adjacent sections of a long polypeptide chain.

19.21 The secondary structure of beta-amyloid protein in the brain of a normal person is alpha-helical, whereas the secondary structure in the brain of a person with Alzheimer's disease changes to beta-pleated sheet.

19.23 An α helix consists of only one peptide chain twisted into a helix. The structure of collagen consists of three helices twisted around each other.

19.25 a. The two cysteines have —SH groups, which react to form a disulfide bond.
   b. Aspartate is acidic and lysine is basic; an ionic attraction called a salt bridge, is formed between the —COO^- in the R group of Asp and the —NH_3^+ in the R group of Lys.
   c. Serine has a polar —OH group that can form a hydrogen bond with the carboxyl group of aspartate.
   d. Two leucines have R groups that are hydrocarbons and nonpolar. They would have a hydrophobic interaction.

19.27 a. Leucine and valine are likely to be found on the inside of the protein structure because they have nonpolar R groups that are hydrophobic.
   b. Cysteine and aspartate would be on the outside of the protein because they have R groups that are polar.
   c. The order of the amino acids (the primary structure) provides the R groups that interact to determine the tertiary structure of the protein.

19.29 a. Disulfide bonds and salt bridges join different sections of the protein chain to give a three-dimensional shape. Disulfide bonds and salt bridges are important in the tertiary and quaternary structures.
   b. Peptide bonds join the amino acid building blocks in the primary structure of a polypeptide.
   c. Hydrogen bonds that hold adjacent polypeptide chains together are found in the secondary structures of β-pleated sheets.
   d. In the secondary structure of α helices, hydrogen bonding occurs between amino acids in the same polypeptide to give a coiled shape to the protein.

19.31 The products of complete hydrolysis would be the amino acids glycine, alanine, and serine.

19.33 The dipeptides that could be produced by partial hydrolysis of this tetrapeptide are His–Met (HM), Met–Gly (MG), and Gly–Val (GV).

19.35 Hydrolysis splits the peptide linkages in the primary structure.

19.37 a. Placing an egg in boiling water denatures the proteins of the egg because the heat disrupts hydrogen bonds and hydrophobic interactions.
   b. The alcohol on the swab denatures the proteins of any bacteria present on the surface of the skin by forming hydrogen bonds and disrupting hydrophobic interactions.
   c. The heat from an autoclave will denature the proteins of any bacteria on the surgical instruments by disrupting hydrogen bonds and hydrophobic interactions.
   d. Heat will denature the surrounding proteins to close the wound by disrupting hydrogen bonds and hydrophobic interactions.

19.39 $CH_4N_2O_2$

19.41 $21 \; \cancel{lb} \times \dfrac{1 \; \cancel{kg}}{2.20 \; \cancel{lb}} \times \dfrac{20. \; \text{mg hydroxyurea}}{1 \; \cancel{\text{kg body weight}}} = 190 \; \text{mg of hydroxyurea (2 SFs)}$

19.43 a. The amino acids in aspartame are aspartate and phenylalanine.
   b. The dipeptide in aspartame would be named aspartylphenylalanine.

19.45 a. Asparagine and serine are both polar neutral amino acids; their R groups can interact by hydrogen bonding.

  b. Aspartate is a polar acidic amino acid, and lysine is a polar basic amino acid; their R groups can interact by forming a salt bridge.

  c. The polar neutral amino acid cysteine contains the —SH group; two cysteines can form a disulfide bond.

  d. Leucine and alanine are both nonpolar amino acids; their R groups can have a hydrophobic interaction.

19.47 a. Yes, a combination of rice and garbanzo beans provides all the essential amino acids; garbanzo beans contain the lysine missing in rice.

  b. No, a combination of lima beans and cornmeal does not provide all the essential amino acids; both are deficient in the amino acid tryptophan.

  c. No, a combination of garbanzo beans and lima beans does not provide all the essential amino acids; both are deficient in the amino acid tryptophan.

19.49 The possible primary structures of a tripeptide of one valine and two serines are Val–Ser–Ser (VSS), Ser–Val–Ser (SVS), and Ser–Ser–Val (SSV).

19.51 a.

$$\overset{+}{H_3N}-\overset{\underset{H}{|}}{\overset{\overset{\displaystyle CH_2-OH}{|}}{C}}-\overset{O}{\overset{||}{C}}-\overset{\underset{H}{|}}{N}-\overset{\underset{H}{|}}{\overset{\overset{\displaystyle CH_2-CH_2-CH_2-CH_2-\overset{+}{N}H_3}{|}}{C}}-\overset{O}{\overset{||}{C}}-\overset{\underset{H}{|}}{N}-\overset{\underset{H}{|}}{\overset{\overset{\displaystyle CH_2-COO^-}{|}}{C}}-C-O^-$$

  b. This segment contains polar R groups, which would be more likely to be found on the surface of a protein where they can hydrogen bond with water.

19.53 a. Collagen is found in connective tissue, blood vessels, skin, tendons, ligaments, the cornea of the eye, and cartilage.

  b. Collagen consists of three helical polypeptides twisted together into a triple helix.

19.55 A polypeptide with a high content of amino acids with large R groups, like His, Met, and Leu, would be expected to have more α-helical than β-pleated sheet sections. In an α helix, these large R groups could extend toward the outside of the helix, away from the central corkscrew shape.

19.57 Serine is a polar amino acid, whereas valine is nonpolar. Serine would likely move to the outside surface of the protein where it can form hydrogen bonds with water. However, valine, which is nonpolar, would likely be pushed to the center of the tertiary structure where it can be stabilized by forming hydrophobic interactions.

19.59 The acid from the lemon juice and the mechanical whipping (agitation) of the egg white denature the proteins, which turn into solids as meringue.

19.61 a. The secondary structure of a protein depends on hydrogen bonds between atoms in the protein backbone to form a helix or a pleated sheet; the tertiary structure is determined by the interactions of R groups such as disulfide bonds, hydrogen bonds, and salt bridges, and determines the three-dimensional shape of the protein.

  b. Nonessential amino acids can be synthesized by the body; essential amino acids must be supplied by the diet.

    c. Polar amino acids have hydrophilic R groups, whereas nonpolar amino acids have hydrophobic R groups.

    d. Dipeptides contain two amino acids, whereas tripeptides contain three amino acids.

19.63 a. Valine has a nonpolar R group and would be found in hydrophobic regions.

    b. Lysine and aspartate have polar R groups and would be found in hydrophilic regions.

    c. Lysine and aspartate have polar R groups containing —COO⁻ and —NH₃⁺ groups which can form hydrogen bonds.

    d. Lysine is a polar basic amino acid, and aspartate is a polar acidic amino acid; they can form salt bridges.

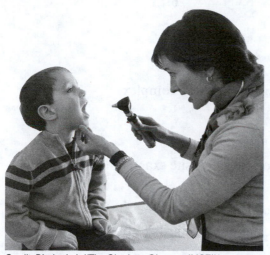

Noah's mother learns that Noah is lactose intolerant, which means that he lacks the lactase enzyme needed to completely digest lactose in milk and milk-based products. Under the guidance of Noah's physician assistant, Emma, Noah's mother limits the amount of dairy products in Noah's diet by using lactose-free milk and milk products or soy milk. Because calcium is usually provided by milk, Noah's mother adds nonmilk products high in calcium such as rhubarb and spinach to his diet. She gives him lactase if he consumes a milk-based product. Noah no longer suffers from abdominal pain, gas, or diarrhea. What is the substrate for the enzyme lactase?

Credit: Diedra Laird/The Charlotte Observer/MCT/Newscom

LOOKING AHEAD

20.1 Enzymes and Enzyme Action
20.2 Classification of Enzymes
20.3 Factors Affecting Enzyme Activity

20.4 Regulation of Enzyme Activity
20.5 Enzyme Inhibition

20.6 Enzyme Cofactors and Vitamins

 The Health icon indicates a question that is related to health and medicine.

20.1 Enzymes and Enzyme Action

Learning Goal: Describe enzymes and their roles in enzyme-catalyzed reactions.

REVIEW

Rates of Reaction (10.1)

- Enzymes are proteins that act as biological catalysts.
- Enzymes accelerate the rate of biological reactions by lowering the activation energy of a reaction.
- Within the structure of an enzyme, there is a small pocket called the active site, which has a specific shape that fits a specific substrate.
- In the induced-fit model, a substrate and a flexible active site adjust their shapes to form an enzyme–substrate complex in which the reaction of the substrate is catalyzed to give product(s).

♦ **Learning Exercise 20.1A**

Indicate whether each of the following is a characteristic of an enzyme: yes (Y) or no (N)

CORE CHEMISTRY SKILL

Describing Enzyme Action

1. _____ is a biological catalyst

2. _____ does not change the equilibrium position of a reaction

3. _____ greatly increases the rate of a cellular reaction

4. _____ is needed for every reaction that takes place in the cell

5. _____ catalyzes at a faster rate at lower temperatures

6. _____ lowers the activation energy of a biological reaction

7. _____ increases the rate of the forward, but not the reverse, reaction

Answers   **1.** Y   **2.** Y   **3.** Y   **4.** Y   **5.** N   **6.** Y   **7.** N

◆ **Learning Exercise 20.1B**

Match the following terms with the correct description:

    a. active site   **b.** substrate   **c.** enzyme–substrate complex
    d. lock-and-key   **e.** induced-fit

1. _____ the combination of an enzyme with a substrate

2. _____ a model of enzyme action in which the rigid shape of the active site exactly fits the shape of the substrate

3. _____ has a tertiary structure that fits the structure of the active site

4. _____ a model of enzyme action in which the shape of the active site adjusts to fit the shape of a substrate

5. _____ the portion of an enzyme that binds to the substrate and catalyzes the reaction

Answers   **1.** c   **2.** d   **3.** b   **4.** e   **5.** a

◆ **Learning Exercise 20.1C**

Write an equation to illustrate each of the following:

1. the formation of an enzyme–substrate complex

2. the conversion of an enzyme–substrate complex to product

3. the general equation for the formation of a product from an enzyme and substrate

Answers   **1.** E + S $\rightleftarrows$ ES complex   **2.** ES complex $\longrightarrow$ EP complex $\longrightarrow$ E + P
               3. E + S $\rightleftarrows$ ES complex $\longrightarrow$ EP complex $\longrightarrow$ E + P

20.2 Classification of Enzymes

Learning Goal: Classify enzymes and give their names.

- The names of many enzymes describe the compound or the reaction that is catalyzed and end with *ase*.
- Enzymes are classified by the type of reaction they catalyze: oxidation–reduction, transfer of groups, hydrolysis, the addition or removal of a group, isomerization, or the joining of two smaller molecules.

♦ Learning Exercise 20.2

Match the classification for enzymes with each of the following types of reactions:

a. oxidoreductase **b.** transferase **c.** hydrolase **d.** lyase **e.** isomerase **f.** ligase

CORE CHEMISTRY SKILL
Classifying Enzymes

1. _____ combines small molecules using energy from ATP

2. _____ transfers phosphate groups

3. _____ hydrolyzes a disaccharide into two glucose units

4. _____ converts a substrate to an isomer of the substrate

5. _____ adds hydrogen to a substrate

6. _____ removes H_2O from a substrate

7. _____ adds oxygen to a substrate

8. _____ converts a cis structure to a trans structure

Answers **1.** f **2.** b **3.** c **4.** e **5.** a **6.** d **7.** a **8.** e

20.3 Factors Affecting Enzyme Activity

Learning Goal: Describe the effects that changes of temperature, pH, enzyme concentration, and substrate concentration have on enzyme activity.

- Enzymes are most effective at their optimum temperature and pH.
- The rate of an enzyme-catalyzed reaction decreases at temperatures and pH above or below the optimum.
- An increase in enzyme concentration increases the rate of an enzyme-catalyzed reaction.
- An increase in substrate concentration increases the reaction rate of an enzyme-catalyzed reaction until all of the enzyme molecules combine with substrate.

♦ Learning Exercise 20.3A

CORE CHEMISTRY SKILL
Identifying Factors Affecting Enzyme Activity

An enzyme has an optimum pH of 7 and an optimum temperature of 37 °C.

a. Draw a graph to represent the activity of the enzyme at the pH values shown. Indicate the optimum pH.

b. Draw a graph to represent the activity of the enzyme at the temperatures shown. Indicate the optimum temperature.

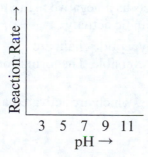

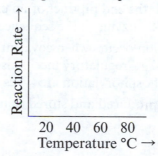

Answers

a.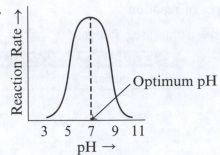

b.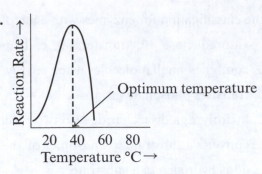

♦ **Learning Exercise 20.3B**

Urease has an optimum pH of 7.0. Indicate the effect of each of the following on the reaction rate of a urease-catalyzed reaction:

    a. increases   **b.** decreases   **c.** doesn't change

1. _____ adding more substrate when an excess of enzyme is present

2. _____ running the reaction at pH 9.0

3. _____ lowering the temperature to 0 °C

4. _____ running the reaction at 85 °C

5. _____ increasing the concentration of substrate when all of the enzyme molecules are bonded to substrate

6. _____ adjusting pH to its optimum

Answers   **1.** a   **2.** b   **3.** b   **4.** b   **5.** c   **6.** a

20.4 Regulation of Enzyme Activity

Learning Goal: Describe the role of allosteric enzymes, feedback control, and covalent modification in regulating enzyme activity.

- When allosteric enzymes bind regulator molecules on a different part of the enzyme, there is a change in the shape of the enzyme and the active site. A positive regulator speeds up a reaction, and a negative regulator slows down a reaction.

- In feedback control, the end product of an enzyme-catalyzed sequence acts as a negative regulator and binds to the first enzyme in the sequence, which slows the rate of catalytic activity.

- Covalent modification occurs when covalent bonds to a group on the polypeptide chain are formed or broken. As a regulatory mechanism, covalent modification is reversible. The formation of zymogens and phosphorylation are types of covalent modification.

- Many enzymes are produced and stored as inactive forms called *zymogens*, which are activated at a later time.

♦ **Learning Exercise 20.4**

Match the following terms with the correct phrase:

      a. zymogen   **b.** allosteric enzyme   **c.** covalent modification
      d. negative regulator   **e.** feedback control   **f.** positive regulator

1. _____ an enzyme that binds molecules at a site that is not the active site to increase the rate of enzyme activity

2. _____ the binding of the end product of a reaction sequence to the first enzyme in the pathway

3. _____ a molecule that slows down a reaction by preventing proper binding to the substrate

4. _____ an inactive form of an enzyme that is activated by removing a peptide section

5. _____ a molecule that binds at a site different than the active site to speed up the reaction

6. _____ the addition or removal of a phosphate group that changes catalytic activity

Answers   **1.** b   **2.** e   **3.** d   **4.** a   **5.** f   **6.** c

20.5 Enzyme Inhibition

Learning Goal: Describe reversible and irreversible inhibition. Describe competitive and noncompetitive inhibition.

- An inhibitor reduces the activity of an enzyme or makes it inactive.
- A competitive inhibitor has a structure similar to the substrate and competes for the active site. When the active site is occupied by a competitive inhibitor, the enzyme cannot catalyze the reaction of the substrate.
- A noncompetitive inhibitor attaches elsewhere on the enzyme, changing the shape of both the enzyme and the active site. As long as the noncompetitive inhibitor is attached to the enzyme, the altered active site cannot bind or catalyze substrate.
- A reversible inhibitor causes a loss of enzymatic activity that can be reversed.
- An irreversible inhibitor forms a permanent covalent bond with an amino acid side group in the active site of the enzyme, which permanently prevents enzymatic activity.

Determining Type of Enzyme Inhibition	
STEP 1	Compare the structure of the inhibitor to that of the substrate.
STEP 2	Describe the characteristics of the inhibitor.
STEP 3	Assign the type of inhibition based on the characteristics.

♦ **Learning Exercise 20.5**

Identify each of the following as characteristic of competitive inhibition (C), noncompetitive inhibition (N), or irreversible inhibition (I):

1. _____ An inhibitor binds to the surface of the enzyme away from the active site.

2. _____ An inhibitor resembling the substrate molecule blocks the active site on the enzyme.

3. _____ An inhibitor causes permanent damage to the enzyme with a total loss of biological activity.

4. _____ The action of this inhibitor can be reversed by adding more substrate.

5. _____ The action of this inhibitor is not reversed by adding more substrate, but can be reversed when chemical reagents remove the inhibitor.

6. _____ Sulfanilamide stops bacterial infections because its structure is similar to PABA (*p*-aminobenzoic acid), which is essential for bacterial growth.

Answers   **1.** N   **2.** C   **3.** I   **4.** C   **5.** N   **6.** C

20.6 Enzyme Cofactors and Vitamins

Learning Goal: Describe the types of cofactors found in enzymes.

- Simple enzymes are biologically active as a protein only, whereas other enzymes require a cofactor.
- A cofactor may be a metal ion, such as Cu^{2+} or Fe^{2+}, or an organic compound called a coenzyme, usually a vitamin.
- Vitamins are organic molecules that are essential for proper health.
- Vitamins must be obtained from the diet because they are not synthesized in the body.
- Vitamins B and C are water-soluble vitamins; vitamins A, D, E, and K are fat-soluble vitamins.
- Many water-soluble vitamins function as coenzymes or precursors of enzymes.

Key Terms for Sections 20.1 to 20.6

Match each of the following key terms with the correct description:

a. lock-and-key model   **b.** vitamin   **c.** inhibitor   **d.** enzyme   **e.** active site

1. _____ the portion of an enzyme structure where a substrate undergoes reaction

2. _____ a protein that catalyzes a biological reaction in the cells

3. _____ a model of enzyme action in which the substrate exactly fits the shape of the active site of an enzyme

4. _____ a substance that makes an enzyme inactive by interfering with its ability to react with a substrate

5. _____ an organic compound essential for normal health and growth that must be obtained from the diet

Answers   **1.** e   **2.** d   **3.** a   **4.** c   **5.** b

♦ **Learning Exercise 20.6A**

> **CORE CHEMISTRY SKILL**
> Describing the Role of Cofactors

Indicate whether each enzyme described below does or does not require a cofactor:

1. an enzyme consisting only of protein _____

2. an enzyme requiring magnesium ion for activity _____

3. an enzyme containing a sugar group _____

4. an enzyme that gives only amino acids upon hydrolysis _____

5. an enzyme that requires zinc ions for activity _____

Answers   **1.** does not require a cofactor   **2.** requires a cofactor   **3.** requires a cofactor
          4. does not require a cofactor   **5.** requires a cofactor

♦ **Learning Exercise 20.6B**

Identify the water-soluble vitamin (**a** to **d**) associated with each of the following:

    a. pantothenic acid (B_5)   **b.** riboflavin (B_2)   **c.** niacin (B_3)   **d.** ascorbic acid (C)

1. _____ collagen formation   **2.** _____ FAD and FMN

3. _____ NAD^+ **4.** _____ coenzyme A

Answers   **1.** d   **2.** b   **3.** c   **4.** a

♦ **Learning Exercise 20.6C**

Identify the fat-soluble vitamin (**a** to **d**) associated with each of the following:

    a. vitamin A   **b.** vitamin D   **c.** vitamin E   **d.** vitamin K

1. _____ blood clotting   **2.** _____ vision

3. _____ antioxidant   **4.** _____ absorption of calcium

Answers   **1.** d   **2.** a   **3.** c   **4.** b

Checklist for Chapter 20

You are ready to take the Practice Test for Chapter 20. Be sure you have accomplished the following learning goals for this chapter. If not, review the Section listed at the end of the goal. Then apply your new skills and understanding to the Practice Test.

After studying Chapter 20, I can successfully:

_____ Describe the lock-and-key and induced-fit models of enzyme action. (20.1)

_____ Classify enzymes according to the type of reaction they catalyze. (20.2)

_____ Discuss the effect of changes in temperature, pH, concentration of enzyme, and concentration of substrate on enzyme action. (20.3)

_____ Describe the reversible and irreversible inhibition of enzymes. (20.4)

_____ Discuss allosteric enzymes, feedback control, and covalent modification in the regulation of enzyme activity. (20.5)

_____ Identify the types of cofactors that are necessary for enzyme activity. (20.6)

_____ Describe the functions of vitamins as coenzymes. (20.6)

Practice Test for Chapter 20

The chapter Sections to review are shown in parentheses at the end of each question.

1. Enzymes (20.1)
    A. are biological catalysts   **B.** are polysaccharides   **C.** are insoluble in water
    D. always contain a cofactor   **E.** are named with an *ose* ending

For questions 2 through 6, select answers from the following (E = *enzyme*; S = *substrate*; P = *product*): (20.1)
    A. S $\longrightarrow$ P   **B.** EP complex $\longrightarrow$ E + P   **C.** E + S $\longrightarrow$ ES complex
    D. ES complex $\longrightarrow$ EP complex   **E.** EP complex $\longrightarrow$ ES complex

2. _____ the enzymatic reaction occurring at the active site

3. _____ the release of product from the enzyme

4. _____ the first step in the enzyme reaction

5. _____ the formation of the enzyme–substrate complex

6. _____ the final step in the enzyme reaction

For questions 7 through 11, match each enzyme with the reaction it catalyzes: (20.2)

    A. decarboxylase   **B.** isomerase   **C.** dehydrogenase

    D. lipase   **E.** sucrase

7. _____ $CH_3-\underset{\underset{OH}{|}}{CH}-\underset{\underset{\|}{O}}{C}-OH \longrightarrow CH_3-\underset{\underset{\|}{O}}{C}-\underset{\underset{\|}{O}}{C}-OH$

8. _____ sucrose + $H_2O \longrightarrow$ glucose and fructose

9. _____ $CH_3-\underset{\underset{\|}{O}}{C}-\underset{\underset{\|}{O}}{C}-OH \longrightarrow CH_3-\underset{\underset{\|}{O}}{C}-OH + CO_2$

10. _____ fructose $\longrightarrow$ glucose

11. _____ triacylglycerol + $3H_2O \longrightarrow$ fatty acids and glycerol

For questions 12 through 16, select your answers from the following: (20.3)

    A. increases the rate of reaction   **B.** decreases the rate of reaction

    C. denatures the enzyme, and no reaction occurs

12. _____ setting the reaction tube in a beaker of water at 100 °C

13. _____ adding substrate to the reaction vessel

14. _____ running the reaction at 10 °C

15. _____ adding ethanol to the reaction system

16. _____ adjusting the pH to optimum pH

For questions 17 through 20, match the following terms with the correct phrase: (20.4)

    A. zymogen   **B.** allosteric enzyme   **C.** covalent modification   **D.** feedback control

17. _____ an enzyme that binds molecules at a site that is not the active site to increase the rate of enzyme activity

18. _____ the binding of the end product of a reaction sequence to the first enzyme in the pathway

19. _____ an inactive form of an enzyme that is activated by removing a peptide section

20. _____ the addition or removal of a phosphate group that changes catalytic activity

For questions 21 through 25, identify each description of inhibition as one of the following: (20.5)

    A. competitive   **B.** noncompetitive

21. _____ This alteration affects the overall shape of the enzyme.

22. _____ A molecule that closely resembles the substrate interferes with enzymatic activity.

23. _____ The inhibition can be reversed by increasing substrate concentration.

24. _____ In cases of methanol poisoning, ethanol is given as an antidote.

25. _____ The inhibition is not affected by increased substrate concentration.

For questions 26 through 29, classify each enzyme as requiring a cofactor (C) *or not requiring a cofactor* (N): (20.6)

26. _____ an active enzyme that contains amino acids and a magnesium ion

27. _____ an active enzyme consisting of protein only

28. _____ an active enzyme requiring zinc ion for activation

29. _____ an active enzyme containing vitamin K

Answers to the Practice Test

1. A	**2.** D	**3.** B	**4.** C	**5.** C
6. B	**7.** C	**8.** E	**9.** A	**10.** B
11. D	**12.** C	**13.** A	**14.** B	**15.** C
16. A	**17.** B	**18.** D	**19.** A	**20.** C
21. B	**22.** A	**23.** A	**24.** B	**25.** B
26. C	**27.** N	**28.** C	**29.** C	

Selected Answers and Solutions to Text Problems

20.1 Chemical reactions in the body can occur without enzymes, but the rates are too slow at the relatively mild conditions of normal body temperature and pH. Catalyzed reactions, which are many times faster, provide the amounts of products needed by the cell at a particular time.

20.3 **a.** The equation for an enzyme-catalyzed reaction is:
$$E + S \rightleftharpoons ES \longrightarrow EP \longrightarrow E + P$$
E = enzyme, S = substrate, ES = enzyme–substrate complex
EP = enzyme–product complex, P = products
b. The active site is a region or pocket within the tertiary structure of an enzyme that accepts the substrate, aligns the substrate for reaction, and catalyzes the reaction.

20.5 **a.** An enzyme (2) has a tertiary structure that recognizes the substrate.
b. The enzyme–substrate complex (1) is the combination of an enzyme with the substrate.
c. The substrate (3) has a structure that fits the active site of the enzyme.

20.7 Because α-amylase catalyzes the reaction for a specific type of bond, it shows linkage specificity.

20.9 **a.** Oxidoreductases catalyze oxidation–reduction reactions.
b. Transferases move (or transfer) groups, such as amino or phosphate groups, from one substance to another.
c. Hydrolases use water to split bonds in molecules (hydrolysis) such as carbohydrates, peptides, and lipids.

20.11 **a.** A hydrolase would catalyze the hydrolysis of sucrose.
b. An oxidoreductase would catalyze the addition of oxygen (oxidation).
c. An isomerase would catalyze the conversion of glucose-6-phosphate to fructose-6-phosphate.
d. A transferase would catalyze the transfer of an amino group from one molecule to another.

20.13 **a.** A lyase such as a decarboxylase removes CO_2 from a molecule.
b. The transfer of an amino group to another molecule would be catalyzed by a transferase.

20.15 **a.** Succinate oxidase catalyzes the oxidation of succinate.
b. Glutamine synthetase catalyzes the combination of glutamate and ammonia to form glutamine.
c. Alcohol dehydrogenase removes 2H from an alcohol.

20.17 From the graph, the optimum pH values for the enzymes are approximately: pepsin, pH 2; sucrase pH 6; trypsin, pH 8.

20.19 **a.** Decreasing the substrate concentration decreases the rate of reaction.
b. The reaction will slow or stop because the enzyme will be denatured at low pH.
c. The reaction will slow or stop because the high temperature will denature the enzyme.
d. The reaction will go faster as long as there are polypeptides to react.

20.21 When a regulator molecule binds to an enzyme's allosteric site, the shape of the active site is altered to either bind the substrate more efficiently (in the case of a positive regulator) or less efficiently (in the case of a negative regulator), thereby increasing or decreasing the rate of the reaction.

20.23 In feedback control, the product binds to an enzyme at or near the beginning of a series and changes the shape of the active site. If the active site can no longer bind the substrate effectively, the reaction will stop.

20.25 In order to conserve material and energy and to prevent clotting in the bloodstream, thrombin is produced from prothrombin only at a wound site.

20.27 **a.** (3) covalent modification **b.** (2) a zymogen
c. (1) an allosteric enzyme

20.29 **a.** PSA is the abbreviation for the enzyme prostate-specific antigen.
    b. CK is the abbreviation for the enzyme creatine kinase.
    c. CE is the abbreviation for the enzyme cholinesterase.

20.31 Isoenzymes are slightly different forms of an enzyme that catalyze the same reaction in different organs and tissues of the body.

20.33 The enzymes CK and LDH might be present in the blood serum if the patient has had a heart attack.

20.35 **a.** If the inhibitor has a structure similar to the structure of the substrate, the inhibitor is competitive.
    b. If adding more substrate cannot reverse the effect of the inhibitor, the inhibitor is noncompetitive.
    c. If the inhibitor competes with the substrate for the active site, it is a competitive inhibitor.
    d. If the structure of the inhibitor is not similar to the structure of the substrate, the inhibitor is noncompetitive.
    e. If adding more substrate reverses the inhibition, the inhibitor is competitive.

20.37 **a.** Methanol has the condensed structural formula CH_3-OH, whereas ethanol is CH_3-CH_2-OH.
    b. Ethanol has a structure similar to methanol and could compete for the active site.
    c. Ethanol is a competitive inhibitor.

20.39 **a.** An enzyme that requires vitamin B_1 (thiamine) is active with a cofactor.
    b. An enzyme that needs Zn^{2+} for catalytic activity is active with a cofactor.
    c. If the active form of an enzyme consists of only polypeptide chains, a cofactor is not required.

20.41 **a.** Pantothenic acid (vitamin B_5) is part of coenzyme A.
    b. Folic acid is the vitamin that is a component of tetrahydrofolate (THF).
    c. Niacin (vitamin B_3) is a component of NAD^+.

20.43 **a.** A deficiency of vitamin D (cholecalciferol) can lead to rickets.
    b. A deficiency of vitamin C (ascorbic acid) can lead to scurvy.
    c. A deficiency of vitamin B_3 (niacin) can lead to pellagra.

20.45 Vitamin B_6 is a water-soluble vitamin, which means that each day, any excess of vitamin B_6 is eliminated from the body.

20.47 **a.** At 60 min, the baseline level of breath hydrogen is approximately 20 ppm.
    b. At 60 min, the level of hydrogen in Noah's breath is approximately 40 ppm.

20.49 Heating the milk in the hot chocolate will denature the protein enzyme, lactase. Because the enzyme is no longer active, Noah will suffer from the symptoms of lactose intolerance.

20.51 **a.** As a protease, chymosin catalyzes reactions in which the peptide bonds in a protein are hydrolyzed.
    b. Chymosin fits into the hydrolase category of enzymes.

20.53 **a.** The oxidation of a glycol to an aldehyde and carboxylic acid is catalyzed by an oxidoreductase.
    b. Ethanol competes with ethylene glycol for the active site of the alcohol dehydrogenase enzyme, saturating the active site, which allows ethylene glycol to be removed from the body without producing oxalic acid.

20.55 The many different reactions that take place in cells require different enzymes because enzymes react with only a certain type of substrate.

20.57 Enzymes are catalysts that are usually proteins and that function only at mild temperature and pH. Catalysts used in chemistry laboratories are usually inorganic materials that function at high temperatures and in strongly acidic or basic conditions.

20.59 Enzymes lower the activation energy of a reaction, thereby requiring less energy to convert reactants to products. They also bind the substrate in the active site of the enzyme in a position that is more favorable for the reaction to proceed.

20.61 In the lock-and-key model, the substrate sucrose fits the exact shape of the active site in the enzyme sucrase, but lactose does not.

20.63 **a.** The disaccharide lactose is a substrate (S).
   b. The suffix *ase* in lactase indicates that it is an enzyme (E).
   c. The suffix *ase* in lipase indicates that it is an enzyme (E).
   d. Trypsin is an enzyme that hydrolyzes polypeptides (E).
   e. Pyruvate is a substrate (S).
   f. The suffix *ase* in transaminase indicates that it is an enzyme (E).

20.65 **a.** Urea is the substrate for urease.
   b. Succinate is the substrate for succinate dehydrogenase.
   c. Aspartate is the substrate for aspartate transaminase.
   d. Phenylalanine is the substrate for phenylalanine hydroxylase.

20.67 **a.** The transfer of an acyl group is catalyzed by a transferase.
   b. Oxidases are classified as oxidoreductases.
   c. A lipase, which uses water to split ester bonds in lipids, is classified as a hydrolase.
   d. A decarboxylase is classified as a lyase.

20.69 The optimum temperature for an enzyme is the temperature at which the enzyme is fully active and most effective.

20.71 **a.** The rate of catalysis will slow and stop as a high temperature denatures the enzyme.
   b. The rate of the catalyzed reaction will slow as the temperature is lowered.
   c. The enzyme will not be functional at pH 2.0.

20.73 **a.** An enzyme is saturated if adding more substrate does not increase the rate of reaction.
   b. When increasing the substrate concentration increases the rate of reaction, the enzyme is not saturated.

20.75 An allosteric enzyme contains sites for regulators that alter the structure of the active site of the enzyme, which speeds up or slows down the rate of the catalyzed reaction.

20.77 In feedback control, the end product of the reaction pathway binds to the enzyme to slow down or stop a reaction at or near the beginning of the reaction pathway. It acts as a negative regulator.

20.79 In reversible inhibition, the inhibitor can dissociate from the enzyme, whereas in irreversible inhibition, the inhibitor forms a strong covalent bond with the enzyme and does not dissociate. Irreversible inhibitors act as poisons to enzymes.

20.81 **a.** (3) An irreversible inhibitor forms a covalent bond with an R group in the active site.
   b. (1) A competitive inhibitor has a structure similar to the substrate.
   c. (1) The addition of more substrate reverses the inhibition of a competitive inhibitor.
   d. (2) A noncompetitive inhibitor bonds to the surface of the enzyme, causing a change in the shape of the enzyme and active site.

20.83 **a** and **c** describe enzymes that require cofactors.

20.85 **a.** When pepsinogen enters the stomach, the low pH cleaves a peptide from its protein chain to form pepsin.
   b. An active protease would digest the proteins of the stomach rather than the proteins in the foods.

20.87 A heart attack may be the cause. Normally the enzymes LDH and CK are present only in low levels in the blood.

20.89 When designing an inhibitor, the substrate of an enzyme is usually known even if the structure of the enzyme is not. It is easier to design a molecule that resembles the substrate (competitive) than to find an inhibitor that binds to a second site on an enzyme (noncompetitive).

20.91 **a.** Antibiotics such as ampicillin are irreversible inhibitors.
  b. Antibiotics such as ampicillin inhibit enzymes needed to form cell walls in bacteria, not humans.

20.93 **a.** Coenzyme A (3) requires pantothenic acid (B_5).
  b. NAD^+ (1) requires niacin (B_3).
  c. TPP (2) requires thiamine (B_1).

20.95 A vitamin combines with an enzyme only when the enzyme needs to catalyze a reaction. When the enzyme is not needed, the vitamin dissociates for use by other enzymes in the cell.

20.97 **a.** A deficiency of niacin can lead to pellagra (3).
  b. A deficiency of vitamin A can lead to night blindness (1).
  c. A deficiency of vitamin D can lead to weak bone structure (2).

20.99 **a.** The reactants are lactose and water, and the products are glucose and galactose.
  b.

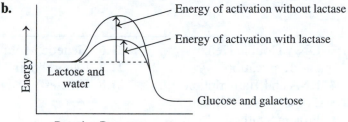

  c. By lowering the energy of activation, the enzyme furnishes a lower energy pathway by which the reaction can take place.

20.101 **a.** In this reaction, oxygen is added to an aldehyde. The enzyme that catalyzes this reaction would be an oxidoreductase.
  b. In this reaction, a dipeptide is hydrolyzed. The enzyme that catalyzes this reaction would be a hydrolase.
  c. In this reaction, water is added to a double bond. The enzyme that catalyzes the addition of a group without hydrolysis would be a lyase.

20.103 Cadmium would be a noncompetitive inhibitor because increasing the substrate and cofactor has no effect on the rate. This implies that the cadmium is binding to another site on the enzyme.

21
Nucleic Acids and Protein Synthesis

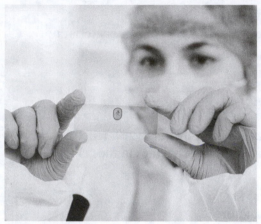

Credit: Alex Traksel/Shutterstock

Ellen, who was diagnosed with breast cancer, has a lumpectomy. Tissue samples from the tumor and lymph nodes are prepared for a pathologist by placing a thin section, 0.001 mm thick, on a microscope slide. The tissues, which test positive for estrogen receptor, confirm that Ellen's breast cancer tumors require estrogen for their growth. The bonding of estrogen to the estrogen receptor increases the production of mammary cells and potential mutations that can lead to cancer. The tissues are also tested for altered genes BRCA1 and BRCA2. The normal genes suppress tumor growth, whereas the mutated genes lose their ability to suppress tumor growth, which increases the risk for breast cancer. Ellen's test results for a BRCA mutation are negative. How does tamoxifen reduce the risk of breast cancer?

LOOKING AHEAD

21.1 Components of Nucleic Acids

21.2 Primary Structure of Nucleic Acids

21.3 DNA Double Helix and Replication

21.4 RNA and Transcription

21.5 The Genetic Code and Protein Synthesis

21.6 Genetic Mutations

21.7 Recombinant DNA

21.8 Viruses

 The Health icon indicates a question that is related to health and medicine.

21.1   Components of Nucleic Acids

Learning Goal: Describe the bases and ribose sugars that make up the nucleic acids DNA and RNA.

- Nucleic acids, such as deoxyribonucleic acid (DNA) and ribonucleic acid (RNA), are unbranched chains of nucleotides.
- A nucleoside is a combination of a pentose sugar and a base.
- A nucleotide is composed of three parts: a pentose sugar, a base, and a phosphate group.
- In DNA, the sugar is deoxyribose, and the base can be adenine, thymine, guanine, or cytosine.
- In RNA, the sugar is ribose, and the base can be adenine, uracil, guanine, or cytosine.

♦ **Learning Exercise 21.1A**

1. Write the names and abbreviations for the bases in each of the following:

    DNA _____   RNA _____

2. Write the name of the sugar in each of the following:

    DNA _____   RNA _____

Answers   1. DNA: adenine (A), thymine (T), guanine (G), cytosine (C)
            RNA: adenine (A), uracil (U), guanine (G), cytosine (C)
          2. DNA: deoxyribose; RNA: ribose

◆ **Learning Exercise 21.1B**

Name each of the following, and classify it as a purine or a pyrimidine:

a.

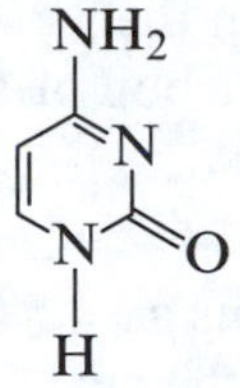

b.

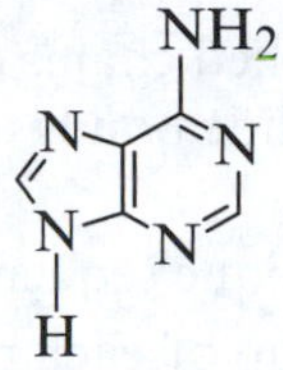

c.

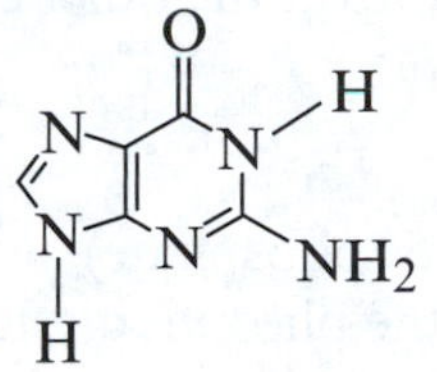

d.

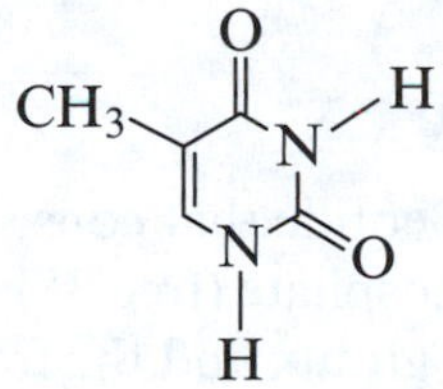

Answers **a.** cytosine; pyrimidine **b.** adenine; purine
c. guanine; purine **d.** thymine; pyrimidine

◆ **Learning Exercise 21.1C**

Identify the nucleic acid (DNA or RNA) in which each of the following is found:

1. _____ adenosine monophosphate 2. _____ guanosine monophosphate

3. _____ dCMP 4. _____ cytidine monophosphate

5. _____ deoxythymidine monophosphate 6. _____ UMP

7. _____ dGMP 8. _____ deoxyadenosine monophosphate

Answers **1.** RNA **2.** RNA **3.** DNA **4.** RNA
5. DNA **6.** RNA **7.** DNA **8.** DNA

◆ **Learning Exercise 21.1D**

Draw the condensed structural formula for deoxythymidine monophosphate. Indicate the 5′ and the 3′ carbon atoms on the sugar.

Answer

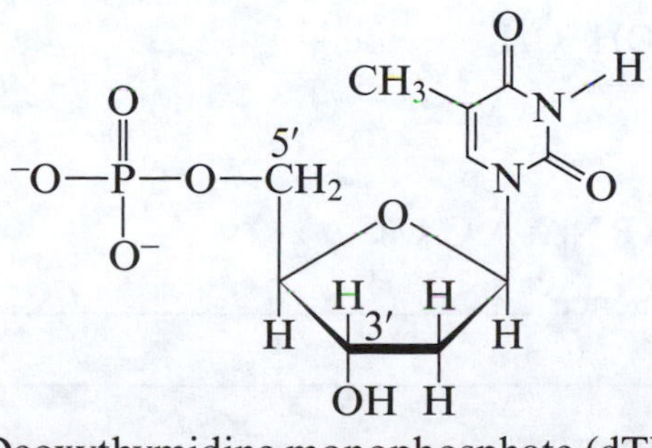

Deoxythymidine monophosphate (dTMP)

21.2 Primary Structure of Nucleic Acids

Learning Goal: Describe the primary structures of RNA and DNA.

- Nucleic acids are linear unbranched chains of nucleotides in which the —OH group on the 3′ carbon of a sugar in one nucleotide bonds to the phosphate group attached to the 5′ carbon of a sugar in the adjacent nucleotide.
- Each nucleic acid has its own unique sequence of bases known as its primary structure.
- In a nucleic acid, the 3′ OH group of each ribose in RNA or deoxyribose in DNA forms a phosphodiester linkage to the phosphate group of the 5′ carbon atom of the sugar in the next nucleotide to give a backbone of alternating sugar and phosphate groups.
- There is a free 5′ phosphate at one end of the nucleic acid and a free 3′ OH group at the other end.

♦ **Learning Exercise 21.2A**

Draw the structure of a dinucleotide that consists of uridine monophosphate (free 5′ phosphate) bonded to guanosine monophosphate (free 3′ hydroxyl). Identify each nucleotide, the phosphodiester linkage, the free 5′ phosphate group, and the free 3′ hydroxyl group.

Answer

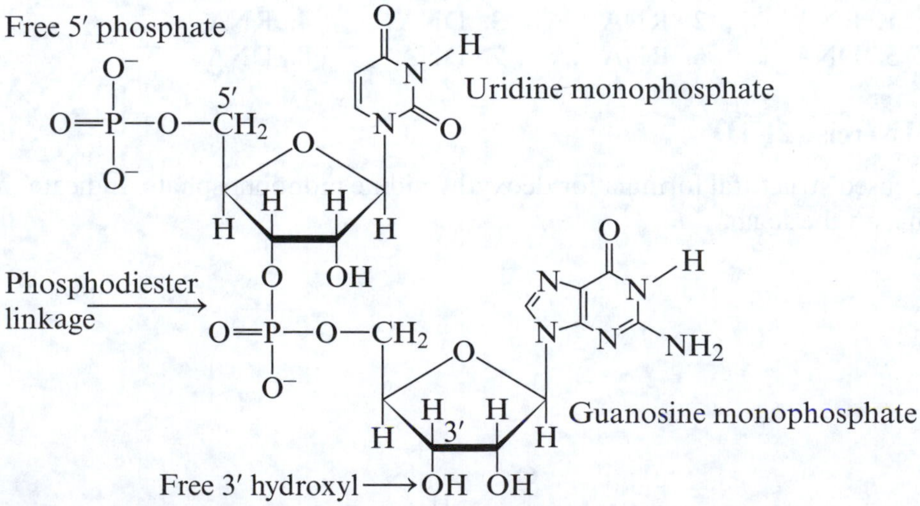

♦ **Learning Exercise 21.2B**

Consider the following sequence of nucleotides in RNA: A G U C

1. Write the names of the nucleotides in this sequence. _____

_____

2. Which nucleotide has the free 5′ phosphate group? _____

3. Which nucleotide has the free 3′ hydroxyl group? _____

Answers   **1.** adenosine monophosphate, guanosine monophosphate,
                    uridine monophosphate, cytidine monophosphate
              2. adenosine monophosphate   **3.** cytidine monophosphate

21.3   DNA Double Helix and Replication

Learning Goal: Describe the double helix of DNA; describe the process of DNA replication.

- A DNA molecule consists of two strands of nucleotides wound around each other like a spiral staircase.
- The two strands in DNA are held together by hydrogen bonds between complementary base pairs, A with T and G with C.
- One DNA strand runs in the 5′ to 3′ direction, and the other strand runs in the 3′ to 5′ direction.
- During replication, the two DNA strands separate, then DNA polymerase makes new DNA strands along each of the original DNA strands that serve as templates.
- Complementary base pairing ensures the correct pairing of bases to give identical copies of the original DNA.

♦ **Learning Exercise 21.3A**

Complete the following statements:

1. The structure of the two strands of nucleotides in DNA is called a _____.

2. In one strand of DNA, the sugar–phosphate backbone runs in the 5′ to 3′ direction, whereas the opposite strand goes in the _____ direction.

3. On the DNA strands, the free phosphate group is at the _____ end and the free hydroxyl group is at the _____ end.

4. The only combinations of base pairs that connect the two DNA strands are _____ and _____.

5. The base pairs along one DNA strand are _____ to the base pairs on the opposite strand.

Answers   **1.** double helix   **2.** 3′ to 5′   **3.** 5′; 3′   **4.** AT; GC   **5.** complementary

♦ **Learning Exercise 21.3B**

Write the complementary base sequence for the each of following segments of a strand of DNA:

1. 5′ A A A T T T C C C G G G 3′

> **CORE CHEMISTRY SKILL**
> Writing the Complementary
> DNA Strand

2. 5′ A T G C T T G G C T C C 3′

3. 5′ G C G C T C A A A T G C 3′

Answers   **1.** 3′ T T T A A A G G G C C C 5′   **2.** 3′ T A C G A A C C G A G G 5′
              3. 3′ C G C G A G T T T A C G 5′

♦ Learning Exercise 21.3C

Describe the process by which DNA replicates to produce identical copies.

Answer   In the replication process, the bases in each strand of the separated parent DNA are paired with their complementary bases. Because each complementary base is specific for a base in the parent strand of DNA, the new DNA strands are exact copies of the original strands of DNA.

♦ Learning Exercise 21.3D

Match each of the following terms with the correct description:

    a. replication forks   **b.** Okazaki fragments   **c.** DNA polymerase
    d. helicase   **e.** leading strand   **f.** lagging strand

1. _____ the enzyme that catalyzes the unwinding of a section of the DNA double helix

2. _____ the points in open sections of DNA where replication begins

3. _____ the enzyme that catalyzes the formation of phosphodiester linkages between nucleotides

4. _____ the short segments produced in the formation of the 3′ to 5′ daughter DNA strand

5. _____ the new DNA strand that grows in the 5′ to 3′ direction during the formation of daughter DNA

6. _____ the new DNA strand that is synthesized in the 3′ to 5′ direction

Answers   **1.** d   **2.** a   **3.** c   **4.** b   **5.** e   **6.** f

21.4   RNA and Transcription

Learning Goal: Identify the different types of RNA; describe the synthesis of mRNA.

- The three main types of RNA differ by their function in the cell: ribosomal RNA (rRNA) makes up most of the structure of the ribosomes and is the site of protein synthesis; messenger RNA (mRNA) carries information from the DNA to the ribosomes; and transfer RNA (tRNA) places the correct amino acids in the protein.

- Transcription is the process by which genetic information for the synthesis of a protein is copied from a gene in DNA to make mRNA.

- The bases in the mRNA are complementary to the DNA, except U in RNA is paired with A in DNA.

- In eukaryotes, the noncoding introns in a pre-mRNA are removed, and the exons that code for proteins are spliced together before the RNA leaves the nucleus.

- The production of mRNA occurs when certain proteins are needed in the cell.

- Transcription factors bind to the promoter to activate the RNA polymerase for the transcription of a particular mRNA.

- The RNA polymerase enzyme moves along an unwound section of DNA in a 3′ to 5′ direction.

♦ **Learning Exercise 21.4A**

Match each of the following characteristics with mRNA, tRNA, rRNA, pre-mRNA, or spliceosome:

1. _____ is most abundant in a cell

2. _____ contains both exons and introns

3. _____ has the shortest chain of nucleotides

4. _____ removes introns from a pre-mRNA

5. _____ carries information from DNA to the ribosomes for protein synthesis

6. _____ is the major component of ribosomes

7. _____ carries specific amino acids to the ribosome for protein synthesis

8. _____ consists of a large and a small subunit

Answers **1.** rRNA **2.** pre-mRNA **3.** tRNA **4.** spliceosome
5. mRNA **6.** rRNA **7.** tRNA **8.** rRNA

♦ **Learning Exercise 21.4B**

Fill in the blanks with a word or phrase that answers each of the following questions:

1. Where in the cell does transcription take place? _____

2. How many strands of a DNA molecule are involved in transcription? _____

3. What is the name of the sections in genes that code for proteins? _____

4. What is the name of the sections in genes that do not code for proteins? _____

5. What are the abbreviations for the four nucleotides in mRNA? _____

6. What is the corresponding section of an mRNA produced from each of the following?

   a. 3′ C A T T C G G T A 5′　　　　　　**b.** 3′ G T A C C T A A C G T C C G 5′

Answers **1.** nucleus **2.** one **3.** exons **4.** introns **5.** A, U, G, C
6. a. 5′ G U A A G C C A U 3′　　**b.** 5′ C A U G G A U U G C A G G C 3′

21.5 The Genetic Code and Protein Synthesis

Learning Goal: Use the genetic code to write the amino acid sequence for a segment of mRNA.

- The genetic code consists of a sequence of three nucleotides (triplets) called codons that specify the amino acids in a protein.
- The 64 codons for 20 amino acids allow several codons for most amino acids.
- The codon AUG signals the start of transcription, and codons UAG, UGA, and UAA signal the end of transcription.
- Proteins are synthesized at the ribosomes in a translation process that includes three steps: initiation, chain elongation, and termination.
- During translation, the appropriate tRNA brings the correct amino acid to the ribosome, where the amino acid is bonded by a peptide bond to the growing peptide chain.
- When the polypeptide is released, it takes on its secondary and tertiary structures to become a functional protein in the cell.

♦ Learning Exercise 21.5A

CORE CHEMISTRY SKILL
Writing the Amino Acid for an mRNA Codon

Give the three-letter and one-letter abbreviations for each of the amino acids coded for by the following mRNA codons:

1. UUU ______________ **2.** AGG ______________

3. AGC ______________ **4.** GAC ______________

5. GGA ______________ **6.** ACA ______________

7. AUG ______________ **8.** CUC ______________

9. CAU ______________ **10.** GUU ______________

Answers **1.** Phe, F **2.** Arg, R **3.** Ser, S **4.** Asp, D **5.** Gly, G **6.** Thr, T **7.** Start/Met, M **8.** Leu, L **9.** His, H **10.** Val, V

♦ Learning Exercise 21.5B

Match the components **a** to **e** of the translation process with the correct description **1** to **5** below:

a. initiation **b.** activation **c.** anticodon **d.** translocation **e.** termination

1. ____ the three bases in tRNA that complement a codon on the mRNA
2. ____ the combining of an amino acid with a specific tRNA
3. ____ the placement of methionine on the large ribosomal subunit
4. ____ the shift of the ribosome from one codon on mRNA to the next
5. ____ the process that occurs when the ribosome reaches a UAA, UGA, or UAG codon on mRNA

Answers **1.** c **2.** b **3.** a **4.** d **5.** e

♦ Learning Exercise 21.5C

Write the mRNA that forms for each of the following template sections of DNA. For each codon in the mRNA, write the three-letter abbreviation of the amino acid that would be placed in the protein by a tRNA.

1. DNA: 3′ ACG TCA CCA CGC 5′

mRNA: ________ ________ ________ ________

amino acid: ________ ________ ________ ________

2. DNA: 3′ ATG GCC TTT GGC AAC 5′

mRNA: ________ ________ ________ ________ ________

amino acid: ________ ________ ________ ________ ________

Answers   **1.** mRNA: 5′ UGC AGU GGU GCG 3′
amino acid: Cys   Ser   Gly   Ala

2. mRNA: 5′ UAC CGG AAA CCG UUG 3′
amino acid: Tyr   Arg   Lys   Pro   Leu

♦ **Learning Exercise 21.5D**

A segment of DNA that codes for a protein contains 270 nucleotides. How many amino acids would be present in the protein for this DNA segment?

Answer   Assuming that the entire segment codes for a protein, there would be 90 (270 ÷ 3) amino acids in the protein produced.

21.6   Genetic Mutations

Learning Goal: Identify the type of change in DNA for a point mutation, a deletion mutation, and an insertion mutation.

* A genetic mutation is a change of one or more bases in the DNA sequence that may alter the structure and ability of the resulting protein to function properly.
* In a point mutation, one base is changed, which may result in a different amino acid being placed in the protein.
* In an insertion or deletion mutation, the insertion or deletion of one base changes all of the codons that follow, which leads to a different sequence of amino acids from the point of the mutation.
* Mutations result from X-rays, overexposure to UV light, chemicals called mutagens, and some viruses.

♦ **Learning Exercise 21.6**

Consider the following segment of a DNA template strand:
3′ AAT CCC GGG 5′

> **REVIEW**
> Identifying the Primary, Secondary, Tertiary, and Quaternary Structures of Proteins (19.2, 19.3, 19.4)
> Describing Enzyme Action (20.1)

1. Write the mRNA produced.

_____   _____   _____

2. Write the amino acid order for the mRNA codons.

_____   _____   _____

3. A point mutation replaces the thymine in this DNA template strand with guanine. Write the mRNA produced.

_____   _____   _____

4. Write the new amino acid order.

_____   _____   _____

5. Why is this mutation called a point mutation?

6. How is a point mutation different from a deletion mutation?

7. What are some possible causes of genetic mutations?

Answers
1. 5′ UUA GGG CCC 3′ **2.** Leu–Gly–Pro

3. 5′ UUC GGG CCC 3′ **4.** Phe–Gly–Pro

5. One base is replaced with another.

6. In a point mutation, only one codon is affected and only one amino acid may be different. In a deletion mutation, all of the codons that follow the mutation are shifted by one base, which causes a different amino acid order in the remaining sequence of the protein.

7. X-rays, UV light, chemicals called mutagens, and some viruses are possible causes of genetic mutations.

21.7 Recombinant DNA

Learning Goal: Describe the preparation and uses of recombinant DNA.

- Recombinant DNA is synthesized by opening a piece of DNA and inserting a DNA section from another source.
- Much of the work with recombinant DNA is done with small circular DNA molecules called plasmids found in *E. coli* bacteria.
- Recombinant DNA is used to produce large numbers of copies of foreign DNA that are useful in genetic engineering techniques.
- Polymerase chain reaction (PCR) is a process which makes it possible to produce multiple copies of a gene in a short time.

♦ **Learning Exercise 21.7**

Match each of the following terms with the correct description:

   a. plasmids   **b.** restriction enzymes   **c.** polymerase chain reaction
   d. recombinant DNA

1. _____ a synthetic form of DNA that contains a piece of foreign DNA

2. _____ small, circular, DNA molecules found in *E. coli* bacteria

3. _____ a process that makes multiple copies of DNA in a short amount of time

4. _____ enzymes that cut open the DNA strands in plasmids

Answers   **1.** d   **2.** a   **3.** c   **4.** b

21.8   Viruses

Learning Goal: Describe the methods by which a virus infects a cell.

- Viruses are small particles of 3 to 200 genes that cannot replicate unless they invade a host cell.
- A viral infection involves using the host cell's machinery to replicate the viral nucleic acid.
- A retrovirus contains RNA as its genetic material.

Key Terms for Sections 21.1 to 21.8

Match each of the following key terms with the correct description:

    a. DNA   **b.** RNA   **c.** double helix   **d.** mutation
    e. virus   **f.** recombinant DNA   **g.** transcription

1. _____ the formation of mRNA to carry genetic information from DNA to allow for protein synthesis

2. _____ the genetic material that contains the bases adenine, cytosine, guanine, and thymine

3. _____ the shape of DNA with a sugar–phosphate backbone and base pairs linked in the center

4. _____ the combination of DNA from different organisms which forms new DNA

5. _____ a small particle containing DNA or RNA in a protein coat that requires a host cell for replication

6. _____ a change in the DNA base sequence that may alter the shape and function of a protein

7. _____ a type of nucleic acid with a single strand containing the nucleotides adenine, cytosine, guanine, and uracil

Answers   **1.** g   **2.** a   **3.** c   **4.** f   **5.** e   **6.** d   **7.** b

♦ **Learning Exercise 21.8**

Match each of the following terms with the correct description:

    a. host cell   **b.** retrovirus   **c.** vaccine   **d.** protease   **e.** virus

1. _____ the enzyme inhibited by drugs that prevent the synthesis of viral proteins

2. _____ a small, disease-causing particle that contains either DNA or RNA as its genetic material

3. _____ a type of virus that must use reverse transcriptase to make a viral DNA

4. _____ required by viruses to replicate

5. _____ inactive form of a virus that boosts the immune response by causing the body to produce antibodies

Answers   **1.** d   **2.** e   **3.** b   **4.** a   **5.** c

Checklist for Chapter 21

You are ready to take the Practice Test for Chapter 21. Be sure you have accomplished the following learning goals for this chapter. If not, review the Section listed at the end of the goal. Then apply your new skills and understanding to the Practice Test.

After studying Chapter 21, I can successfully:

_____ Identify the components of the nucleic acids RNA and DNA. (21.1)

_____ Describe the nucleotides contained in DNA and RNA. (21.1)

_____ Describe the primary structure of nucleic acids. (21.2)

_____ Describe the structures of RNA and DNA; show the relationship between the bases in the double helix. (21.3)

_____ Explain the process of DNA replication. (21.3)

_____ Describe the structure and characteristics of the three types of RNA. (21.4)

_____ Describe the synthesis of mRNA (transcription). (21.4)

_____ Describe the function of the codons in the genetic code. (21.5)

_____ Describe the role of translation in protein synthesis. (21.5)

_____ Describe some ways in which DNA is altered to cause mutations. (21.6)

_____ Describe the process used to prepare recombinant DNA. (21.7)

_____ Explain how retroviruses use reverse transcription to synthesize DNA. (21.8)

Practice Test for Chapter 21

The chapter Sections to review are shown in parentheses at the end of each question.

1. A nucleotide contains (21.1)
   - **A.** a base
   - **B.** a base and a sugar
   - **C.** a phosphate group and a sugar
   - **D.** a base and deoxyribose
   - **E.** a base, a sugar, and a phosphate group

2. For the nucleic acid segment UGCA: (21.2)
   - **A.** it is from DNA
   - **B.** A has a free 5′ phosphate
   - **C.** C has a free 3′ OH group
   - **D.** U has a free 3′ OH group
   - **E.** U has a free 5′ phosphate

3. The double helix in DNA is held together by (21.3)
   - **A.** hydrogen bonds
   - **B.** ester linkages
   - **C.** peptide bonds
   - **D.** salt bridges
   - **E.** disulfide bonds

4. The process of producing DNA in the nucleus is called (21.3)
   - **A.** complementation
   - **B.** replication
   - **C.** translation
   - **D.** transcription
   - **E.** mutation

5. Which type of molecule carries amino acids to the ribosomes? (21.4)
   - **A.** DNA
   - **B.** mRNA
   - **C.** tRNA
   - **D.** rRNA
   - **E.** protein

6. Which type of molecule determines protein structure in protein synthesis? (21.4)
   - **A.** DNA
   - **B.** mRNA
   - **C.** tRNA
   - **D.** rRNA
   - **E.** ribosomes

For questions 7 through 15, select answers from the following nucleic acids: (21.1, 21.2, 21.3, 21.4)
   - **A.** DNA
   - **B.** mRNA
   - **C.** tRNA
   - **D.** rRNA

7. _____ a major component of the ribosomes

8. _____ a double helix consisting of two chains of nucleotides held together by hydrogen bonds between bases

9. _____ a nucleic acid that uses deoxyribose as the sugar

10. _____ a nucleic acid produced in the nucleus that migrates to the ribosomes to direct the formation of a protein

11. _____ places the proper amino acid into the peptide chain

12. _____ contains the bases adenine, cytosine, guanine, and thymine

13. _____ contains the codons for the amino acid order

14. _____ contains a triplet called an anticodon loop

15. _____ replicated during cellular division

For questions 16 through 20, select answers from the following: (21.3, 21.4, 21.5)

A. 5′ A G C C T A 3′
⋮ ⋮ ⋮ ⋮ ⋮ ⋮
3′ T C G G A T 5′

B. 5′ A U U G C U C 3′

C. 5′ A G T U G U 3′
⋮ ⋮ ⋮ ⋮ ⋮ ⋮
3′ T C A A C A 5′

D. G U A

E. 5′ A T G T A T 3′

16. _____ a section of an mRNA

17. _____ a strand of DNA that is not possible

18. _____ a codon

19. _____ a section from a DNA molecule

20. _____ a single strand that would not be possible for mRNA

For questions 21 through 25, indicate the correct order of protein synthesis: (21.4, 21.5)

A. tRNA assembles the amino acids at the ribosomes.
B. DNA forms a complementary copy of itself called mRNA.
C. Protein is formed and breaks away.
D. tRNA picks up specific amino acids.
E. mRNA goes to the ribosomes.

21. _____ first step

22. _____ second step

23. _____ third step

24. _____ fourth step

25. _____ fifth step

For questions 26 through 28, match each term with the correct description: (21.4, 21.6)

A. mutation

B. intron

C. exon

26. _____ an error in the transmission of the base sequence of DNA

27. _____ a section of DNA that codes for the synthesis of protein

28. _____ a section of DNA that does not code for the synthesis of protein

29. A mutation that occurs by elimination of one base in a DNA sequence is a(n) _____. (21.6)
  A. insertion mutation   **B.** viral mutation   **C.** double mutation
  D. deletion mutation   **E.** point mutation

30. A mutation that occurs by substitution of one base for another in a DNA sequence is a(n) _____. (21.6)
  A. insertion mutation   **B.** viral mutation   **C.** double mutation
  D. deletion mutation   **E.** point mutation

31. Small circular pieces of DNA used in recombinant DNA are called _____. (21.7)
  A. genes   **B.** plasmids   **C.** bacteria   **D.** viruses   **E.** tumors

32. Small living particles that cannot replicate without a host cell are called _____. (21.8)
  A. genes   **B.** plasmids   **C.** bacteria   **D.** viruses   **E.** tumors

Answers to the Practice Test

1. E	**2.** E	**3.** A	**4.** B	**5.** C
6. A	**7.** D	**8.** A	**9.** A	**10.** B
11. C	**12.** A	**13.** B	**14.** C	**15.** A
16. B, D	**17.** C	**18.** D	**19.** A	**20.** E
21. B	**22.** E	**23.** D	**24.** A	**25.** C
26. A	**27.** C	**28.** B	**29.** D	**30.** E
31. B	**32.** D			

Selected Answers and Solutions to Text Problems

21.1 Purine bases (e.g., adenine, guanine) have a double-ring structure; pyrimidines (e.g., cytosine, thymine, uracil) have a single ring.
  a. Thymine is a pyrimidine base.
  b. This double-ring base is the purine adenine.

21.3 DNA contains two purines, adenine (A) and guanine (G), and two pyrimidines, cytosine (C) and thymine (T). RNA contains the same bases, except thymine (T) is replaced by the pyrimidine uracil (U).
  a. Thymine is a base present in DNA only.
  b. Adenine is a base present in both DNA and RNA.

21.5 Nucleotides contain a base, a sugar, and a phosphate group. The nucleotides found in DNA would all contain the sugar deoxyribose. The four nucleotides are deoxyadenosine monophosphate (dAMP), deoxyguanosine monophosphate (dGMP), deoxycytidine monophosphate (dCMP), and deoxythymidine monophosphate (dTMP).

21.7 **a.** Adenosine is a nucleoside.
  b. Deoxycytidine is a nucleoside.
  c. Uridine is a nucleoside.
  d. Cytidine monophosphate is a nucleotide.

21.9 **a.** Phosphate is present in both DNA and RNA.
  b. Ribose is present in RNA only.
  c. Deoxycytidine monophosphate is present in DNA only.
  d. UMP is present in RNA only.

21.11

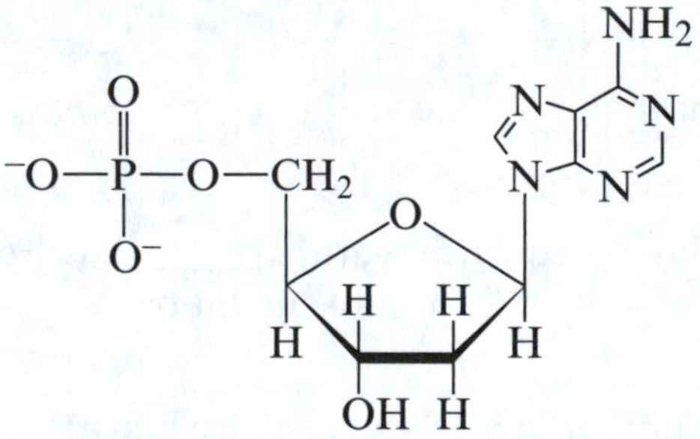

21.13

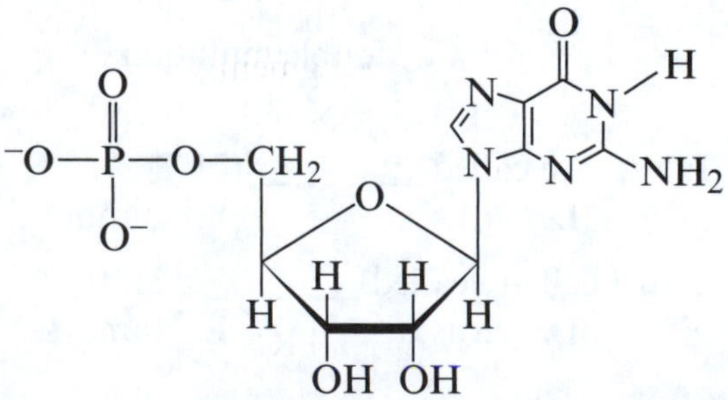

21.15 The nucleotides in nucleic acids are held together by phosphodiester linkages between the 3′ OH group of a sugar (ribose or deoxyribose) and the phosphate group on the 5′ carbon of another sugar.

21.17 The backbone of a nucleic acid consists of alternating sugar and phosphate groups.

  —sugar—phosphate—sugar—phosphate—

21.19 The free 5′ end is determined by the free phosphate group on the 5′ carbon of ribose or deoxyribose of a nucleic acid.

21.21

Guanine (G)

Cytosine (C)

21.23 Structural features of DNA include that it is shaped like a double helix, it contains a sugar–phosphate backbone, the nitrogen-containing bases are hydrogen bonded between strands, and the strands run in opposite directions. A forms two hydrogen bonds to T, and G forms three hydrogen bonds to C.

21.25 The two DNA strands are held together by hydrogen bonds between the complementary bases in each strand.

21.27 a. Since T pairs with A, if one strand of DNA has the sequence 5′ A A A A A A 3′, the second strand would be 3′ T T T T T T 5′.
   b. Since C pairs with G, if one strand of DNA has the sequence 5′ G G G G G G 3′, the second strand would be 3′ C C C C C C 5′.
   c. Since T pairs with A, and C pairs with G, if one strand of DNA has the sequence 5′ A G T C C A G G T 3′, the second strand would be 3′ T C A G G T C C A 5′.
   d. Since T pairs with A, and C pairs with G, if one strand of DNA has the sequence 5′ C T G T A T A C G T T A 3′, the second strand would be 3′ G A C A T A T G C A A T 5′.

21.29 The enzyme helicase unwinds the DNA helix by breaking the hydrogen bonds between the complementary bases so that the parent DNA strands can be replicated.

21.31 Once the DNA strands separate, DNA polymerase pairs each of the bases with its complementary base and produces two exact copies of the original DNA.

21.33 a. Helicase (2) unwinds the DNA helix.
   b. Primase (3) synthesizes primers at the replication fork.
   c. The replication fork (4) is the point in DNA where nucleotides add to the daughter DNA strand.
   d. Daughter DNA is synthesized from short sections called Okazaki fragments on the lagging strand (1).

21.35 a. 5′ C G A G G T A C 3′
   b. No; since the template strand runs from 3′ to 5′, the new DNA segment can be synthesized continuously in the 5′ to 3′ direction.

21.37 The three major types of RNA are messenger RNA (mRNA), ribosomal RNA (rRNA), and transfer RNA (tRNA).

21.39 A ribosome consists of a small subunit and a large subunit that each contain rRNA combined with proteins.

21.41 In transcription, the sequence of nucleotides on a DNA template strand is used to produce the base sequence of a messenger RNA. The DNA unwinds, and one strand is copied as complementary bases are placed in the mRNA molecule. In RNA, U (uracil) is paired with A in DNA.

21.43 To form mRNA, the bases in the DNA template strand are paired with their complementary bases: G with C, C with G, T with A, and A with U. The strand of mRNA would have the following sequence: 5′ GGC UUC CAA GUG 3′.

21.45 In eukaryotic cells, genes contain sections called exons that code for proteins and sections called introns that do not code for proteins.

21.47 5′ GGC GGA UCG 3′

21.49 A transcription factor is a molecule that binds to DNA and regulates the binding of RNA polymerase, which is needed for transcription.

21.51 An activator binds to a section of DNA known as an enhancer, making it easier for RNA polymerase to bind to the promoter and increasing the rate of transcription of a particular gene.

21.53 A codon is a three-base sequence (triplet) in mRNA that codes for a specific amino acid in a protein.

21.55 a. The codon CCA in mRNA codes for the amino acid proline (Pro).
  b. The codon AAC in mRNA codes for the amino acid asparagine (Asn).
  c. The codon GGU in mRNA codes for the amino acid glycine (Gly).
  d. The codon AGG in mRNA codes for the amino acid arginine (Arg).

21.57 At the beginning of an mRNA, the codon AUG signals the start of protein synthesis; thereafter, the AUG codon specifies the amino acid methionine.

21.59 A codon is a base triplet in the mRNA template. An anticodon is the complementary triplet on a tRNA for a specific amino acid.

21.61 The new amino acid is joined by a peptide bond to the growing peptide chain. The ribosome moves to the next codon, which hydrogen bonds to a tRNA carrying the next amino acid.

21.63 The three steps in translation are initiation, chain elongation, and termination.

21.65 a. The codons ACC, ACA, and ACU in mRNA all code for threonine: Thr–Thr–Thr, TTT.
  b. The codons UUU and UUC code for phenylalanine, and CCG and CCA both code for proline: Phe–Pro–Phe–Pro, FPFP.
  c. The codon UAC codes for tyrosine, GGG for glycine, AGA for arginine, and UGU for cysteine: Tyr–Gly–Arg–Cys, YGRC.

21.67 a. The mRNA sequence would be: 5′ CGA UAU GGU UUU 3′.
  b. The tRNA triplet anticodons would be: GCU, AUA, CCA, and AAA.
  c. From the table of mRNA codons, the amino acids would be: Arg–Tyr–Gly–Phe, RYGF.

21.69 a. 5′ AAA CAC UUG GUU GUG GAC 3′
  b. Lys–His–Leu–Val–Val–Asp, KHLVVD

21.71 In a point mutation, a base in DNA is replaced by a different base.

21.73 In a mutation caused by the deletion of a base, all the codons from the mutation onward are changed, which changes the order of amino acids in the rest of the polypeptide chain.

21.75 The normal triplet TTT in DNA forms the codon AAA in mRNA, which codes for lysine. The mutation TTC in DNA forms the codon AAG in mRNA, which also codes for lysine. Thus, there is no effect on the amino acid sequence.

21.77 a. Thr–Ser–Arg–Val is the amino acid sequence produced by normal DNA.
  b. Thr–Thr–Arg–Val is the amino acid sequence produced by a mutation.
  c. Thr–Ser–Gly–Val is the amino acid sequence produced by a mutation.
  d. Thr–STOP; protein synthesis would terminate early. If this mutation occurs early in the formation of the polypeptide, the resulting protein will probably be nonfunctional.
  e. The new protein will contain the sequence Asp–Ile–Thr–Gly.
  f. The new protein will contain the sequence His–His–Gly.

21.79 a. GCC and GCA both code for alanine.
  b. A vital ionic interaction in the tertiary structure of hemoglobin cannot be formed when the polar acidic glutamate is replaced by valine, which is nonpolar. The resulting hemoglobin is malformed and less capable of carrying oxygen.

21.81 *E. coli* bacterial cells contain several small circular plasmids of DNA that can be isolated easily. After the recombinant DNA is formed, *E. coli* multiply rapidly, producing many copies of the recombinant DNA in a relatively short time.

21.83 *E. coli* can be soaked in a detergent solution that disrupts the plasma membrane and releases the cell contents, including the plasmids, which are then collected.

21.85 When a gene has been cut out of donor DNA with a restriction enzyme, it is mixed with plasmids that have been cut by the same enzyme. When mixed together, the sticky ends of the donor DNA fragments bond with the sticky ends of the plasmid DNA to form a recombinant DNA.

21.87 In DNA fingerprinting, restriction enzymes cut a sample of DNA into fragments, which are sorted by size by gel electrophoresis. A radioactive probe that adheres to specific DNA sequences exposes an X-ray film placed over the gel and creates a pattern of dark and light bands known as a DNA fingerprint.

21.89 Asp–Arg–Val–Tyr–Ile–His–Pro–Phe

21.91 A virus contains either DNA or RNA, but not both, inside a protein coat.

21.93 a. An RNA-containing virus must make viral DNA from the RNA to produce proteins for the protein coat, which allows the virus to replicate and leave the cell to infect new cells.
  b. A virus that uses reverse transcription is called a retrovirus.

21.95 Nucleoside analogs such as AZT and ddI are similar to the nucleosides required to make viral DNA in reverse transcription. When they are incorporated into viral DNA, the lack of a hydroxyl group on the $3'$ carbon in the sugar prevents the formation of the sugar–phosphate bonds and stops the replication of the virus.

21.97 Estrogen receptors are molecules on cell surfaces that bind to estrogen. The bonding of estrogen and the estrogen receptor to DNA increases the production of mammary cells, which can lead to mutations and cancer.

21.99 Tamoxifen blocks the binding of estrogen to the estrogen receptor, which prevents the growth of cancers that are estrogen positive.

21.101 a. $5'$ ACC UUA AUA GAC GAG AAG CGC $3'$
  b. Thr–Leu–Ile–Asp–Glu–Lys–Arg, TLIDEKR
  c. There is no change, Asp is still the fourth amino acid.

21.103 a.

Parent strand: | A | G | G | T | C | G | C | C | T |

New strand: | T | C | C | A | G | C | G | G | A |

  b.

| A | G | G | U | C | G | C | C | U |

  c. (Arg)–(Ser)–(Pro)

21.105 DNA contains two purines, adenine (A) and guanine (G), and two pyrimidines, cytosine (C) and thymine (T). RNA contains the same bases, except thymine (T) is replaced by the pyrimidine uracil (U).
  a. Cytosine is a pyrimidine base.   **b.** Adenine is a purine base.
  c. Uracil is a pyrimidine base.   **d.** Thymine is a pyrimidine base.
  e. Guanine is a purine base.

21.107 a. Deoxythymidine contains the base thymine and the sugar deoxyribose.
  b. Adenosine contains the base adenine and the sugar ribose.
  c. Cytidine contains the base cytosine and the sugar ribose.
  d. Deoxyguanosine contains the base guanine and the sugar deoxyribose.

21.109 Thymine and uracil are both pyrimidines, but thymine has a methyl group on carbon 5.

21.111

21.113 Both RNA and DNA are polymers of nucleotides connected through phosphodiester linkages between alternating sugar and phosphate groups, with bases extending out from each sugar.

21.115 Because A bonds with T, DNA containing 28% A will also have 28% T. Thus the sum of A + T is 56%, which leaves 44% divided equally between G and C: 22% G and 22% C.

21.117 There are two hydrogen bonds between A and T in DNA.

21.119 **a.** 3′ C T G A A T C C G 5′
 b. 5′ A C G T T T G A T C G A 3′
 c. 3′ T A G C T A G C T A G C 5′

21.121 DNA polymerase synthesizes the leading strand continuously in the 5′ to 3′ direction. The lagging strand is synthesized in small segments called Okazaki fragments because DNA polymerase can only work in the 5′ to 3′ direction.

21.123 One strand of the parent DNA is found in each of the two copies of the daughter DNA molecule.

21.125 **a.** Transfer RNA (tRNA) is the smallest type of RNA.
 b. Ribosomal RNA (rRNA) makes up the highest percentage of RNA in the cell.
 c. Messenger RNA (mRNA) carries genetic information from the nucleus to the ribosomes.

21.127 **a.** ACU, ACC, ACA, and ACG are all codons for the amino acid threonine.
 b. UCU, UCC, UCA, UCG, AGU, and AGC are all codons for the amino acid serine.
 c. UGU and UGC are codons for the amino acid cysteine.

21.129 **a.** AAG codes for lysine.   **b.** AUU codes for isoleucine.
 c. CGA codes for arginine.

21.131 The anticodon on tRNA consists of the three complementary bases to the codon in mRNA.
 a. UCG   **b.** AUA   **c.** GGU

21.133 Using the genetic code, the codons indicate the following amino acid sequence:
 START–Tyr–Gly–Gly–Phe–Leu–STOP

21.135 Glu–Phe–Arg–His–Asp–Ser–Gly–Tyr–Glu–Val

21.137 The codon for each amino acid contains three nucleotides, plus the start and stop codon consisting of three nucleotides each, which makes a minimum total of $(9 \times 3) + 3 + 3 = 33$ nucleotides.

21.139 A DNA virus attaches to a cell and injects viral DNA that uses the host cell to produce copies of the DNA to make viral RNA. A retrovirus injects viral RNA from which complementary DNA is produced by reverse transcription.

21.141 **a.** DNA polymerase is involved in the replication of DNA (1).
 b. mRNA is synthesized from nuclear DNA during transcription (2).
 c. Some viruses use reverse transcription (5).
 d. Restriction enzymes are used to make recombinant DNA (4).
 e. tRNA molecules bond to codons during translation (3).

Selected Answers to Combining Ideas from Chapters 19 to 21

CI.41 a. The functional groups in aromasin are ketone, alkene, and cycloalkene.
 b. The functional groups in testosterone are ketone, cycloalkene, and alcohol.
 c. The functional groups in estrogen are phenol and alcohol.
 d. The enzyme aromatase converts a six-carbon cycloalkene into an aromatic ring.
 e. Aromasin has a steroid structure similar to that of testosterone, which binds with aromatase at the active site.
 f. Molar mass of aromasin ($C_{20}H_{24}O_2$)
  $= 20(12.01 \text{ g}) + 24(1.008 \text{ g}) + 2(16.00 \text{ g}) = 296.4 \text{ g/mole (4 SFs)}$

 g. $30 \text{ tablets} \times \dfrac{25 \text{ mg aromasin}}{1 \text{ tablet}} \times \dfrac{1 \text{ g aromasin}}{1000 \text{ mg aromasin}} \times \dfrac{1 \text{ mole aromasin}}{296.4 \text{ g aromasin}}$

  $= 2.5 \times 10^{-3} \text{ mole of aromasin (2 SFs)}$

CI.43 a. Glu–His–Trp–Ser–Tyr–Gly–Leu–Arg–Pro–Gly
 b. EHWSYGLRPG
 c. a point mutation
 d. Serine is replaced with phenylalanine.
 e. Serine has a polar —OH group, whereas phenylalanine is nonpolar. The polar serine would most likely have been on the outside of the normal peptide, whereas the phenylalanine will most likely move to the inside of the mutated GnRF. This could change the tertiary structure of the decapeptide, and possibly cause a loss in biological activity.
 f. feedback control

CI.45 a. 5′ UGU AUA AAA CGU UUA AAA CGU ACG 3′
 b. The respective anticodons on tRNA would be ACA, UAU, UUU, GCA, AAU, UUU, GCA, and UGC.
 c. Cys–Ile–Lys–Arg–Leu–Lys–Arg–Thr

22

Metabolic Pathways for Carbohydrates

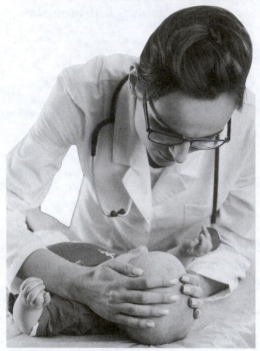

Philip, who is 8 months old, is in the hepatology unit with symptoms that include convulsions, lethargy, difficulty breathing, and diarrhea. Blood tests indicate that Philip has very low blood glucose, low blood pH, high lactate levels, and high liver glycogen. He also has a doll-like face from fat deposits and an extended abdomen caused by a enlarged liver. Because of the large amount of glycogen in his liver, Philip is diagnosed with a glycogen storage disease (GSD) known as von Gierke's disease in which a genetic disorder produces a defective form of the enzyme glucose-6-phosphatase. As a result, glycogen in his liver cannot be degraded to glucose for energy and is stored in the liver. Because Philip cannot produce sufficient glucose, he is hypoglycemic and weak. After nurses began to feed him with small amounts of glucose or starch by nasogastric tube to raise his blood glucose levels, he has shown a rapid improvement in physical behavior. During the day, he is fed cornstarch, which is easily digested to provide small, continuous, amounts of glucose in the blood. Why did Philip's condition require him to be fed cornstarch every few hours?

Credit: KatarzynaBialasiewicz/Getty Images

LOOKING AHEAD

22.1 Metabolism and Energy
22.2 Important Coenzymes in Metabolic Pathways
22.3 Digestion of Carbohydrates

22.4 Glycolysis: Oxidation of Glucose
22.5 Pathways for Pyruvate

22.6 Glycogen Synthesis and Degradation
22.7 Gluconeogenesis: Glucose Synthesis

 The Health icon indicates a question that is related to health and medicine.

22.1 Metabolism and Energy

Learning Goal: Describe the three stages of catabolism, the structure of ATP, and the role of ATP.

- Metabolism includes all of the chemical reactions that provide energy and substances for maintenance of life and for cell growth.
- Catabolic reactions degrade large molecules to produce energy.
- Anabolic reactions utilize energy in the cell to build large molecules for the cell.
- The three stages of catabolism are digestion of food, degradation of larger molecules into smaller groups such as the acetyl group, and the oxidation of the acetyl groups to release energy for ATP synthesis.
- In cells, different organelles contain the enzymes and coenzymes for the various catabolic and anabolic reactions.
- Energy obtained from catabolic reactions is stored primarily in adenosine triphosphate (ATP), a high-energy compound.
- The hydrolysis of ATP, which releases energy, is linked with many anabolic reactions in the cell.

◆ **Learning Exercise 22.1A**

Match each of the following components of a cell with its description or function:

    a. lysosomes   **b.** ribosomes   **c.** mitochondria
    d. cytosol   **e.** cell membrane

1. _____ separates the contents of a cell from the external environment

2. _____ contain enzymes that catalyze energy-producing reactions

3. _____ the fluid of the cell, which is an aqueous solution of electrolytes and enzymes

4. _____ are the sites of protein synthesis

5. _____ contain hydrolytic enzymes that digest old cell structures

Answers   **1.** e   **2.** c   **3.** d   **4.** b   **5.** a

◆ **Learning Exercise 22.1B**

Identify the stages of metabolism for each of the following processes:

    a. Stage 1   **b.** Stage 2   **c.** Stage 3

1. _____ involves the oxidation of two-carbon acetyl CoA

2. _____ polysaccharides undergo digestion to form monosaccharides, such as glucose

3. _____ digestion products such as glucose are degraded to two- or three-carbon compounds

Answers   **1.** c   **2.** a   **3.** b

◆ **Learning Exercise 22.1C**

Fill in the blanks with one of the following terms: phosphate group(s), adenine, hydrolysis, ribose, high-energy, P_i, energy, anabolic, $ATP + H_2O \longrightarrow ADP + P_i + energy$ (7.3 kcal/mole)

The ATP molecule is composed of the nitrogen base **(1)** _____, a **(2)** _____ sugar, and three **(3)** _____. ATP undergoes **(4)** _____, which cleaves a **(5)** _____ and releases **(6)** _____. For this reason, ATP is called a **(7)** _____ compound. The resulting phosphate group, called inorganic phosphate, is abbreviated as **(8)** _____. The equation for the hydrolysis of ATP is written as **(9)** _____. The energy from ATP is linked to cellular reactions that are **(10)** _____.

Answers   **1.** adenine   **2.** ribose   **3.** phosphate groups   **4.** hydrolysis
                5. phosphate group   **6.** energy   **7.** high-energy   **8.** P_i
                9. $ATP + H_2O \longrightarrow ADP + P_i + energy$ (7.3 kcal/mole)   **10.** anabolic

22.2   Important Coenzymes in Metabolic Pathways

Learning Goal: Describe the components and functions of the coenzymes NAD$^+$, NADP$^+$, FAD, and coenzyme A.

> **REVIEW**
> Describing the Role of Cofactors (20.6)

- Coenzymes such as NAD$^+$, NADP$^+$, and FAD pick up hydrogen ions and electrons during oxidative processes.

- Coenzyme A is a coenzyme that carries acetyl (two-carbon) groups produced when glucose, fatty acids, and amino acids are degraded.

♦ **Learning Exercise 22.2**

Select the coenzyme(s) that match(es) each of the following descriptions:

    a. NAD^+ **b.** NADH   **c.** $NADP^+$ **d.** FAD   **e.** $FADH_2$ **f.** coenzyme A

1. _____ participates in reactions that convert a C=O group to a hydroxyl group

> **CORE CHEMISTRY SKILL**
> Identifying Important Coenzymes in Metabolism

2. _____ contains riboflavin (vitamin B_2)

3. _____ is the oxidized form of nicotinamide adenine dinucleotide

4. _____ contains the vitamin niacin

5. _____ is the oxidized form of flavin adenine dinucleotide

6. _____ contains the vitamin pantothenic acid, ADP, and aminoethanethiol

7. _____ participates in oxidation reactions that produce a carbon–carbon (C=C) double bond

8. _____ transfers acyl groups such as the two-carbon acetyl group

9. _____ is the reduced form of flavin adenine dinucleotide

10. _____ is the oxidized form of NADPH

Answers	**1.** b	**2.** d, e	**3.** a	**4.** a, b, c	**5.** d
	6. f	**7.** d	**8.** f	**9.** e	**10.** c

22.3 Digestion of Carbohydrates

Learning Goal: Give the sites and products of the digestion of carbohydrates.

- Digestion is a series of reactions that break down large molecules into smaller molecules that can be absorbed and used by the cells.
- The end products of digestion of carbohydrates are the monosaccharides glucose, fructose, and galactose.

♦ **Learning Exercise 22.3**

Complete the table by providing sites, enzymes, and products for the digestion of carbohydrates:

Carbohydrate	Digestion Site(s)	Enzyme	Products
1. Amylose			
2. Amylopectin			
3. Maltose			
4. Lactose			
5. Sucrose			

Answers

Carbohydrate	Digestion Site(s)	Enzyme	Products
1. Amylose	a. mouth b. small intestine (mucosa)	a. salivary amylase b. pancreatic amylase	a. smaller polysaccharides (dextrins), some maltose and glucose b. maltose, glucose
2. Amylopectin	a. mouth b. small intestine (mucosa)	a. salivary amylase b. pancreatic amylase	a. smaller polysaccharides (dextrins), some maltose and glucose b. maltose, glucose
3. Maltose	small intestine (mucosa)	maltase	glucose and glucose
4. Lactose	small intestine (mucosa)	lactase	glucose and galactose
5. Sucrose	small intestine (mucosa)	sucrase	glucose and fructose

22.4 Glycolysis: Oxidation of Glucose

Learning Goal: Describe the conversion of glucose to pyruvate in glycolysis.

REVIEW

Classifying Enzymes (20.2)

Identifying Factors Affecting Enzyme Activity (20.3)

- Glycolysis, which occurs in the cytosol, consists of 10 reactions that degrade glucose (six carbons) to two pyruvate molecules (three carbons each).
- The overall series of reactions yields two molecules of the reduced coenzyme NADH and two ATP.
- The activity of three enzymes—hexokinase, phosphofructokinase, and pyruvate kinase—can be regulated to respond to cellular requirements for ATP and/or the end products of glycolysis.
- The pentose phosphate pathway utilizes glucose-6-phosphate to supply NADPH and pentose sugars for the biosynthesis of nucleotides, fatty acids, cholesterol, and amino acids.

♦ **Learning Exercise 22.4**

Match each of the following terms associated with glycolysis with the best description(s):

CORE CHEMISTRY SKILL

Identifying the Compounds in Glycolysis

a. 2 NADH	**b.** anaerobic	**c.** glucose	**d.** two pyruvate
e. 2 NADP$^+$	**f.** energy generated	**g.** 2 ATP	**h.** 4 ATP

1. _____ the starting material for glycolysis

2. _____ reduced in the pentose phosphate pathway

3. _____ operates without oxygen

4. _____ net ATP energy produced

5. _____ number of reduced coenzymes produced

6. _____ reactions 6 to 10 of glycolysis

7. _____ end products of glycolysis

8. _____ number of ATP required

Answers　　**1.** c　　**2.** e　　**3.** b　　**4.** g

　　　　　　　5. a　　**6.** f　　**7.** d, a　　**8.** g

22.5   Pathways for Pyruvate

Learning Goal: Give the conditions for the conversion of pyruvate to lactate, ethanol, and acetyl coenzyme A.

- Under aerobic conditions, pyruvate is oxidized in the mitochondria to acetyl CoA, which enters the citric acid cycle.

- In the absence of oxygen, pyruvate is reduced to lactate and NAD^+ is regenerated for the continuation of glycolysis, whereas microorganisms such as yeast reduce pyruvate to ethanol, a process known as fermentation.

♦ Learning Exercise 22.5A

Fill in the blanks with the following terms:

lactate	NAD^+	fermentation
NADH	anaerobic	acetyl CoA

When oxygen is available during glycolysis, the three-carbon pyruvate may be oxidized to form
(1) _____ + CO_2. The coenzyme **(2)** _____ is reduced to **(3)** _____. Under
(4) _____ conditions, pyruvate is reduced to **(5)** _____. In yeast, pyruvate forms
ethanol in a process known as **(6)** _____.

Answers   **1.** acetyl CoA   **2.** NAD^+ **3.** NADH
                 4. anaerobic   **5.** lactate   **6.** fermentation

♦ Learning Exercise 22.5B

Explain how the formation of lactate from pyruvate during anaerobic conditions allows glycolysis to continue.

Answer   Under anaerobic conditions, the oxidation of pyruvate to acetyl CoA to regenerate
                  NAD^+ cannot take place. Pyruvate is reduced to lactate using NADH in the cytosol
                  and regenerating NAD^+

22.6   Glycogen Synthesis and Degradation

Learning Goal: Describe the synthesis and breakdown of glycogen.

- Glycogen, the storage form of glucose, is synthesized when blood glucose levels are high.

- In the *glycogenesis* pathway, glucose units are added to a glycogen chain.

- *Glycogenolysis*, the breakdown of glycogen, occurs when blood glucose levels are depleted and glucose is required for energy by muscles and the brain.

♦ Learning Exercise 22.6

Associate each of the following descriptions with the correct
pathway in glycogen metabolism:

> **CORE CHEMISTRY SKILL**
>
> Identifying the Compounds and
> Enzymes in Glycogenesis and
> Glycogenolysis

a. glycogenesis   **b.** glycogenolysis

1. ____ breakdown of glycogen to glucose
2. ____ activated by glucagon
3. ____ starting material is glucose-6-phosphate
4. ____ synthesis of glycogen from glucose
5. ____ activated by insulin
6. ____ UDP activates glucose

Answers   1. b   2. b   3. a   4. a   5. a   6. a

22.7   Gluconeogenesis: Glucose Synthesis

Learning Goal: Describe how glucose is synthesized from molecules from noncarbohydrate sources.

- In *gluconeogenesis*, glucose is synthesized from noncarbohydrate compounds such as lactate, pyruvate, citric acid cycle intermediates, and the carbon atoms of amino acids.
- Most of the enzymes used in gluconeogenesis are the same as the enzymes used in glycolysis except for hexokinase, phosphofructokinase, and pyruvate kinase.
- In the *Cori cycle*, lactate formed in the muscles is transported to the liver, where it is converted to pyruvate and then to glucose, which is used again by the muscles.

Key Terms for Sections 22.1 to 22.7

Match each of the following key terms with the correct description:

   a. ATP   **b.** glycogen   **c.** glycolysis
   d. catabolic reaction   **e.** gluconeogenesis   **f.** mitochondria

1. ____ the storage form of glucose in the muscle and liver

2. ____ a metabolic reaction that produces energy for the cell by degrading large molecules

3. ____ a high-energy compound produced from energy-releasing processes that provides energy for energy-requiring reactions

4. ____ the synthesis of glucose from noncarbohydrate compounds

5. ____ the degradation reactions of glucose that yield two pyruvate molecules

6. ____ the part of the cell where energy-producing reactions take place

Answers   1. b   2. d   3. a   4. e   5. c   6. f

♦ **Learning Exercise 22.7**

Match each of the following terms with the correct description:

   a. gluconeogenesis   **b.** pyruvate   **c.** pyruvate kinase
   d. pyruvate carboxylase   **e.** Cori cycle

1. ____ an enzyme in glycolysis that cannot be used in gluconeogenesis

2. ____ a typical noncarbohydrate source of carbon atoms for glucose synthesis

3. ____ a process whereby lactate produced in muscle is used for glucose synthesis in the liver

4. ____ a metabolic pathway that converts noncarbohydrate sources to glucose

5. _____ an enzyme used in gluconeogenesis that is not used in glycolysis

6. _____ a metabolic pathway that is activated when glycogen reserves are depleted

Answers   **1.** c   **2.** b   **3.** e   **4.** a   **5.** d   **6.** a

Checklist for Chapter 22

You are ready to take the Practice Test for Chapter 22. Be sure you have accomplished the following learning goals for this chapter. If not, review the Section listed at the end of the goal. Then apply your new skills and understanding to the Practice Test.

After studying Chapter 22, I can successfully:

_____ Associate catabolic and anabolic reactions with organelles in the cell. (22.1)

_____ Describe the role of ATP in catabolic and anabolic reactions. (22.1)

_____ Describe the coenzymes NAD^+, $NADP^+$, FAD, and coenzyme A. (22.2)

_____ Describe the sites, enzymes, and products of the digestion of carbohydrates. (22.3)

_____ Describe the conversion of glucose to pyruvate in glycolysis. (22.4)

_____ Give the conditions for the conversion of pyruvate to lactate, ethanol, and acetyl coenzyme A. (22.5)

_____ Describe the formation and breakdown of glycogen. (22.6)

_____ Describe the reactions in which noncarbohydrate sources are used to synthesize glucose. (22.7)

Practice Test for Chapter 22

The chapter Sections to review are shown in parentheses at the end of each question.

1. The main function of the mitochondria is (22.1, 22.5)
   A. energy production   **B.** protein synthesis   **C.** glycolysis
   D. genetic instructions   **E.** waste disposal

2. ATP is a(n) (22.1)
   A. nucleotide unit in RNA and DNA   **B.** end product of glycogenolysis
   C. end product of transamination   **D.** enzyme
   E. energy storage molecule

For questions 3 through 7, match each term or phrase with the correct cellular component: (22.1)

   A. mitochondria   **B.** lysosomes   **C.** cytosol
   D. ribosomes   **E.** cell membrane

3. _____ protein synthesis   **4.** _____ fluid part of the cell

5. _____ separates cell contents from external fluids   **6.** _____ energy-producing reactions

7. _____ degradation by hydrolytic enzymes of old cell structures

For questions 8 through 14, match each description with the correct coenzyme: (22.2)

   A. NAD^+ **B.** NADH   **C.** FAD   **D.** $FADH_2$ **E.** coenzyme A

8. _____ converts a hydroxyl group to a $C{=}O$ group

9. _____ reduced form of nicotinamide adenine dinucleotide

10. _____ oxidized form of flavin adenine dinucleotide

11. _____ contains the vitamin pantothenic acid, ADP, and aminoethanethiol

12. _____ participates in oxidation reactions that produce a carbon–carbon (C=C) double bond

13. _____ transfers acyl groups such as the two-carbon acetyl group

14. _____ reduced form of flavin adenine dinucleotide

Match each of the substances in questions 15 through 18, with one or more of the correct enzymes and the end product(s) of its digestion: (22.3)

A. maltase	**B.** lucose	**C.** fructose	**D.** sucrase
E. galactose	**F.** lactase	**G.** pancreatic amylase	

15. _____ sucrose   16. _____ lactose

17. _____ amylose   18. _____ maltose

19. Glycolysis (22.4)
    A. requires oxygen for the catabolism of glucose
    B. represents the aerobic sequence for glucose anabolism and ATP production
    C. represents the splitting off of glucose residues from glycogen
    D. represents the anaerobic catabolism of glucose to pyruvate
    E. produces acetyl units and ATP as end products

20. Which does *not* appear in the glycolysis pathway? (22.4)
    A. dihydroxyacetone phosphate   B. pyruvate   C. NAD$^+$
    D. acetyl CoA   E. lactate

For questions 21 through 23, answer the following for glycolysis: (22.4)

21. _____ number of ATP invested for the oxidation of one glucose molecule

22. _____ number of ATP (net) produced from one glucose molecule

23. _____ number of NADH produced from the degradation of one glucose molecule

For questions 24 through 28, match each description with one of the following metabolic pathways: (22.4, 22.6, 22.7)

A. glycolysis	**B.** glycogenolysis	**C.** gluconeogenesis
D. glycogenesis	**E.** fermentation	

24. _____ conversion of pyruvate to alcohol

25. _____ breakdown of glucose to pyruvate

26. _____ formation of glycogen

27. _____ synthesis of glucose

28. _____ breakdown of glycogen to glucose

Answers to the Practice Test

1. A	**2.** E	**3.** D	**4.** C	**5.** E
6. A	**7.** B	**8.** A	**9.** B	**10.** C
11. E	**12.** C	**13.** E	**14.** D	**15.** D, B, C
16. F, B, E	**17.** G, A, B	**18.** A, B	**19.** D	**20.** D, E
21. 2	**22.** 2	**23.** 2	**24.** E	**25.** A
26. D	**27.** C	**28.** B		

Selected Answers and Solutions to Text Problems

22.1 The digestion of polysaccharides takes place in stage 1.

22.3 In metabolism, a catabolic reaction breaks apart large molecules, releasing energy.

22.5 **a.** The synthesis of large molecules requires energy and involves anabolic reactions.
b. The addition of phosphate to glucose requires energy and is an anabolic process.
c. The breakdown of ATP releases energy and is a catabolic reaction.
d. The breakdown of large molecules involves catabolic reactions.

22.7 When ATP is hydrolyzed, sufficient energy is released for many energy-requiring processes in the cell.

22.9 **a.** PEP $\longrightarrow$ pyruvate + P_i + 14.8 kcal/mole
b. ADP + P_i + 7.3 kcal/mole $\longrightarrow$ ATP
c. Overall: PEP + ADP $\longrightarrow$ ATP + pyruvate + 7.5 kcal/mole

22.11 $2100 \text{ kJ} \times \dfrac{1000 \text{ J}}{1 \text{ kJ}} \times \dfrac{1 \text{ cal}}{4.184 \text{ J}} \times \dfrac{1 \text{ kcal}}{1000 \text{ cal}} \times \dfrac{1 \text{ mole ATP}}{7.3 \text{ kcal}} \times \dfrac{507 \text{ g ATP}}{1 \text{ mole ATP}}$
$= 3.5 \times 10^4 \text{ g of ATP (2 SFs)}$

22.13 **a.** Pantothenic acid is a component of coenzyme A.
b. Niacin is the vitamin component of NAD^+ and $NADP^+$.
c. Ribitol is the sugar alcohol that is a component of riboflavin in FAD.

22.15 In biochemical systems, oxidation is usually accompanied by the gain of oxygen or loss of hydrogen. Loss of oxygen or gain of hydrogen usually accompanies reduction.
a. The reduced form of NAD^+ is abbreviated NADH.
b. The oxidized form of $FADH_2$ is abbreviated FAD.

22.17 When a carbon–carbon double bond is formed, the coenzyme that gains hydrogen is FAD.

22.19 Hydrolysis is the main reaction involved in the digestion of carbohydrates.

22.21 **a.** Lactose + H_2O $\longrightarrow$ galactose + glucose
b. Sucrose + H_2O $\longrightarrow$ glucose + fructose
c. Maltose + H_2O $\longrightarrow$ glucose + glucose

22.23 Glucose is the starting compound of glycolysis.

22.25 In the initial steps of glycolysis, ATP molecules are required to add phosphate groups to glucose (phosphorylation reactions).

22.27 When fructose-1,6-bisphosphate splits, the three-carbon intermediates glyceraldehyde-3-phosphate and dihydroxyacetone phosphate are formed.

22.29 ATP is produced directly in glycolysis in two places. In reaction 7, a phosphate group from 1,3-bisphosphoglycerate is transferred to ADP and yields ATP. In reaction 10, a phosphate group from phosphoenolpyruvate is transferred directly to ADP to yield another ATP molecule.

22.31 **a.** In glycolysis, phosphorylation is catalyzed by the enzyme hexokinase in reaction 1 and by phosphofructokinase in reaction 3.
b. In glycolysis, direct transfer of a phosphate group is catalyzed by the enzyme phosphoglycerate kinase in reaction 7 and by pyruvate kinase in reaction 10.

22.33 **a.** One ATP is required in the phosphorylation of glucose to glucose-6-phosphate.
b. One NADH is produced in the conversion of each glyceraldehyde-3-phosphate to 1,3-bisphosphoglycerate.
c. Two ATP and two NADH are produced when glucose is converted to pyruvate.

22.35 **a.** In reaction 1 of glycolysis, a hexokinase uses ATP to phosphorylate glucose.
   b. In reactions 7 and 10 of glycolysis, phosphate groups are transferred from 1,3-bisphospho-glycerate and phosphoenolpyruvate directly to ADP to produce ATP.
   c. In reaction 4 of glycolysis, the six-carbon molecule fructose-1,6-bisphosphate is split into two three-carbon molecules, glyceraldehyde-3-phosphate and dihydroxyacetone phosphate.

22.37 In a series of enzymatic reactions, galactose is phosphorylated to yield glucose-1-phosphate, which is converted to glucose-6-phosphate, which enters glycolysis in reaction 2. In the muscles, fructose is phosphorylated to fructose-6-phosphate, which enters glycolysis in reaction 3. In the liver, fructose is phosphorylated to fructose-1-phosphate, which is converted to glyceraldehyde-3-phosphate, which enters glycolysis at reaction 6.

22.39 **a.** Low levels of ATP will activate phosphofructokinase and increase the rate of glycolysis.
   b. High levels of ATP will inhibit phosphofructokinase and slow or stop glycolysis.

22.41 Glucose-6-phosphate is the initial substrate for the pentose phosphate pathway.

22.43 The pentose phosphate pathway produces NADPH required for anabolic reactions.

22.45 Sucrose contains the monosaccharides glucose and fructose. Once hydrolyzed during digestion, glucose directly enters glycolysis at step 1, and fructose enters glycolysis either at step 3 in the muscle and kidney or step 6 in the liver.

22.47 The disaccharide lactose can be hydrolyzed into glucose and galactose. Patients with galactosemia lack one of the enzymes needed to metabolize galactose, so galactose and its by-products can build up to toxic levels if products containing lactose are eaten.

22.49 A cell converts pyruvate to acetyl CoA only under aerobic conditions; there must be sufficient oxygen available.

22.51 The oxidation of pyruvate converts NAD^+ to NADH and produces acetyl CoA and CO_2.

$$CH_3-\overset{\overset{O}{\|}}{C}-COO^- + HS-CoA + NAD^+ \xrightarrow{\underset{\text{dehydrogenase}}{\text{Pyruvate}}} CH_3-\overset{\overset{O}{\|}}{C}-S-CoA + CO_2 + NADH$$

Pyruvate   Acetyl CoA

22.53 During fermentation, the three-carbon compound pyruvate is reduced to ethanol while decarboxylation removes one carbon as carbon dioxide, CO_2.

22.55 **a.** (3) ethanol
   b. (1) acetyl CoA
   c. (2) lactate
   d. (1) acetyl CoA, (3) ethanol

22.57 During strenuous exercise, under anaerobic conditions, pyruvate is converted to lactate. The accumulation of lactate and the corresponding drop in pH causes the muscles to tire and become sore.

22.59 $$CH_3-\overset{\overset{O}{\|}}{C}-\overset{\overset{O}{\|}}{C}-O^- + NADH + H^+ \longrightarrow CH_3-\overset{\overset{OH}{|}}{CH}-\overset{\overset{O}{\|}}{C}-O^- + NAD^+$$

Pyruvate   Lactate

22.61 Glycogenesis is the synthesis of glycogen from glucose molecules.

22.63 Muscle cells break down glycogen to glucose-6-phosphate, which enters glycolysis.

22.65 Glycogen phosphorylase cleaves the glycosidic bonds at the ends of glycogen chains to remove glucose monomers as glucose-1-phosphate.

22.67 **a.** activates   **b.** inhibits   **c.** inhibits

22.69 The movement of glucose out of the bloodstream and into the cells is regulated by insulin, whereas fructose is not. Fructose can move into the cells and be degraded in glycolysis.

22.71 When the glycogen phosphorylase enzyme in the liver is defective, glycogen in the liver cannot be broken down to glucose, which leads to hypoglycemia, muscle weakness, and exercise intolerance.

22.73 When there are no glycogen stores remaining in the liver, gluconeogenesis synthesizes glucose from noncarbohydrate compounds such as pyruvate and lactate.

22.75 The enzymes in glycolysis that are also used in their reverse directions for gluconeogenesis are phosphoglucose isomerase, aldolase, triose phosphate isomerase, glyceraldehyde-3-phosphate dehydrogenase, phosphoglycerate kinase, phosphoglycerate mutase, and enolase.

22.77 **a.** Low glucose levels activate glucose synthesis (gluconeogenesis).
 b. Glucagon, produced when glucose levels are low, activates gluconeogenesis.
 c. Insulin, produced when glucose levels are high, inhibits gluconeogenesis.
 d. AMP inhibits gluconeogenesis.

22.79 People on low carbohydrate diets do not store normal amounts of glycogen and therefore lose this weight and the water weight used to hydrolyze the glycogen.

22.81 Glucocorticoids increase gluconeogenesis levels in the liver.

22.83 Some of the symptoms of von Gierke's disease are: liver enlargement due to glycogen accumulation, and hypoglycemia, which can lead to a shortened life expectancy.

22.85 Glycogen cannot be degraded completely to glucose, causing low blood glucose. Glycogen accumulates as stored glycogen in the liver.

22.87 $2.5 \text{ h} \times \dfrac{350 \text{ kcal}}{1 \text{ h}} \times \dfrac{1 \text{ mole ATP}}{7.3 \text{ kcal}} = 120 \text{ moles of ATP (2 SFs)}$

22.89 **a.** Pyrophosphorylase is utilized in (1) glycogenesis but not glycogenolysis.
 b. Phosphoglucomutase is utilized in (3) both glycogenesis and glycogenolysis.
 c. Glycogen phosphorylase is utilized in (2) glycogenolysis but not glycogenesis.

22.91 **a.** Glucose-6-phosphatase is utilized in (2) gluconeogenesis but not glycolysis.
 b. Aldolase is utilized in (3) both glycolysis and gluconeogenesis.
 c. Phosphofructokinase is utilized in (1) glycolysis but not gluconeogenesis.
 d. Phosphoglycerate mutase is utilized in (3) both glycolysis and gluconeogenesis.

22.93 Metabolism includes all the reactions in cells that provide energy and material for cell growth.

22.95 Stage 1 involves the digestion of large food polymers, such as polysaccharides and proteins.

22.97 A eukaryotic cell has a nucleus.

22.99 ATP is the abbreviation for adenosine triphosphate.

22.101 $\text{ATP} + H_2O \longrightarrow \text{ADP} + P_i + 7.3 \text{ kcal/mole (or 31 kJ/mole)}$

22.103 The reduced form is $FADH_2$; the oxidized form is FAD.

22.105 The reduced form is NADH; the oxidized form is NAD.

22.107 **a.** The reduced form of FAD is abbreviated $FADH_2$.
 b. The reduced form of NAD^+ is abbreviated $NADH + H^+$.
 c. The reduced form of $NADP^+$ is abbreviated $NADPH + H^+$.

22.109 Lactose undergoes digestion in the small intestine to yield glucose and galactose.

22.111 Glucose is the reactant and pyruvate is the product of glycolysis.

22.113 a. Reactions 1 and 3 of glycolysis involve phosphorylation of hexoses with ATP.
  b. Reactions 7 and 10 of glycolysis involve direct substrate phosphorylation that generates ATP.

22.115 Reaction 4, catalyzed by aldolase, converts fructose-1,6-bisphosphate into two three-carbon intermediates.

22.117 Phosphoglucose isomerase converts glucose-6-phosphate to the isomer fructose-6-phosphate.

22.119 Pyruvate is converted to lactate when oxygen is not present in the cell (anaerobic conditions) to regenerate NAD^+ for glycolysis.

22.121 Phosphofructokinase is an allosteric enzyme that is activated by high levels of AMP and ADP because the cell needs to produce more ATP. When ATP levels are high due to a decrease in energy needs, ATP inhibits phosphofructokinase, which reduces its catalysis of fructose-6-phosphate.

22.123 The rate of glycogenolysis increases when blood glucose levels are low and glucagon has been secreted, which accelerates the breakdown of glycogen.

22.125 The breakdown of glycogen (glycogenolysis) in the liver produces glucose.

22.127 a. A low blood glucose level increases the rate of glycogenolysis in the liver.
  b. Insulin, secreted when glucose levels are high, decreases the rate of glycogenolysis in the liver.
  c. Glucagon, secreted when glucose levels are low, increases the rate of glycogenolysis in the liver.
  d. High levels of ATP decrease the rate of glycogenolysis in the liver.

22.129 a. High blood glucose levels decrease the rate of gluconeogenesis.
  b. Insulin, secreted when glucose levels are high, decreases the rate of gluconeogenesis.
  c. Glucagon, secreted when glucose levels are low, increases the rate of gluconeogenesis.
  d. High levels of ATP decrease the rate of gluconeogenesis.

22.131 The cells in the liver, but not skeletal muscle, contain a phosphatase enzyme needed to convert glucose-6-phosphate to free glucose that can diffuse through cell membranes into the bloodstream. Glucose-6-phosphate, which is the end product of glycogenolysis in muscle cells, cannot diffuse easily across cell membranes.

22.133 Insulin increases the rate of glycogenesis and glycolysis and decreases the rate of glycogenolysis. Glucagon decreases the rate of glycogenesis and glycolysis and increases the rate of glycogenolysis.

22.135 The Cori cycle is a cyclic process that involves the flow of lactate and glucose between muscle and the liver. Lactate produced during anaerobic exercise is transported to the liver where it is oxidized to pyruvate, which is used to synthesize glucose. Glucose enters the bloodstream and returns to the muscle to rebuild glycogen stores.

22.137 a. $2.0 \times 10^3 \text{ kcal} \times \dfrac{1 \text{ mole ATP}}{7.3 \text{ kcal}} \times \dfrac{507 \text{ g ATP}}{1 \text{ mole ATP}} \times \dfrac{1 \text{ kg ATP}}{1000 \text{ g ATP}} = 140 \text{ kg of ATP (2 SFs)}$

  b. $250 \text{ g ATP} \times \dfrac{1 \text{ mole ATP}}{507 \text{ g ATP}} \times \dfrac{7.3 \text{ kcal}}{1 \text{ mole ATP}} \times \dfrac{1 \text{ day}}{2.0 \times 10^3 \text{ kcal}} \times \dfrac{24 \text{ h}}{1 \text{ day}} \times \dfrac{60 \text{ min}}{1 \text{ h}}$

  $= 2.6 \text{ min (2 SFs)}$

23
Metabolism and Energy Production

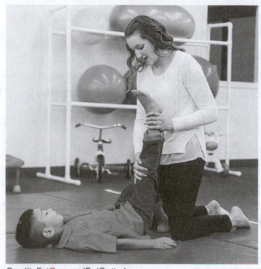

Brian is a four-year-old who has difficulty with balance and coordination. His pediatrician diagnoses Brian with mitochondrial myopathy, which causes fatigue and muscle weakness. The pediatrician refers Brian to Kate, a physical therapist, who specializes in muscular therapy for children. Kate works with Brian and his parents to develop a program of diet and exercises that will improve the quality of Brian's life. Why is oxygen required during aerobic exercise?

Credit: FatCamera/E+/Getty Images

LOOKING AHEAD

23.1 The Citric Acid Cycle   **23.2** Electron Transport and ATP   **23.3** ATP Energy from Glucose

 The Health icon indicates a question that is related to health and medicine.

23.1   The Citric Acid Cycle

Learning Goal: Describe the oxidation of acetyl CoA in the citric acid cycle.

- Under aerobic conditions, pyruvate is oxidized in the mitochondria to acetyl CoA, which enters the citric acid cycle.
- At the start of the citric acid cycle, acetyl CoA combines with oxaloacetate to yield citrate.
- In one turn of the citric acid cycle, the oxidation of acetyl CoA yields 2 CO_2, 1 GTP, 3 NADH, and 1 $FADH_2$. The phosphorylation of ADP by GTP yields ATP.

♦ **Learning Exercise 23.1A**

Match each enzyme (**a** to **h**) with the equation(s) (**1** to **9**):

a. isocitrate dehydrogenase

b. α-ketoglutarate dehydrogenase

c. fumarase

d. succinate dehydrogenase

e. malate dehydrogenase

f. aconitase

g. succinyl CoA synthetase

h. citrate synthase

1. ___ acetyl CoA + oxaloacetate ⟶ citrate   2. ___ citrate ⟶ isocitrate

3. ___ isocitrate ⟶ α-ketoglutarate   4. ___ α-ketoglutarate ⟶ succinyl CoA

5. ___ succinyl CoA ⟶ succinate   6. ___ succinate ⟶ fumarate

7. ___ fumarate ⟶ malate   8. ___ malate ⟶ oxaloacetate

9. ___ allosteric enzymes that regulate the citric acid cycle

Answers   **1.** h   **2.** f   **3.** a   **4.** b   **5.** g   **6.** d   **7.** c   **8.** e   **9.** a, b

♦ **Learning Exercise 23.1B**

For each of the following steps in the citric acid cycle, indicate whether oxidation occurs (yes/no) and identify any coenzyme or direct phosphorylation product produced ($NADH + H^+$, $FADH_2$, GTP):

Step in Citric Acid Cycle	Oxidation	Coenzyme
1. acetyl CoA + oxaloacetate ⟶ citrate	_____	_____
2. citrate ⟶ isocitrate	_____	_____
3. isocitrate ⟶ α-ketoglutarate	_____	_____
4. α-ketoglutarate ⟶ succinyl CoA	_____	_____
5. succinyl CoA ⟶ succinate	_____	_____
6. succinate ⟶ fumarate	_____	_____
7. fumarate ⟶ malate	_____	_____
8. malate ⟶ oxaloacetate	_____	_____

Answers   **1.** no   **2.** no   **3.** yes, $NADH + H^+$ **4.** yes, $NADH + H^+$
5. no, GTP   **6.** yes, $FADH_2$ **7.** no   **8.** yes, $NADH + H^+$

23.2   Electron Transport and ATP

Learning Goal: Describe the transfer of hydrogen ions and electrons in electron transport and the process of oxidative phosphorylation in ATP synthesis.

- The reduced coenzymes from glycolysis and the citric acid cycle are oxidized to NAD^+ and FAD by transferring hydrogen ions and electrons to electron transport.

- In electron transport, electrons are transferred to electron carriers, including coenzyme Q and cytochrome *c*.

- The final acceptor, O_2, combines with hydrogen ions and electrons to yield H_2O.

- The flow of electrons in electron transport pumps hydrogen ions across the inner mitochondrial membrane, which produces a high-energy hydrogen ion gradient that provides energy for the synthesis of ATP.

- The process of using the energy of electron transport to synthesize ATP is called oxidative phosphorylation.

◆ Learning Exercise 23.2A

Write the oxidized and reduced forms of each of the following electron carriers:

1. coenzyme Q   oxidized _____   reduced _____

2. cytochrome c oxidized _____   reduced _____

3. flavin adenine dinucleotide   oxidized _____   reduced _____

4. nicotinamide adenine dinucleotide   oxidized _____   reduced _____

Answers
1. oxidized: CoQ   reduced: $CoQH_2$
2. oxidized: cyt c (Fe^{3+})   reduced: cyt c (Fe^{2+})
3. oxidized: FAD   reduced: $FADH_2$
4. oxidized: NAD^+ reduced: $NADH + H^+$

◆ Learning Exercise 23.2B

1. Write an equation for the transfer of hydrogen from $FADH_2$ to CoQ.

2. What is the function of coenzyme Q in electron transport?

3. What are the end products of electron transport?

Answers
1. $FADH_2 + CoQ \longrightarrow FAD + CoQH_2$
2. CoQ accepts hydrogen atoms from NADH or $FADH_2$. From $CoQH_2$, the hydrogen atoms are separated into hydrogen ions and electrons, with the electrons being passed on to cytochrome c.
3. H_2O and ATP

◆ Learning Exercise 23.2C

Match the following terms with the correct description below:

    a. oxidative phosphorylation   b. ATP synthase
    c. hydrogen ion pump   d. hydrogen ion gradient

1. ____ the complexes I, III, and IV, through which hydrogen ions move out of the matrix into the intermembrane space

2. ____ the protein complex through which hydrogen ions flow from the intermembrane space back to the matrix to generate energy for ATP synthesis

3. ____ use of energy from electron transport to form a hydrogen ion gradient that drives ATP synthesis

4. ____ the accumulation of hydrogen ions in the intermembrane space that lowers pH

Answers   1. c   2. b   3. a   4. d

23.3   ATP Energy from Glucose

Learning Goal: Account for the ATP produced by the complete oxidation of glucose.

- The oxidation of NADH yields 2.5 ATP molecules, whereas the oxidation of $FADH_2$ yields 1.5 ATP molecules.
- The complete oxidation of glucose yields a total of 32 ATP from direct phosphorylation and the oxidation of the reduced coenzymes NADH and $FADH_2$ from electron transport and oxidative phosphorylation.

Key Terms for Sections 23.1 to 23.3

Match each of the following key terms with the correct description:

    a. citric acid cycle   **b.** oxidative phosphorylation   **c.** coenzyme Q
    d. cytochrome *c*   **e.** hydrogen ion pump

1. _____ a mobile carrier that passes electrons from NADH and $FADH_2$ to cytochrome *c* in complex III

2. _____ a mobile carrier that transfers electrons from $CoQH_2$ to oxygen

3. _____ moves hydrogen ions from the matrix into the intermembrane space

4. _____ the synthesis of ATP from ADP and P_i using energy generated from electron transport

5. _____ oxidation reactions that convert acetyl CoA to CO_2, producing reduced coenzymes for energy production via electron transport

Answers   **1.** c   **2.** d   **3.** e   **4.** b   **5.** a

> **CORE CHEMISTRY SKILL**
> Calculating the ATP Produced from Glucose

♦ **Learning Exercise 23.3**

Complete the following:

Substrate	Reaction	Products	Amount of ATP Produced
1. Glucose	glycolysis (aerobic)		
2. Pyruvate	oxidation		
3. Acetyl CoA	citric acid cycle		
4. Glucose	complete oxidation		

Answers

Substrate	Reaction	Products	Amount of ATP Produced
1. Glucose	glycolysis (aerobic)	2 pyruvate	7
2. Pyruvate	oxidation	acetyl CoA + CO_2	2.5
3. Acetyl CoA	citric acid cycle	$2CO_2$	10
4. Glucose	complete oxidation	$6CO_2 + 6H_2O$	32

Checklist for Chapter 23

You are ready to take the Practice Test for Chapter 23. Be sure you have accomplished the following learning goals for this chapter. If not, review the Section listed at the end of the goal. Then apply your new skills and understanding to the Practice Test.

After studying Chapter 23, I can successfully:

_____ Describe the oxidation of acetyl CoA in the citric acid cycle. (23.1)

_____ Identify the electron carriers in electron transport. (23.2)

_____ Describe the process of electron transport. (23.2)

_____ Explain oxidative phosphorylation whereby ATP synthesis is linked to the energy of electron transport and a hydrogen ion gradient. (23.2)

_____ Account for the ATP produced by the complete oxidation of glucose. (23.3)

Practice Test for Chapter 23

The chapter Sections to review are shown in parentheses at the end of each question.

1. Which is true of the citric acid cycle? (23.1)
   A. Acetyl CoA is converted to CO_2 and H_2O.
   B. Oxaloacetate combines with acetyl units to form citric acid.
   C. The coenzymes are NAD^+ and FAD.
   D. ATP is produced by direct phosphorylation.
   E. All of the above.

For questions 2 through 6, match the description with each of the following: (23.1)

   A. malate   B. fumarate   C. succinate
   D. citrate   E. oxaloacetate

2. _____ This compound is formed when oxaloacetate combines with acetyl CoA.

3. _____ H_2O adds to the double bond of this compound to form malate.

4. _____ FAD removes hydrogen from this compound to form a double bond.

5. _____ This compound is formed when the hydroxyl group in malate is oxidized.

6. _____ This compound is regenerated in the citric acid cycle.

7. One turn of the citric acid cycle produces (23.1, 23.3)
   A. 3 NADH   B. 3 NADH, 1 $FADH_2$ C. 3 $FADH_2$, 1 NADH, 1 ATP
   D. 3 NADH, 1 $FADH_2$, 1 ATP   E. 1 NADH, 1 $FADH_2$, 1 ATP

8. The citric acid cycle is activated by (23.1)
   A. high ATP levels   B. NADH   C. high ADP levels
   D. low ATP levels   E. succinyl CoA

9. The end product(s) of electron transport is (are) (23.2)
   A. H_2O + ATP   B. CO_2 + H_2O C. NH_3 + CO_2 + H_2O
   D. H_2 + O_2 E. urea

10. How many complexes in electron transport provide electrons for ATP synthesis? (23.2)
    A. none   B. 1   C. 2   D. 3   E. 4

11. Electron transport (23.2, 23.3)
    A. produces most of the ATP in the body   B. carries oxygen to the cells
    C. produces CO_2 + H_2O D. is involved only in the citric acid cycle
    E. operates during fermentation

For questions 12 through 20, match the metabolic processes with the correct component of electron transport: (23.2)

A. NAD$^+$ **B.** FAD   **C.** CoQ   **D.** cytochrome c **E.** ATP synthase

12. _____ a mobile carrier that picks up electrons from FADH$_2$

13. _____ the coenzyme that accepts hydrogen atoms from NADH

14. _____ a mobile carrier that picks up electrons from CoQH$_2$

15. _____ the coenzyme used to remove hydrogen atoms from two adjacent carbon atoms to form carbon–carbon double bonds

16. _____ protein complex that returns hydrogen ions to the matrix to generate ATP

17. _____ coenzyme that contains flavin

18. _____ the reduced form of this coenzyme that generates 2.5 molecules of ATP

19. _____ the reduced form of this coenzyme that generates 1.5 molecules of ATP

20. _____ electron acceptor containing iron

For questions 21 through 25, indicate the number of ATP produced: (23.3)

A. 2.5 ATP   **B.** 5 ATP   **C.** 7 ATP   **D.** 10 ATP   **E.** 32 ATP

21. _____ from one turn of the citric acid cycle (acetyl CoA $\longrightarrow$ 2CO$_2$)

22. _____ from complete reaction of glucose (glucose + 6O$_2$ $\longrightarrow$ 6CO$_2$ + 6H$_2$O)

23. _____ produced for each NADH that enters electron transport

24. _____ produced from glycolysis (glucose + O$_2$ $\longrightarrow$ 2 pyruvate + 2H$_2$O)

25. _____ from oxidation of 2 pyruvate (2 pyruvate $\longrightarrow$ 2 acetyl CoA + 2CO$_2$)

Answers to the Practice Test

1. E	**2.** D	**3.** B	**4.** C	**5.** E
6. E	**7.** D	**8.** C, D	**9.** A	**10.** D
11. A	**12.** C	**13.** C	**14.** D	**15.** B
16. E	**17.** B	**18.** A	**19.** B	**20.** D
21. D	**22.** E	**23.** A	**24.** C	**25.** B

Selected Answers and Solutions to Text Problems

23.1 The citric acid cycle is also known as the Krebs cycle and the tricarboxylic acid cycle.

23.3 One turn of the citric acid cycle converts 1 acetyl CoA to $2CO_2$, $3NADH + 3H^+$, $1FADH_2$, 1GTP (ATP), and $1HS$—CoA.

23.5 **a.** Two reactions, reactions 3 and 4, involve oxidation and decarboxylation which reduces the length of the carbon chain by one carbon in each reaction.
  b. There is a dehydration in reaction 2. Citrate is dehydrated to aconitate, which is then hydrated to isocitrate.
  c. NAD^+ is reduced by the oxidation reactions 3, 4, and 8 of the citric acid cycle.

23.7 **a.** The six-carbon compounds in the citric acid cycle are citrate and isocitrate.
  b. Decarboxylation reactions remove carbon atoms as CO_2, which reduces the number of carbon atoms in the chain (reactions 3 and 4).
  c. The five-carbon compound in the citric acid cycle is α-ketoglutarate.
  d. Secondary alcohols are oxidized in reactions 3 and 8.

23.9 **a.** Citrate synthase joins acetyl CoA to oxaloacetate.
  b. Succinate dehydrogenase and aconitase form a carbon–carbon double bond.
  c. Fumarase adds water to the double bond in fumarate.

23.11 a. NAD^+ accepts a hydrogen from the oxidation and decarboxylation of isocitrate.
  b. FAD accepts two hydrogens from the oxidation of succinate.

23.13 Citrate synthase, isocitrate dehydrogenase, and α-ketoglutarate dehydrogenase are allosteric enzymes, which can be modified to increase or decrease the flow of materials through the citric acid cycle.

23.15 High levels of ADP means there are low levels of ATP. To provide more ATP for the cell, the reaction rate of the citric acid cycle increases.

23.17 The citric acid cycle plays a central role in aerobic metabolism, and it is likely that defects or deficiencies in citric acid cycle enzymes would be fatal to a developing fetus.

23.19 Since malate builds up, the deficient enzyme is malate dehydrogenase.

23.21 Cyt c (Fe^{3+}) is the abbreviation for the oxidized form of cytochrome c.

23.23 a. The loss of H^+ and $2\,e^-$ is oxidation.   **b.** The gain of $2H^+$ and $2\,e^-$ is reduction.

23.25 NADH molecules provide the hydrogen ions and electrons for electron transport at complex I.

23.27 Their order in electron transport is: $FADH_2$, CoQ, cytochrome c (Fe^{3+}).

23.29 The mobile carrier coenzyme Q transfers electrons from complex I to complex III.

23.31 When NADH transfers electrons to complex I, NAD^+ is produced (oxidation).

23.33 a. $NADH + H^+ + \underline{CoQ} \longrightarrow \underline{NAD^+} + CoQH_2$
  b. $CoQH_2 + 2cyt\ c\ (Fe^{3+}) \longrightarrow CoQ + \underline{2cyt\ c\ (Fe^{2+})} + \underline{2H^+}$

23.35 In oxidative phosphorylation, the energy from the oxidation reactions in electron transport is used to synthesize ATP from ADP and P_i.

23.37 As hydrogen ions return to the lower-energy environment in the matrix, they pass through ATP synthase, releasing energy to drive the synthesis of ATP.

23.39 Glycolysis and the citric acid cycle produce the reduced coenzymes NADH and $FADH_2$, which enter electron transport and are oxidized to provide energy for the synthesis of ATP.

23.41 If CN^- binds to the heme of a cytochrome, electrons cannot reduce Fe^{3+} to Fe^{2+} and will not be transported, stopping electron transport.

23.43 Because dinitrophenol induces thermogenesis and inhibits ATP synthase, it raises body temperature to unsafe levels and can cause constant sweating.

23.45 Two. In glycolysis, one glucose molecule forms two pyruvate, which oxidize to give 2 acetyl CoA to enter the citric acid cycle.

23.47 a. 2.5 ATP are produced from the oxidation of NADH in electron transport.
    b. 7 ATP are produced in glycolysis when glucose degrades to two pyruvate molecules.
    c. 5 ATP are produced when two pyruvate molecules are oxidized to 2 acetyl CoA and $2CO_2$.

23.49 Normally cells are 33% efficient at converting the energy stored in glucose to ATP energy. The cells described in the problem are 24% efficient:

$$25 \; \cancel{\text{g glucose}} \times \frac{1 \; \cancel{\text{mole glucose}}}{180.2 \; \cancel{\text{g glucose}}} \times \frac{32 \text{ moles ATP}}{1 \; \cancel{\text{mole glucose}}} \times \frac{24\% \; \cancel{\text{efficiency}}}{33\% \; \cancel{\text{efficiency}}}$$

$$= 3.2 \text{ moles of ATP (2 SFs)}$$

23.51 a. accumulates: NADH; not produced in sufficient quantity: NAD^+
    b. accumulates: $CoQH_2$; not produced in sufficient quantity: CoQ
    c. accumulates: electrons, H^+, and O_2; not produced in sufficient quantity: H_2O
    d. accumulates: ADP, P_i, and H^+; not produced in sufficient quantity: ATP

23.53 a. Succinate is part of the citric acid cycle.
    b. $CoQH_2$ is part of electron transport.
    c. FAD is part of both the citric acid cycle and electron transport.
    d. Cyt c (Fe^{2+}) is part of electron transport.
    e. Citrate is part of the citric acid cycle.

23.55 citrate $\longrightarrow$ isocitrate
    succinyl CoA $\longrightarrow$ succinate
    malate $\longrightarrow$ oxaloacetate

23.57 a. reactant: citrate   product: isocitrate
    b. reactant: succinate   product: fumarate
    c. reactant: fumarate   product: malate

23.59 a. Aconitase uses H_2O.
    b. Succinate dehydrogenase uses FAD.
    c. Isocitrate dehydrogenase uses NAD^+.

23.61 a. Aconitase catalyzes a hydration (4) reaction.
    b. Succinate dehydrogenase catalyzes an oxidation (1) reaction.
    c. Isocitrate dehydrogenase catalyzes an oxidation (1) and decarboxylation (2) reaction.

23.63 A maximum of 2.5 ATP can be produced by energy released when electrons flow from NADH to oxygen (O_2).

23.65 The oxidation reactions of the citric acid cycle produce a source of reduced coenzymes for electron transport and ATP synthesis.

23.67 The oxidized coenzymes NAD^+ and FAD needed for the citric acid cycle are regenerated by electron transport, which requires oxygen.

23.69 a. Citrate and isocitrate are six-carbon compounds in the citric acid cycle.
    b. α-Ketoglutarate is a five-carbon compound.
    c. The compounds α-ketoglutarate, succinyl CoA, and oxaloacetate have keto groups.

23.71 a. In reaction 4, α-ketoglutarate, a five-carbon keto acid, is decarboxylated.
  b. In reactions 2 and 7, carbon–carbon double bonds in aconitate and fumarate are hydrated.
  c. NAD^+ is reduced in reactions 3, 4, and 8.
  d. In reactions 3 and 8, secondary hydroxyl groups in isocitrate and malate are oxidized.

23.73 a. NAD^+ is the coenzyme for the oxidation of a secondary hydroxyl group in isocitrate to a keto group in α-ketoglutarate.
  b. The coenzymes NAD^+ and CoA are needed in the oxidation and decarboxylation of α-ketoglutarate to succinyl CoA.

23.75 a. High levels of NADH inhibit citrate synthase, isocitrate dehydrogenase, and α-ketoglutarate dehydrogenase to slow the rate of the citric acid cycle.
  b. High levels of ATP inhibit citrate synthase and isocitrate dehydrogenase to slow the rate of the citric acid cycle.

23.77 The transfer of electrons by complexes I, III, and IV provides energy to pump hydrogen ions out of the matrix and into the intermembrane space.

23.79 a. Amytal and rotenone block electron flow from complex I to coenzyme Q.
  b. Antimycin A blocks the flow of electrons from complex III to cytochrome c.
  c. Cyanide and carbon monoxide block electron flow from cytochrome c to complex IV.

23.81 In the chemiosmotic model, energy is released as hydrogen ions flow through ATP synthase back to the mitochondrial matrix and is utilized for the synthesis of ATP.

23.83 In the intermembrane space, there is a higher concentration of hydrogen ions, which reduces the pH and forms an electrochemical gradient. As a result, hydrogen ions flow into the matrix, where the H^+ concentration is lower and the pH is higher.

23.85 As hydrogen ions from the hydrogen ion gradient move through the ATP synthase to return to the matrix, energy is released and used to synthesize ATP by ATP synthase.

23.87 ATP synthase is a protein complex that spans the inner mitochondrial membrane and protrudes into the mitochondrial matrix.

23.89 A hibernating bear has more brown fat because it can be used during the winter for heat rather than ATP energy.

23.91 The oxidation of glucose to pyruvate produces 7 ATP, whereas the oxidation of glucose to CO_2 and H_2O produces a maximum of 32 ATP.

23.93 a. $7 \text{ moles ATP} \times \dfrac{7.3 \text{ kcal}}{1 \text{ mole ATP}} = 51 \text{ kcal (2 SFs) (from glycolysis)}$

  b. $5 \text{ moles ATP} \times \dfrac{7.3 \text{ kcal}}{1 \text{ mole ATP}} = 37 \text{ kcal (2 SFs) (2 pyruvate molecules to 2 acetyl CoA)}$

  c. $20 \text{ moles ATP} \times \dfrac{7.3 \text{ kcal}}{1 \text{ mole ATP}} = 150 \text{ kcal (2 SFs) (2 acetyl CoA in the citric acid cycle)}$

  d. $32 \text{ moles ATP} \times \dfrac{7.3 \text{ kcal}}{1 \text{ mole ATP}} = 230 \text{ kcal (2 SFs) (complete oxidation of glucose to } CO_2 \text{ and } H_2O)$

23.95 If the combustion of glucose produces 690 kcal in a calorimeter, but only 230 kcal (from 32 ATP) in cells, the efficiency of glucose use in the cells is 230 kcal/690 kcal or 33%.

23.97 a. weight $= 115 \text{ lb}$; height $= 5 \text{ ft } 4 \text{ in.} = (5 \times 12) + 4 = 64 \text{ in.}$; age $= 21 \text{ yr}$
  BMR $= 655 + (4.35 \times 115) + (4.70 \times 64) - (4.70 \times 21)$
  $= 1360 \text{ kcal (rounded to nearest 10 kcal)}$

  b. $1360 \text{ kcal} \times \dfrac{1 \text{ mole ATP}}{7.3 \text{ kcal}} \times \dfrac{507 \text{ g ATP}}{1 \text{ mole ATP}} \times \dfrac{1 \text{ kg ATP}}{1000 \text{ g ATP}} = 94 \text{ kg of ATP (2 SFs)}$

Metabolic Pathways for Lipids and Amino Acids

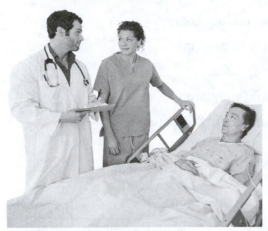

Credit: Tyler Olson/Fotolia

Luke is a paramedic. Recently, blood work from a physical examination indicate a plasma cholesterol level of 256 mg/dL, which is elevated. Luke's doctor orders a liver profile that shows elevated liver enzymes: aspartate transaminase (AST) of 226 Units/L (the normal AST is 5 to 50 Units/L) and alanine transaminase (ALT) of 282 Units/L (the normal ALT is 5 to 35 Units/L).

During Luke's career as a paramedic, he was exposed several times to blood and was accidentally stuck by a needle containing infected blood. A hepatitis profile shows that Luke is positive for antibodies to both hepatitis B and C. His doctor diagnoses Luke with chronic hepatitis C virus (HCV) infection. Hepatitis C is an infection caused by a virus that attacks the liver and leads to inflammation. Hepatitis C is usually passed by contact with contaminated blood or by needles shared during illegal drug use. Most people infected with the hepatitis C virus have no symptoms. As part of his treatment, Luke attends a class given by a public health nurse on living with hepatitis C. How many ATP molecules are produced from the beta oxidation of capric acid, the 10-carbon saturated fatty acid?

LOOKING AHEAD

 The Health icon indicates a question that is related to health and medicine.

24.1 Digestion of Triacylglycerols

Learning Goal: Describe the process by which triacylglycerols are digested.

- Dietary fats begin digestion in the small intestine, where they are emulsified by bile salts.
- Pancreatic lipases catalyze the hydrolysis of triacylglycerols to yield monoacylglycerols and free fatty acids.
- Triacylglycerols reform in the intestinal lining, where they combine with proteins to form chylomicrons for transport through the lymphatic system and bloodstream.
- In the cells, triacylglycerols are hydrolyzed to glycerol and fatty acids, which can be used for energy.

♦ **Learning Exercise 24.1**

Match each of the following terms with the correct description:

    a. chylomicrons   **b.** lipases   **c.** fat utilization

    d. monoacylglycerols and fatty acids   **e.** emulsification

1. _____ the hydrolysis of triacylglycerols in adipose tissues to produce energy

2. _____ lipoproteins formed when triacylglycerols are coated with proteins

3. _____ the breakup of fat globules in the small intestine by bile salts

4. _____ enzymes released from the pancreas that hydrolyze triacylglycerols

5. _____ the products of lipase hydrolysis of triacylglycerols in the small intestine

Answers   **1.** c   **2.** a   **3.** e   **4.** b   **5.** d

24.2   Oxidation of Fatty Acids

Learning Goal: Describe the metabolic pathway of β oxidation.

> REVIEW
> Identifying Fatty Acids (17.2)

- When needed for energy, fatty acids bond to coenzyme A for transport to the mitochondria, where they undergo β oxidation.
- In β oxidation, an acyl chain is oxidized to yield a shortened fatty acid, acetyl CoA, and the reduced coenzymes NADH and $FADH_2$.

♦ **Learning Exercise 24.2A**

Match each of the following terms with the correct description:

    a. activation   **b.** carnitine   **c.** β oxidation

    d. NAD^+ and FAD   **e.** mitochondria

1. _____ coenzymes needed for β oxidation

2. _____ site in the cell where β oxidation of fatty acids takes place

3. _____ combination of a fatty acid with HS—CoA to form fatty acyl CoA

4. _____ carrier that moves the acyl group into the mitochondrial matrix

5. _____ the sequential removal of two-carbon sections from fatty acids

Answers   **1.** d   **2.** e   **3.** a   **4.** b   **5.** c

♦ **Learning Exercise 24.2B**

1. Write the equation for the activation of myristic (C_{14}) acid: $CH_3-(CH_2)_{12}-\overset{\displaystyle O}{\overset{\|}{C}}-OH$

2. Write the equation for the first oxidation of myristoyl CoA.

3. Write the equation for the hydration of the double bond.

4. Write the equation for the second oxidation of myristoyl CoA.

5. a. How many cycles of β oxidation are needed for complete oxidation?

    b. How many acetyl CoA units will be produced by complete oxidation?

1. $CH_3-(CH_2)_{12}-\overset{\overset{\displaystyle O}{\|}}{C}-OH \ + \ HS-CoA \ + \ ATP \longrightarrow$

        $CH_3-(CH_2)_{12}-\overset{\overset{\displaystyle O}{\|}}{C}-S-CoA \ + \ AMP \ + \ 2P_i \ + \ H_2O$

2. $CH_3-(CH_2)_{12}-\overset{\overset{\displaystyle O}{\|}}{C}-S-CoA \ + \ FAD \longrightarrow$

        $CH_3-(CH_2)_{10}-\overset{\overset{\displaystyle H}{|}}{C}=\overset{\overset{\displaystyle H \ \ O}{| \ \ \|}}{\underset{\underset{\displaystyle H}{|}}{C}}-C-S-CoA \ + \ FADH_2$

3. $CH_3-(CH_2)_{10}-\overset{\overset{\displaystyle H}{|}}{\underset{\underset{\displaystyle H}{|}}{C}}=\overset{\overset{\displaystyle H \ \ O}{| \ \ \|}}{C}-C-S-CoA \ + \ H_2O \longrightarrow$

        $CH_3-(CH_2)_{10}-\overset{\overset{\displaystyle OH}{|}}{C}H-CH_2-\overset{\overset{\displaystyle O}{\|}}{C}-S-CoA$

4. $CH_3-(CH_2)_{10}-\overset{\overset{\displaystyle OH}{|}}{C}H-CH_2-\overset{\overset{\displaystyle O}{\|}}{C}-S-CoA \ + \ NAD^+ \longrightarrow$

        $CH_3-(CH_2)_{10}-\overset{\overset{\displaystyle O}{\|}}{C}-CH_2-\overset{\overset{\displaystyle O}{\|}}{C}-S-CoA \ + \ NADH+H^+$

5. a. 6 cycles   **b.** 7 acetyl CoA units

24.3 ATP and Fatty Acid Oxidation

Learning Goal: Calculate the total ATP produced by the complete oxidation of a fatty acid.

REVIEW

Identifying Important Coenzymes in Metabolism (22.2)

- The energy obtained from a particular fatty acid depends on the number of carbon atoms.
- Two ATP are required for activation. Then each acetyl CoA produces 10 ATP via the citric acid cycle; each NADH gives 2.5 ATP and each $FADH_2$ gives 1.5 ATP from electron transport.

♦ **Learning Exercise 24.3**

Lauric acid is a 12-carbon saturated fatty acid:

CORE CHEMISTRY SKILL

Calculating the ATP from Fatty Acid Oxidation (β Oxidation)

$$CH_3-(CH_2)_{10}-\overset{\overset{\displaystyle O}{\|}}{C}-OH$$

1. How many ATP are needed for activation? _____

2. How many cycles of β oxidation are needed for complete oxidation? _____

3. How many NADH and $FADH_2$ are produced during β oxidation? _____

4. How many acetyl CoA units are produced? _____

5. What is the total number of ATP produced from the citric acid cycle and electron transport? _____

Answers

1. 2 ATP
2. 5 cycles
3. Five cycles produce 5 NADH and 5 $FADH_2$.
4. 6 acetyl CoA units
5. 5 NADH × 2.5 ATP/NADH = 12.5 ATP; 5 $FADH_2$ × 1.5 ATP/$FADH_2$ = 7.5 ATP; 6 acetyl CoA × 10 ATP/acetyl CoA = 60 ATP; Total ATP = 12.5 ATP + 7.5 ATP + 60 ATP − 2 ATP (for activation) = 78 ATP

24.4 Ketogenesis and Ketone Bodies

Learning Goal: Describe the pathway of ketogenesis.

- When the oxidation of large amounts of fatty acids causes high levels of acetyl CoA, the acetyl CoA undergoes ketogenesis.
- Two molecules of acetyl CoA form acetoacetyl CoA, which is converted to the ketone bodies acetoacetate, 3-hydroxybutyrate, and acetone.

♦ **Learning Exercise 24.4**

Match each of the following terms with the correct description:

  a. ketone bodies   **b.** ketogenesis   **c.** ketosis

  d. liver   **e.** ketoacidosis

  1. _____ high levels of ketone bodies in the blood

  2. _____ a metabolic pathway that produces ketone bodies

  3. _____ 3-hydroxybutyrate, acetoacetate, and acetone

  4. _____ the condition whereby ketone bodies lower the blood pH below 7.4

  5. _____ site where ketone bodies form

Answers   **1.** c   **2.** b   **3.** a   **4.** e   **5.** d

24.5   Fatty Acid Synthesis

Learning Goal: Describe the synthesis of fatty acids from acetyl CoA.

- When all energy needs have been met and glycogen stores are full, excess acetyl CoA is used to synthesize fatty acids that are stored in the adipose tissue.
- Two-carbon acetyl CoA molecules link together to form palmitic (C_{16}) acid and other fatty acids.

♦ **Learning Exercise 24.5**

Indicate whether each of the following is characteristic of lipogenesis (L) or β oxidation (O):

  1. _____ occurs in the matrix of mitochondria   **2.** _____ occurs in the cytosol of mitochondria

  3. _____ activated by insulin   **4.** _____ activated by glucagon

  5. _____ starts with fatty acids   **6.** _____ starts with acetyl CoA units

  7. _____ produces fatty acids   **8.** _____ produces acetyl CoA units

  9. _____ requires $NADPH + H^+$ **10.** _____ requires FAD and NAD^+ coenzymes

  11. _____ activated by HS—CoA   **12.** _____ activated by HS—ACP

Answers   **1.** O   **2.** L   **3.** L   **4.** O
                5. O   **6.** L   **7.** L   **8.** O
                9. L   **10.** O   **11.** O   **12.** L

24.6   Degradation of Proteins and Amino Acids

Learning Goal: Describe the hydrolysis of dietary protein and the reactions of transamination and oxidative deamination in the degradation of amino acids.

- Proteins begin digestion in the stomach, where HCl denatures proteins and activates protease enzymes that hydrolyze peptide bonds.
- In the small intestine, trypsin and chymotrypsin complete the hydrolysis of peptides to amino acids.
- Amino acids are normally used for protein synthesis.
- Amino acids are degraded by transferring an amino group from an amino acid to an α-keto acid to yield a different amino acid and α-keto acid.
- In oxidative deamination, the amino group in glutamate is removed as an ammonium ion, NH_4^+.

♦ **Learning Exercise 24.6A**

Match each of the following terms with the correct description:

    a. nitrogen-containing compounds   **b.** protein turnover   **c.** stomach
    d. nitrogen balance   **e.** small intestine

1. _____ HCl activates enzymes that hydrolyze peptide bonds in proteins.

2. _____ Trypsin and chymotrypsin convert peptides to amino acids.

3. _____ These compounds include amino acids, amino alcohols, proteins, hormones, and nucleic acids.

4. _____ The process of synthesizing proteins and breaking them down.

5. _____ The amount of protein hydrolyzed is equal to the amount of protein used in the body.

Answers   **1.** c   **2.** e   **3.** a   **4.** b   **5.** d

♦ **Learning Exercise 24.6B**

Match each of the following descriptions with transamination (T) or oxidative deamination (D):

1. _____ produces an ammonium ion, NH_4^+

2. _____ transfers an amino group to an α-keto acid

3. _____ usually involves the degradation of glutamate

4. _____ requires NAD^+

5. _____ produces another amino acid and α-keto acid

6. _____ usually produces α-ketoglutarate

Answers   **1.** D   **2.** T   **3.** D   **4.** D   **5.** T   **6.** D

♦ **Learning Exercise 24.6C**

1. Write an equation for the transamination reaction of serine and oxaloacetate.

2. Write an equation for the oxidative deamination of glutamate.

Answers

1. $HO-CH_2-\overset{\overset{+}{N}H_3}{\underset{|}{C}H}-COO^- + {}^-OOC-\overset{\overset{O}{\|}}{C}-CH_2-COO^- \xrightarrow{\text{Aminotransferase}}$

$HO-CH_2-\overset{\overset{O}{\|}}{C}-COO^- + {}^-OOC-\overset{\overset{+}{N}H_3}{\underset{|}{C}H}-CH_2-COO^-$

2. ${}^-OOC-\overset{\overset{+}{N}H_3}{\underset{|}{C}H}-CH_2-CH_2-COO^- + NAD^+ + H_2O \xrightarrow{\text{Glutamate dehydrogenase}}$

${}^-OOC-\overset{\overset{O}{\|}}{C}-CH_2-CH_2-COO^- + NH_4^+ + NADH + H^+$

24.7 Urea Cycle

Learning Goal: Describe the formation of urea from an ammonium ion.

- The ammonium ion, NH_4^+, from amino acid degradation is toxic if allowed to accumulate.
- The urea cycle converts ammonium ion to urea, which forms urine in the kidneys.

♦ **Learning Exercise 24.7**

Arrange the following reactions in the order they occur in the urea cycle:

Reaction 1. _____   **Reaction 2.** _____   **Reaction 3.** _____   **Reaction 4.** _____

   a. Argininosuccinate is split to yield arginine and fumarate.

   b. Aspartate condenses with citrulline to yield argininosuccinate.

   c. Arginine is hydrolyzed to yield urea and regenerates ornithine.

   d. Ornithine combines with the carbamoyl group from carbamoyl phosphate.

Answers   **Reaction 1.** d   **Reaction 2.** b   **Reaction 3.** a   **Reaction 4.** c

24.8 Fates of the Carbon Atoms from Amino Acids

Learning Goal: Describe where carbon atoms from amino acids enter the citric acid cycle or other pathways.

- α-Keto acids resulting from transamination can be used as intermediates in the citric acid cycle or in the synthesis of lipids or glucose or oxidized for energy.

REVIEW

Identifying the Compounds in Glycolysis (22.4)

Identifying the Compounds and Enzymes in Glycogenesis and Glycogenolysis (22.6)

Describing the Reactions in the Citric Acid Cycle (23.1)

Calculating the ATP Produced from Glucose (23.3)

♦ **Learning Exercise 24.8**

Match each of the following terms with the correct description:

   a. glucogenic   b. ketogenic   c. oxaloacetate
   d. acetyl CoA   e. α-ketoglutarate   f. pyruvate

1. _____ amino acids that generate pyruvate or oxaloacetate, which can be used to synthesize glucose

2. _____ intermediate produced from carbon atoms of alanine and serine

3. _____ intermediate produced from carbon atoms of glutamine and glutamate

4. _____ intermediate produced from carbon atoms of aspartate and asparagine

5. _____ amino acids that generate compounds that can produce ketone bodies

6. _____ intermediate produced from carbon atoms of leucine and isoleucine

Answers   **1.** a   **2.** f   **3.** e   **4.** c   **5.** b   **6.** d

24.9   Synthesis of Amino Acids

Learning Goal: Describe how some nonessential amino acids are synthesized from intermediates in the citric acid cycle and other metabolic pathways.

- Humans synthesize only 11 amino acids. The other 9, called essential amino acids, must be obtained from the diet.
- Nonessential amino acids are synthesized when an amino group from glutamate is transferred to an alpha keto acid obtained from glycolysis or the citric acid cycle.

Key Terms for Sections 24.1 to 24.9

Match each of the following key terms with the correct description:

a. ketone bodies   b. oxidative deamination   c. lipogenesis
d. ketoacidosis   e. β oxidation

1. _____ a reaction cycle that oxidizes fatty acids by removing acetyl CoA units

2. _____ the synthesis of fatty acids by linking two-carbon acetyl units

3. _____ a condition in which high levels of ketone bodies lower blood pH

4. _____ the products of ketogenesis: acetoacetate, 3-hydroxybutyrate, and acetone

5. _____ an ammonium ion is produced when an amino group is removed from glutamate

Answers   **1.** e   **2.** c   **3.** d   **4.** a   **5.** b

> **CORE CHEMISTRY SKILL**
> Distinguishing Anabolic and Catabolic Pathways

♦ **Learning Exercise 24.9**

Match each of the following terms with the correct description:

a. essential amino acids   b. nonessential amino acids
c. transamination   d. phenylketonuria (PKU)

1. _____ amino acids synthesized in humans

2. _____ a genetic condition in which phenylalanine is not converted to tyrosine

3. _____ amino acids that must be supplied by the diet

4. _____ reaction that produces some nonessential amino acids

Answers   **1.** b   **2.** d   **3.** a   **4.** c

Checklist for Chapter 24

You are ready to take the Practice Test for Chapter 24. Be sure you have accomplished the following learning goals for this chapter. If not, review the Section listed at the end of the goal. Then apply your new skills and understanding to the Practice Test.

After studying Chapter 24, I can successfully:

_____ Describe the sites, enzymes, and products for the digestion of triacylglycerols. (24.1)

_____ Describe the oxidation of fatty acids via β oxidation. (24.2)

_____ Calculate the number of ATP produced by the complete oxidation of a fatty acid. (24.3)

_____ Explain ketogenesis and the conditions in the cell that form ketone bodies. (24.4)

_____ Describe the biosynthesis of fatty acids from acetyl CoA. (24.5)

_____ Describe the sites, enzymes, and products for the digestion of dietary proteins. (24.6)

_____ Explain the role of transamination and oxidative deamination in degrading amino acids. (24.6)

_____ Describe the formation of urea from ammonium ion. (24.7)

_____ Explain how carbon atoms from amino acids are prepared to enter the citric acid cycle or other pathways. (24.8)

_____ Show how nonessential amino acids are synthesized from substances used in the citric acid cycle and other pathways. (24.9)

Practice Test for Chapter 24

The chapter Sections to review are shown in parentheses at the end of each question.

1. The digestion of triacylglycerols takes place in the _____ by enzymes called _____. (24.1)
   A. small intestine; proteases
   B. stomach; lipases
   C. stomach; protcases
   D. small intestine; lipases
   E. all of these

2. The products of the digestion of triacylglycerols are (24.1)
   A. fatty acids
   B. monoacylglycerols
   C. glycerol
   D. diacylglycerols
   E. all of these

3. The function of bile salts in the digestion of fats is (24.1)
   A. emulsification
   B. hydration
   C. dehydration
   D. oxidation
   E. reduction

4. Chylomicrons formed in the intestinal lining (24.1)
   A. are lipoproteins
   B. are triacylglycerols coated with proteins
   C. transport fats into the lymphatic system and bloodstream
   D. carry triacylglycerols to the cells of the heart, muscle, and adipose tissues
   E. all of these

5. Glycerol obtained from the hydrolysis of triacylglycerols enters glycolysis when converted to (24.1)
   A. glucose
   B. fatty acids
   C. dihydroxyacetone phosphate
   D. pyruvate
   E. glycerol-3-phosphate

6. Fatty acids are prepared for β oxidation by forming (24.2)
   A. carnitine
   B. acyl carnitine
   C. acetyl CoA
   D. acyl CoA
   E. pyruvate

7. The fatty acids in the β oxidation cycle do *not* undergo (24.2)

A. reduction   B. hydration   C. dehydrogenation

D. oxidation   E. formation of acetyl CoA

For questions 8 through 11, consider the β oxidation of palmitic (C_{16}) acid: (24.2, 24.3)

8. The number of β oxidation cycles required to oxidize palmitic (C_{16}) acid is

A. 16   B. 9   C. 8   D. 7   E. 6

9. The number of acetyl CoA units produced by the β oxidation of palmitic (C_{16}) acid is

A. 16   B. 9   C. 8   D. 7   E. 6

10. The number of NADH and $FADH_2$ produced by the β oxidation of palmitic (C_{16}) acid is

A. 16   B. 9   C. 8   D. 7   E. 6

11. The total ATP produced by the β oxidation of palmitic (C_{16}) acid is

A. 96   B. 106   C. 126   D. 134   E. 136

12. The oxidation of large amounts of fatty acids can produce (24.4)

A. ketone bodies   B. glucose   C. low pH level in the blood

D. acetone   E. pyruvate

13. The metabolic pathway of lipogenesis requires (24.5)

A. fatty acids   B. acetyl CoA   C. FAD and NAD^+

D. glucagon   E. ketone bodies

14. The digestion of proteins takes place in the _____ by enzymes called _____. (24.6)

A. small intestine; proteases   B. stomach; lipases   C. stomach; proteases

D. small intestine; lipases   E. stomach and small intestine; proteases

15. The process of transamination (24.6)

A. is part of the citric acid cycle   B. converts α-amino acids to β-keto acids

C. produces new amino acids   D. is not used in the metabolism of amino acids

E. is part of the β oxidation of fats

16. The oxidative deamination of glutamate produces (24.6)

A. a new amino acid   B. a new α-keto acid   C. ammonia, NH_3

D. ammonium ion, NH_4^+ E. urea

17. The purpose of the urea cycle in the liver is to (24.7)

A. synthesize urea

B. convert urea to ammonium ion, NH_4^+

C. convert ammonium ion, NH_4^+, to urea

D. synthesize new amino acids

E. take part in the β oxidation of fats

18. The urea cycle begins with the conversion of NH_4^+ to (24.7)

A. aspartate   B. carbamoyl phosphate

C. citrulline   D. argininosuccinate

E. urea

19. The carbon atoms from a ketogenic amino acid can be used to (24.8)

A. synthesize ketone bodies

B. convert urea to ammonium ion, NH_4^+

C. synthesize fatty acids

D. produce energy

E. synthesize proteins

20. The carbon atoms from various amino acids can be used (24.8)
  A. as intermediates in the citric acid cycle
  B. in the formation of pyruvate
  C. in the synthesis of glucose
  D. in the formation of ketone bodies
  E. for all of these

21. Essential amino acids (24.9)
  A. are not synthesized by humans
  B. are required in the diet
  C. are excreted if in excess
  D. include leucine, lysine, and valine
  E. all of these

22. Phenylketonuria is a condition (24.9)
  A. that is abbreviated as PKU
  B. in which a person does not synthesize tyrosine
  C. that can be detected at birth
  D. that can cause severe mental retardation if untreated
  E. all of these

Answers to the Practice Test

1. D	**2.** E	**3.** A	**4.** E	**5.** C
6. D	**7.** A	**8.** D	**9.** C	**10.** D
11. B	**12.** A, C, D	**13.** B	**14.** E	**15.** C
16. B, D	**17.** C	**18.** B	**19.** A, C	**20.** E
21. E	**22.** E			

Selected Answers and Solutions to Text Problems

24.1 The bile salts emulsify fat so that it forms small fat globules for hydrolysis by pancreatic lipase.

24.3 Fats are released from fat stores when blood glucose and glycogen stores are depleted.

24.5 Glycerol is converted to glycerol-3-phosphate, and then to dihydroxyacetone phosphate, an intermediate of glycolysis.

24.7 The enzyme glycerol kinase would be defective.

24.9 Fatty acids are activated in the cytosol.

24.11 The coenzymes FAD, NAD^+, and HS—CoA are required for β oxidation.

24.13 The designation β-carbon is based on the common names of carboxylic acids in which the α-carbon is the carbon adjacent to the carboxyl group.

a.
$$CH_3-CH_2-CH_2-CH_2-CH_2-\underset{\beta}{CH_2}-CH_2-\overset{\overset{\textstyle O}{\|}}{C}-S-CoA$$

b.

24.15 a.

b.
$$CH_3-(CH_2)_6-\underset{\underset{\textstyle H}{|}}{\overset{\overset{\textstyle H}{|}}{C}}=C-\overset{\overset{\textstyle O}{\|}}{C}-S-CoA$$

24.17 a. and b.
$$CH_3-(CH_2)_4-\underset{\beta}{CH_2}-\underset{\alpha}{CH_2}-\overset{\overset{\textstyle O}{\|}}{C}-S-CoA$$

c. Three β oxidation cycles are needed.

d. Four acetyl CoA units are produced from a C_8 fatty acid.

24.19 The hydrolysis of ATP to AMP involves the hydrolysis of ATP to ADP, and then ADP to AMP, which provides the same amount of energy as the hydrolysis of two ATP to two ADP.

24.21

Number of Carbon Atoms	Number of β Oxidation Cycles	Number of Acetyl CoA	Number of NADH	Number of FADH$_2$
22	10	11	10	10

a. A C_{22} fatty acid will go through 10 β oxidation cycles.

b. The β oxidation of a chain of 22 carbon atoms produces 11 acetyl CoA units.

c. **ATP Production from Behenic Acid (C$_{22}$)**

Activation	-2 ATP
$11 \text{ acetyl CoA} \times \dfrac{10 \text{ ATP}}{\text{acetyl CoA}}$ (citric acid cycle)	110 ATP
$10 \text{ NADH} \times \dfrac{2.5 \text{ ATP}}{\text{NADH}}$ (electron transport)	25 ATP
$10 \text{ FADH}_2 \times \dfrac{1.5 \text{ ATP}}{\text{FADH}_2}$ (electron transport)	15 ATP
Total	**148 ATP**

24.23

Number of Carbon Atoms	Number of β Oxidation Cycles	Number of Acetyl CoA	Number of NADH	Number of FADH$_2$
18	8	9	8	7

a. A C$_{18}$ fatty acid will go through eight β oxidation cycles.

b. The β oxidation of a chain of 18 carbon atoms produces nine acetyl CoA units.

c. Note that the number of FADH$_2$ produced from the β oxidation cycles is one less than the number of NADH produced because one cis double bond must be isomerized to a trans double bond.

ATP Production from Oleic Acid (C$_{18}$)

Activation	-2 ATP
9 ~~acetyl CoA~~ $\times \dfrac{10\text{ ATP}}{\text{acetyl CoA}}$ (citric acid cycle)	90 ATP
8 ~~NADH~~ $\times \dfrac{2.5\text{ ATP}}{\text{NADH}}$ (electron transport)	20 ATP
7 ~~FADH$_2$~~ $\times \dfrac{1.5\text{ ATP}}{\text{FADH$_2$}}$ (electron transport)	10.5 ATP
Total	118.5 ATP

24.25 Hypoglycin A is metabolized in the body to produce a toxin that inhibits acyl CoA dehydrogenase, the first enzyme in the β oxidation pathway.

24.27 Ketogenesis is the synthesis of ketone bodies from excess acetyl CoA from fatty acid oxidation, which occurs when glucose is not available for energy, particularly in starvation, low-carbohydrate diets, fasting, alcoholism, and diabetes.

24.29 Acetoacetate undergoes reduction using NADH + H$^+$ to yield 3-hydroxybutyrate.

24.31 Ketoacidosis is a condition characterized by a drop in blood pH values, excessive urination, strong thirst, vomiting, shortness of breath, fatigue, and confusion.

24.33 Fatty acid synthesis primarily occurs in the cytosol of cells in adipose tissue.

24.35 The reactants for the first step of fatty acid synthesis are malonyl ACP and acetyl ACP.

24.37 a. (3) Malonyl CoA transacylase converts malonyl CoA to malonyl ACP.
    b. (1) Acetyl CoA carboxylase combines acetyl CoA with bicarbonate to yield malonyl CoA.
    c. (2) Acetyl CoA transacylase converts acetyl CoA to acetyl ACP.

24.39 a. A C$_{10}$ fatty acid requires the formation of 4 malonyl ACP, which uses 4 HCO$_3^-$.
    b. Four ATP are required to produce 4 malonyl CoA.
    c. Five acetyl CoA molecules are needed to make 1 acetyl ACP and 4 malonyl ACP.
    d. A C$_{10}$ fatty acid requires 4 malonyl ACP and 1 acetyl ACP.
    e. A C$_{10}$ fatty acid chain requires 4 cycles with 2 NADPH/cycle or a total of 8 NADPH.
    f. The four cycles remove a total of 4 CO$_2$.

24.41 a. Reaction 4, the reduction of the double bond to a single bond, is catalyzed by enoyl ACP reductase.
    b. If fatty acid synthesis is inhibited, the bacteria cannot form cell membranes and will not thrive.

24.43 The digestion of proteins begins in the stomach and is completed in the small intestine.

24.45 Hormones, heme, purines and pyrimidines for nucleotides, proteins, nonessential amino acids, amino alcohols, and neurotransmitters require nitrogen obtained from amino acids.

24.47 The reactants are an amino acid and an α-keto acid, and the products are a new amino acid and a new α-keto acid.

24.49 In transamination, an amino group is transferred from an amino acid to an α-keto acid, creating a new amino acid and a new α-keto acid.

a.
$$H-\overset{\overset{\displaystyle O}{\|}}{C}-COO^-$$

b.
$$HS-CH_2-\overset{\overset{\displaystyle O}{\|}}{C}-COO^-$$

c.
$$CH_3-\overset{\overset{\displaystyle CH_3}{|}}{CH}-\overset{\overset{\displaystyle O}{\|}}{C}-COO^-$$

24.51 In oxidative deamination, the amino group in an amino acid such as glutamate is removed as an ammonium ion. The reaction requires NAD^+.

$$^-OOC-\overset{\overset{\displaystyle \overset{+}{N}H_3}{|}}{CH}-CH_2-CH_2-COO^- + H_2O + NAD^+ \xrightarrow[\text{dehydrogenase}]{\text{Glutamate}}$$
Glutamate

$$^-OOC-\overset{\overset{\displaystyle O}{\|}}{C}-CH_2-CH_2-COO^- + NH_4^+ + NADH + H^+$$
α-Ketoglutarate

24.53
$$H_2N-\overset{\overset{\displaystyle O}{\|}}{C}-NH_2$$

24.55 The carbon atom in urea is obtained from CO_2.

24.57 The body converts NH_4^+ to urea because NH_4^+ is toxic if allowed to accumulate.

24.59 If there is a deficiency of argininosuccinate synthetase, citrulline will accumulate.

24.61 Glucogenic amino acids can be used to synthesize glucose.

24.63 a. Carbon atoms from alanine form the citric acid cycle intermediate oxaloacetate.
    b. Carbon atoms from asparagine form the citric acid cycle intermediate oxaloacetate.
    c. Carbon atoms from valine form the citric acid cycle intermediate succinyl CoA.
    d. Carbon atoms from glutamine form the citric acid cycle intermediate α-ketoglutarate.

24.65 The amino acids that humans can synthesize are called nonessential amino acids.

24.67 Glutamine synthetase catalyzes the addition of $-NH_3^+$ to glutamate to form glutamine using energy from the hydrolysis of ATP.

24.69 PKU is the abbreviation for **p**henyl**k**eton**u**ria.

24.71
$$CH_3-\overset{\overset{\displaystyle O}{\|}}{C}-COO^- + {}^-OOC-\overset{\overset{\displaystyle \overset{+}{N}H_3}{|}}{CH}-CH_2-CH_2-COO^-$$
     Pyruvate   Glutamate

24.73 a. and **b.**

Number of Carbon Atoms	Number of β Oxidation Cycles	Number of Acetyl CoA	Number of NADH	Number of FADH$_2$
12	5	6	5	5

    c. Five cycles of β oxidation are needed.
    d. Six acetyl CoA units are produced.

e. ATP Production from Lauric Acid (C_{12})

Activation	−2 ATP
6 ~~acetyl CoA~~ $\times \dfrac{10\ \text{ATP}}{\text{acetyl CoA}}$ (citric acid cycle)	60 ATP
5 ~~NADH~~ $\times \dfrac{2.5\ \text{ATP}}{\text{NADH}}$ (electron transport)	12.5 ATP
5 ~~FADH$_2$~~ $\times \dfrac{1.5\ \text{ATP}}{\text{FADH}_2}$ (electron transport)	7.5 ATP
Total	78 ATP

24.75 Triacylglycerols are hydrolyzed to monoacylglycerols and fatty acids in the small intestine, which reform as triacylglycerols in the intestinal lining for transport as lipoproteins to the tissues.

24.77 Fats can be stored in unlimited amounts in adipose tissue compared to the limited storage of carbohydrates as glycogen.

24.79 The fatty acids cannot diffuse across the blood–brain barrier.

24.81 a. Glycerol is converted to glycerol-3-phosphate and then to dihydroxyacetone phosphate, which can enter glycolysis or gluconeogenesis.
b. Activation of fatty acids occurs in the cytosol.
c. The energy cost is equal to two ATP.
d. Only acyl CoA can move into the intermembrane space for transport by carnitine into the matrix.

24.83 a. NAD^+ is involved in β oxidation.
b. β Oxidation occurs in the mitochondrial matrix.
c. Malonyl ACP is involved in fatty acid synthesis.
d. Cleavage of a two-carbon acetyl group occurs in β oxidation.
e. Acyl carrier protein is involved in fatty acid synthesis.
f. Acetyl CoA carboxylase is involved in fatty acid synthesis.

24.85 a. High blood glucose stimulates fatty acid synthesis.
b. Secretion of glucagon stimulates fatty acid oxidation.

24.87 Ammonium ion is toxic if allowed to accumulate.

24.89 a. Citrulline reacts with aspartate in the urea cycle.
b. Carbamoyl phosphate reacts with ornithine in the urea cycle.

24.91 a. Carbon atoms from serine form the citric acid cycle intermediate oxaloacetate.
b. Carbon atoms from lysine form the citric acid cycle intermediate acetyl CoA.
c. Carbon atoms from methionine form the citric acid cycle intermediate succinyl CoA.
d. Carbon atoms from glutamate form the citric acid cycle intermediate α-ketoglutarate.

24.93 Serine is degraded to pyruvate, which is oxidized to acetyl CoA. The oxidation produces $NADH + H^+$, which can be oxidized to provide the energy to synthesize 2.5 ATP. From one turn of the citric acid cycle, the oxidation of the acetyl CoA eventually produces 10 ATP. Thus, serine can provide a total of 12.5 ATP.

24.95 a. first oxidation **b.** cleavage **c.** activation

24.97 a.
$$CH_3\overset{}{\underset{|}{CH}}-\overset{+}{\underset{|}{NH_3}}\ \ $$

$$CH_3-CH-CH-\overset{O}{\overset{\|}{C}}-O^- \quad \text{Valine}$$

b.
$$CH_3-CH_2-\overset{CH_3}{\underset{|}{CH}}-\overset{\overset{+}{NH_3}}{\underset{|}{CH}}-\overset{O}{\overset{\|}{C}}-O^- \quad \text{Isoleucine}$$

Selected Answers to Combining Ideas from Chapters 22 to 24

CI.47 a. GTP is part of the citric acid cycle.
   b. $CoQH_2$ is part of electron transport.
   c. $FADH_2$ is part of both the citric acid cycle and electron transport.
   d. Cyt *c* is part of electron transport.
   e. Succinate dehydrogenase is part of both the citric acid cycle and electron transport.
   f. Complex I is part of electron transport.
   g. Isocitrate is part of the citric acid cycle.
   h. NAD^+ is part of both the citric acid cycle and electron transport.

CI.49 a. The components of acetyl CoA are an acetyl group, aminoethanethiol, pantothenic acid (vitamin B_5), and phosphorylated ADP.
   b. Coenzyme A carries an acetyl group to the citric acid cycle for oxidation.
   c. The acetyl group links to the sulfur atom ($-S-$) in the aminoethanethiol part of CoA.
   d. Molar mass of acetyl CoA ($C_{23}H_{38}N_7O_{17}P_3S$)

$$= 23(12.01 \text{ g}) + 38(1.008 \text{ g}) + 7(14.01 \text{ g}) + 17(16.00 \text{ g}) + 3(30.97 \text{ g}) + 32.07 \text{ g}$$

$$= 810. \text{ g/mole (3 SFs)}$$

   e. $1.0 \text{ mg acetyl CoA} \times \dfrac{1 \text{ g acetyl CoA}}{1000 \text{ mg acetyl CoA}} \times \dfrac{1 \text{ mole acetyl CoA}}{810. \text{ g acetyl CoA}} \times \dfrac{10 \text{ moles ATP}}{1 \text{ mole acetyl CoA}}$

$$= 1.2 \times 10^{-5} \text{ mole of ATP (2 SFs)}$$

CI.51 a.

   b. Molar mass of glyceryl tripalmitate ($C_{51}H_{98}O_6$)
   $= 51(12.01 \text{ g}) + 98(1.008 \text{ g}) + 6(16.00 \text{ g}) = 807 \text{ g/mole (3 SFs)}$

   c.

Number of Carbon Atoms	Number of β Oxidation Cycles	Number of Acetyl CoA	Number of NADH	Number of FADH$_2$
16	7	8	7	7

ATP Production from Palmitic Acid (C$_{16}$)

Activation	−2 ATP
$8 \text{ acetyl CoA} \times \dfrac{10 \text{ ATP}}{\text{acetyl CoA}}$ (citric acid cycle)	80 ATP
$7 \text{ NADH} \times \dfrac{2.5 \text{ ATP}}{\text{NADH}}$ (electron transport)	17.5 ATP
$7 \text{ FADH}_2 \times \dfrac{1.5 \text{ ATP}}{\text{FADH}_2}$ (electron transport)	10.5 ATP
Total	106 ATP

∴ 1 mole of palmitic acid will yield 106 moles of ATP.

d. $0.50 \text{ oz butter} \times \dfrac{1 \text{ lb}}{16 \text{ oz}} \times \dfrac{454 \text{ g}}{1 \text{ lb}} \times \dfrac{80. \text{ g glyceryl tripalmitate}}{100. \text{ g butter}} \times \dfrac{1 \text{ mole glyceryl tripalmitate}}{807 \text{ g glyceryl tripalmitate}}$

$\times \dfrac{3 \text{ moles palmitic acid}}{1 \text{ mole glyceryl tripalmitate}} \times \dfrac{106 \text{ moles ATP}}{1 \text{ mole palmitic acid}} \times \dfrac{7.3 \text{ kcal}}{1 \text{ mole ATP}} = 33 \text{ kcal (2 SFs)}$

e. $45 \text{ min} \times \dfrac{1 \text{ h}}{60 \text{ min}} \times \dfrac{750 \text{ kcal}}{1 \text{ h}} \times \dfrac{0.50 \text{ oz butter}}{33 \text{ kcal}} = 8.5 \text{ oz of butter (2 SFs)}$

CI.53 a. The disaccharide maltose will produce more ATP per mole than the monosaccharide glucose.
  b. The C_{18} fatty acid stearic acid will produce more ATP per mole than the C_{14} fatty acid myristic acid.
  c. The six-carbon molecule glucose will produce more ATP per mole than two two-carbon molecules of acetyl CoA.